Springer Series in
SOLID-STATE SCIENCES 155

Springer Series in
SOLID-STATE SCIENCES

Series Editors:
M. Cardona   P. Fulde   K. von Klitzing   R. Merlin   H.-J. Queisser   H. Störmer

The Springer Series in Solid-State Sciences consists of fundamental scientific books prepared by leading researchers in the field. They strive to communicate, in a systematic and comprehensive way, the basic principles as well as new developments in theoretical and experimental solid-state physics.

138 **Phase Separation in Soft Matter Physics**
Micellar Solutions, Microemulsions, Critical Phenomena
By P.K. Khabibullaev and A.A. Saidov

139 **Optical Response of Nanostructures**
Microscopic Nonlocal Theory
By K. Cho

140 **Fractal Concepts in Condensed Matter Physics**
By T. Nakayama and K. Yakubo

141 **Excitons in Low-Dimensional Semiconductors**
Theory, Numerical Methods, Applications   By S. Glutsch

142 **Two-Dimensional Coulomb Liquids and Solids**
By Y. Monarkha and K. Kono

143 **X-Ray Multiple-Wave Diffraction**
Theory and Application
By S.-L. Chang

144 **Physics of Transition Metal Oxides**
By S. Maekawa, T. Tohyama, S.E. Barnes, S. Ishihara, W. Koshibae, and G. Khaliullin

145 **Point-Contact Spectroscopy**
By Y.G. Naidyuk and I.K. Yanson

146 **Optics of Semiconductors and Their Nanostructures**
Editors: H. Kalt and M. Hetterich

147 **Electron Scattering in Solid Matter**
A Theoretical and Computational Treatise
By J. Zabloudil, R. Hammerling, L. Szunyogh, and P. Weinberger

148 **Physical Acoustics in the Solid State**
By B. Lthi

149 **Solitary Waves in Complex Dispersive Media**
Theory · Simulation · Applications
By V.Yu. Belashov and S.V. Vladimirov

150 **Topology in Condensed Matter**
Editor: M.I. Monastyrsky

151 **Particle Penetration and Radiation Effects**
By P. Sigmund

152 **Magnetism**
From Fundamentals to Nanoscale Dynamics
By H.C. Siegmann and J. Sthr

153 **Quantum Chemistry of Solids**
The LCAO First Principles Treatment of Crystals
By R.A. Evarestov

154 **Low-Dimensional Molecular Metals**
By N. Toyota, M. Lang and J. Mller

155 **Diffusion in Solids**
Fundamentals, Methods, Materials, Diffusion-Controlled Processes
By H. Mehrer

Volumes 91–137 are listed at the end of the book.

Helmut Mehrer

Diffusion in Solids

Fundamentals, Methods, Materials,
Diffusion-Controlled Processes

With 267 Figures and 27 Tables

 Springer

Professor Dr. Helmut Mehrer
Universität Münster
Institut für Materialphysik
Wilhelm-Klemm-Str. 10
48149 Münster
Germany

Series Editors:

Professor Dr., Dres. h.c. Manuel Cardona
Professor Dr., Dres. h.c. Peter Fulde*
Professor Dr., Dres. h.c. Klaus von Klitzing
Professor Dr., Dres. h.c. Hans-Joachim Queisser
Max-Planck-Institut für Festkörperforschung, Heisenbergstrasse 1, 70569 Stuttgart, Germany
* Max-Planck-Institut für Physik komplexer Systeme, Nöthnitzer Strasse 38
  01187 Dresden, Germany

Professor Dr. Roberto Merlin
Department of Physics, 5000 East University, University of Michigan
Ann Arbor, MI 48109-1120, USA

Professor Dr. Horst Störmer
Dept. Phys. and Dept. Appl. Physics, Columbia University, New York, NY 10027 and
Bell Labs., Lucent Technologies, Murray Hill, NJ 07974, USA

ISSN 0171-1873
ISBN 978-3-642-09070-7   e-ISBN 978-3-540-71488-0

This work is subject to copyright. All rights are reserved, whether the whole or part of the material is concerned, specifically the rights of translation, reprinting, reuse of illustrations, recitation, broadcasting, reproduction on microfilm or in any other way, and storage in data banks. Duplication of this publication or parts thereof is permitted only under the provisions of the German Copyright Law of September 9, 1965, in its current version, and permission for use must always be obtained from Springer. Violations are liable for prosecution under the German Copyright Law.

Springer is a part of Springer Science+Business Media

springer.com

© Springer-Verlag Berlin Heidelberg 2010

The use of general descriptive names, registered names, trademarks, etc. in this publication does not imply, even in the absence of a specific statement, that such names are exempt from the relevant protective laws and regulations and therefore free for general use.

Cover design: eStudioCalamar S.L., F. Steinen-Broo, Girona, Spain

For Karin and my family, who wonder what I did all the time.

In particular I express my enduring thanks to my wife Karin. Without her patience, understanding, and love I could not have completed this book.

Preface

Diffusion is the transport of matter from one point to another by thermal motion of atoms or molecules. It is relatively fast in gases, slow in liquids, and very slow in solids. Diffusion plays a key rôle in many processes as diverse as intermixing of gases and liquids, permeation of atoms or molecules through membranes, evaporation of liquids, drying of timber, doping silicon wafers to make semiconductor devices, and transport of thermal neutrons in nuclear power reactors. Rates of important chemical reactions are limited by how fast diffusion can bring reactants together or deliver them to reaction sites on enzymes or other catalysts.

Diffusion in solid materials is the subject of this book. Already in ancient times reactions in the solid state such as surface hardening of steels were in use, which according to our present knowledge involves the diffusion of carbon atoms in the crystal lattice of iron. Nevertheless, until the end of the nineteenth century the paradigm *'Corpora non agunt nisi fluida'* was widely accepted by the scientific community. It was mainly due to the pioneering work of William Roberts-Austen and Georg von Hevesy that this paradigm had to be abandoned.

Diffusion in solids is fundamental in the art and science of materials and thus an important topic of solid-state physics, physical chemistry, physical metallurgy, and materials science. Diffusion processes are relevant for the kinetics of many microstructural changes that occur during preparation, processing, and heat treatment of materials. Typical examples are nucleation of new phases, diffusive phase transformations, precipitation and dissolution of a second phase, homogenisation of alloys, recrystallisation, high-temperature creep, and thermal oxidation. Diffusion and electrical conduction in ionic conductors are closely related phenomena. Direct technological applications of diffusion concern, e.g., doping during fabrication of microelectronic devices, solid electrolytes for batteries and fuel cells, surface hardening of steel through carburisation or nitridation, diffusion bonding, and sintering.

Appreciable diffusion in solids mostly takes place at temperatures well above room temperature. Knowledge of diffusion is therefore particularly important for scientists who design materials for elevated temperatures and for engineers who build equipment for operation at such temperatures. However, processes connected with diffusion at room temperature pose problems, too.

Creep, atmospheric corrosion, and embrittlement of solders are among the more prominent of those. With the downscaling of microelectronic circuits to nanometer dimensions, diffusion and electromigration in these circuits must be taken into account.

A deeper knowledge about diffusion requires information on the position of atoms and how they move in solids. The atomic mechanisms of diffusion in crystalline solids are closely connected with defects. Point defects such as vacancies or interstitials are the simplest defects and often mediate diffusion in crystals. Dislocations, grain-boundaries, phase boundaries, and free surfaces are other types of defects. They can act as high-diffusivity paths (diffusion short circuits), because the mobility of atoms along such defects is usually much higher than in the lattice. In solids with structural disorder such as glasses or crystals with highly disordered sublattices the concept of defects is no longer useful. Nevertheless, diffusion is fundamental for transport of matter and for ionic conduction in disordered materials.

The content of this book is divided into seven parts. After a historical introduction and a diffusion bibliography, *Part I* introduces basic concepts of diffusion in solid matter such as continuum description, random walk theory, point defects, atomic mechanisms, correlation effects, dependence of diffusion on temperature, pressure and isotope mass, diffusion with driving forces , and some remarks about the relation between diffusion and thermodynamics of irreversible processes. The necessary background is a course in solid-state physics. In *Part II* we describe experimental methods for the determination of diffusion coefficients in solid matter. Direct methods based on Fick's laws and indirect methods such as anelastic relaxation, internal friction, nuclear magnetic relaxation, Mössbauer spectroscopy, quasielastic neutron scattering, impedance spectroscopy, and spreading resistance measurements are treated. In further parts we provide access to information on diffusion in various types of materials such as metals, intermetallics and quasicrystalline alloys (*Part III*), semiconductors (*Part IV*), ionic materials including fast ion conductors (*Part V*), metallic and oxide glasses (*Part VI*). Finally, rapid diffusion paths such as grain-boundary diffusion and diffusion in nanomaterials are considered (*Part VII*). Although these parts cannot replace a comprehensive data collection, typical up-to-date resources available on diffusion for various types of materials are noted.

A thorough understanding of diffusion in materials is crucial for materials development and engineering. Graduate students in solid state physics, physical metallurgy, physical and inorganic chemistry, and geophysical materials will benefit from this book as will physicists, chemists and metallurgists, working in academia and industry.

Münster, May 2007   Helmut Mehrer

Acknowledgements

Like any author of a scientific book, I am indebted to previous writers on diffusion and allied subjects. In particular I acknowledge the encouragement of colleagues and friends who provided invaluable assistance during the preparation of the book. Prof. Hartmut Bracht, Prof. Klaus Funke, and Dr. Nikolaas Stolwijk, all from the University of Münster, have read many chapters as they unfolded. Prof. Gabor Erdelyi, Debrecen, Hungary, Prof. Andry Gusak, Cherkassy, Ukraine, Dr. J.N. Mundy, USA, and Prof. Malcom Ingram, University of Aberdeen, Great Britain, stayed in my University as guest scientist for some time and have also reviewed several chapters of the book. Dr. M. Hirscher, Max-Planck-Institut für Metallforschung, Stuttgart, provided comments for the chapter on hydrogen diffusion. Prof. Gerhard Wilde from the University of Münster made useful suggestions for the chapter on nanomaterials. Prof. Graeme Murch, Newcastle, Australia, did a great job in reading the whole book, polishing my English, and providing many helpful suggestions.

To my collaborators Dr. Serguei Divinski, Dr. Arpad Imre, Dr. Halgard Staesche and my PhD students and postdocs Dr. Robert Galler, Dr. Marcel Salamon, Dr. Serguei Peteline, Dr. Eugene Tanguep Nijokep, Dr. Stephan Voss, and to my secretary Sylvia Gurnik I owe many thanks for critically reading parts of the book. Dr. Arpad Imre was a great help in preparing most figures. The contributions and constructive criticisms of all these persons were most helpful.

Contents

1 **History and Bibliography of Diffusion** 1
   1.1 Pioneers and Landmarks of Diffusion 2
   References 16
   1.2 Bibliography of Solid-State Diffusion 18

Part I Fundamentals of Diffusion

2 **Continuum Theory of Diffusion** 27
   2.1 Fick's Laws in Isotropic Media 27
       2.1.1 Fick's First Law 28
       2.1.2 Equation of Continuity 29
       2.1.3 Fick's Second Law – the 'Diffusion Equation' 30
   2.2 Diffusion Equation in Various Coordinates 31
   2.3 Fick's Laws in Anisotropic Media 33
   References 35

3 **Solutions of the Diffusion Equation** 37
   3.1 Steady-State Diffusion 37
   3.2 Non-Steady-State Diffusion in one Dimension 39
       3.2.1 Thin-Film Solution 39
       3.2.2 Extended Initial Distribution
              and Constant Surface Concentration 41
       3.2.3 Method of Laplace Transformation 45
       3.2.4 Diffusion in a Plane Sheet – Separation of Variables .. 47
       3.2.5 Radial Diffusion in a Cylinder 50
       3.2.6 Radial Diffusion in a Sphere 51
   3.3 Point Source in one, two, and three Dimensions 52
   References 53

4 **Random Walk Theory and Atomic Jump Process** 55
   4.1 Random Walk and Diffusion 56
       4.1.1 A Simplified Model 56
       4.1.2 Einstein-Smoluchowski Relation 58
       4.1.3 Random Walk on a Lattice 60

		4.1.4 Correlation Factor	62
	4.2	Atomic Jump Process	64
	References		66

5 Point Defects in Crystals ... 69
5.1 Pure Metals ... 70
5.1.1 Vacancies ... 70
5.1.2 Divacancies ... 72
5.1.3 Determination of Vacancy Properties ... 74
5.1.4 Self-Interstitials ... 79
5.2 Substitutional Binary Alloys ... 80
5.2.1 Vacancies in Dilute Alloys ... 81
5.2.2 Vacancies in Concentrated Alloys ... 82
5.3 Ionic Compounds ... 83
5.3.1 Frenkel Disorder ... 84
5.3.2 Schottky Disorder ... 85
5.4 Intermetallics ... 86
5.5 Semiconductors ... 88
References ... 91

6 Diffusion Mechanisms ... 95
6.1 Interstitial Mechanism ... 95
6.2 Collective Mechanisms ... 97
6.3 Vacancy Mechanism ... 98
6.4 Divacancy Mechanism ... 100
6.5 Interstitialcy Mechanism ... 100
6.6 Interstitial-substitutional Exchange Mechanisms ... 102
References ... 103

7 Correlation in Solid-State Diffusion ... 105
7.1 Interstitial Mechanism ... 107
7.2 Interstitialcy Mechanism ... 107
7.3 Vacancy Mechanism of Self-diffusion ... 108
7.3.1 A 'Rule of Thumb' ... 108
7.3.2 Vacancy-tracer Encounters ... 109
7.3.3 Spatial and Temporal Correlation ... 112
7.3.4 Calculation of Correlation Factors ... 112
7.4 Correlation Factors of Self-diffusion ... 115
7.5 Vacancy-mediated Solute Diffusion ... 116
7.5.1 Face-Centered Cubic Solvents ... 117
7.5.2 Body-Centered Cubic Solvents ... 120
7.5.3 Diamond Structure Solvents ... 121
7.6 Concluding Remarks ... 122
References ... 124

Contents XIII

8 Dependence of Diffusion on Temperature and Pressure ... 127
- 8.1 Temperature Dependence ... 127
  - 8.1.1 The Arrhenius Relation ... 127
  - 8.1.2 Activation Parameters – Examples ... 130
- 8.2 Pressure Dependence ... 132
  - 8.2.1 Activation Volumes of Self-diffusion ... 135
  - 8.2.2 Activation Volumes of Solute Diffusion ... 139
  - 8.2.3 Activation Volumes of Ionic Crystals ... 140
- 8.3 Correlations between Diffusion and Bulk Properties ... 141
  - 8.3.1 Melting Properties and Diffusion ... 141
  - 8.3.2 Activation Parameters and Elastic Constants ... 146
  - 8.3.3 Use of Correlations ... 147
- References ... 147

9 Isotope Effect of Diffusion ... 151
- 9.1 Single-jump Mechanisms ... 151
- 9.2 Collective Mechanisms ... 155
- 9.3 Isotope Effect Experiments ... 155
- References ... 159

10 Interdiffusion and Kirkendall Effect ... 161
- 10.1 Interdiffusion ... 161
  - 10.1.1 Boltzmann Transformation ... 162
  - 10.1.2 Boltzmann-Matano Method ... 163
  - 10.1.3 Sauer-Freise Method ... 166
- 10.2 Intrinsic Diffusion and Kirkendall Effect ... 168
- 10.3 Darken Equations ... 170
- 10.4 Darken-Manning Equations ... 172
- 10.5 Microstructural Stability of the Kirkendall Plane ... 173
- References ... 176

11 Diffusion and External Driving Forces ... 179
- 11.1 Overview ... 179
- 11.2 Fick's Equations with Drift ... 181
- 11.3 Nernst-Einstein Relation ... 182
- 11.4 Nernst-Einstein Relation for Ionic Conductors and Haven Ratio ... 184
- 11.5 Nernst-Planck Equation – Interdiffusion in Ionic Crystals ... 186
- 11.6 Nernst-Planck Equation *versus* Darken Equation ... 188
- References ... 189

12 Irreversible Thermodynamics and Diffusion ... 191
- 12.1 General Remarks ... 191
- 12.2 Phenomenological Equations of Isothermal Diffusion ... 193
  - 12.2.1 Tracer Self-Diffusion in Element Crystals ... 193

            12.2.2 Diffusion in Binary Alloys 195
      12.3 The Phenomenological Coefficients........................ 199
            12.3.1 Phenomenological Coefficients, Tracer Diffusivities,
                   and Jump Models 202
            12.3.2 Sum Rules – Relations
                   between Phenomenological Coefficients 204
      References 205

Part II Experimental Methods

13 Direct Diffusion Studies 209
      13.1 Direct *versus* Indirect Methods........................ 209
      13.2 The Various Diffusion Coefficients 212
            13.2.1 Tracer Diffusion Coefficients 212
            13.2.2 Interdiffusion and Intrinsic Diffusion Coefficients 214
      13.3 Tracer Diffusion Experiments 215
            13.3.1 Profile Analysis by Serial Sectioning 217
            13.3.2 Residual Activity Method 222
      13.4 Isotopically Controlled Heterostructures 223
      13.5 Secondary Ion Mass Spectrometry (SIMS) 224
      13.6 Electron Microprobe Analysis (EMPA) 227
      13.7 Auger-Electron Spectroscopy (AES) 230
      13.8 Ion-beam Analysis: RBS and NRA........................... 231
      References 234

14 Mechanical Spectroscopy 237
      14.1 General Remarks... 237
      14.2 Anelasticity and Internal Friction...................... 239
      14.3 Techniques of Mechanical Spectroscopy................... 242
      14.4 Examples of Diffusion-related Anelasticty............... 244
            14.4.1 Snoek Effect (Snoek Relaxation) 244
            14.4.2 Zener Effect (Zener Relaxation) 247
            14.4.3 Gorski Effect (Gorski Relaxation) 248
            14.4.4 Mechanical Loss in Ion-conducting Glasses......... 249
      14.5 Magnetic Relaxation 250
      References 251

15 Nuclear Methods 253
      15.1 General Remarks... 253
      15.2 Nuclear Magnetic Relaxation (NMR) 253
            15.2.1 Fundamentals of NMR 254
            15.2.2 Direct Diffusion Measurement
                   by Field-Gradient NMR............................. 256
            15.2.3 NMR Relaxation Methods 258

    15.3 Mössbauer Spectroscopy (MBS) 264
    15.4 Quasielastic Neutron Scattering (QENS) 269
          15.4.1 Examples of QENS studies 278
          15.4.2 Advantages and Limitations of MBS and QENS 279
    References 281

16 Electrical Methods 285
    16.1 Impedance Spectroscopy 285
    16.2 Spreading Resistance Profiling 290
    References 293

Part III Diffusion in Metallic Materials

17 Self-diffusion in Metals 297
    17.1 General Remarks 297
    17.2 Cubic Metals 299
          17.2.1 FCC Metals – Empirical Facts 299
          17.2.2 BCC Metals – Empirical Facts 301
          17.2.3 Monovacancy Interpretation 302
          17.2.4 Mono- and Divacancy Interpretation 303
    17.3 Hexagonal Close-Packed and Tetragonal Metals 306
    17.4 Metals with Phase Transitions 308
    References 311

18 Diffusion of Interstitial Solutes in Metals 313
    18.1 'Heavy' Interstitial Solutes C, N, and O 313
          18.1.1 General Remarks 313
          18.1.2 Experimental Methods 314
          18.1.3 Interstitial Diffusion in Dilute Interstial Alloys 316
    18.2 Hydrogen Diffusion in Metals 317
          18.2.1 General Remarks 317
          18.2.2 Experimental Methods 318
          18.2.3 Examples of Hydrogen Diffusion.................... 320
          18.2.4 Non-Classical Isotope Effects 323
    References 324

19 Diffusion in Dilute Substitutional Alloys 327
    19.1 Diffusion of Impurities 327
          19.1.1 'Normal' Impurity Diffusion 327
          19.1.2 Impurity Diffusion in Al........................... 332
    19.2 Impurity Diffusion in 'Open' Metals –
          Dissociative Mechanism 333
    19.3 Solute Diffusion and Solvent Diffusion in Alloys 336
    References 338

XVI Contents

20 Diffusion in Binary Intermetallics 341
20.1 General Remarks 341
20.2 Influence of Order-Disorder Transitions 344
20.3 B2 Intermetallics 346
    20.3.1 Diffusion Mechanisms in B2 Phases 347
    20.3.2 Example B2 NiAl 351
    20.3.3 Example B2 Fe-Al 353
20.4 $L1_2$ Intermetallics 355
20.5 $D0_3$ Intermetallics 357
20.6 Uniaxial Intermetallics 360
    20.6.1 $L1_0$ Intermetallics 360
    20.6.2 Molybdenum Disilicide ($C11_b$ structure) 362
20.7 Laves Phases 364
20.8 The Cu_3Au Rule 366
References 367

21 Diffusion in Quasicrystalline Alloys 371
21.1 General Remarks on Quasicrystals 371
21.2 Diffusion Properties of Quasicrystals 373
    21.2.1 Icosahedral Quasicrystals 374
    21.2.2 Decagonal Quasicrystals 379
References 381

Part IV Diffusion in Semiconductors

22 General Remarks on Semiconductors 385
22.1 'Semiconductor Age' and Diffusion 386
22.2 Specific Features of Semiconductor Diffusion 389
References 392

23 Self-diffusion in Elemental Semiconductors 395
23.1 Intrinsic Point Defects and Diffusion 396
23.2 Germanium 398
23.3 Silicon 402
References 406

24 Foreign-Atom Diffusion in Silicon and Germanium 409
24.1 Solubility and Site Occupancy 409
24.2 Diffusivities and Diffusion Modes 412
    24.2.1 Interstitial Diffusion 414
    24.2.2 Dopant Diffusion 416
    24.2.3 Diffusion of Hybrid Foreign Elements 420
24.3 Self- and Foreign Atom Diffusion – a Summary 421
References 422

25 Interstitial-Substitutional Diffusion ... 425
- 25.1 Combined Dissociative and Kick-out Diffusion ... 425
  - 25.1.1 Diffusion Limited by the Flow of Intrinsic Defects ... 427
  - 25.1.2 Diffusion Limited by the Flow of Interstitial Solutes ... 429
  - 25.1.3 Numerical Analysis of an Intermediate Case ... 430
- 25.2 Kick-out Mechanism ... 431
  - 25.2.1 Basic Equations and two Solutions ... 431
  - 25.2.2 Examples of Kick-Out Diffusion ... 434
- 25.3 Dissociative Mechanism ... 439
  - 25.3.1 Basic Equations ... 439
  - 25.3.2 Examples of Dissociative Diffusion ... 440
- References ... 445

Part V Diffusion and Conduction in Ionic Materials

26 Ionic Crystals ... 449
- 26.1 General Remarks ... 449
- 26.2 Point Defects in Ionic Crystals ... 451
  - 26.2.1 Intrinsic Defects ... 452
  - 26.2.2 Extrinsic Defects ... 454
- 26.3 Methods for the Study of Defect and Transport Properties ... 456
- 26.4 Alkali Halides ... 458
  - 26.4.1 Defect Motion, Tracer Self-diffusion, and Ionic Conduction ... 458
  - 26.4.2 Example NaCl ... 462
  - 26.4.3 Common Features of Alkali Halides ... 467
- 26.5 Silver Halides AgCl and AgBr ... 468
  - 26.5.1 Self-diffusion and Ionic Conduction ... 469
  - 26.5.2 Doping Effects ... 471
- References ... 473

27 Fast Ion Conductors ... 475
- 27.1 Fast Silver-Ion Conductors ... 477
  - 27.1.1 AgI and related Simple Anion Structures ... 477
  - 27.1.2 $RbAg_4I_5$ and related Compounds ... 479
- 27.2 PbF_2 and other Halide Ion Conductors ... 480
- 27.3 Stabilised Zirconia and related Oxide Ion Conductors ... 481
- 27.4 Perovskite Oxide Ion Conductors ... 482
- 27.5 Sodium β-Alumina and related Materials ... 482
- 27.6 Lithium Ion Conductors ... 484
- 27.7 Polymer Electrolytes ... 485
- References ... 488

Part VI Diffusion in Glasses

28 The Glassy State 493
    28.1 What is a Glass? 493
    28.2 Volume-Temperature Diagram 494
    28.3 Temperature-Time-Transformation Diagram 496
    28.4 Glass Families 498
    References 501

29 Diffusion in Metallic Glasses 503
    29.1 General Remarks 503
    29.2 Structural Relaxation and Diffusion 506
    29.3 Diffusion Properties of Metallic Glasses 509
    29.4 Diffusion and Viscosity in Glass-forming Alloys 517
    References 518

30 Diffusion and Ionic Conduction in Oxide Glasses 521
    30.1 General Remarks 521
    30.2 Experimental Methods 526
    30.3 Gas Permeation 529
    30.4 Examples of Diffusion and Ionic Conduction 530
    References 542

Part VII Diffusion along High-Diffusivity Paths and in Nanomaterials

31 High-diffusivity Paths in Metals 547
    31.1 General Remarks 547
    31.2 Diffusion Spectrum 548
    31.3 Empirical Rules for Grain-Boundary Diffusion 549
    31.4 Lattice Diffusion and Microstructural Defects 551
    References 552

32 Grain-Boundary Diffusion 553
    32.1 General Remarks 553
    32.2 Grain Boundaries 554
        32.2.1 Low- and High-Angle Grain Boundaries 555
        32.2.2 Special High-Angle Boundaries 557
    32.3 Diffusion along an Isolated Boundary (Fisher Model) 559
    32.4 Diffusion Kinetics in Polycrystals 568
        32.4.1 Type A Kinetics Regime 568
        32.4.2 Type B Kinetics Regime 570
        32.4.3 Type C Kinetics Regime 574

       32.5 Grain-Boundary Diffusion and Segregation 576
       32.6 Atomic Mechanisms of Grain-Boundary Diffusion 579
       References 580

33 Dislocation Pipe Diffusion 583
       33.1 Dislocation Pipe Model 584
       33.2 Solutions for Mean Thin Layer Concentrations 586
       References 591

34 Diffusion in Nanocrystalline Materials 593
       34.1 General Remarks 593
       34.2 Synthesis of Nanocrystalline Materials 594
              34.2.1 Powder Processing 594
              34.2.2 Heavy Plastic Deformation 596
              34.2.3 Chemical and Related Synthesis Methods 598
              34.2.4 Devitrification of Amorphous Precursors 598
       34.3 Diffusion in Poly- and Nanocrystals 599
              34.3.1 Grain Size and Diffusion Regimes 599
              34.3.2 Effective Diffusivities in Poly- and Nanocrystals 604
       34.4 Diffusion in Nanocrystalline Metals 606
              34.4.1 General Remarks 606
              34.4.2 Structural Relaxation and Grain Growth 607
              34.4.3 Nanomaterials with Bimodal Grain Structure 608
              34.4.4 Grain Boundary Triple Junctions 612
       34.5 Diffusion and Ionic Conduction in Nanocrystalline Ceramics . 612
       References 618

Index 639

1 History and Bibliography of Diffusion

If a droplet of ink is placed without stirring at the bottom of a bottle filled with water, the colour will slowly spread through the bottle. At first, it will be concentrated near the bottom. After a few days, it will penetrate upwards a few centimeters. After several days, the solution will be coloured homogeneously. The process responsible for the movement of the coloured material is diffusion. Diffusion is caused by the BROWNian motion of atoms or molecules that leads to complete mixing. In gases, diffusion progresses at a rate of centimeters per second; in liquids, its rate is typically fractions of millimeters per second; in solids, diffusion is a fairly slow process and the rate of diffusion decreases strongly with decreasing temperature: near the melting temperature of a metal a typical rate is about one micrometer per second; near half of the melting temperature it is only of the order of nanometers per second.

The science of diffusion in solids had its beginnings in the 19th century, although the blacksmiths and metal artisans of antiquity already used the phenomenon to make such objects as swords of steel, gilded copper or bronze wares. Diffusion science is based on several corner stones. The most important ones are: (i) The continuum theory of diffusion originated from work of the German scientist *Adolf Fick*, who was inspired by elegant experiments on diffusion in gases and of salt in water performed by *Thomas Graham* in Scotland. (ii) The BROWNian motion was detected by the Scotish botanist *Robert Brown*. He observed small particles suspended in water migrating in an erratic fashion. This phenomenon was interpreted many decades later by *Albert Einstein*. He realised that the 'dance' described by *Brown* was a random walk driven by the collisions between particles and the water molecules. His theory provided the statistical cornerstone of diffusion and bridged the gap between mechanics and thermodynamics. It was verified in beautiful experiments by the French Nobel laureate *Jean Baptiste Perrin*. (iii) The atomistics of solid-state diffusion had to wait for the birth date of solid-state physics heralded by the experiments of *Max von Laue*. Equally important was the perception of the Russian and German scientists *Jakov Frenkel* and *Walter Schottky* that point defects play an important rôle for properties of crystalline substances, most notably for those controlling diffusion and the many properties that stem from it.

This chapter is not meant to be a systematic history of diffusion science. It is devoted in its first section to some major landmarks and eminent people in the field. The second section contains information about bibliography of diffusion in textbooks, monographs, conference proceedings, and data collections.

1.1 Pioneers and Landmarks of Diffusion

Establishment of the diffusion law: Experimental studies of diffusion were probably performed for the first time by *Thomas Graham* (1805–1869). Graham was born in Glasgow. His father was a successful textile manufacturer. He wished his son to enter the Church of Scotland. Defying his father's wishes he studied natural sciences, developed a strong interest in chemistry and became professor of chemistry in 1830 at the Andersonian Institute (now Strathclyde University) in Glasgow. Later he became professor of chemistry at several colleges including the Royal College of Science and Technology and the University of London in 1837. Graham helped to found the Chemical Society of London and became its first president. In 1854 Graham succeeded Sir John Herschel as Master of the Mint in London following the tradition – established by Sir Isaac Newton – of distinguished scientists occupying the post.

Graham is one of the founders of physical chemistry and he discovered the medical method of 'dialysis'. He initiated the quantitative study of diffusion in gases, largely conducted in the years of 1828 to 1833 [1, 2]. In one of his articles he explicitly stated what we now call Graham's law: *'The diffusion or spontaneous intermixture of two gases is effected by an interchange in position of indefinitely minute volumes of the gases, which volumes are not of equal magnitude, being, in the case of each gas, inversely proportional to the square root of the density of that gas.'* The crucial point about Graham's work on diffusion in gases was that it could be understood by the kinetic theory of gases developed by Maxwell and Clausius shortly after the middle of the 19th century. Graham's law can be attributed to the equipartition of kinetic energies between molecules with different molecular masses. In this way diffusion was connected with the thermal motion of atoms or molecules, and the idea of the mean free path entered science. Graham also extended his studies to diffusion of salts in liquids [3] and to the uptake of hydrogen in metals. He showed that diffusion in liquids was at least several thousand times slower than in gases.

The next major advance in the field of diffusion came from the work done by *Adolf Eugen Fick* (1829–1901). He was born in Kassel, Germany, as the youngest of five siblings. His father, a civil engineer, was a superintendent of buildings. During his secondary schooling, Adolf Fick was delighted by mathematics, especially by the work of Fourier and Poisson. He entered the University of Marburg with the intention to specialise in mathematics, but switched to medicine on the advice of an elder brother, a professor of anatomy. He got his doctorate with a thesis on *'Visual Errors due to Astigmatism'*. He spent the years from 1852 to 1868 at the University of Zürich, Switzerland, in various positions. After sixteen years in Zürich he accepted a chair in physiology in Würzburg, Germany.

Graham's work on diffusion of salts in water stimulated Fick to develop a mathematical framework to describe the phenomena of diffusion using the analogy between Fourier's law of thermal conduction and diffusion [4, 5]. Fick signed his papers on diffusion as 'Demonstrator of Anatomy, Zürich'. They were published in high-ranking journals. His approach was a phenomenological one and uses a continuum description. Nowadays, we would call his theory a 'linear response' approach. Fick is even better known in medicine. He published a well-rounded monography on *'Medical Physics'* [6] and a textbook on *'The Anatomy of Sense Organs'*. He became an outstanding person in the small group of nineteenth century physiologists who applied concepts and methods of physics to the study of living organisms, and thereby laid the foundations of modern physiology. Fick's vital contribution to the field of diffusion was to define the diffusion coefficient and to measure it for diffusion of salt in water. Mathematical solutions of Fick's equations began with the nineteenth century luminaries *Jozef Stephan* [7] and *Franz Neumann*, who were among the first to recognise the significance of boundary conditions for solutions of the diffusion equation.

Roberts-Austen – **discovery of solid-state diffusion:** *William Chandler Roberts-Austen* (1843–1902) graduated from the Royal School of Mines, London, in 1865 and became personal assistant to Graham at the Mint. After Graham's death in 1869 Roberts-Austen became 'Chemist and Assayer of the Mint', a position he occupied until his death. He was appointed professor of metallurgy at the Royal School of Mines in 1880 and was knighted in 1899 by Queen Victoria. He was a man of wide interests, with charm, and an understanding of people, which made him very popular. He conducted studies on the effects of impurities on the physical properties of pure metals and alloys and became a world authority on the technical aspects of minting coins. His work had many practical and industrial applications. Austenite – a non-magnetic solution of carbon in iron – is named after Sir Roberts-Austen.

He records his devotion to diffusion research as follows [8]: *'... My long connection with Graham's researches made it almost a duty to attempt to extend his work on liquid diffusion to metals.'* Roberts-Austen perfected the technique for measuring high temperatures adopting Le Chatelier's platinum-based thermocouples and studied the diffusion of gold, platinum, and rhodium in liquid lead; of gold, silver, and lead in liquid tin; and of gold in bismuth. These solvents were selected because of their relatively low melting temperatures. The solidified samples were sectioned and the diffused species determined in each section using the high precision assaying techniques developed for use in the Mint. Typically six or seven sections were taken and diffusion coefficients determined. Even more importantly, Roberts-Austen applied these techniques to the study of gold diffusion in solid lead as well. It is interesting to observe that the values of the diffusion coefficient of gold in lead reported by him are very close to those determined by modern techniques using radioactive isotopes. The choice of the system gold in lead was really fortunate. Nowadays, we know that the diffusion of noble metals in lead is exceptionally fast in comparison to most other diffusion processes in solids.

Arrhenius law of solid-state diffusion: The most surprising omission in Roberts-Austen's work is any discussion of the temperature dependence of the diffusion coefficient. Historically, the temperature dependence of reaction rates, diffusivities etc., now generally referred to as 'Arrhenius law', is named after the Swedish scientist *Svante August Arrhenius* (1859–1927). Arrhenius got a doctorship in chemistry in Uppsala, Sweden, in 1884 with a thesis about *electrolytic dissociation*. He was awarded a travel fellowship which enabled him to work with Ostwald in Riga, now Latvia, and with Kohlrausch in Würzburg, Germany. He also cooperated with Boltzmann in Graz, Austria, in 1887 and with Van't Hoff in Amsterdam, The Netherlands, in 1888. He was appointed for a chair in chemistry at the University of Stockholm in 1891. He abandoned this position in 1905 to become director of the Nobel Institute of Physical Chemistry. Arrhenius was awarded the 1903 Nobel prize in chemistry *for his theory of electrolytic dissociation*. It appears that the Arrhenius law for chemical reactions was proposed by the Dutch scientist *Jacobus Hendrik van't Hoff* (1852–1921), the first Nobel laureate in chemistry (1901). The suggestion that the diffusivity in solids should obey that law was apparently made by *Dushman and Langmuir* in 1922 [9].

Von Hevesy – the first measurements of self-diffusion: The idea of self-diffusion was already introduced by Maxwell, when treating the rate of diffusion of gases. The first attempts to measure self-diffusion in condensed matter were those of *Georg Karl von Hevesy* (1885–1966), who studied self-diffusion in liquid [10] and in solid lead [11] by using a natural radioactive isotope of lead. Von Hevesy had a fascinating scientific career. He was born in Budapest, Austria-Hungary, and studied at the Universities of Budapest, Berlin and Freiburg. He did research work in physical chemistry at the ETH in Zürich, with Fritz Haber in Karlsruhe, with Ernest Rutherford in Manchester, and with Fritz Paneth in Vienna. After World War I he teached for six months at the University of Budapest, and from 1920 to 1926 he worked with Niels Bohr at the University of Copenhagen. Together with the Dutch physicist Dirk Coster he discovered the new element 'hafnium' among the oreg of zirconium. He was professor at the University of Freiburg, Germany, from 1926 to 1934. During his eight years in Freiburg he initiated work with radiotracers in solids and in animal tissues. Fleeing from the Nazis in Germany, he moved to the Niels Bohr Institute in Copenhagen in 1934 and from there to Stockholm. In 1944 the Swedish Royal Academy of Sciences awarded him the Nobel prize in Chemistry of the year 1943 for '... *his work on the use of isotopes as tracers in the study of chemical processes.*' He became a Swedish citizen and was appointed professor of organic chemistry at the University of Stockholm in 1959. Von Hevesy, who married Pia Riis, daughter of a Danish ship owner, had four children, died in Freiburg, Germany. Von Hevesy is also the founder of radioistope applications in nuclear medicine. For example, the hospital of the author's University has a station called 'von Hevesy station'. Wolfgang Seith, who collaborated with von Hevesy in Freiburg, was appointed as the first professor in physical chemistry at the author's University in Münster, Germany.

Brownian motion: The phenomenon of irregular motion of small particles suspended in a liquid had been known for a long time. It had been discovered by the Scottish scientist *Robert Brown* (1773–1858). Brown was the son of an episcopalian priest. He studied medicine at Edinburgh University, but did not obtain his degree. At the age of twenty-one he enlisted in a newly raised Scottish regiment. At that time he already knew that his true interests lay not in medicine but in botany, and he already had acquired some reputation as a botanist. On a visit to London in 1798 to recruit for his regiment, he met the botanist Sir Joseph Banks, president of the Royal Society, who recommended Brown to the Admiralty for the post of a naturalist aboard a ship. The ship was to embark on a surveying voyage at the coasts of Australia. Brown made extensive plant collections in Australia and it took him about 5 years to classify approximately 3900 species he had gathered, almost all of which were new to science. By that time, Robert Brown was already a renowned botanist. Much later *Charles Darwin* referred to him as '... *princeps botanicorum*'. In addition to collecting and classifying, Brown made several important discoveries. Perhaps the most celebrated one by biologists is his discovery that plant cells have a nucleus.

Robert Brown is best known in science for his description of the random movement of small particles in liquid suspension, first described in a pamphlet entitled *'A brief account of microscopical observations in the months June, July and August 1827 on the particles contained in pollen ...'*, which was originally intended for private circulation, but was reprinted in the archival literature shortly after its appearance [12]. Brown investigated the way in which pollen acted during impregnation. A plant he studied under the microscope was *Clarkia pulchella*, a wildflower found in the Pacific Northwest of the United States. The pollen of this plant contains granules varying from about five to six micrometers in linear dimension. It is these granules, not the whole pollen grains, upon which Brown made his observation. He wrote '... *While examining the form of these particles immersed in water, I observed many of them very evidently in motion ... These motions were such as to satisfy me, after frequently repeated observation, that they arose neither from currents in the fluid, nor from its gradual evaporation, but belonged to the particle itself*'. The inherent, incessant motion of small particles is nowadays called BROWNian motion in honour of Robert Brown.

Einstein's and Smoluchowski's theory of Brownian motion: In the period between 1829 and about 1900 not much progress was made in the understanding of BROWNian motion, although developments in the theory of heat and kinetic theory stimulated new experiments and conjectures. It is striking that the founders and developers of kinetic theory, Maxwell, Boltzmann, and Clausius, never published anything on BROWNian motion. The reason for the lack in progress was that the major studies of that period focused on the particle velocities. Measurements of the particle velocities gave puzzling results. The path of a small particle, on the length scales available from observations in a microscope, is an extremely erratic curve. In modern language we would say that it is a fractal. Such curves are differentiable almost nowhere. Consequently, the particles whose trajectories they represent have no velocity, as usually defined. It was not until the work of *Einstein* and *Smoluchowski* that it was understood that the velocity is not a useful thing to measure in this context.

Albert Einstein (1879–1955), born in Ulm, Germany, is certainly the best known physicist of the twentieth century, perhaps even of all time. In the year of 1905, he published four papers that at once raised him to the rank of a physicist of the highest caliber: the photon hypothesis to explain the photo effect, for which he received the Nobel prize in physics in 1922 for the year 1921, his first paper on BROWNian motion, and his two first papers on relativity theory. At that time Einstein was employed at the *'The Eidgenössische Amt für Geistiges Eigentum'* in Bern, Switzerland. He did not receive the doctoral degree until the following year, 1906. Interestingly, his thesis was on none of the above problems, but rather concerned with the determination of the dimensions of molecules. His first paper on BROWNian motion was entitled *'Die von der molekularkinetischen Theorie der Wärme geforderte Bewegung von in ruhenden Flüssigkeiten suspendierten Teilchen'* [13]. A second paper was entitled *'Zur Theorie der Brownschen Bewegung'* [14]. Einstein published two additional short papers on this topic [15, 16], but these were of relatively minor interest. Einstein was the first to understand, contrary to many scientists of his time, that the basic quantity is not the velocity but the mean-square displacement of particles. He related the mean-square displacement to the diffusion coefficient.

The Polish physicist and mountaineer *Marian Smoluchowski* (1872–1917) was born in Vienna, Austria. During his lifetime, Poland was not an independent country; it was partitioned between Russia, Prussia, and Austria. Marian Smoluchowski entered the University of Vienna and studied physics under *Joseph Stephan* and *Franz Exner*. He was impressed by the work of Ludwig Boltzmann. In his later life he was called *'der geistige Nachfolger Boltzmanns'* (the intellectual successor of Boltzmann). He got his PhD and his 'venia legendi' from the University of Vienna and was appointed full professor at Lvov University (now Ukraine) in 1903. He accepted a chair in physics at the Jagellonian University at Cracow in 1913, when he was a wellknown physicist of worldwide recognition. Smoluchowski also served as president of the Polish Tatra Society and received the *'Silberne Edelweiss'* from the German and Austrian Alpine Society, an award given to distinguished alpinists.

Smoluchowski's interest for molecular statistics led him already around 1900 to consider BROWNian motion. He did publish his results not before 1906 [17, 18], under the impetus of Einstein's first paper. Smoluchowski later studied BROWNian motion for particles under the influence of an external force [19, 20]. Einstein's and Smoluchowski's scientific paths crossed again, when both considered the theory of the scattering of light near the critical state of a fluid, the critical opalescence. Smoluchowski died as a result of a dysentery epidemic, aggravated by wartime conditions in 1917. Einstein wrote a sympathetic obituary for him with special reference to Smoluchowski's interest in fluctuations [21].

Atomic reality – Perrin's experiments: The idea that matter was made up of atoms was already postulated by Demokrit of Abdeira, an ancient Greek philosopher, who lived about four hundred years before Christ. However, an experimental proof had to wait for more than two millennia. The concept of atoms and molecules took strong hold of the scientific community since the time of English scientist *John Dalton* (1766–1844). It was also shown that the ideas of the Italian scientist *Amadeo Avogadro* (1776–1856) could be used to construct a table of atomic weights, a central idea of chemistry and physics. Most scientists were willing to accept atoms as real, since the facts of chemistry and the kinetic theory of gases provided strong indirect evidence. Yet there were famous sceptics. Perhaps the most prominent ones were the German physical chemist and Nobel laureate *Wilhelm Ostwald* (1853–1932) and the Austrian physicist *Ernst Mach* (1938–1916). They agreed that atomic theory was a useful way of summarising experience. However, the lack of direct experimental verification led them to maintain their scepticism against atomic theory with great vigour.

The Einstein-Smoluchowski theory of Brownian motion provided ammunition for the atomists. This theory explains the incessant motion of small particles by fluctuations, which seems to violate the second law of thermodynamics. The question remained, what fluctuates? Clearly, fluctuations can be explained on the basis of atoms and/or molecules that collide with a Brownian particle and push it around. The key question was then, what is the experimental evidence that the Einstein-Smoluchowski theory is quantitatively correct? The answer had to wait for experiments of the French scientist *Jean Baptiste Perrin* (1870–1942), a convinced atomist. The experiments were difficult. In order to study the dependence of the mean-square displacement on the particle radius, it was necessary to prepare monodisperse suspensions. The experiments of Perrin were successful and showed agreement with the Einstein-Smoluchowski theory [22, 23]. He and his students continued refining the work and in 1909 Perrin published a long paper on his own and his students' research [24]. He became an energetic advocate for the reality of atoms and received the 1926 Nobel prize in physics '... *for his work on the discontinuous structure of matter* ...'.

Crystalline solids and atomic defects: Solid-state physics was born when *Max von Laue* (1879–1960) detected diffraction of X-rays on crystals. His experiments demonstrated that solid matter usually occurs in three-dimensional periodic arrangements of atoms. His discovery, published in 1912 together with *Friedrich* and *Knipping*, was awarded with the 1914 Nobel prize in physics.

However, the ideal crystal of Max von Laue is a 'dead' crystal. Solid-state diffusion and many other properties require deviations from ideality. The Russian physicist *Jakov Il'ich Frenkel* (1894–1952) was the first to introduce the concept of disorder in the field of solid-state physics. He suggested that thermal agitation causes transitions of atoms from their regular lattice sites into interstitial positions leaving behind lattice vacancies [25]. This kind of disorder is now called Frenkel disorder and consists of pairs of vacant lattice sites (vacancies) and lattice atoms on interstitial sites of the host crystal (self-interstitials). Only a few years later, *Wagner and Schottky* [26] generalised the concept of disorder and treated disorder in binary compounds considering the occurrence of vacancies, self-interstititals and antisite defects on both sublattices. Nowadays, it is common wisdom that atomic defects are necessary to mediate diffusion in crystals. The German physicist *Walter Schottky* (1886–1976) taught at the universities of Rostock and Würzburg, Germany, and worked in the research laboratories of Siemens. He had a strong influence on the development of telecommunication. Among Schottky's many achievements a major one was the development of a theory for the rectifying behaviour of metal-semiconductor contact, which revolutionised semiconductor technology. Since 1973 the German Physical Society decorates outstanding achievements of young German scientists in solid-state physics with the 'Walter-Schottky award'.

Kirkendall effect: A further cornerstone of solid-state diffusion comes from the work of *Ernest Kirkendall* (1914–2005). In the 1940s, it was still a widespread belief that atomic diffusion in metals takes place via direct exchange or ring mechanisms. This would suggest that in binary alloys the two components should have the same coefficient of self-diffusion. Kirkendall and coworkers observed the inequality of copper and zinc diffusion during interdiffusion between brass and copper, since the interface between the two different phases moves [27–29]. The direction of the mass flow was such as might be expected if zinc diffuses out of the brass more rapidly than copper diffuses in. Such phenomena have been observed in the meantime in many other binary alloys. The movement of inert markers placed at the initial interface of a diffusion couple is now called the *Kirkendall effect*. Kirkendall's discovery, which took the scientific world about ten years to be appreciated, is nowadays taken as evidence for a vacancy mechanism of diffusion in metals and alloys. Kirkendall left research in 1947 and served as secretary of the American Institute of Mining, Metallurgical and Petroleum Engineers. He then became a manager at the United Engineering Trustees and concluded his career as a vice president of the American Iron and Steel Institute.

Thermodynamics of irreversible processes: The Norwegian Nobel laureate in chemistry of 1968 *Lars Onsager* (1903–1976) had widespread interests, which include colloids, dielectrics, order-disorder transitions, hydrodynamics, thermodynamics, and statistical mechanics. His work had a great impact on the 'Thermodynamics of Irreversible Processes'. He received the Nobel prize for the reciprocity theorem, which is named after him. This theorem states that the matrix of phenomenological coefficients, which relate fluxes and generalised forces of transport theory, is symmetric. The non-diagonal terms of the Onsager matrix also include cross-phenomena, such as the influence of a gradient in concentration of one species upon the flow of another one or the effect of a temperature gradient upon the flow of various atomic species, both of which can be significant for diffusion processes.

Solid-state diffusion after World War II: The first period of solid-state diffusion under the guidance of Roberts-Austen, von Hevesy, Frenkel, and Schottky was followed by a period which started in the mid 1930s, when 'artificial' radioactive isotopes, produced in accelerators, became available. Soon after World War II nuclear reactors became additional sources of radioisotopes. This period saw first measurements of self-diffusion on elements other than lead. Examples are self-diffusion of gold [30, 31], copper [32], silver [33], zinc [34], and α-iron [35]. In all these experiments the temperature dependence of diffusion was adequately described by the Arrhenius law, which by about 1950 had become an accepted 'law of nature'.

It is hardly possible to review the following decades, since the field has grown explosively. This period is characterised by the extensive use of radioactive isotopes produced in nuclear reactors and accelerators, the study of the dependence of diffusion on the tracer mass (isotope effect), and of diffusion under hydrostatic pressure. Great improvements in the precision of diffusion measurements and in the accessible temperature ranges were achieved by using refined profiling techniques such as electron microprobe analysis, sputter sectioning, secondary ion mass spectroscopy, Rutherford back-scattering, and nuclear reaction analysis. Methods not directly based on Fick's law to study atomic motion such as the anelastic or magnetic after-effect, internal friction, and impedance spectroscopy for ion-conducting materials were developed and widely applied. Completely new approaches making use of nuclear methods such as nuclear magnetic relaxation (NMR) [36], Mössbauer spectroscopy (MBS), and quasielastic neutron scattering (QENS) have been successfully applied to diffusion problems.

Whereas diffusion on solid surfaces nowadays can be recorded by means of scanning tunnelling microscopy, the motion of atoms inside a solid is still difficult to observe in a direct manner. Nevertheless, diffusion occurs and it is the consequence of a large number of atomic or molecular jumps. The mathematics of the random-walk problem allows one to go back and forth between the diffusion coefficient and the jump distances and jump rates of the diffusing atoms. Once the diffusion coefficient was interpreted in this way, it was only a question of time before attempts were made to understand the measured values in terms of atomistic diffusion mechanisms.

The past decades have seen a tremendous increase in the application of computer modeling and simulation methods to diffusion processes in materials. Along with continuum modeling aimed at describing complex diffusion problems by differential equations, atomic-level modeling such as ab-initio calculations, molecular dynamics studies, and Monte Carlo simulations, play an increasingly important rôle as means of gaining fundamental insights into diffusion processes.

Grain-boundary diffusion: By 1950, the fact that grain-boundary diffusion exists had been well documented by autoradiographic images [37], from which the ratio of grain-boundary to lattice-diffusion coefficients in metals was estimated to be a few orders of magnitude [38]. *Fisher* published his now classical paper presenting the first theoretical model of grain-boundary diffusion in 1951 [39]. That pioneering paper, together with concurrent experimental work by *Hoffman and Turnbull* (1915–2007) [40], initiated the whole area of quantitative studies of grain-boundary diffusion in solids. Nowadays, grain-boundary diffusion is well recognised to be a transport phenomenon of great fundamental interest and of technical importance in normal polycrystals and in particular in nanomaterials.

Distinguished scientists of solid-state diffusion: In what follows some people are mentioned, who have made or still make significant contributions to the field of solid-state diffusion. The author is well aware that such an attempt is necessarily incomplete and perhaps biased by personal flavour.

Wilhelm Jost (1903–1988) was a professor of physical chemistry at the University of Göttingen, Germany. He had a very profound knowledge of diffusion not only for solids but also for liquids and gases. His textbook 'Diffusion in Solids, Liquids and Gases', which appeared for the first time in 1952 [41], is still today a useful source of information. Although the author of the present book never had the chance to meet Wilhelm Jost, it is obvious that Jost was one of the few people who overlooked the whole field of diffusion, irrespective whether diffusion in condensed matter or in gases is concerned.

John Bardeen (1908–1991) and *C. Herring*, both from the Bell Telephone Laboratories, Murray Hill, New Jersey, USA, recognised in 1951 that diffusion of atoms in a crystal by a vacancy mechanism is correlated [42]. After this pioneering work it was soon appreciated that correlation effects play an important rôle for any solid-state diffusion process, when point defects act as diffusion vehicles. Nowadays, a number of methods are available for the calculation of correlation factors. Correlation factors of self-diffusion in elements with cubic lattices are usually numbers characteristic for a given diffusion mechanism. Correlation factors of foreign atom diffusion are temperature dependent and thus contribute to the activation enthalpy of foreign atom diffusion. It may be interesting to mention that John Bardeen is one of the very few scientists, who received the Nobel prize twice. Schockley, Bardeen, and Brattain were awarded for their studies of semiconducors and for the development of the transition in 1956. Bardeen, Cooper, and Schriefer received the 1972 Nobel price for the so-called BCS theory of superconductivity.

Yakov E. Geguzin (1918–1987) was born in the town of Donetsk, now Ukraine. He graduated from Gor'kii State University at Kharkov, Ukraine. After years of industrial and scientific work in solid-state physics he became professor at the Kharkov University. He founded the Department of Crystal Physics, which he headed till his death. The main scientific areas of Geguzin were diffusion and mass transfer in crystals. He carried our pioneering studies of surface diffusion, diffusion and mass transfer in the bulk and on the surface of metals and ionic crystals, interdiffusion and accompanying effects in binary metal and ionic systems. He was a bright person, a master not only to realise experiments but also to tell of them. His enthusiasm combined with his talent for physics attracted many students. His passion is reflected in numerous scientific and popular books, which include topics such as defects in metals, physics of sintering, diffusion processes on crystal surfaces, and an essay on diffusion in crystals [43].

Norman Peterson (1934–1985) was an experimentalist of the highest calibre and a very active and lively person. His radiotracer diffusion studies performed together with *Steven Rothman, John Mundy, Himanshu Jain* and other members of the materials science group of the Argonne National Laboratory, Illinois, USA, set new standards for high precision measurements of tracer diffusivities in solids. Gaussian penetration profiles of lattice diffusion over more than three orders of magnitude in tracer concentration were often reported. This high precision allowed the detection of small deviations from Arrhenius behaviour of self-diffusion, e.g., in fcc metals, which could be attributed to the simultaneous action of monovacancy and divacancy mechanisms. The high precision was also a prerequisite for successful isotope effect experiments of tracer diffusion, which contributed a lot to the interpretation of diffusion mechanisms. Furthermore, the high precision permitted reliable studies of grain-boundary diffusion in poly- and bi-crystals with tracer techniques. The author of this book collaborated with Norman Peterson, when Peterson spent a sabbatical in Stuttgart, Germany, as a Humboldt fellow. The author and his groups either at the University of Stuttgart, Germany, until 1984 or from then at the University of Münster, Germany, struggled hard to fulfill 'Peterson standards' in own tracer diffusion experiments.

John Manning (1933–2005) had strong interests in the 'Diffusion Kinetics of Atoms in Crystals', as evidenced by the title of his book [44]. He received his PhD from the University of Illinois, Urbana, USA. Then, he joined the metals physics group at the National Bureau of Standards (NBS/NIST) in Washington. Later, he was the chief of the group until his retirement. He also led the Diffusion in Metals Data Center together with *Dan Butrymowics* and *Michael Read*. The obituary published by NIST has the following very rightful statement: *'His papers have explained the significance of the correlation factor and brought about an appreciation of its importance in a variety of diffusion phenomena'*. The author of this book met John Manning on several conferences, Manning was a great listener and a strong advocate, fair, honest, friendly, courteous, kind and above all a gentleman.

Paul Shewmon is professor emeritus in the Department of Materials Science and Engineering at the Ohio State Univeristy, USA. He studied at the University of Illinois and at the Carnegie Mellon University, where he received his PhD. Prior to becoming a professor at the Ohio State University he served among other positions as director of the Materials Science Division of the Argonne National Laboratory, Illinois, and as director of the Division of Materials Research for the National Science Foundation of the United States. Shewmon is an outstanding materials scientists of the United States. He has also written a beautiful textbook on 'Diffusion in Solids', which is still today usefull to introduce students into the field. It appeared first in 1963 and in slightly revised form in 1989 [45].

The diffusion community owes many enlightening contributions to the British theoretician *Alan B. Lidiard* from AEA Technology Harwell and the Department of Theoretical Chemistry, University of Oxford, GB. He co-authored the textbook 'Atomic Transport in Solids' together with *A.R. Allnatt* from the Department of Chemistry, University of Western Ontario, Canada [46]. Their book provides the fundamental statistical theory of atomic transport in crystals, that is the means by which processes occurring at the atomic level are related to macroscopic transport coefficients and other observable quantities. Alan Lidiard is also the father of the so-called 'five-frequency model' [47]. This model provides a theoretical framework for solute and solvent diffusion in dilute alloys and permits to calculate correlation factors for solute and solvent diffusion. It has been also successfully applied to foreign atom diffusion in ionic crystals.

Jean Philibert, a retired professor of the University Paris-sud, France, is an active member and highly respected senior scientist of the international diffusion community. Graduate students in solid-state physics, physical metallurgy, physical and inorganic chemistry, and geophysical materials as well as physicists, metallurgists in science and industrial laboratories benefit from his comprehensive textbook 'Atom Movements – Diffusion and Mass Transport in Solids', which was translated from the French-language book of 1985 by *Steven J. Rothman*, then senior scientist at the Argonne National Laboratory, Illinois, USA [48]. *David Lazarus*, then a professor at the University of Illinois, Urbana, USA, wrote in the preface to Philiberts book: *'This is a work of love by a scientist who understands the field thoroughly and deeply, from its fundamental atomistic aspects to the most practical of its 'real-world' applications.'* The author of the present book often consulted Philibert's book and enjoyed Jean Philibert's well-rounded contributions to scientific discussions during conferences.

Graeme Murch, head of the theoretical diffusion group at the University of Newcastle, Australia, serves the international diffusion community in many respects. He is an expert in computer modeling of diffusion processes and has a deep knowledge of irreversible thermodynamics and diffusion. He authored and co-authored chapters in several specialised books on diffusion, stand-alone chapters on diffusion in solids, and a chapter about interdiffusion in a data collection [69]. He also edited books on certain aspects of diffusion. Graeme Murch is since many years the editor-in-chief of the international journal 'Defect and Diffusion Forum'. This journal is an important platform of the solid-state diffusion community. The proceedings of many international diffusion conferences have been published in this journal.

Other people, who serve or served the diffusion community with great success, can be mentioned only shortly. Many of them were also involved in the laborious and time-consuming organisation of international conferences in the field of diffusion:

The Russian scientists *Semjon Klotsman*, the retired chief of the diffusion group in Jekaterinburg, Russia, and *Boris Bokstein*, head of the thermodynamics and physical chemistry group at the Moscow Institute of Steels and Alloys, Moscow, Russia, organised stimulating international conferences on special topics of solid-state diffusion.

Deszö Beke, head of the solid-state physics department at the University of Debrecen, Hungary, and his group contribute significantly to the field and organised several conferences. The author of this book has a very good remembrance to DIMETA-82 [49], which took place at lake Balaton, Hungary, in 1982. This conference was one of the very first occasions where diffusion experts from western and eastern countries could participate and exchange experience in a fruitful manner, although the 'iron curtain' still did exist. DIMETA-82 was the starting ignition for a series of international conferences on diffusion in materials. These were: DIMETA-88 once more organised by Beke and his group at lake Balaton, Hungary [50]; DIMAT-92 organised by *Masahiro Koiwa and Hideo Nakajima* in Kyoto, Japan [51]; DIMAT-96 organised by the author of this book and his group in Nordkirchen near Münster, Germany [52]; DIMAT-2000 organised by *Yves Limoge and J.L. Bocquet* in Paris, France [53]; DIMAT-2004 organised by *Marek Danielewski* and colleagues in Cracow, the old capital of Poland [54].

Devendra Gupta, retired senior scientist from the IBM research laboratories in Yorktown Heights, New York, USA, was one of the pioneers of grain-boundary and dislocation diffusion studies in thin films. He organised symposia on 'Diffusion in Ordered Alloys' and on 'Diffusion in Amorphous Materials' and co-edited the proceedings [55, 56]. Gupta also edited a very useful book on 'Diffusion Processes in Advanced Technological Materials', which appeared in 2005 [57].

Yuri Mishin, professor at the Computational Materials Science group of Georg Mason University, Fairfax, Virginia, USA, is an expert in grain-boundary diffusion and in computer modeling of diffusion processes. He co-authored a book on 'Fundamentals of Grain and Interphase Boundary Diffusion' [58] and organised various symposia, e.g., one on 'Diffusion Mechanisms in Crystalline Materials' [59].

Frans van Loo, retired professor of physical chemistry at the Technical University of Eindhoven, The Netherlands, is one of the few experts in multiphase diffusion and of diffusion in ternary systems. He is also a distinguished expert in Kirkendall effect studies. Van Loo and his group have made significant contributions to the question of microstructural stability of the Kirkendall plane. It was demonstrated experimentally that binary systems with stable, unstable, and even with several Kirkendall planes exist.

Mysore Dayananda is professor of the School of Engineering of Purdue University, West Lafayette, Indiana, USA. His research interests mainly concern interdiffusion, multiphase diffusion and diffusion in ternary alloys. Dayananda has also organised several specialised diffusion symposia and co-edited the proceedings [60, 61].

The 150th anniversary of the laws of Fick and the 100th anniversary of Einstein's theory of Brownian motion was celebrated on two conferences. One conference was organised by *Jörg Kärger*, University of Leipzig, Germany, and *Paul Heitjans*, University of Hannover, Germany, at Leipzig in 2005. It was was devoted to the 'Fundamentals of Diffusion' [62]. Heitjans and Kärger also edited a superb text on diffusion, in which experts cover various topics concerning methods, materials and models [63]. The anniversaries were also celebrated during a conference in Moscow, Russia, organised by *Boris Bokstein and Boris Straumal* with the topics 'Diffusion in Solids – Past, Present and Future' [64].

Andreas Öchsner, professor at the University of Aveiro, Portugal, organised a first international conference on 'Diffusion in Solids and Liquids (DSL2005)' in 2005 [65]. The interesting idea of this conference was, to bring diffusion experts from solid-state and liquid-state diffusion together again. Obviously, this idea was successful since many participants also attended DSL2006 only one year later [66].

Diffusion research at the University of Münster, Germany: Finally, one might mention, that the field of solid-state diffusion has a long tradition at the University of Münster, Germany – the author's university. *Wolfgang Seith* (1900–1955), who had been a coworker of Georg von Hevesy at the University of Freiburg, Germany, was full professor of physical chemistry at the

University of Münster from 1937 until his early death in 1955. He established diffusion research in Münster under aggravated war-time and post-war conditions. He also authored an early textbook on 'Diffusion in Metallen', which appeared in 1939 [66]. A revised edition of this book was published in 1955 and co-authored by Seith's associate Heumann [67]. *Theodor Heumann* (1914–2002) was full professor and director of the 'Institut für Metallforschung' at the University of Münster from 1958 until his retirement in 1982. Among other topics, he continued research in diffusion, introduced radiotracer techniques and electron microprobe analysis together with his associate *Christian Herzig*. As professor emeritus Heumann wrote a new book on 'Diffusion in Metallen', which appeared in 1992 [68]. Its German edition was translated to Japanese language by *S.-I. Fujikawa*. The Japanese edition appeared in 2006.

The author of the present book, *Helmut Mehrer*, was the head of a diffusion group at the University of Stuttgart, Germany, since 1974. He was then appointed full professor and successor on Heumann's chair at the University of Münster in 1984 and retired in 2005. Diffusion was reinforced as one of the major research topics of the institute. In addition to metals, further classes of materials have been investigated and additional techniques applied. These topics have been pursued by the author and his colleagues *Christian Herzig, Nicolaas Stolwijk, Hartmut Bracht,* and *Serguei Divinski*. The name of the institute was changed into 'Institut für Materialphysik' in accordance with the wider spectrum of materials in focus. Metals, intermetallic compounds, metallic glasses, quasicrystals, elemental and compound semiconductors, and ion-conducting glasses and polymers have been investigated. Lattice diffusion has been mainly studied by tracer techniques using mechanical and/or sputter-sectioning techniques and in cooperation with other groups by SIMS profiling. Interdiffusion and multi-phase diffusion was studied by electron microprobe analysis. The pressure and mass dependence of diffusion has been investigated with radiotracer techniques on metals, metallic and oxide glasses. Grain-boundary diffusion and segregation into grain boundaries has been picked up as a further topic. Ionic conduction studied by impedance spectroscopy combined with element-specific tracer measurements, provided additional insight into mass and charge transport in ion-conducting oxide glasses and polymer electrolytes. Numerical modeling of diffusion processes has been applied to obtain a better understanding of experimental data. A data collection on diffusion in metals and alloys was edited in 1990 [69], DIMAT-96 was organised in 1996 and the conference proceedings were edited [52].

Further reading on history of diffusion: An essay on the early history of solid-state diffusion has been given by *L. W. Barr* in a paper on *'The origin of quantitative diffusion measurements in solids. A centenary view'* [71]. *Jean Philibert* has written a paper on *'One and a Half Century of Diffusion: Fick, Einstein, before and beyond'* [72]. Remarks about the more recent history can be found in an article of *Steven Rothman* [70], *Masahiro Koiwa* [73], and

Alfred Seeger [74]. Readers interested in the history of diffusion mechanisms of solid-state diffusion may benefit from *C. Tuijn's* article on *'History of models for solid-state diffusion'* [75]. Steven Rothman ends his personal view of diffusion research with the conclusion that *'... Diffusion is alive and well'*.

References

1. T. Graham, Quaterly Journal of Science, Literature and Art **27**, 74 (1829)
2. T. Graham, Philos. Mag. **2**, 175, 222, 351 (1833)
3. T. Graham, Philos. Trans. of the Roy. Soc. of London **140**, 1 (1950)
4. A.E. Fick, Annalen der Physik und Chemie **94**, 59 (1855)
5. A.E. Fick, Philos. Mag. **10**, 30 (1855)
6. A.E. Fick, Gesammelte Abhandlungen, Würzburg (1903)
7. J. Stephan, Sitzungsberichte d. Kaiserl. Akad. d. Wissenschaften II **79**, 161 (1879)
8. W.C. Roberts-Austen, Phil. Trans. Roy. Soc. A **187**, 383 (1896)
9. S. Dushman and I. Langmuir, Phys. Rev. **20**, 113 (1922)
10. J. Groh, G. von Hevesy, Ann. Physik **63**, 85 (1920)
11. J. Groh, G. von Hevesy, Ann. Physik **65**, 216 (1921)
12. R. Brown, Edin. New. Phil. J **5**, 358–371 (1828); Edin. J. Sci. **1**, 314 (1829)
13. A. Einstein, Annalen der Physik **17**, 549 (1905)
14. A. Einstein, Annalen der Physik **19**, 371 (1906)
15. A. Einstein, Z. für Elektrochemie **13**, 98 (1907)
16. A. Einstein, Z. für Elektrochemie **14**, 235 (1908)
17. M. van Smoluchowski, Annalen der Physik **21**, 756 (1906)
18. M. van Smoluchowski, Physikalische Zeitschrift **17**, 557 (1916)
19. M. van Smoluchowski, Bull. Int. de l'Acad. de Cracovie, Classe de Sci. math.-nat. **A**, 418 (1913)
20. M. van Smoluchowski, Annalen der Physik **48**, 1103 (1915)
21. A. Einstein, Naturwissenschaften **50**, 107 (1917)
22. J. Perrin, C.R. Acad. Sci. Paris **147**, 475 (1908)
23. J. Perrin, C.R. Acad. Sci. Paris **147**, 530 (1908)
24. J. Perrin, Ann. de Chim. et de Phys. **18**, 1 (1909)
25. J.I. Frenkel, Z. Physik **35**, 652 (1926)
26. C. Wagner, W. Schottky, Z. Phys. Chem. B **11**, 163 (1930)
27. E.O. Kirkendall, L. Thomassen, C. Upthegrove, Trans. AIME **133**, 186 (1939)
28. E.O. Kirkendall, Trans. AIME **147**, 104 (1942)
29. A.D. Smigelskas, E.O. Kirkendall, Trans. AIME **171**, 130 (1947)
30. A.M. Sagrubskij, Phys. Z. Sowjetunion **12**, 118 (1937)
31. H.A.C. McKay, Trans. Faraday Soc. **34**, 845 (1938)
32. B.V. Rollin, Phys. Rev. **55**, 231 (1939)
33. W.A. Johnson, Trans. Americ. Inst. Min. Met. Engrs. **143**, 107 (1941)
34. P.H. Miller, R.R. Banks, Phys. Rev. **61**, 648 (1942)
35. C.E. Birchenall, R.F. Mehl, J. Appl. Phys. **19**, 217 (1948)
36. N. Bloembergen, E.H. Purcell, and R.V. Pound, Phys. Rev. **73**, 674 (1948)
37. R.S. Barnes, Nature **166**, 1032 (1950)
38. A.D. Le Claire, Philos. Mag. **42**, 468 (1951)

39. J.C. Fisher, J. Appl. Phys. **22**, 74 (1951)
40. R.E. Hoffman, D. Turnbull, J. Appl. Phys. **22**, 634 (1951)
41. W. Jost, *Diffusion in Solids, Liquids, and Gases*, Academic Press, New York, 1952
42. J. Bardeen, C. Herring, in: *Atom Movements*, ASM Cleveland, p. 87, 1951
43. Y.E. Geguzin, German edition: *Grundzüge der Diffusion in Kristallen*, VEB Verlag für Grundstoffindustrie, Leipzig, 1977
44. J.R. Manning, *Diffusion Kinetics of Atoms in Crystals*, van Norstrand Comp., 1968
45. P.G. Shewmon, *Diffusion in Solids*, 1st edition, MacGraw Hill Book Company, 1963; 2nd edition, The Minerals, Metals & Materials Society, Warrendale, USA, 1989
46. A.R. Allnatt, A.B. Lidiard, *Atomic Transport in Solids*, Cambridge University Press, 1991
47. A.B. Lidiard, Philos. Mag. **40**, 1218 (1955)
48. J. Philibert, *Atom Movements – Diffusion and Mass Transport in Solids*, Les Editions de Physique, Les Ulis, Cedex A, France, 1991
49. *DIMETA-82, Diffusion in Metals and Alloys*, F.J. Kedves, D.L. Beke (Eds.), Defect and Diffusion Monograph Series No. 7, Trans Tech Publications, Switzerland, 1983
50. *DIMETA-88, Diffusion in Metals and Alloys*, F.J. Kedves, D.L. Beke (Eds.), Defect and Diffusion Forum **66–69**, 1989
51. *DIMAT-92, Diffusion in Materials*, M. Koiwa, K. Hirano, H. Nakajima, T. Okada (Eds.), Trans Tech Publications, Zürich, Switzerland, 1993; also Defect and Diffusion Forum **95–98**, 1993
52. *DIMAT-96, Diffusion in Materials*, H. Mehrer, Chr. Herzig, N.A. Stolwijk, H. Bracht (Eds.), Scitec Publications, Zürich-Uetikon, Switzerland, 1997; also Defect and Diffusion Forum **143–147**, 1997
53. *DIMAT-2000, Diffusion in Materials*, Y. Limoge, J.L.Bocquet (Eds.), Scitec Publications, Zürich-Uetikon, Switzerland, 2001; also Defect and Diffusion Forum **194–199**, 2001
54. *DIMAT-2004, Diffusion in Materials*, M. Danielewski, R. Filipek, R. Kozubski, W. Kucza, P. Zieba (Eds.), Trans Tech Publications, Zürich-Uetikon, Switzerland, 2005; also Defect and Diffusion Forum **237–240**, 2005
55. B. Fultz, R.W. Cahn, D. Gupta (Eds.), *Diffusion in Ordered Alloys*, The Minerals, Metals & Materials Society, Warrendale, Pennsylvania, USA, 1993
56. H. Jain, D. Gupta (Eds.), *Diffusion in Amorphous Materials*, The Minerals, Metals & Materials Society, Warrendale, Pennsylvania, 1993
57. D. Gupta (Ed.), *Diffusion Processes in Advanced Technological Materials*, William Andrew, Inc., 2005
58. I. Kaur, Y. Mishin, W. Gust, *Fundamentals of Grain and Interphase Boundary Diffusion*, John Wiley & Sons, Ltd., 1995
59. Y. Mishin, G. Vogl, N. Cowern, R. Catlow, R. Farkas (Eds.), *Diffusion Mechanism in Crystalline Materials*, Mat. Res. Soc. Symp. Proc. Vol. 527, Materials Research Society, Warrendale, Pennsylvania, USA, 1997
60. D. Gupta, A.D. Romig, M.A. Dayananda (Eds.), *Diffusion in High Technological Materials*, Trans Tech Publications, Aedermannsdorf, Switzerland, 1988
61. A.D. Romig, M.A. Dayanada (Eds.), *Diffusion Analysis and Applications*, The Minerals, Metals & Materials Society, Warrendale, Pennsylvania, 1989

62. J. Kärger, F. Grindberg, P. Heitjans (Eds.), *Diffusion Fundamentals* – Leipzig 2005, Leipziger Universitätsverlag GmbH, 2005
63. P. Heitjans, J. Kärger (Eds.), *Diffusion in Condensed Matter – Methods, Materials, Models*, Springer-Verlag, 2005
64. B.S. Bokstein, B.B. Straumal (Eds.), *Diffusion in Solids – Past, Present, and Future*, Trans Tech Publications, Ltd., Switzerland, 2006; also Defect and Diffusion Forum **249**, 2006
65. A. Öchsner, J. Gracio, F. Barlat (Eds.), *First International Conference on Diffusion in Solids and Liquids – DSL 2005*, Centre for Mechanical Technology and Automation and Department of Mechanical Engineering, University of Aveiro, Portugal, Editura MEDIAMIRA, Cluj-Napoca, 2006
66. W. Seith, *Diffusion in Metallen*, Verlag Julius Spriger, 1939
67. W. Seith, Th. Heumann, *Diffusion in Metallen*, Springer-Verlag, 1955
68. Th. Heumann, *Diffusion in Metallen*, Springer-Verlag, 1992; Japanese language edition 2006 translated by S.-I. Fujikawa
69. H. Mehrer (Vol. Ed.), *Diffusion in Solid Metals and Alloys*, Landolt-Börnstein, Numerical Data and Functional Relationships in Science and Technology, New Series, Group III: Crystal and Solid State Physics, Vol. 26, Springer-Verlag, 1990
70. S.J. Rothman, Defect and Diffusion Forum **99–100**, 1 (1993)
71. L.W. Barr, Defect and Diffusion Forum **143–147**, 3 (1997); see also [52]
72. J. Philibert, in: *Diffusion Fundamentals – Leipzig 2005*, Universitätsverlag Leipzig 2005, p.8; see also [62]
73. M. Koiwa, in: *Proc. of PRIMCN -3*, Honolulu, Hawai, July 1998
74. A. Seeger, Defect and Diffusion Forum **143–147**, 21 (1997); see also [52]
75. C. Tuijn, Defect and Diffusion Forum **143–147**, 11 (1997); see also [52]

1.2 Bibliography of Solid-State Diffusion

In this section, we list diffusion-related bibliography from the past four or five decades. Textbooks on diffusion in solids and some books that are devoted to the mathematics of diffusion are supplemented by monographs and/or books on specific topics or materials, and by stand-alone chapters on diffusion. Conference proceedings of international conferences on diffusion in solids and comprehensive collections of diffusion data complete the bibliography. The literature is ordered in each section according to the year of publication.

General Textbooks

R.M. Barrer, *Diffusion in and through Solids*, Cambridge, The Syndics of the Cambridge University Press, first printed 1941, reprinted with corrections 1951

L.A. Girifalco, *Atomic Migration in Crystals*, Blaisdell Publ. Comp., New York, 1964

W. Jost, *Diffusion in Solids, Liquids, Gases*, Academic Press, Inc., New York, 1952, 4th printing with addendum, 1965

Y. Adda and J. Philibert *La Diffusion dans les Solides*, 2 volumes, Presses Universitaires de France, 1966

J.R. Manning, *Diffusion Kinetics of Atoms in Crystals*, D. van Norstrand Company, Inc., Princeton, 1968

C.P. Flynn, *Point Defects and Diffusion*, Clarendon Press, Oxford, 1972

J.P. Stark, *Solid-State Diffusion*, John Wiley & Sons, New York, 1976

S. Mrowec, *Defects and Diffusion – an Introduction*, Materials Science Monographs, Vol. 5, Elsevier, Amsterdam, 1980

R.J. Borg and G.J. Dienes, *An Introduction to Solid-State Diffusion*, Academic Press, Inc., 1988

P.G. Shewmon, *Diffusion in Solids*, 1st edition, MacGraw-Hill Book Company, Inc., 1963; 2nd edition, The Minerals, Metals & Materials Society, Warrendale, USA, 1989

J.S. Kirkaldy and D.J. Young, *Diffusion in the Condensed State*, The Institute of Metals, London, 1987

J. Philibert, *Atom Movement – Diffusion and Mass Transport in Solids*, Les Editions de Physique, Les Ulis, Cedex A, France, 1991

A.R. Allnatt and A.B. Lidiard, *Atomic Transport in Solids*, Cambridge University Press, 1993

D.S. Wilkinson, *Mass Transport in Solids and Liquids*, Cambridge University Press, 2000

M.E. Glicksman, *Diffusion in Solids – Field Theory, Solid-State Principles and Applications*, John Wiley & Sons, Inc., 2000

Mathematics of Diffusion

H.S. Carslaw and J.C. Jaeger, *Conduction of Heat in Solids*, Clarendon Press, Oxford, 1959

J. Crank, *The Mathematics of Diffusion*, 2nd edition, Oxford University Press, 1975

R. Ghez, *A Primer of Diffusion Problems*, John Wiley & Sons, Inc., 1988

J. Crank, *Free and Moving Boundary Problems*, Oxford University Press, Oxford, 1984; reprinted in 1988, 1996

R.M. Mazo, *Brownian Motion – Fluctuations, Dynamics, and Applications*, Clarendon Press, Oxford, 2002

Specialised Books and Monographs on Solid-State Diffusion

W. Seith and Th. Heumann, *Diffusion in Metallen*, Springer-Verlag, Berlin, 1955

B.I. Boltaks, *Diffusion in Semiconductors*, translated from Russian by J.I. Carasso, Infosearch Ltd., London, 1963

G.R. Schulze, *Diffusion in metallsichen Werkstofffen*, VEB Verlag für Grundstoffindustrie, Leipzig, 1970

G. Neumann and G.M. Neumann, *Surface Self-diffusion of Metals*, Diffusion and Defect Monograph Series No. 1, edited by Y. Adda, A.D. Le Claire, L.M. Slifkin, F.H. Wöhlbier, Trans Tech SA, Switzerland, 1972

D. Shaw (Ed.), *Atomic Diffusion in Semiconductors*, Plenum Press, New York, 1973

J.N. Pratt and P.G.R. Sellors, *Electrotransport in Metals and Alloys*, Trans Tech Publications, Zürich, 1973

G. Frischat, *Ionic Diffusion in Oxide Glasses*, Diffusion and Defect Monograph Series No. 3/4, edited by Y. Adda, A.D. Le Claire, L.M. Slifkin, F.H. Wöhlbier, Trans Tech SA, Switzerland, 1973

B. Tuck, *Introduction to Diffusion in Semiconductors*, IEE Monograph Series 16, Inst. Electr. Eng., 1974

A.S. Nowick, J.J. Burton (Eds.), *Diffusion in Solids – Recent Developments*, Academic Press, Inc. 1975

H. Wever, *Elektro- und Thermotransport in Metallen*, Johann Ambrosius Barth, Leipzig, 1975

G.E. Murch, *Atomic Diffusion Theory in Highly Defective Solids*, Diffusion and Defect Monograph Series No. 6, edited by Y. Adda, A.D. Le Claire, L.M. Slifkin, F.H. Wöhlbier, Trans Tech SA, Switzerland, 1980

L.N. Larikov, V.V. Geichenko, and V.M. Fal'chenko, *Diffusion Processes in Ordered Alloys*, Kiev 1975, English translation published by Oxonian Press, New Dehli, 1981

G.E. Murch and A.S. Nowick (Eds.), *Diffusion in Crystalline Solids*, Academic Press, Inc., 1984

G.B. Fedorov and E.A. Smirnov, *Diffusion in Reactor Materials*, Trans Tech Publications, Zürich, Switzerland, 1984

B. Tuck, *Diffusion in III-V Semiconductors*, A. Hilger, London, 1988

W.R. Vieth, *Diffusion in and through Polymers – Principles and Application*, Carl Hanser Verlag, Munich, 1991

G.E. Murch (Ed.), *Diffusion in Solids – Unsolved Problems*, Trans Tech Publications, Ltd., Zürich, Switzerland, 1992

Th. Heumann, *Diffusion in Metallen*, Springer-Verlag, 1992

I. Kaur, Y. Mishin, and W. Gust, *Fundamentals of Grain and Interphase Boundary Diffusion*, John Wiley & Sons, Ltd., 1995

H. Schmalzried, *Chemical Kinetics of Solids*, VCH Verlagsgesellschaft mbH, Weinheim, Germany, 1995

E.L. Cussler, *Diffusion – Mass Transfer in Fluid Systems*, Cambridge University Press, 1997

J. Kärger, P. Heitjans, and R. Haberlandt (Eds.), *Diffusion in Condensed Matter*, Friedr. Vieweg & Sohn Verlagsgesellschaft mbH, Braunschweig/Wiesbaden, Germany, 1998

D.S. Wilkinson, *Mass Transport in Solids and Liquids*, Cambridge University Press, 2000

V.I. Dybkov, *Reaction Diffusion and Solid State Chemical Kinetics*, The IPMS Publications, Kyiv, Ukraine, 2002

G. Neumann and C. Tuijn, *Impurity Diffusion in Metals*, Scitec Publications Ltd, Zürich-Uetikon, Switzerland, 2002

R.H. Doremus, *Diffusion of Reactive Molecules in Solids and Melts*, John Wiley and Sons, Inc., 2002

D.L. Beke (Ed.) *Nanodiffusion*, Special Issue of J. of Metastable and Nanocrystalline Materials **19**, 2004

A. Gusak, *Diffusion, Reactions, Coarsening – Some New Ideas*, Cherkassy National University, 2004

J. Maier, *Physical Chemistry of Ionic Materials – Ions and Electrons*, J. Wiley & Sons, Ltd., 2004

D. Gupta (Ed.), *Diffusion Processes in Advanced Technological Materials*, William Andrew, Inc., 2005

P. Heitjans, J. Kärger (Eds.), *Diffusion in Condensed Matter – Methods, Materials, Models*, Springer-Verlag, 2005

Y. Iijima (Ed.), *Diffusion Study in Japan 2006*, Research Signpost, Kerala, India, 2006

Stand-alone Chapters on Diffusion in Solids

R.E. Howard and A.B. Lidiard, *Matter Transport in Solids*, Reports on Progress in Physics **27**, 161 (1964)

A.D. Le Claire, *Diffusion*, in: *Treatise in Solid State Chemistry, Vol. 4, Reactivity of Solids*, edited by N.B. Hannay, Plenum Press, 1975

S.J. Rothman, *The Measurement of Tracer Diffusion Coefficients in Solids*, in: *Diffusion in Crystalline Solids*, edited by G.E. Murch and A.S. Nowick, Academic Press, Orlando, Fl, 1984

G.E. Murch, *Diffusion Kinetics in Solids*, Ch. 3 in *Phase Transformations in Materials*, G. Kostorz (Ed.), Wiley-VCh Verlag GmbH, Weinheim, Germany, 2001

J. L Bocquet, G. Brebec, and Y. Limoge, *Diffusion in Metals and Alloys*, Ch. 7 in *Physical Metallurgy*, 4th edition, R.W. Cahn and P. Haasen (Eds.), Elsevier Science BV, 1996

H. Mehrer, *Diffusion in Metals*, in: *Diffusion in Condensed Matter*, edited by J. Kärger, P. Heitjans, and R. Haberlandt, Friedr. Vieweg & Sohn Verlagsgesellschaft mbH, Braunschweig/Wiesbaden, Germany, 1998

H. Mehrer, *Diffusion: Introduction and Case Studies in Metals and Binary Alloys*, Ch. 1 in: *Diffusion in Condensed Matter – Methods, Materials, Models*, Springer-Verlag, 2005. Braunschweig/Wiesbaden, Germany, 1998

Conference Proceedings

J.A. Wheeler, Jr. and F.R. Winslow (Eds.), *Diffusion in Body-Centered Cubic Metals*, American Society for Metals, Metals Park, Ohio, 1965

J.N. Sherwood, A.V. Chadwick, W.M. Muir, and F.L. Swinton (Eds.), *Diffusion Processes*, 2 volumes, Gordon and Breach Science Publishers, London, 1971

H.I. Aaronson (seminar coordinator), *Diffusion*, American Society for Metals, Metals Park, Ohio, 1973

P. Vashista, J.N. Mundy, and G.K. Shenoy (Eds.), *Fast Ion Transport in Solids – Electrodes and Electrolytes*, Elsevier North-Holland, Inc., 1979

F.J. Kedves and D.L. Beke (Eds.), *DIMETA-82 – Diffusion in Metals and Alloys*, Defect and Diffusion Monograph Series No. 7, Trans Tech Publications, Switzerland, 1983

G.E. Murch, H.K. Birnbaum, and J.R. Cost (Eds.), *Nontraditional Methods in Diffusion*, The Metallurgical Society of AIME, Warrendale, Pennsylvania, USA, 1984

D. Gupta, A.D. Romig, and M.A. Dayananda (Eds.), *Diffusion in High Technology Materials*, Trans Tech Publications, Aedermannsdorf, Switzerland, 1988

F.J. Kedves and D.L. Beke (Eds.), *DIMETA-88 – Diffusion in Metals and Alloys*, Defect and Diffusion Forum **66–69**, 1989

A.L. Laskar, J.L. Bocquet, G. Brebec, and C. Monty (Eds.), *Diffusion in Materials*, NATO ASI Series, Kluwer Academic Publishers, The Netherlands, 1989

A.D. Romig, Jr. and M.A. Dayananda (Eds.), *Diffusion Analysis and Applications*, The Minerals, Metals & Materials Society, Warrendale, Pennsylvania, 1989
J. Nowotny (Ed.), *Diffusion in Solids and High Temperature Oxidation of Metals*, Trans Tech Publications, 1992
B. Fultz, R.W. Cahn, and D. Gupta (Eds.), *Diffusion in Ordered Alloys*, The Minerals, Metals & Materials Society, Warrendale, Pennsylvania, USA, 1993
M. Koiwa, K. Hirano, H. Nakajima, and T. Okada (Eds.), *Diffusion in Materials – DIMAT-92*, 2 volumes, Trans Tech Publications, Zürich, Switzerland, 1993; and Defect and Diffusion Forum **95–98**, 1993
H. Jain and D. Gupta (Eds.), *Diffusion in Amorphous Materials*, The Minerals, Metals & Materials Society, Warrendale, Pennsylvania, USA, 1993
J. Jedlinki (Ed.), *Diffusion and Reactions – from Basics to Applications*, Scitec Publications, Ltd., Switzerland, 1995; also Solid State Phenomena **41**, 1995
D.L. Beke, I.A. Szab (Eds.), *Diffusion and Stresses*, Scitec Publications Ltd., Zürich-Uetikon, Switzerland; also: Defect and Diffusion Forum **129–130**, 1996
H. Mehrer, Chr. Herzig, N.A. Stolwijk, and H. Bracht (Eds.), *Diffusion in Materials – DIMAT-96*, 2 volumes, Scitec Publications, Ltd., Zürich-Uetikon, Switzerland, 1997; also Defect and Diffusion Forum **143–147**, 1997
Y. Mishin, G. Vogl, N. Cowern, R. Catlow, and D. Farkas (Eds.), *Diffusion Mechanisms in Crystalline Materials*, Mat. Res. Soc. Symp. Proc. Vol. 527, Materials Research Society, Warrendale, Pennsylvania, USA, 1997
M. Danielewski (Ed.), *Diffusion and Reactions*, Scitec Publications, Ltd., Zürich-Uetikon, Switzerland, 2000
Y. Limoge and J.L. Bocquet (Eds.), *Diffusion in Materials – DIMAT-2000*, 2 volumes, Scitec Publications, Ltd., Zürich-Uetikon, Switzerland, 2001; also: Defect and Diffusion Forum **194–199**, 2001
B.S. Bokstein and B.B. Straumal (Eds.), *Diffusion, Segregation and Stresses in Materials*, Scitec Publications, Ltd., Zürich-Uetikon, Switzerland, 2003; also: Defect and Diffusion Forum **216–217**, 2003
M. Danielewski, R. Filipek, R. Kozubski, W. Kucza, P. Zieba, Z. Zurec (Eds.), *Diffusion in Materials – DIMAT-2004*, 2 volumes, Tans Tech Publications, Ltd., Zürich-Uetikon, Switzerland, 2005; also: Defect and Diffusion Forum **237–240**, 2005
A. Öchsner, J. Grácio, F. Barlat (Eds.), *First International Conference on Diffusion in Solids and Liquids – DSL 2005*, Centre for Mechanical Technology and Automation and Department of Mechanical Engineering, University of Aveiro, Editura MEDIAMIRA, Cluj-Napoca, 2005
J. Kärger, F. Grindberg, P. Heitjans (Eds.), *Diffusion Fundamentals – Leipzig 2005*, Leipziger Universitätsverlag GmbH 2005
B.S. Bokstein, B.B. Straumal (Eds.), *Diffusion in Solids – Past, Present and Future*, Trans Tech Publications, Ltd., Switzerland, 2006; also: Defect and Diffusion Forum **249**, 2006
A. Öchsner, J. Grácio (Eds.), *Diffusion in Solids and Liquids – DSL 2006*, Proc. of 2nd Int. Conf. on Diffusion in Solids and Liquids, Mass Transfer-Heat Transfer-Microstructure and Properties, Areiro, Portugal, 2006; also: Defect and Diffusion Forum **258–260**, 2006
J. Čermak, I. Stloukal (Eds.), *Diffusion and Thermodynamics of Materials – DT 2006*, Proc. of 9th Seminar on Diffusion and Thermodynamics of Materials, Brno, Czech Republik, 2006; also: Defect and Diffusion Forum **263**, 2007

Compilations of Diffusion Data

I. Kaur, W. Gust, L. Kozma, *Handbook of Grain and Interphase Boundary Diffusion Data*, 2 volumes, Ziegler Press, Stuttgart, 1989

H. Mehrer (Vol. Ed.), *Diffusion in Solid Metals and Alloys*, Landolt-Börnstein, New Series, Group III, Vol. 26, Springer-Verlag, 1990

D.L. Beke (Vol. Ed.), *Diffusion in Semiconductors and Non-Metallic Solids*, Subvolume A, *Diffusion in Semiconductors*, Landolt-Börnstein, New Series, Group III, Vol. 33, Springer-Verlag, 1998

D.L. Beke (Vol. Ed.), *Diffusion in Semiconductors and Non-Metallic Solids*, Subvolume B1, *Diffusion in Non-Metallic Solids (Part 1)*, Landolt-Börnstein, New Series, Group III, Vol. 33, Springer-Verlag, 1999

Diffusion and Defect Data, Journal of Abstracts, published by Trans Tech Publications, Aedermannsdorf, Switzerland, 1974–2003

Part I

Fundamentals of Diffusion

2 Continuum Theory of Diffusion

The equations governing diffusion processes are Fick's laws. These laws represent a continuum description and are purely phenomenological. The original work of ADOLF FICK appeared in 1855 [1] and described a salt-water system undergoing diffusion. Fick introduced the concept of the diffusion coefficient and suggested a linear response between the concentration gradient and the mixing of salt and water. Already in 1807 JOSEF FOURIER had developed an analogous relation between the flow of heat and the temperature gradient [2]. Fick's laws describe the diffusive transport of matter as an empirical fact without claiming that it derives from basic concepts. It is, however, indicative of the power of Fick's continuum description that all subsequent developments have in no way affected the validity of his approach. A deeper physical understanding of diffusion in solids is based on random walk theory and on the atomic mechanisms of diffusion, which are treated later in this book.

2.1 Fick's Laws in Isotropic Media

In an isotropic medium, physical and chemical properties are independent of direction, whereas in anisotropic media properties depend on the direction considered. Diffusion is isotropic in gases, most liquids, in glassy solids, in polycrystalline materials without texture, in cubic crystals and in icosahedral quasicrystals. In isotropic materials the diffusivity (introduced below) is a scalar quantity. Numerous engineering materials have cubic structures. Examples are face-centered cubic metals (Cu, Ag, Au, Al, Pb, Ni, ...), body-centered cubic metals (V, Nb, Ta, Cr, Mo, W, β-Ti, β-Zr, ...), α-Fe and ferritic steels, which are body-centered cubic, and austenitic steels which are face-centered cubic. All of these important materials, and vastly more of their alloys, share cubic symmetry and exhibit scalar diffusivities. The elemental semiconductors Si and Ge crystallise in the diamond structure which is cubic. Many compound semiconductors occur in the cubic zinc blende structure. Many ionic crystals such as alkali halides and many oxides are cubic or have cubic modifications. Diffusion is anisotropic in non-cubic crystals and in some quasicrystals. Anisotropic diffusion is discussed in Sect. 2.3.

2.1.1 Fick's First Law

Let us first consider the flux of diffusing particles in one dimension (x-direction) illustrated in Fig. 2.1. The particles can be atoms, molecules, or ions. Fick's first law for an isotropic medium can be written as

$$J_x = -D\frac{\partial C}{\partial x}. \qquad (2.1)$$

Here J_x is the flux of particles (diffusion flux) and C their number density (concentration). The negative sign in Eq. (2.1) indicates opposite directions of diffusion flux and concentration gradient. Diffusion is a process which leads to an equalisation of concentration. The factor of proportionality, D, is denoted as the *diffusion coefficient* or as the *diffusivity* of the species considered.

Units: The diffusion flux is expressed in number of particles (or moles) traversing a unit area per unit time and the concentration in number of particles per unit volume. Thus the diffusivity D has the dimension of *length^2 per time* and bears the units [cm^2 s^{-1}] or [m^2 s^{-1}].

Fick's first law in three dimensions: Fick's first law is easily generalised to three dimensions using a vector notation:

$$\boldsymbol{J} = -D\boldsymbol{\nabla}C. \qquad (2.2)$$

The vector of the diffusion flux $\boldsymbol{J}$ is directed opposite in direction to the concentration gradient vector $\boldsymbol{\nabla}C$. The *nabla* symbol, $\boldsymbol{\nabla}$, is used to express the vector operation on the right-hand side of Eq. (2.2). The nabla operator acts on the scalar concentration field $C(x,y,z,t)$ and produces the concentration-gradient field $\boldsymbol{\nabla}C$. The concentration-gradient vector always points in that direction for which the concentration field undergoes the most rapid increase,

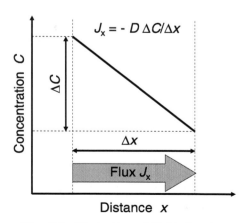

Fig. 2.1. Illustration of Fick's first law

and its magnitude equals the maximum rate of increase of concentration at the point. For an isotropic medium the diffusion flux is antiparallel to the concentration gradient.

Equations (2.1) and (2.2) represent the simplest form of Fick's first law. Complications leading to modifications of Eq. (2.2) may arise from anisotropy, concentration dependence of D, chemical reactions of the diffusing particles, external fields, and high-diffuasivity paths. Anisotropy is considered in Sect. 2.3. Further complications are treated in later chapters of this book.

Analogous equations: As already mentioned Fick's first law is formally equivalent to *Fourier's law* of heat flow

$$\boldsymbol{J}_q = -\kappa \boldsymbol{\nabla} T ,$$

where $\boldsymbol{J}_q$ is the flux of heat, T the temperature field, and κ the thermal conductivity. It is also analogous to *Ohm's law*

$$\boldsymbol{J}_e = -\sigma \boldsymbol{\nabla} V ,$$

where $\boldsymbol{J}_e$ is the electric current density, V the electrostatic potential, and σ the electrical conductivity. Fick's law describes the transport of particles, Fourier's law the transport of heat, and Ohm's law the transport of electric charge.

2.1.2 Equation of Continuity

Usually, in diffusion processes the number of diffusing particles is conserved[1]. For a diffusing species which obeys a *conservation law* an *equation of continuity* can be formulated. To this end, let us choose an aribitrary point P located at (x, y, z) and a test volume of size $\Delta x, \Delta y$, and Δz (Fig. 2.2). The diffusion flux $\boldsymbol{J}$ and its components J_x, J_y, J_z vary across the test volume. If the sum of the fluxes leaving and entering the test volume do not balance, a net accumulation (or loss) must occur. This material balance can be expressed as

$$\text{inflow - outflow} = \text{accumulation (or loss) rate}.$$

The flux components can be substituted into this equation to yield

$$[J_x(P) - J_x(P + \Delta x)]\Delta y \Delta z +$$
$$[J_y(P) - J_y(P + \Delta y)]\Delta x \Delta z +$$
$$[J_z(P) - J_z(P + \Delta z)]\Delta x \Delta y \ = \text{accumulation (or loss) rate}.$$

---

[1] This implies that the diffusing species neither undergoes reactions nor exchanges with internal sources or sinks. Sources and sinks are important for intrinsic point defects. Reactions of the diffusing species with intrinsic point defects can be important as well. Such complications are treated later in the relevant chapters.

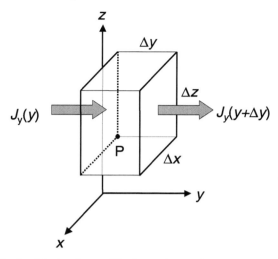

Fig. 2.2. Infinitesimal test volume. The in- and outgoing y-components of the diffusion flux are indicated by arrows. The other components (not shown) are analogous

Using Taylor expansions of the flux components up to their linear terms, the expressions in square brackets can be replaced by $\Delta x \partial J_x/\partial x$, $\Delta y \partial J_y/\partial y$, and $\Delta z \partial J_z/\partial z$, respectively. This yields

$$-\left[\frac{\partial J_x}{\partial x} + \frac{\partial J_y}{\partial y} + \frac{\partial J_z}{\partial z}\right]\Delta x \Delta y \Delta z = \frac{\partial C}{\partial t}\Delta x \Delta y \Delta z, \qquad (2.3)$$

where the accumulation (or loss) rate in the test volume is expressed in terms of the partial time derivative of the concentration. For infinitesimal size of the test volume Eq. (2.3) can be written in compact form by introducing the vector operation *divergence* $\nabla \cdot$, which acts on the vector of the diffusion flux:

$$-\nabla \cdot J = \frac{\partial C}{\partial t}. \qquad (2.4)$$

Equation (2.4) is denoted as the *continuity equation*.

2.1.3 Fick's Second Law – the 'Diffusion Equation'

Fick's first law Eq. (2.2) and the equation of continuity (2.4) can be combined to give an equation which is called *Fick's second law* or sometimes also the *diffusion equation*:

$$\frac{\partial C}{\partial t} = \nabla \cdot (D\nabla C). \qquad (2.5)$$

From a mathematical viewpoint Fick's second law is a second-order partial differential equation. It is non-linear if D depends on concentration, which is, for example, the case when diffusion occurs in a chemical composition gradient. The composition-dependent diffusivity is usually denoted as the *interdiffusion coefficient*. For arbitrary composition dependence $D(C)$, Eq. (2.5) usually cannot be solved analytically. The strategy to deal with interdiffusion is described in Chap. 10.

If the diffusivity is independent of concentration, which is the case for tracer diffusion in chemically homogenous systems or for diffusion in ideal solid solutions, Eq. (2.5) simplifies to

$$\frac{\partial C}{\partial t} = D \Delta C, \qquad (2.6)$$

where Δ denotes the *Laplace operator*. This form of Fick's second law is sometimes also called the *linear diffusion equation*. It is a linear second-order partial differential equation for the concentration field $C(x, y, z, t)$. One can strive for solutions of this equation, if boundary and initial conditions are formulated. Some solutions are considered in Chap. 3.

Analogous equations: If one combines Fourier's law for the conduction of heat with an equation for the conservation of heat energy, assuming a constant thermal conductivity κ, one arrives at

$$\frac{\partial T}{\partial t} = \frac{\kappa}{\rho C_V} \Delta T,$$

where $T(x, y, z, t)$ is the temperature field, ρ the mass density, and C_V the specific heat for constant volume. This equation for time-dependent heat conduction is mathematically identical with the linear diffusion equation.

The time-dependent Schrödinger equation for free particles can be written in a similar way:

$$\frac{\partial \Psi}{\partial t} = \frac{i}{\hbar} \left(-\frac{\hbar^2}{2m} \right) \Delta \Psi.$$

Here $\Psi(x, y, z, t)$ denotes the wave function, $\hbar$ the Planck constant divided by 2π, and i the imaginary unit. Similar mathematical concepts such as the method of separation of variables can be used to solve diffusion and Schrödinger equations. We note, however, that C is a function with real values, whereas the wave function Ψ is a function with a real and an imaginary part.

2.2 Diffusion Equation in Various Coordinates

As already mentioned, Fick's second law for constant diffusivity is a linear second-order partial differential equation. The Laplacian operator on the right-hand side of Eq. (2.6) has different representations in different coordinate systems (Fig. 2.3). Using these representations we get for isotropic diffusion the following forms of the linear diffusion equation [3, 4].

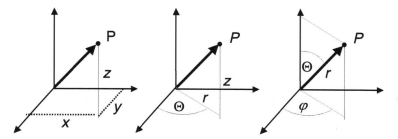

Fig. 2.3. Cartesian (*left*), cylindrical (*middle*), and spherical (*right*) coordinates

Cartesian coordinates x, y, z:

$$\frac{\partial C}{\partial t} = D\left(\frac{\partial^2 C}{\partial x^2} + \frac{\partial^2 C}{\partial y^2} + \frac{\partial^2 C}{\partial z^2}\right); \quad (2.7)$$

Cylindrical coordinates r, Θ, z:

$$\frac{\partial C}{\partial t} = \frac{D}{r}\left[\frac{\partial}{\partial r}\left(r\frac{\partial C}{\partial r}\right) + \frac{\partial}{\partial \Theta}\left(\frac{1}{r}\frac{\partial C}{\partial \Theta}\right) + \frac{\partial}{\partial z}\left(r\frac{\partial C}{\partial z}\right)\right]; \quad (2.8)$$

Spherical coordinates r, Θ, φ:

$$\frac{\partial C}{\partial t} = \frac{D}{r^2}\left[\frac{\partial}{\partial r}\left(r^2\frac{\partial C}{\partial r}\right) + \frac{1}{\sin\Theta}\frac{\partial}{\partial \Theta}\left(\sin\Theta\frac{\partial C}{\partial \Theta}\right) + \frac{1}{\sin^2\Theta}\frac{\partial^2 C}{\partial \varphi^2}\right]$$

$$= D\left[\frac{\partial^2 C}{\partial r^2} + \frac{2}{r}\frac{\partial C}{\partial r} + \frac{1}{r^2 \sin^2\Theta}\frac{\partial^2 C}{\partial \varphi^2} + \frac{1}{r^2}\frac{\partial^2 C}{\partial \Theta^2} + \frac{1}{r^2}\cot\Theta\frac{\partial C}{\partial \Theta}\right]. \quad (2.9)$$

Experimental diffusion studies often use simple geometric settings, which impose special symmetries on the diffusion field. In the following we mention some special symmetries:

Linear flow in x-direction is a special case of Eq. (2.7), if $\partial/\partial y = \partial/\partial z = 0$:

$$\frac{\partial C}{\partial t} = D\frac{\partial^2 C}{\partial x^2}. \quad (2.10)$$

Axial flow in r-direction is a special case of Eq. (2.8), if $\partial/\partial z = \partial/\partial \Theta = 0$:

$$\frac{\partial C}{\partial t} = D\left(\frac{\partial^2 C}{\partial r^2} + \frac{1}{r}\frac{\partial C}{\partial r}\right). \quad (2.11)$$

Spherical flow in r-direction is a special case of Eq. (2.9), if $\partial/\partial\phi = \partial/\partial\Theta = 0$:
$$\frac{\partial C}{\partial t} = D\left(\frac{\partial^2 C}{\partial r^2} + \frac{2}{r}\frac{\partial C}{\partial r}\right). \tag{2.12}$$

Such symmetries are conducive to analytical solutions, of which some are discussed in Chap. 3.

2.3 Fick's Laws in Anisotropic Media

Aniosotropic media have different diffusion properties in different directions. Anisotropy is encountered, for example, in non-cubic single crystals, composite materials, textured polycrystals, and decagonal quasicrystals. Anisotropy affects the directional relationship between the vectors of the diffusion flux and of the concentration gradient. For such media, for arbitrary directions the direction of the diffusion flux at an arbitrary is not normal to the surface of constant concentration. The generalisation of Fick's first law for anisotropic media is
$$\boldsymbol{J} = -\mathbf{D}\nabla C. \tag{2.13}$$

Application of Neumann's principle [5] shows that the diffusivity is a second-rank tensor $\mathbf{D}$. Furthermore, as a consequence of Onsager's reciprocity relations from the thermodynamics of irreversible processes (see, e.g., [3, 6–8] and Chap. 12) the diffusivity tensor is symmetric. Any symmetric second-rank tensor can be transformed to its three orthogonal *principal axes*. The diffusivity tensor then takes the form
$$\mathbf{D} = \begin{pmatrix} D_1 & 0 & 0 \\ 0 & D_2 & 0 \\ 0 & 0 & D_3 \end{pmatrix},$$

where D_1, D_2, and D_3 are called the *principal diffusion coefficients* or the *principal diffusivities* (self-diffusivities, solute diffusivities, ...). There are thus not more than three coefficients of diffusion. There are, however, always $p \leq 6$ independent parameters; the $p-3$ others define the orientations of the principal axes. The number p varies according to the symmetry of the crystal system as indicated in Table 2.1.

If x_1, x_2, x_3 denote the principal diffusion axes and J_1, J_2, J_3 the pertinent components of the diffusion flux, Eq. (2.13) can be written as
$$J_1 = -D_1 \frac{\partial C}{\partial x_1},$$
$$J_2 = -D_2 \frac{\partial C}{\partial x_2},$$
$$J_3 = -D_3 \frac{\partial C}{\partial x_3}. \tag{2.14}$$

Table 2.1. Number of parameters, p, decribing the principal diffusivities plus the orientations of principal axes

System	triclinic	monoclinic	orthorhombic	hexagonal tetragonal rhombohedral (or trigonal)	cubic
p	6	4	3 principal axes and crystal axes coincide	2 one principal axis parallel crystal axis	1 isotropic

These equations imply that the diffusion flux $\boldsymbol{J}$ and the concentration gradient $\boldsymbol{\nabla} C$ usually point in different directions.

Let us describe a selected diffusion direction by its angles $\Theta_1, \Theta_2, \Theta_3$ with respect to the principal diffusion axes (Fig. 2.4) and introduce the direction cosines of the diffusion direction by

$$\alpha_1 \equiv \cos\Theta_1, \quad \alpha_2 \equiv \cos\Theta_2, \quad \alpha_3 \equiv \cos\Theta_3. \tag{2.15}$$

Then the diffusion coefficient for that direction, $D(\alpha_1, \alpha_2, \alpha_3)$, can be written as

$$D(\alpha_1, \alpha_2, \alpha_3) = \alpha_1^2 D_1 + \alpha_2^2 D_2 + \alpha_3^2 D_3. \tag{2.16}$$

Equation (2.16) shows that for given principal axes, anisotropic diffusion can be completely described by the principal diffusion coefficients.

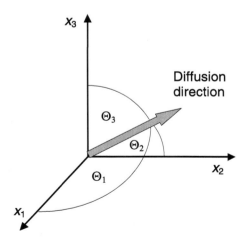

Fig. 2.4. Diffusion direction in a single-crystal with principal diffusion axes x_1, x_2, x_3

For crystals with **triclinic, monoclinic, and orthorhombic** symmetry all three principal diffusivities are different:

$$D_1 \neq D_2 \neq D_3. \tag{2.17}$$

Among these crystal systems only for crystals with *orthorhombic* symmetry the principal axes of diffusion do coincide with the axes of crystallographic symmetry.

For **uniaxial** materials, such as **trigonal, tetragonal, and hexagonal** crystals and **decagonal or octagonal** quasicrystals, with their unique axis parallel to the x_3-axis we have

$$D_1 = D_2 \neq D_3. \tag{2.18}$$

For uniaxial materials Eq. (2.16) reduces to

$$D(\Theta) = D_1 \sin^2 \Theta + D_3 \cos^2 \Theta, \tag{2.19}$$

where Θ denotes the angle between diffusion direction and the crystal axis. For *cubic crystals* and *icosahedral quasicrystals*

$$D_1 = D_2 = D_3 \equiv D$$

and the diffusivity tensor reduces to a scalar quantity (see above).

The majority of experiments for the measurement of diffusion coefficients in single crystals are designed in such a way that the flow is one-dimensional. Diffusion is one-dimensional if a concentration gradient exists only in the x-direction and both, C and $\partial C/\partial x$, are everywhere independent of y and z. Then the diffusivity depends on the crystallographic direction of the flow. If the direction of diffusion is chosen parallel to one of the principal axis (x_1, or x_2, or x_3) the diffusivity coincides with one of the principal diffusivities D_1, or D_2, or D_3. For an arbitrary direction, the measured D is given by Eq. (2.16).

For uniaxial materials the diffusivity $D(\Theta)$ is measured when the crystal or quasicrystal is cut in such a way that an angle Θ occurs between the normal of the front face and the crystal axis. For a full characterisation of the diffusivity tensor in crystals with orthorhombic or lower symmetry measurements in three independent directions are necessary. For uniaxial crystals two measurements in independent directions suffice. For cubic crystals one measurement in an arbitrary direction is sufficient.

References

1. A. Fick, Annalen der Phyik und Chemie **94**, 59 (1855); Philos. Mag. **10**, 30 (1855)

2. J.B.J. Fourier, *The Analytical Theory of Heat*, translated by A. Freeman, University Press, Cambridge, 1978
3. J. Crank, *The Mathematics of Diffusion*, 2nd edition, Oxford University Press, 1975
4. I.N. Bronstein, K.A. Semendjajew, *Taschenbuch der Mathematik*, 9. Auflage, Verlag Harri Deutsch, Zürich & Frankfurt, 1969
5. J.F. Nye, *Physical Properties of Crystals: their Representation by Tensors and Matrices*, Clarendon Press, Oxford, 1957
6. S.R. de Groot, P. Mazur, *Thermodynamics of Irreversible Processes*, North-Holland Publ. Comp., 1952
7. J. Philibert, *Atom Movement – Diffusion and Mass Transport in Solids*, Les Editions de Physique, Les Ulis, Cedex A, France, 1991
8. M.E. Glicksman, *Diffusion in Solids – Field Theory, Solid-State Principles and Applications*, John Wiley & Sons, Inc., 2000

3 Solutions of the Diffusion Equation

The aim of this chapter is to give the reader a feeling for properties of the diffusion equation and to acquaint her/him with frequently encountered solutions. No attempt is made to achieve completeness or full rigour. Solutions of Eq. (2.6), giving the concentration as a function of time and position, can be obtained by various means once the boundary and initial conditions have been specified. In certain cases, the conditions are geometrically highly symmetric. Then it is possible to obtain explicit analytic solutions. Such solutions comprise either Gaussians, error functions and related integrals, or they are given in the form of Fourier series.

Experiments are often designed to satisfy simple initial and boundary conditions (see Chap. 13). In what follows, we limit ourselves to a few simple cases. First, we consider solutions of steady-state diffusion for linear, axial, and spherical flow. Then, we describe examples of non-steady state diffusion in one dimension. A powerful method of solution, which is mentioned briefly, employs the Laplace transform. We end this chapter with a few remarks about instantaneous point sources in one, two, and three dimensions.

For more comprehensive treatments of the mathematics of diffusion we refer to the textbooks of CRANK [1], JOST [2], GHEZ [3] and GLICKSMAN [4]. As mentioned already, the conduction of heat can be described by an analogous equation. Solutions of this equation have been developed for many practical cases of heat flow and are collected in the book of CARSLAW AND JAEGER [5]. By replacing T with C and D with the corresponding thermal property these solution can be used for diffusion problems as well. In many other cases, numerical methods must be used to solve diffusion problems. Describing numerical procedures is beyond the scope of this book. Useful hints can be found in the literature, e.g., in [1, 3, 4, 6, 7].

3.1 Steady-State Diffusion

At steady state, there is no change of concentration with time. Steady-state diffusion is characterised by the condition

$$\frac{\partial C}{\partial t} = 0. \tag{3.1}$$

For the special geometrical settings mentioned in Sect. 2.2, this leads to different *stationary concentration distributions*:

For *linear flow* we get from Eqs. (2.10) and (3.1)

$$D\frac{\partial^2 C}{\partial x^2} = 0 \quad \text{and} \quad C(x) = a + Ax, \tag{3.2}$$

where a and A in Eq. (3.2) denote constants. A constant concentration gradient and a linear distribution of concentration is established under linear flow steady-state conditions, if the diffusion coefficient is a constant.

For *axial flow* substitution of Eq. (3.1) into Eq. (2.8) gives

$$\frac{\partial}{\partial r}\left(r\frac{\partial C}{\partial r}\right) = 0 \quad \text{and} \quad C(r) = B\ln r + b, \tag{3.3}$$

where B and b denote constants.

For *spherical flow* substitution of Eq. (3.1) into Eq. (2.9) gives

$$\frac{\partial}{\partial r}\left(r^2\frac{\partial C}{\partial r}\right) = 0 \quad \text{and} \quad C(r) = \frac{C_a}{r} + C_b. \tag{3.4}$$

C_a and C_b in Eq. (3.4) denote constants.

Permeation through membranes: The passage of gases or vapours through membranes is called *permeation*. A well-known example is diffusion of hydrogen through palladium membranes. A steady state can be established in permeation experiments after a certain transient time (see Sect. 3.2.4). Based on Eqs. (3.2), (3.3), and (3.4) a number of examples are easy to formulate and are useful in permeation studies of diffusion:

Planar Membrane: If δ is the thickness, q the cross section of a planar membrane, and C_1 and C_2 the concentrations at $x = 0$ and $x = \delta$, we get from Eq. (3.2)

$$C(x) = C_1 + \frac{C_2 - C_1}{\delta}x; \quad J = qD\frac{C_1 - C_2}{\delta}. \tag{3.5}$$

If J, C_1, and C_2 are measured in an experiment, the diffusion coefficient can be determined from Eq. (3.5).

Hollow cylinder: Consider a hollow cylinder, which extends from an inner radius r_1 to an outer radius r_2. If at r_1 and r_2 the stationary concentrations C_1 and C_2 are maintained, we get from Eq. (3.3)

$$C(r) = C_1 + \frac{C_1 - C_2}{\ln(r_1/r_2)}\ln\frac{r}{r_1}. \tag{3.6}$$

Spherical shell: If the shell extends from an inner radius r_1 to an outer radius r_2, and if at r_1 and r_2 the stationary concentrations C_1 and C_2 are maintained, we get from Eq. (3.4)

$$C(r) = \frac{C_1 r_1 - C_2 r_2}{r_1 - r_2} + \frac{(C_1 - C_2)}{\left(\frac{1}{r_1} - \frac{1}{r_2}\right)}\frac{1}{r}. \tag{3.7}$$

For the geometrical conditions treated above, it is also possible to solve the steady-state equations, if the diffusion coefficient is not a constant [8]. Solutions for concentration-dependent and position-dependent diffusivities can be found, e.g., in the textbook of JOST [2].

3.2 Non-Steady-State Diffusion in one Dimension

3.2.1 Thin-Film Solution

An initial condition at $t = 0$, which is encountered in many one-dimensional diffusion problems, is the following:

$$C(x,0) = M\,\delta(x)\,. \tag{3.8}$$

The diffusing species (diffusant) is deposited at the plane $x = 0$ and allowed to spread for $t > 0$. M denotes the number of diffusing particles per unit area and $\delta(x)$ the Dirac delta function. This initial condition is also called *instantaneous planar source*.

Sandwich geometry: If the diffusant (or diffuser) is allowed to spread into two material bodies occupying the half-spaces $0 < x < \infty$ and $-\infty < x < 0$, which have equal and constant diffusivity, the solution of Eq. (2.10) is

$$C(x,t) = \frac{M}{2\sqrt{\pi Dt}} \exp\left(-\frac{x^2}{4Dt}\right). \tag{3.9}$$

Thin-film geometry: If the diffuser is deposited initially onto the surface of a sample and spreads into one half-space, the solution is

$$C(x,t) = \frac{M}{\sqrt{\pi Dt}} \exp\left(-\frac{x^2}{4Dt}\right). \tag{3.10}$$

These solutions are also denoted as *Gaussian solutions*. Note that Eqs. (3.9) and (3.10) differ by a factor of 2. Equation (3.10) is illustrated in Fig. 3.1 and some of its further properties in Fig. 3.2.

The quantity $2\sqrt{Dt}$ is a characteristic *diffusion length*, which occurs frequently in diffusion problems. Salient properties of Eq. (3.9) are the following:

1. The diffusion process is subject to the conservation of the integral number of diffusing particles, which for Eq. (3.9) reads

$$\int_{-\infty}^{+\infty} \frac{M}{2\sqrt{\pi Dt}} \exp\left(-\frac{x^2}{4Dt}\right) dx = \int_{-\infty}^{+\infty} M\delta(x) dx = M\,. \tag{3.11}$$

2. $C(x,t)$ and $\partial^2 C/\partial x^2$ are *even* functions of x. $\partial C/\partial x$ is an *odd* function of x.

40   3 Solutions of the Diffusion Equation

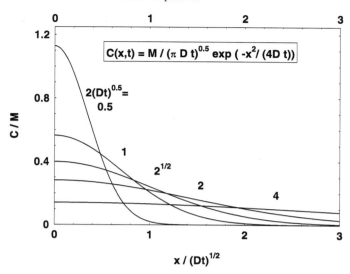

Fig. 3.1. Gaussian solution of the diffusion equation for various values of the diffusion length $2\sqrt{Dt}$

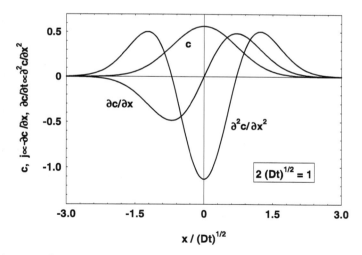

Fig. 3.2. Gaussian solution of the diffusion equation and its derivatives

3. The diffusion flux, $J = -D\partial C/\partial x$, is an *odd* function of x. It is zero at the plane $x = 0$.
4. According to the diffusion equation the rate of accumulation of the diffusing species $\partial C/\partial t$ is an *even* function of x. It is negative for small $|x|$ und positive for large $|x|$.

The tracer method for the experimental determination of diffusivities exploits these properties (see Chap. 13). The Gaussian solutions are also applicable if the thickness of the deposited layer is very small with respect to the diffusion length.

3.2.2 Extended Initial Distribution and Constant Surface Concentration

So far, we have considered solutions of the diffusion equation when the diffusant is initially concentrated in a very thin layer. Experiments are also often designed in such a way that the diffusant is distributed over a finite region. In practice, the diffusant concentration is often kept constant at the surface of the sample. This is, for example, the case during carburisation or nitridation experiments of metals. The linearity of the diffusion equation permits the use of the 'principle of superposition' to produce new solutions for different geometric arrangements of the sources. In the following, we consider examples which exploit this possibility.

Diffusion Couple: Let us suppose that the diffusant has an initial distribution at $t = 0$ which is given by:

$$C = C_0 \quad \text{for} \quad x < 0 \quad \text{and} \quad C = 0 \quad \text{for} \quad x > 0. \tag{3.12}$$

This situation holds, for example, when two semi-infinite bars differing in composition (e.g., a dilute alloy and the pure solvent material) are joined end to end at the plane $x = 0$ to form a diffusion couple. The initial distribution can be interpreted as a continuous distribution of instantaneous, planar sources of infinitesimal strength $dM = C_0 d\xi$ at position ξ spread uniformly along the left-hand bar, i.e. for $x < 0$. A unit length of the left-hand bar initially contains $M = C_0 \cdot 1$ diffusing particles per unit area. Initially, the right-hand bar contains no diffusant, so one can ignore contributions from source points $\xi > 0$. The solution of this diffusion problem, $C(x,t)$, may be thought as the sum, or integral, of all the infinitesimal responses resulting from the continuous spatial distribution of instantaneous source releases from positions $\xi < 0$. The total response occurring at any plane x at some later time t is given by the superposition

$$C(x,t) = C_0 \int_{-\infty}^{0} \frac{\exp\left[-(x-\xi)^2/4Dt\right]}{2\sqrt{\pi Dt}} d\xi = \frac{C_0}{\sqrt{\pi}} \int_{x/2\sqrt{Dt}}^{\infty} \exp(-\eta^2) d\eta. \tag{3.13}$$

Here we used the variable substitution $\eta \equiv (x-\xi)/2\sqrt{Dt}$. The right-hand side of Eq. (3.13) may be split and rearranged as

$$C(x,t) = \frac{C_0}{2}\left[\frac{2}{\sqrt{\pi}}\int_0^\infty \exp(-\eta^2)d\eta - \frac{2}{\sqrt{\pi}}\int_0^{x/2\sqrt{Dt}} \exp(-\eta^2)d\eta\right]. \tag{3.14}$$

It is convenient to introduce the *error function*[1]

$$\text{erf}(z) \equiv \frac{2}{\sqrt{\pi}} \int_0^z \exp(-\eta^2) d\eta, \qquad (3.15)$$

which is a standard mathematical function. Some properties of erf(z) and useful approximations are discussed below. Introducing the error function we get

$$C(x,t) = \frac{C_0}{2} \left[\text{erf}(\infty) - \text{erf}\left(\frac{x}{2\sqrt{Dt}}\right) \right] \equiv \frac{C_0}{2} \text{erfc}\left(\frac{x}{2\sqrt{Dt}}\right), \qquad (3.16)$$

where the abbreviation

$$\text{erfc}(z) \equiv 1 - \text{erf}(z) \qquad (3.17)$$

is denoted as the *complementary error function*. Like the thin-film solution, Eq. (3.16) is applicable when the diffusivity is constant. Equation (3.16) is sometimes called the *Grube-Jedele solution*.

Diffusion with Constant Surface Concentration: Let us suppose that the concentration at $x = 0$ is maintained at concentration $C_s = C_0/2$. The Grube-Jedele solution Eq. (3.16) maintains the concentration in the midplane of the diffusion couple. This property can be exploited to construct the diffusion solution for a semi-infinite medium, the free end of which is continuously exposed to a fixed concentration C_s:

$$C = C_s \text{erfc}\left(\frac{x}{2\sqrt{Dt}}\right). \qquad (3.18)$$

The quantity of material which diffuses into the solid per unit area is:

$$M(t) = 2C_s \sqrt{Dt/\pi}. \qquad (3.19)$$

Equation (3.18) is illustrated in Fig. 3.3. The behaviour of this solution reveals several general features of diffusion problems in infinite or semi-infinite media, where the initial concentration at the boundary equals some constant for all time: The concentration field $C(x,t)$ in these cases may be expressed with

---

[1] The *probability integral* introduced by GAUSS is defined as

$$\Phi(a) \equiv \frac{2}{\sqrt{2\pi}} \int_0^a \exp(-\eta^2/2) d\eta.$$

The error function and the probability integral are related via

$$\text{erf}(z) = \Phi(\sqrt{2}z).$$

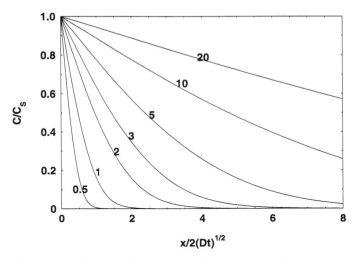

Fig. 3.3. Solution of the diffusion equation for constant surface concentration C_s and for various values of the diffusion length $2\sqrt{Dt}$

a single variable $z = x/2\sqrt{Dt}$, which is a special combination of space-time field variables. The quantity z is sometimes called a similarity variable which captures both, the spatial and temporal features of the concentration field. Similarity scaling is extremely useful in applying the diffusion solution to diverse situations. For example, if the average diffusion length is increased by a factor of ten, the product of the diffusivity times the diffusion time would have to increase by a factor of 100 to return to the same value of z.

Applications of Eq. (3.18) concern, e.g., carburisation or nitridation of metals, where in-diffusion of C or N into a metal occurs from an atmosphere, which maintains a constant surface concentration. Other examples concern in-diffusion of foreign atoms, which have a limited solubility, C_s, in a matrix.

Diffusion from a Slab Source: In this arrangement a slab of width $2h$ having a uniform initial concentration C_0 of the diffusant is joined to two half-spaces which, in an experiment may be realised as two bars of the pure material. If the slab and the two bars have the same diffusivity, the diffusion field can be expressed by an integral of the source distribution

$$C(x,t) = \frac{C_0}{2\sqrt{\pi Dt}} \int_{-h}^{+h} \exp\left[-\frac{(x-\xi)^2}{4Dt}\right] d\xi. \tag{3.20}$$

This expression can be manipulated into standard form and written as

$$C(x,t) = \frac{C_0}{2}\left[\operatorname{erf}\left(\frac{x+h}{2\sqrt{Dt}}\right) + \operatorname{erf}\left(\frac{x-h}{2\sqrt{Dt}}\right)\right]. \tag{3.21}$$

44   3 Solutions of the Diffusion Equation

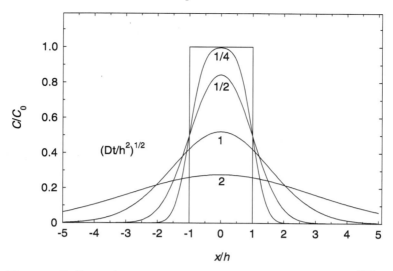

Fig. 3.4. Diffusion from a slab of width $2h$ for various values of $\sqrt{Dt}/h$

The normalised concentration field, $C(x/h, t)/C_0$, resulting from Eq. (3.21) is shown in Fig. 3.4 for various values of $\sqrt{Dt}/h$.

Error Function and Approximations: The error function defined in Eq. (3.15) is an odd function and for large arguments $|z|$ approaches asymptotically ± 1:

$$\mathrm{erf}(-z) = \mathrm{erf}(z), \quad \mathrm{erf}(\pm\infty) = \pm 1, \quad \mathrm{erf}(0) = 0. \tag{3.22}$$

The complementary error function defined in Eq. (3.17) has the following asymtotic properties:

$$\mathrm{erfc}(-\infty) = 2, \quad \mathrm{erfc}(+\infty) = 0, \quad \mathrm{erfc}(0) = 1. \tag{3.23}$$

Tables of the error function are available in the literature, e.g., in [4, 9–11].

Detailed calculations cannot be performed just relying on tabular data. For advanced computations and for graphing one needs, instead, numerical estimates for the error function. Approximations are available in commercial mathematics software. In the following, we mention several useful expressions:

1. For small arguments, $|z| < 1$, the error function is obtained to arbitrary accuracy from its Taylor expansion [10] as

$$\mathrm{erf}(z) = \frac{2}{\sqrt{\pi}} \left[z - \frac{z^3}{(3 \times 1)!} + \frac{z^5}{(5 \times 2)!} - \frac{z^7}{(7 \times 3)!} + \cdots \right]. \tag{3.24}$$

2. For large arguments, $z \gg 1$, it is approximated by its asymptotic form

$$\mathrm{erf}(z) = 1 - \frac{\exp(-z^2)}{2\sqrt{\pi}}\left(1 - \frac{1}{2z^2} + \dots\right). \tag{3.25}$$

3. A convenient rational expression reported in [11] is the following:

$$\mathrm{erf}(z) = 1 - \frac{1}{(1 + 0.278393z + 0.230389z^2 + 0.000972z^3 + 0.078108z^4)^4} + \epsilon(z). \tag{3.26}$$

This expression works for $z > 0$ with an associated error $\epsilon(z)$ less than 5×10^{-4}.

3.2.3 Method of Laplace Transformation

The Laplace transformation is a mathematical procedure, which is useful for various problems in mathematical physics. Application of the Laplace transformation to the diffusion equation removes the time variable, leaving an ordinary differential equation, the solution of which yields the transform of the concentration field. This is then interpreted to give an expression for the concentration in terms of space variables and time, satisfying the initial and boundary conditions. Here we deal only with an application to the one-dimensional diffusion equation, the aim being to describe rather than to justify the procedure.

The solution of many problems in diffusion by this method calls for no knowledge beyond ordinary calculus. For more difficult problems the theory of functions of a complex variable must be used. No attempt is made here to explain problems of this kind, although solutions obtained in this way are quoted, e.g., in the chapter on grain-boundary diffusion. Fuller accounts of the method and applications can be found in the textbooks of CRANK [1], CARSLAW AND JAEGER [5], CHURCHILL [12] and others.

Definition of the Laplace Transform: The Laplace transform $\bar{f}(p)$ of a known function $f(t)$ for positive values of t is defined as

$$\bar{f}(p) = \int_0^\infty \exp(-pt) f(t) \mathrm{d}t. \tag{3.27}$$

p is a number sufficiently large to make the integral Eq. (3.27) converge. It may be a complex number whose real part is sufficiently large, but in the following discussion it suffices to think of it in terms of a real positive number.

Laplace transforms are common functions and readily constructed by carrying out the integration in Eq. (3.27) as in the following examples:

$$f(t) = 1, \qquad \bar{f}(p) = \int_0^\infty \exp(-pt)\mathrm{d}t = \frac{1}{p}, \qquad (3.28)$$

$$f(t) = \exp(\alpha t), \qquad \bar{f}(p) = \int_0^\infty \exp(-pt)\exp(\alpha t)\mathrm{d}t = \frac{1}{p-\alpha}, \qquad (3.29)$$

$$f(t) = \sin(\omega t), \qquad \bar{f}(p) = \int_0^\infty \exp(-pt)\sin(\omega t)\mathrm{d}t = \frac{\omega}{p^2+\omega^2}. \qquad (3.30)$$

Semi-infinite Medium: As an application of the Laplace transform, we consider diffusion in a semi-infinite medium, $x > 0$, when the surface is kept at a constant concentration C_s. We need a solution of Fick's equation satisfying this boundary condition and the initial condition $C = 0$ at $t = 0$ for $x > 0$. On multiplying both sides of Fick's second law Eq. (2.6) by $\exp(-pt)$ and integrating, we obtain

$$D\int_0^\infty \exp(-pt)\frac{\partial^2 C}{\partial x^2}\mathrm{d}t = \int_0^\infty \exp(-pt)\frac{\partial C}{\partial t}\mathrm{d}t. \qquad (3.31)$$

By interchanging the orders of differentiation and integration, the left-hand term is then

$$D\int_0^\infty \exp(-pt)\frac{\partial^2 C}{\partial x^2}\mathrm{d}t = D\frac{\partial^2}{\partial x^2}\int_0^\infty C\exp(-pt)\mathrm{d}t = D\frac{\partial^2 \bar{C}}{\partial x^2}. \qquad (3.32)$$

Integrating the right-hand term of Eq. (3.31) by parts, we have

$$\int_0^\infty \exp(-pt)\frac{\partial C}{\partial t}\mathrm{d}t = [C\exp(-pt)]_0^\infty + p\int_0^\infty C\exp(-pt)\mathrm{d}t = p\bar{C}, \qquad (3.33)$$

since the term in brackets vanishes by virtue of the initial condition and through the exponential factor. Thus Fick's second equation transforms to

$$D\frac{\partial^2 \bar{C}}{\partial x^2} = p\bar{C}. \qquad (3.34)$$

The Laplace transformation reduces Fick's second law from a partial differential equation to the ordinary differential equation Eq. (3.34). By treating the boundary condition at $x = 0$ in the same way, we obtain

$$\bar{C} = \int_0^\infty C_s \exp(-pt)\mathrm{d}t = \frac{C_s}{p}. \qquad (3.35)$$

3.2 Non-Steady-State Diffusion in one Dimension   47

The solution of Eq. (3.34), which satisfies the boundary condition and for which $\bar{C}$ remains finite for large x is

$$\bar{C} = \frac{C_s}{p} \exp\left(\sqrt{\frac{p}{D}}\right) x. \tag{3.36}$$

Reference to a table of Laplace transforms [1] shows that the function whose transform is given by Eq. (3.36) is the complementary error function

$$C = C_s \operatorname{erfc}\left(\frac{x}{2\sqrt{Dt}}\right). \tag{3.37}$$

We recognise that this is the solution given already in Eq. (3.16).

3.2.4 Diffusion in a Plane Sheet – Separation of Variables

Separation of variables is a mathematical method, which is useful for the solution of partial differential equations and can also be applied to diffusion problems. It is particularly suitable for solutions of Fick's law for finite systems by assuming that the concentration field can be expressed in terms of a periodic function in space and a time-dependent function. We illustrate this method below for the problem of diffusion in a plane sheet.

The starting point is to strive for solutions of Eq. (2.10) trying the 'Ansatz'

$$C(x,t) = X(x)T(t), \tag{3.38}$$

where $X(x)$ and $T(t)$ separately express spatial and temporal functions of x and t, respectively. In the case of linear flow, Fick's second law Eq. (2.10) yields

$$\frac{1}{DT}\frac{dT}{dt} = \frac{1}{X}\frac{d^2X}{dx^2}. \tag{3.39}$$

In this equation the variables are separated. On the left-hand side we have an expression depending on time only, while the right-hand side depends on the distance variable only. Then, both sides must equal the same constant, which for the sake of the subsequent algebra is chosen as $-\lambda^2$:

$$\frac{1}{DT}\frac{\partial T}{\partial t} = \frac{1}{X}\frac{\partial^2 X}{\partial^2 x} \equiv -\lambda^2. \tag{3.40}$$

We then arrive at two ordinary linear differential equations: one is a first-order equation for $T(t)$, the other is a second-order equation for $X(x)$. Solutions to each of these equations are well known:

$$T(t) = T_0 \exp\left(-\lambda^2 Dt\right) \tag{3.41}$$

and

$$X(x) = a \sin(\lambda x) + b \cos(\lambda x), \tag{3.42}$$

48   3 Solutions of the Diffusion Equation

where T_0, a, and b are constants. Inserting Eqs. (3.41) and (3.42) in (3.38) yields a particular solution of the form

$$C(x,t) = [A \sin(\lambda x) + B \cos(\lambda x)] \exp(-\lambda^2 Dt), \tag{3.43}$$

where $A = aT_0$ and $B = bT_0$ are again constants of integration. Since Eq. (2.10) is a linear equation its general solution is obtained by summing solutions of the type of Eq. (3.43). We get

$$C(x,t) = \sum_{n=1}^{\infty} [A_n \sin(\lambda_n x) + B_n \cos(\lambda_n x)] \exp(-\lambda_n^2 Dt), \tag{3.44}$$

where A_n, B_n and λ_n are determined by the initial and boundary conditions for the particular problem. The separation constant $-\lambda^2$ cannot be arbitrary, but must take discrete values. These *eigenvalues* uniquely define the eigenfunctions of which the concentration field $C(x,t)$ is composed.

Out-diffusion from a plane sheet: Let us consider out-diffusion from a plane sheet of thickness L. An example provides out-diffusion of hydrogen from a metal sheet during degassing in vacuum. The diffusing species is initially distributed with constant concentration C_0 and both surfaces of the sheet are kept at zero concentration for times $t > 0$:

Initial condition $C = C_0$, for $0 < x < L$ at $t = 0$
Boundary condition $C = 0$, for $x = 0$ and $x = L$ at $t > 0$.

The boundary conditions demand that

$$B_n = 0 \quad \text{and} \quad \lambda_n = \frac{n\pi}{L}, \quad \text{where} \quad n = 1, 2, 3, \ldots \tag{3.45}$$

The numbers λ_n are the eigenvalues of the plane-sheet problem. Inserting these eigenvalues, Eq. (3.44) reads

$$C(x,t) = \sum_{n=1}^{\infty} A_n \sin\left(\frac{n\pi}{L}x\right) \exp\left(-\frac{n^2\pi^2 D}{L^2}t\right). \tag{3.46}$$

The initial conditions require that

$$C_0 = \sum_{n=1}^{\infty} A_n \sin\left(\frac{n\pi}{L}x\right). \tag{3.47}$$

By multiplying both sides of Eq. (3.47) by $\sin(p\pi x/L)$ and integrating from 0 to L we get

$$\int_0^L \sin\left(\frac{p\pi x}{L}\right) \sin\left(\frac{n\pi x}{L}\right) \mathrm{d}x = 0 \tag{3.48}$$

for $n \neq p$ and $L/2$ for $n = p$. Using these orthogonality relations all terms vanish for which n is even. Thus

$$A_n = \frac{4C_0}{n\pi}; \quad n = 1, 3, 5, \ldots \tag{3.49}$$

The final solution of the problem of out-diffusion from a plane sheet is

$$C(x,t) = \frac{4C_0}{\pi} \sum_{j=0}^{\infty} \frac{1}{2j+1} \sin\left[\frac{(2j+1)\pi}{L}x\right] \exp\left[-\frac{(2j+1)^2\pi^2 D}{L^2}t\right], \tag{3.50}$$

where for convenience $2j + 1$ was substituted for n so that j takes values $0, 1, 2, \ldots$. Each term in Eq. (3.50) corresponds to a term in the *Fourier series* (here a trigonometrical series) by which for $t = 0$ the initial distribution Eq. (3.47) can be represented. Each term is also characterised by a relaxation time

$$\tau_j = \frac{L^2}{(2j+1)^2\pi^2 D}, \quad j = 0, 1, 2, \ldots \tag{3.51}$$

The relaxation times decrease rapidly with increasing j, which implies that the series Eq. (3.50) converges satisfactorily for moderate and large times.

Desorption and Absorption: It is sometimes of interest to consider the average concentration in the sheet, $\bar{C}$, defined as

$$\bar{C}(t) = \frac{1}{L}\int_0^L C(x,t)dx. \tag{3.52}$$

Inserting Eq. (3.50) into Eq. (3.52) yields

$$\frac{\bar{C}(t)}{C_0} = \frac{8}{\pi^2} \sum_{j=0}^{\infty} \frac{1}{(2j+1)^2} \exp\left(-\frac{t}{\tau_j}\right). \tag{3.53}$$

We recognise that for $t \gg \tau_1$ the average concentration decays exponentially with the relaxation time

$$\tau_0 = \frac{L^2}{\pi^2 D}. \tag{3.54}$$

Direct applications of the solution developed above concern degassing of a hydrogen-charged metal sheet in vacuum or decarburisation of a sheet of steel. If we consider the case $t \gg \tau_1$, we get

$$C(x,t) \approx \frac{4C_0}{\pi} \sin\left(\frac{\pi x}{L}\right) \exp\left(-\frac{t}{\tau_0}\right). \tag{3.55}$$

The diffusion flux from both surfaces is then given by

$$|J| = 2D\left(\frac{\partial C}{\partial x}\right)_{x=0} = \frac{8DC_0}{L}\exp\left(-\frac{t}{\tau_0}\right). \tag{3.56}$$

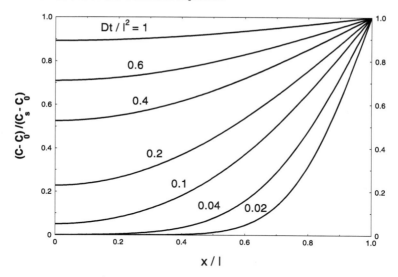

Fig. 3.5. Absorption/desorption of a diffusing species of/from a thin sheet for various values of Dt/l^2

An experimental determination of $|J|$ and/or of the relaxation time τ_0 can be used to measure D.

The solution for a plane sheet with constant surface concentration maintained at C_s and uniform initial concentration C_0 inside the sheet (region $-l < x < +l$) is a straightforward generalisation of Eq. (3.50). We get

$$\frac{C - C_0}{C_s - C_0} = 1 - \frac{4}{\pi} \sum_{j=0}^{\infty} \frac{(-1)^j}{2j+1} \cos\left[\frac{(2j+1)\pi}{2l}x\right] \exp\left[-\frac{(2j+1)^2\pi^2 D}{4l^2}t\right]. \tag{3.57}$$

For $C_s < C_0$ this solution describes *desorption* and for $C_s > C_0$ *absorption*. It is illustrated for various normalised times Dt/l^2 in Fig. 3.5.

3.2.5 Radial Diffusion in a Cylinder

We consider a long circular cylinder, in which the diffusion flux is radial everywhere. Then the concentration is a function of radius r and time t, and the diffusion equation becomes

$$\frac{\partial C}{\partial t} = \frac{1}{r}\frac{\partial}{\partial r}\left(rD\frac{\partial C}{\partial r}\right). \tag{3.58}$$

Following the method of separation of the variables, we see that for constant D

$$C(r,t) = u(r)\exp(-D\alpha^2 t) \tag{3.59}$$

3.2 Non-Steady-State Diffusion in one Dimension

is a solution of Eq. (3.58), provided that u satisfies

$$\frac{\partial^2 u}{\partial r^2} + \frac{1}{r}\frac{\partial u}{\partial r} + \alpha^2 u = 0, \qquad (3.60)$$

which is the *Bessel equation* of order zero. Solutions may be obtained in terms of *Bessel functions*, suitably chosen so that the initial and boundary conditions are satisfied.

Let us suppose that the surface concentration is constant and that the initial distribution of the diffusant is $f(r)$. For a cylinder of radius R, the conditions are:

$$C = C_0, \qquad r = R, \qquad t \geq 0;$$
$$C = f(r), \qquad 0 < r < R, \qquad t = 0.$$

The solution to this problem is [1]

$$C(r,t) = C_0 \left[1 - \frac{2}{R} \sum_{n=1}^{\infty} \frac{1}{\alpha_n} \frac{J_0(r\alpha_n)}{J_1(R\alpha_n)} \exp(-D\alpha_n^2 t) \right]$$
$$+ \frac{2}{R^2} \sum_{n=1}^{\infty} \exp(-D\alpha_n^2 t) \frac{J_0(r\alpha_n)}{J_1^2(R\alpha_n)} \int rf(r) J_0(r\alpha_n) dr. \qquad (3.61)$$

In Eq. (3.61) J_0 is the Bessel function of the first kind and order zero and J_1 the Bessel function of first order. The α_n are the positive roots of $J_0(R\alpha_n) = 0$.

If the concentration is initially uniform throughout the cylinder, we have $f(r) = C_1$ and Eq. (3.61) reduces to

$$\frac{C - C_1}{C_0 - C_1} = 1 - \frac{2}{R} \sum_{n=1}^{\infty} \frac{\exp(-D\alpha_n^2 t) J_0(\alpha_n r)}{\alpha_n J_1(\alpha_n R)}. \qquad (3.62)$$

If $M(t)$ denotes the quantity of diffusant which has entered or left the cylinder in time t and $M(\infty)$ the corresponding quantity at infinite time, we have

$$\frac{M(t)}{M(\infty)} = 1 - \sum_{n=1}^{\infty} \frac{4}{\alpha_n^2 R^2} \exp(-D\alpha_n^2 t). \qquad (3.63)$$

3.2.6 Radial Diffusion in a Sphere

The diffusion equation for a constant diffusivity and radial flux takes the form

$$\frac{\partial C}{\partial t} = D \left(\frac{\partial^2 C}{\partial r^2} + \frac{2}{r} \frac{\partial C}{\partial r} \right). \qquad (3.64)$$

By substituting

$$u(r,t) = C(r,t)r, \qquad (3.65)$$

Eq. (3.64) becomes

$$\frac{\partial u}{\partial t} = D\frac{\partial^2 u}{\partial r^2}. \qquad (3.66)$$

This equation is analogous to linear flow in one dimension. Therefore, solutions of many problems of radial flow in a sphere can be deduced from those of the corresponding linear flow problems.

If we suppose that the sphere is initially at a uniform concentration C_1 and the surface concentration is maintained constant at C_0, the solution is [1]

$$\frac{C - C_1}{C_0 - C_1} = 1 + \frac{2R}{\pi}\sum_{n=1}^{\infty}\frac{(-1)^n}{n}\sin\left(\frac{n\pi r}{R}\right)\exp(-Dn^2\pi^2 t/R^2). \qquad (3.67)$$

The concentration at the centre is given by the limit $r \to 0$, that is by

$$\frac{C - C_1}{C_0 - C_1} = 1 + 2\sum_{n=1}^{\infty}(-1)^n \exp(-Dn^2\pi^2 t/R^2). \qquad (3.68)$$

If $M(t)$ denotes the quantity of diffusant which has entered or left the sphere in time t and $M(\infty)$ the corresponding quantity at infinite time, we have

$$\frac{M(t)}{M(\infty)} = 1 - \frac{6}{\pi^2}\sum_{n=1}^{\infty}\frac{1}{n^2}\exp(-Dn^2\pi^2 t/R^2). \qquad (3.69)$$

The corresponding solutions for small times are

$$\frac{C - C_1}{C_0 - C_1} = \frac{R}{r}\sum_{n=0}^{\infty}\left[\operatorname{erfc}\frac{(2n+1)-r}{2\sqrt{Dt}} - \operatorname{erfc}\frac{(2n+1)+r}{2\sqrt{Dt}}\right] \qquad (3.70)$$

and

$$\frac{M(t)}{M(\infty)} = 6\sqrt{\frac{Dt}{R^2}}\left[\frac{1}{\sqrt{\pi}} + 2\sum_{n=1}^{\infty}\operatorname{ierfc}\frac{nR}{\sqrt{Dt}}\right] - 3\frac{Dt}{R^2}, \qquad (3.71)$$

where ierfc denotes the inverse of the complementary error function.

3.3 Point Source in one, two, and three Dimensions

In the previous section, we have dealt with one-dimensional solutions of the linear diffusion equation. As examples for diffusion in higher dimensions, we consider now diffusion from instantaneous sources in two- and three-dimensional media.

The diffusion response for a point source in three dimensions and for a line source in two dimensions differs from that of the thin-film source in one dimension given by Eq. (3.9). Now we ask for particular solutions of

Fick' second law under spherical or axial symmetry conditions described by Eqs. (2.12) and (2.11). Let us suppose that in the case of spherical flow a point source located at $|\boldsymbol{r}_3| = 0$ releases at time $t = 0$ a fixed number N_3 of diffusing particles into an infinite and isotropic medium. Let us also suppose that in the case of axial flow a line source located at $|\boldsymbol{r}_2| = 0$ releases N_2 diffusing particles into an infinite and isotropic medium. The diffusion flow will be either spherical or axisymmetric, respectively. The concentration fields that develop around instantaneous plane-, line-, and point-sources in one, two, three dimensions, can all be expressed in homologous form by

$$C(\boldsymbol{r}_d, t) = \frac{N_d}{(4\pi Dt)^{d/2}} \exp\left(-\frac{|\boldsymbol{r}_d|^2}{4Dt}\right) \quad (d = 1, 2, 3). \qquad (3.72)$$

In Eq. (3.72) $\boldsymbol{r}_d$ denotes the d-component vector extending from the source located at $\boldsymbol{r}_d = 0$ to the field point, $\boldsymbol{r}_d$, of the concentration field. If the source strength N_d denotes the number of particles in all three dimensions, the diffusion fields predicted by Eq. (3.72) must be expressed in dimensionality-compatible concentration units. These are [*number per length*] for $d = 1$, [*number per length^2*] for $d = 2$, and [*number per length^3*] for $d = 3$. We note that the source solutions are all linear, in the sense that the concentration response is proportional to the initial source strength.

References

1. J. Crank, *The Mathematics of Diffusion*, 2nd edition, Oxford University Press, Oxford, 1975
2. W. Jost, *Diffusion in Solids, Liquids, Gases*, Academic Press, Inc., New York, 1952, 4th printing with addendum 1965
3. R. Ghez, *A Primer of Diffusion Problems*, Wiley and Sons, 1988
4. M.E. Glicksman, *Diffusion in Solids*, John Wiley and Sons, Inc. 2000
5. H.S. Carslaw, J.C. Jaeger, *Conduction of Heat in Solids*, Clarendon Press, Oxford, 1959
6. L. Fox, *Moving Boundary Problems in Heat Flow and Diffusion*, Clarendon Press, Oxford, 1974
7. J. Crank, *Free and Moving Boundary Problems*, Oxford University Press, Oxford, 1984; reprinted in 1988, 1996
8. R.M. Barrer, Proc. Phys. Soc. (London) **58**, 321 (1946)
9. Y. Adda, J. Philibert, *La Diffusion dans les Solides*, 2 volumes, Presses Universitaires de France, 1966
10. I.S. Gradstein, L.M. Ryshik, *Tables of Series, Products, and Integrals*, Verlag MIR, Moscow, 1981
11. A. Milton, I.A. Stegun (Eds.), *Handbook of Mathematical Functions*, Applied Mathematical Series 55, National Bureau of Standards, U.S. Government Printing Office, Washington, DC, 1964
12. R.V. Churchill, *Modern Operational Mathematics in Engineering*, McGraw Hill, new York , 1944

4 Random Walk Theory and Atomic Jump Process

From a microscopic viewpoint, diffusion occurs by the *Brownian motion* of atoms or molecules. As mentioned already in Chap. 1, ALBERT EINSTEIN in 1905 [1] published a theory for the chaotic motion of small particles suspended in a liquid. This phenomenon had been observed by the Scotish botanist ROBERT BROWN more than three quarters of a century earlier in 1827, when he studied the motion of granules from pollen in water. Einstein argued that the motion of mesoscopic particles is due to the presence of molecules in the fluid. He further reasoned that molecules due to their Boltzmann distribution of energy are always subject to thermal movements of a statistical nature. These statistical fluctuations are the source of stochastic motions occurring in matter all the way down to the atomic scale. Einstein related the mean square displacement of particles to the diffusion coefficient. This relation was, almost at the same time, developed by the Polish scientist SMOLUCHOWSKI [2, 3]. It is nowadays called the *Einstein relation* or the *Einstein-Smoluchowski relation*.

In gases, diffusion occurs by free flights of atoms or molecules between their collisions. The individual path lengths of these flights are distributed around some well-defined mean free path. Diffusion in liquids exhibits more subtle atomic motion than gases. Atomic motion in liquids can be described as randomly directed shuffles, each much smaller than the average spacing of atoms in a liquid.

Most solids are crystalline and diffusion occurs by atomic hops in a lattice. The most important point is that a separation of time scales exists between the elementary jump process of particles between neighbouring lattice sites and the succession of steps that lead to macroscopic diffusion. The elementary diffusion jump of an atom on a lattice, for instance, the exchange of a tracer atom with a neighbouring vacancy or the jump of an interstitial atom, has a duration which corresponds to about the reciprocal of the Debye frequency ($\approx 10^{-13}$ s). This process is usually very rapid as compared to the mean residence time of an atom on a lattice site. Hence the problem of diffusion in lattices can be separated into two different tasks:

1. The more or less random walk of particles on a lattice is the first topic of the present chapter. Diffusion in solids results from many individual displacements (jumps) of the diffusing particles. Diffusive jumps are usually

single-atom jumps of fixed length(s), the size of which is of the order of the lattice parameter. In addition, atomic jumps in crystals are frequently mediated by lattice defects such as vacancies and/or self-interstitials. Thus, the diffusivity can be expressed in terms of physical quantities that describe these elementary jump processes. Such quantities are the jump rates, the jump distances of atoms, and the correlation factor (see below).

2. The second topic of this chapter concerns the rate of individual jumps. Jump processes are promoted by thermal activation. Usually an Arrhenius law holds for the jump rate Γ:

$$\Gamma = \nu^0 \exp\left(-\frac{\Delta G}{k_B T}\right) . \qquad (4.1)$$

The prefactor ν^0 denotes an attempt frequency of the order of the Debye frequency of the lattice. ΔG is the Gibbs free energy of activation, k_B the Boltzmann constant, and T the absolute temperature. A detailed treatment can be found in the textbook of FLYNN [4] and in a more recent review of activated processes by HÄNGGI ET AL. [5]. We consider the jump rate in the second part of this section.

4.1 Random Walk and Diffusion

The mathematics of the random-walk problem allows us to go back and forth between the diffusion coefficient defined in Fick's laws and the underlying physical quantities of diffusing atoms. This viewpoint is most exciting since it transforms the study of diffusion from the question how a system will homogenise into a tool for studying the atomic processes involved in a variety of reactions in solids and for studying defects in solids.

4.1.1 A Simplified Model

Before going through a more rigorous treatment of random walks, it may be helpful to study a simple situation: unidirectional diffusion of interstitials in a simple cubic crystal. Let us assume that the diffusing atoms are dissolved in low concentrations and that they move by jumping from an interstitial site to a neighbouring one with a jump length λ (Fig. 4.1). We suppose a concentration gradient along the x-direction and introduce the following definitions:

Γ: jump rate (number of jumps per unit time) from one plane to the neigbouring one,
n_1: number of interstitials per unit area in plane 1,
n_2: number of interstitials per unit area in plane 2.

4.1 Random Walk and Diffusion

Without a driving force, forward and backward hops occur with the same jump rate and the net flux J from plane 1 to 2 is

$$J = \Gamma n_1 - \Gamma n_2. \tag{4.2}$$

The quantities n_1 and n_2 are related to the volume concentrations (number densities) of diffusing atoms via

$$C_1 = \frac{n_1}{\lambda}, \quad C_2 = \frac{n_2}{\lambda}. \tag{4.3}$$

Usually in diffusion studies the concentration field, $C(x,t)$, changes slowly as a function of the distance variable x in terms of interatomic distances. From a Taylor expansion of the concentration-distance function, keeping only the first term (Fig. 4.1), we get

$$C_1 - C_2 = -\lambda \frac{\partial C}{\partial x}. \tag{4.4}$$

Inserting Eqs. (4.3) and (4.4) into Eq. (4.2) we arrive at

$$J = -\lambda^2 \Gamma \frac{\partial C}{\partial x}. \tag{4.5}$$

By comparison with Fick's first law we obtain for the diffusion coefficient

$$D = \Gamma \lambda^2. \tag{4.6}$$

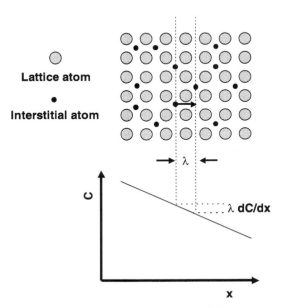

Fig. 4.1. Schematic representation of unidirectional diffusion of atoms in a lattice

Taking into account that in a simple cubic lattice the jump rate of an atom to one of its six nearest-neighbour interstices is related to its total jump rate via $\Gamma_{\text{tot}} = 6\Gamma$, we obtain

$$D = \frac{1}{6}\Gamma_{\text{tot}}\lambda^2 . \tag{4.7}$$

This equation shows that the diffusion coefficient is essentially determined by the product of the jump rate and the jump distance squared. We will show later that this expression is true for any cubic Bravais lattice as long as only nearest-neighbour jumps are considered.

4.1.2 Einstein-Smoluchowski Relation

Let us now consider the random walk of diffusing particles in a more rigorous way. The total displacement $\boldsymbol{R}$ of a particle is composed of many individual displacements $\boldsymbol{r}_i$. Imagine a cloud of diffusing particles starting at time t_0 from the origin and making many individual displacements during the time $t - t_0$. We then ask the question, what is the magnitude characteristic of a random walk after some time $t - t_0 = \tau$? We shall see below that the mean square displacement plays a prominent rôle.

The total displacement of a particle after many individual displacements

$$\boldsymbol{R} = (X, Y, Z) \tag{4.8}$$

is composed of its components X, Y, Z along the x, y, z-axes of the coordinate system and we have

$$R^2 = X^2 + Y^2 + Z^2 . \tag{4.9}$$

To keep the derivation general, the medium is taken not necessarily as isotropic. We concentrate on the X-component of the total displacement and introduce a *distribution function* $W(X, \tau)$. The quantity W denotes the probability that after time τ the particle will have travelled a path with an x-projection X. We assume that W is independent of the choice of the origin and depends only on $\tau = t - t_0$. These assumptions entail that diffusivity and mobility are independent of position and time. Y- and Z-component of the displacement can be treated in analogous way. Fortunately, the precise analytical form of W need not to be known in the following.

Consider now the balance for the number of the diffusing particles (concentration C) located in the plane x at time $t+\tau$. These particles were located in the planes $x - X$ at time t. We thus have

$$C(x, t+\tau) = \sum_X C(x - X, t) W(X, \tau) , \tag{4.10}$$

where the summation must be carried over all values of X. The rate at which the concentration is changing can be found by expanding $C(x, t + \tau)$ and $C(x - X, t)$ around $X = 0$, $\tau = 0$. We get

4.1 Random Walk and Diffusion

$$C(x,t) + \tau \frac{\partial C}{\partial t} + \cdots = \sum_X \left[C(x,t) - X \frac{\partial C}{\partial x} + \frac{X^2}{2} \frac{\partial^2 C}{\partial x^2} + \ldots \right] W(X,\tau). \tag{4.11}$$

The derivatives of C are to be taken at plane x for the time t.

It is convenient to define the nth-moments of X in the usual way:

$$\sum_X W(X,\tau) = 1$$

$$\sum_X X^n W(X,\tau) = \langle X^n \rangle. \tag{4.12}$$

The first expression in Eq. (4.12) states that $W(X,\tau)$ is normalised. The second expression defines the so-called n-th moment $\langle X^n \rangle$ of X. The average values of X^n must be taken over a large number of diffusing particles. In particular, we are be interested in the first and second moment. The second moment $\langle X^2 \rangle$ is also denoted as the *mean square displacement*.

The derivatives $\partial C/\partial t, \partial C/\partial x, \partial^2 C/\partial x^2 \ldots$ have fixed values for time t and position x. For small values of τ, the higher order terms on the left-hand side of Eq. (4.11) are negligible. In addition, because of the nature of diffusion processes, $W(X,\tau)$ becomes more and more localised around $X = 0$ when τ is small. Therefore, for sufficiently small τ terms higher than second order on the right-hand side of Eq. (4.11) can be omitted as well. The terms $C(x,t)$ cancel and we get

$$\frac{\partial C}{\partial t} = -\frac{\langle X \rangle}{\tau} \frac{\partial C}{\partial x} + \frac{\langle X^2 \rangle}{2\tau} \frac{\partial^2 C}{\partial x^2}. \tag{4.13}$$

We recognise that the first term on the right-hand side corresponds to a drift term and the second one to the diffusion term.

In the absence of a driving force, we have $\langle X \rangle = 0$ and Eq. (4.13) reduces to Fick's second law with the diffusion coefficient

$$D_x = \frac{\langle X^2 \rangle}{2\tau}. \tag{4.14}$$

This expression relates the *mean square displacement* in the x-direction with the pertinent component D_x of the diffusion coefficient. Analogous equations hold between the diffusivities D_y, D_z and the mean square displacements in the y- and z-directions:

$$D_y = \frac{\langle Y^2 \rangle}{2\tau}; \quad D_z = \frac{\langle Z^2 \rangle}{2\tau}. \tag{4.15}$$

In an *isotropic medium*, in *cubic crystals*, and in *icosahedral quasicrystals* the displacements in x-, y-, and z-directions are the same. Hence

$$\langle X^2 \rangle = \langle Y^2 \rangle = \langle Z^2 \rangle = \frac{1}{3} \langle R^2 \rangle \tag{4.16}$$

and
$$D = \frac{\langle R^2 \rangle}{6\tau}. \tag{4.17}$$

Equations (4.14) or (4.17) are the relations already mentioned at the entrance of this chapter. They are denoted as the *Einstein relation* or as the *Einstein-Smoluchowski relation*.

4.1.3 Random Walk on a Lattice

In a crystal, the total displacement of an atom is composed of many individual jumps of discrete jump length. For example, in a coordination lattice (coordination number Z) each jump direction will occur with the probability $1/Z$ and the jump length will usually be the nearest-neighbour distance.

According to Fig. 4.2 the individual path of a particle in a sequence of n jumps is the sum

$$\boldsymbol{R} = \sum_{i=1}^{n} \boldsymbol{r}_i \quad \text{or} \quad X = \sum_{i=1}^{n} x_i, \tag{4.18}$$

where $\boldsymbol{r}_i$ denotes jump vectors with x-projections x_i. The squared magnitude of the net displacement is

$$\boldsymbol{R}^2 = \sum_{i=1}^{n} r_i^2 + 2 \sum_{i=1}^{n-1} \sum_{j=i+1}^{n} \boldsymbol{r}_i \boldsymbol{r}_j,$$
$$X^2 = \sum_{i=1}^{n} x_i^2 + 2 \sum_{i=1}^{n-1} \sum_{j=i+1}^{n} x_i x_j. \tag{4.19}$$

If we perform an average over an ensemble of particles, we get

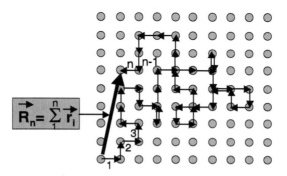

Fig. 4.2. Example for a jump sequence of a particle on a lattice

$$\langle \boldsymbol{R}^2 \rangle = \sum_{i=1}^{n} \langle \boldsymbol{r}_i^2 \rangle + 2 \sum_{i=1}^{n-1} \sum_{j=i+1}^{n} \langle \boldsymbol{r}_i \boldsymbol{r}_j \rangle,$$

$$\langle X^2 \rangle = \sum_{i=1}^{n} \langle x_i^2 \rangle + 2 \sum_{i=1}^{n-1} \sum_{j=i+1}^{n} \langle x_i x_j \rangle. \qquad (4.20)$$

The first term contains squares of the individual jump lengths only. The double sum contains averages between jump i and all subsequent jumps j.

Uncorrelated random walk: Let us consider for the moment a random walker that executes a sequence of jumps in which each individual jump is independent of all prior jumps. Thereby, we deny the 'walker' any memory. Such a jump sequence is sometimes denoted as a Markov sequence (memory free walk) or as an *uncorrelated random walk*. The double sum in Eq. (4.20) contains $n(n-1)/2$ average values of the products $\langle x_i x_j \rangle$ or $\langle \boldsymbol{r}_i \boldsymbol{r}_j \rangle$. These terms contain memory effects also denoted as correlation effects. For a Markov sequence these average values are zero, as for every pair $x_i x_j$ one can find for another particle of the ensemble a pair $x_i x_j$ equal and opposite in sign. Thus, we get from Eq. (4.20) for a random walk without correlation

$$\langle \boldsymbol{R}_{\text{random}}^2 \rangle = \sum_{i=1}^{n} \langle \boldsymbol{r}_i^2 \rangle,$$

$$\langle X_{\text{random}}^2 \rangle = \sum_{i=1}^{n} \langle x_i^2 \rangle. \qquad (4.21)$$

The index 'random' is used to indicate that a *true random walk* is considered with no correlation between jumps.

In a crystal lattice the jump vectors can only take a few definite values. For example, in a *coordination lattice* (coordination number Z), in which nearest-neighbour jumps occur (jump length d with x-projection d_x), Eq. (4.21) reduces to

$$\langle \boldsymbol{R}_{\text{random}}^2 \rangle = \langle n \rangle d^2,$$

$$\langle X_{\text{random}}^2 \rangle = \langle n \rangle d_x^2. \qquad (4.22)$$

Here $\langle n \rangle$ denotes the average number of jumps of a particle. It is useful to introduce the *jump rate* Γ of an atom into one of its Z neighbouring sites via

$$\Gamma \equiv \frac{\langle n \rangle}{Zt}. \qquad (4.23)$$

We then get

$$D = \frac{1}{6} d^2 Z \Gamma = \frac{d^2}{6\bar{\tau}}. \qquad (4.24)$$

Table 4.1. Geometrical properties of cubic Bravais lattices with lattice parameter a

Lattice	Coordination number Z	Jump length d
Primitive cubic	6	a
Body-centered cubic (bcc)	8	$a\sqrt{3}/2$
Face-centered cubic (fcc)	12	$a\sqrt{2}/2$

This equation describes diffusion of interstitial atoms in a dilute interstitial solid solution[1]. The quantity

$$\bar{\tau} = \frac{1}{Z\Gamma} \qquad (4.25)$$

is the *mean residence time of an atom* on a certain site. For cubic Bravais lattices, the jump length d and the lattice parameter a are related to each other as indicated in Table 4.1. Using these parameters we get from Eq. (4.24)

$$D = a^2 \Gamma . \qquad (4.26)$$

4.1.4 Correlation Factor

Random walk theory, to this point, involved a series of independent jumps, each occurring without any memory of the previous jumps. However, several atomic mechanisms of diffusion in crystals entail diffusive motions of atoms which are not free of memory effects. Let us for example consider the vacancy mechanism (see also Chap. 6). If vacancies exchange sites with atoms a memory effect is necessarily involved. Upon exchange, vacancy and 'tagged' atom (tracer) move in opposite directions. Immediately after the exchange the vacancy is for a while available next to the tracer atom, thus increasing the probability for a reverse jump of the tracer. Consequently, the tracer atom does not diffuse as far as expected for a completely random series of jumps. This reduces the efficiency of a tracer walk in the presence of positional memory effects with respect to an uncorrelated random walk.

BARDEEN AND HERRING in 1951 [7, 8] recognised that this can be accounted for by introducing the *correlation factor*

$$f = \lim_{n \to \infty} \frac{\langle \boldsymbol{R}^2 \rangle}{\langle \boldsymbol{R}^2_{\text{random}} \rangle} = 1 + 2 \lim_{n \to \infty} \frac{\sum_{i=1}^{n-1} \sum_{j=i+1}^{n} \langle \boldsymbol{r}_i \boldsymbol{r}_j \rangle}{\sum_{i=1}^{n} \langle \boldsymbol{r}_i^2 \rangle},$$

---

[1] In a non-dilute interstitial solution correlation effects can occur, because some of the neighbouring sites are not available for a jump.

$$f_x = \lim_{n\to\infty} \frac{\langle X^2 \rangle}{\langle X^2_{\text{random}} \rangle} = 1 + 2 \lim_{n\to\infty} \frac{\sum_{i=1}^{n-1}\sum_{j=i+1}^{n} \langle x_i x_j \rangle}{\sum_{i=1}^{n} \langle x_i^2 \rangle}. \qquad (4.27)$$

The correlation factor in Eq. (4.27) equals the sum of two terms: (i) the leading term $+1$, associated with uncorrelated (Markovian) jump sequences and (ii) the double summation contains the correlation between jumps. It has been argued above that for an uncorrelated walk the double summation is zero.

Diffusion in solids is often defect-mediated. Then, successive jumps occur with higher probability in the reverse direction and the contribution of the double sum is negative. Equation (4.27) also shows that one may define the correlation factor as the ratio of the diffusivity of tagged atoms, D^*, and a hypothetical diffusivity arising from uncorrelated jump sequences, D_{random}, via

$$f \equiv \frac{D^*}{D_{\text{random}}}. \qquad (4.28)$$

Correlation effects are important in solid-state diffusion of crystalline materials, whenever diffusion is mediated by a diffusion vehicle (see Table 4.2). Examples for diffusion vehicles are vacancies, vacancy pairs, self-interstitials, etc. An equivalent statement is to say, there must be at least three identifiable 'species' involved in the diffusion process. For example, during tracer diffusion via vacancies in pure crystals, the three participating 'species' are vacancies, host atoms, and tracer atoms. Interstitial diffusion in a dilute interstitial solution is uncorrelated, because no diffusion vehicle is involved.

Let us consider once more diffusion in a cubic Bravais lattice. When correlation occurs, Eq. (4.24) must be replaced by

$$D^* = \frac{1}{6} f d^2 Z \Gamma = f a^2 \Gamma. \qquad (4.29)$$

We will return to the correlation factor in Chap. 7, after having introduced point defects in Chap. 5 and the major mechanisms of diffusion in crystals in Chap. 6.

Table 4.2. Correlation effects of diffusion for crystalline materials

$f = 1$	Markovian jump sequence
	No diffusion vehicle involved: direct interstitial diffusion
$f < 1$	Non-Markovian jump sequence
	Diffusion vehicle involved: vacancy, divacancy, self-interstitial, ... mechanisms

4.2 Atomic Jump Process

In preceding sections, we have considered many atomic jumps on a lattice. Equally important are the rates at which jumps occur. Let us take a closer look to the atomic jump process illustrated in Fig. 4.3. An atom moves into a neighbouring site, which could be either a neighbouring vacancy or an interstitial site. Clearly, the jumping atom has to squeeze between intervening lattice atoms – a process which requires energy. The energy necessary to promote the jump is usually large with respect to the thermal energy $k_B T$. At finite temperatures, atoms in a crystal oscillate around their equilibrium positions. Usually, these oscillations are not violent enough to overcome the barrier and the atom will turn back to its initial position. Occasionally, large displacements result in a successful jump of the diffusing atom. These *activation events* are infrequent relative to the frequencies of the lattice vibrations, which are characterised by the Debye frequency. Typical values of the Debye frequency lie between 10^{12} and $10^{13}\,\text{s}^{-1}$. Once an atom has moved as the result of an activation event, the energy flows away from this atom relatively quickly. The atom becomes deactivated and waits on the average for many lattice vibrations before it jumps again. Thermally activated motion of atoms in a crystal occurs in a series of discrete jumps from one lattice site (or interstitial site) to the next.

The theory of the rate at which atoms move from one site to a neighbouring one was proposed by WERT [9] and has been refined by VINEYARD [10]. Vineyard's approach is based upon the canonical ensemble of statistical mechanics for the distribution of atomic positions and velocities. The jump process can be viewed as occurring in an energy landscape characterised by

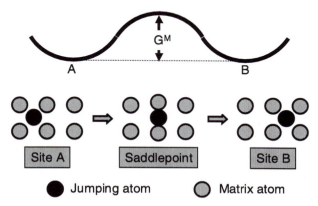

Fig. 4.3. Atomic jump process in a crystalline solid: the black atom moves from an initial configuration (*left*) to a final configuration (*right*) pushing through a saddlepoint configuration (*middle*)

the difference in Gibbs free energy G^M between the saddle-point barrier and the equilibrium position (Fig. 4.3). G^M is denoted as the *Gibbs free energy of migration* (superscript M) of the atom. It can be separated according to

$$G^M = H^M - TS^M, \qquad (4.30)$$

where H^M denotes the *enthalpy of migration* and S^M the *entropy of migration*. Using statistical thermodynamics, VINEYARD [10] has shown that the *jump rate* ω (number of jumps per unit time to a particular neighbouring site) can be written as

$$\omega = \nu^0 \exp\left(-\frac{G^M}{k_B T}\right) = \nu^0 \exp\left(\frac{S^M}{k_B}\right) \exp\left(-\frac{H^M}{k_B T}\right), \qquad (4.31)$$

where ν^0 is called the *attempt frequency*. It is of the order of the Debye frequency. Its rigorous meaning is as follows: ν^0 is the vibration frequency around the equilibrium site but in the direction of the reaction path. The entropy of migration corresponds to the change in lattice vibrations associated with the displacement of the jumping atom from its equilibrium to the saddle point configuration. Vineyard's treatment also provides an expression for the migration entropy:

$$S^M = k_B \left[\sum_{j \neq 0}^{3N-1} \ln\left(\frac{h\nu_j}{k_B T}\right)_A - \sum_{j \neq 0}^{3N-1} \ln\left(\frac{h\nu'_j}{k_B T}\right)_{SP} \right]. \qquad (4.32)$$

The ν_j are the $3N - 1$ normal mode frequencies for vibrations around the equilibrium site A while ν'_j are the frequencies of the system when it is constrained to move within the hyperface which passes through the saddle point (SP) perpendicular to the jump direction.

The theory of the rates at which atoms move from one lattice site to another covers one of the fundamental aspects of diffusion. For most practical purposes, a few general conclusions are important:

1. Atomic migration in a solid is the result of a sequence of localised jumps from one site to another.
2. Atomic jumps in crystals usually occur from one site to a nearest-neighbour site. Molecular dynamic simulations mostly confirm this view. Multiple hops are rare, although their occurrence is indicated in some model substances. Jumps with magnitudes larger than the nearest-neighbour distance are more common on surfaces or in grain boundaries (see Chap. 32).
3. The jump rate ω has an Arrhenius-type dependence on temperature as indicated in Eq. (4.31).
4. The concept of an atomic jump developed above applies to all the diffusion mechanisms discussed in Chap. 6. Of course, the values of $G^M, H^M,$

and S^M depend on the diffusion mechanism and on the material under consideration:

For interstitial diffusion, the quantities G^M, H^M, and S^M pertain to the saddle point separating two interstitial positions. For a dilute interstitial solution, virtually every interstitial solute is surrounded by empty interstitial sites. Thus, for an atom executing a jump the probability to find an empty site next to its starting position is practically unity.

For vacancy-mediated diffusion, G^M, H^M, and S^M pertain to the saddle-point separating the vacant lattice site and the jumping atom on its equilibrium site.

5. There are cases – mostly motion of hydrogen in solids at low temperatures – where a classical treatment is not adequate [16]. However, the end result of different theories including quantum effects is still a movement in a series of distinct jumps from one site to another [17]. For atoms heavier than hydrogen and its isotopes, quantum effects can usually be disregarded.

For more detailed discussions of the problem of thermally activated jumps the reader may consult the textbook of FLYNN [3] and the reviews by FRANKLIN [11], BENNETT [12], JACUCCI [13], HÄNNGI ET AL. [15] PONTIKIS [14] and FLYNN AND STONEHAM [17].

Molecular dynamic calculations become increasingly important and are used to check and to supplement analytical theories. Atomistic computer simulations of diffusion processes have been reviewed, e.g., by MISHIN [18]. Nowadays, simulation methods present a powerful approach to gain fundamental insight into atomic jump processes in materials. The capabilities of simulations have drastically improved due to the development of new simulation methods reinforced by increased computer power. Reliable potentials of atomic interaction have been developed, which allow a quantitative description of point defect properties. Simulation of atomic jump processes have been applied to ordered intermetallic compounds, surface diffusion, grain-boundary diffusion, and other systems. The challenge is to understand, describe, and calculate diffusion coefficients in a particular metal, alloy, or compound. In some – but not all – of the rare cases, when this has been done, the agreement with experiments is encouraging.

References

1. A. Einstein, Annalen der Physik **17**, 549 (1905)
2. M. van Smoluchowski, Annalen der Physik **21**, 756 (1906)
3. M. van Smoluchowski, Z. Phys. **13**, 1069 (1912); and Physikalische Zeitschrift **17**, 557 (1916)
4. C.P. Flynn, *Point Defects and Diffusion*, Clarendon Press, Oxford, 1972
5. P. Hänggi, P. Talkner, M. Borkovec: Rev. Mod. Phys. **62**, 251 (1990)

6. J.R. Manning, *Diffusion Kinetics for Atoms in Crystals*, D. van Norstrand Comp., Inc., 1968
7. J. Bardeen, C. Herring, in: *Atom Movements*, A.S.M. Cleveland, 1951, p. 87
8. J. Bardeen, C. Herring, in: *Imperfections in Nearly Perfect Solids*, W. Shockley (Ed.), Wiley, New York, 1952, p. 262
9. C. Wert, Phys. Rev. **79**, 601 (1950)
10. G. Vineyard, J. Phys. Chem. Sol. **3**, 121 (1957)
11. W.M. Franklin, in: *Diffusion in Solids – Recent Developments*, A.S. Nowick, J.J. Burton (Eds.), Academic Press, Inc., 1975, p.1
12. C.H. Bennett, in: *Diffusion in Solids – Recent Developments*, A.S. Nowick, J.J. Burton (Eds.), Academic Press, Inc., 1975, p.74
13. G. Jacucci, in: *Diffusion in Crystalline Solids*, G.E. Murch, A.S. Nowick (Eds.), Academic Press, Inc., 1984, p.431
14. V. Pontikis, *Thermally Activated Processes*, in: *Diffusion in Materials*, A.L. Laskar, J.L. Bocqut, G. Brebec, C. Monty (Eds.), Kluwer Academic Publishers, Dordrecht, The Netherlands, 1990, p.37
15. P. Hänngi, P. Talkner, M. Borkovec, Rev. Mod. Phys. **62**, 251 (1990)
16. J. Völkl, G. Alefeld, in: *Diffusion in Solids – Recent Developments*, A.S. Nowick, J.J. Burton (Eds.), Academic Press, Inc., 1975
17. C.P. Flynn, A.M. Stoneham, Phys. Rev. **B1**, 3966 (1970)
18. Y. Mishin, *Atomistic Computer Simulation of Diffusion*, in: *Diffusion Processes in Advanced Technological Materials*, D. Gupta (Ed.), William Andrews, Inc., 2005

5 Point Defects in Crystals

The Russian scientist FRENKEL in 1926 [1] was the first author to introduce the concept of point defects (see Chap. 1). He suggested that thermal agitation causes transitions of atoms from their normal lattice sites into interstitial positions leaving behind lattice vacancies. This type of disorder is nowadays denoted as Frenkel disorder and contained already the concepts of vacancies and self-interstitials. Already in the early 1930s WAGNER AND SCHOTTKY [2] treated a fairly general case of disorder in binary AB compounds considering the occurrence of vacancies, self-interstitials, and of antisite defects on both sublattices.

Point defects are important for diffusion processes in crystalline solids. This statement mainly derives from two features: one is the ability of point defects to move through the crystal and to act as 'vehicles for diffusion' of atoms; another is their presence at thermal equilibrium. Of particular interest in this chapter are diffusion-relevant point defects, i.e. defects which are present in appreciable thermal concentrations.

In a defect-free crystal, mass and charge density have the periodicity of the lattice. The creation of a point defect disturbs this periodicity. In metals, the conduction electrons lead to an efficient electronic screening of defects. As a consequence, point defects in metals appear uncharged. In ionic crystals, the formation of a point defect, e.g., a vacancy in one sublattice disturbs the charge neutrality. Charge-preserving defect populations in ionic crystals include Frenkel disorder and Schottky disorder, both of which guarantee global charge neutrality. *Frenkel disorder* implies the formation of equal numbers of vacancies and self-interstitials in one sublattice. *Schottky disorder* consists of corresponding numbers of vacancies in the sublattices of cations and anions. For example, in AB compounds like NaCl composed of cations and anions with equal charges opposite in sign the number of vacancies in both sublattices must be equal to preserve charge neutrality. Point defects in semiconductors introduce electronic energy levels within the band gap and thus can occur in neutral or ionised states, depending on the position of the Fermi level. In what follows, we consider at first point defects in metals and then proceed to ionic crystals and semiconductors.

Nowadays, there is an enormous body of knowledge about point defects from both theoretical and experimental investigations. In this chapter, we provide a brief survey of some features relevant for diffusion. For more comprehensive accounts of the field of point defects in crystals, we refer to the textbooks of FLYNN [3], STONEHAM [4], AGULLO-LOPEZ, CATLOW AND TOWNSEND [5], to a review on defect in metals by WOLLENBERGER [6], and to several conference proceedings [9–12]. For a compilation of data on point defects properties in metals, we refer to a volume edited by ULLMEIER [13]. For semiconductors, data have been assembled by SCHULZ [14], STOLWIJK [15], STOLWIJK AND BRACHT [16], and BRACHT AND STOLWIJK [17]. Properties of point defects in ionic crystals can be found in reviews by BARR AND LIDIARD [7] and FULLER [8] and in the chapters of BENIÈRE [18] and ERDELY [19] of a data collection edited by BEKE.

5.1 Pure Metals

5.1.1 Vacancies

Statistical thermodynamics is a convenient tool to deduce the concentration of lattice vacancies at thermal equilibrium. Let us consider an elemental crystal, which consists of N atoms (Fig. 5.1). We restrict the discussion to metallic elements or to noble gas solids in which the vacancies are in a single electronic state and we suppose (in this subsection) that the concentration is so low that interactions among them can be neglected. At a finite temperature, n_{1V} vacant lattice sites (monovacancies, index $1V$) are formed. The total number of lattice sites then is

$$N' = N + n_{1V}. \tag{5.1}$$

The thermodynamic reason for the occurrence of vacancies is that the Gibbs free energy of the crystal is lowered. The Gibbs free energy $G(p,T)$ of the

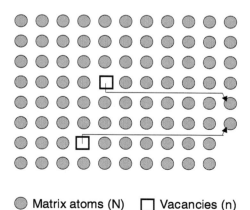

Fig. 5.1. Vacancies in an elemental crystal

crystal at temperature T and pressure p is composed of the Gibbs function of the perfect crystal, $G^0(p,T)$, plus the change in the Gibbs function on forming the actual crystal, ΔG:

$$G(p,T) = G^0(p,T) + \Delta G, \tag{5.2}$$

where

$$\Delta G = n_{1V} G^F_{1V} - T S_{\text{conf}}. \tag{5.3}$$

In Eq. (5.3) the quantity G^F_{1V} represents the *Gibbs free energy of formation* of an isolated vacancy. It corresponds to the work required to create a vacancy by removing an atom from a particular, but arbitrary, lattice site and incorporating it at a surface site ('Halbkristalllage'). Not only surfaces also grain boundaries and dislocations can act as sources or sinks for vacancies. If a vacancy is created, the crystal lattice relaxes around the vacant site and the vibrations of the crystal are also altered. The Gibbs free energy can be decomposed according to

$$G^F_{1V} = H^F_{1V} - T S^F_{1V} \tag{5.4}$$

into the *formation enthalpy* H^F_{1V} and the *formation entropy* S^F_{1V}. The last term on the right-hand side of Eq. (5.3) contains the *configurational entropy* S_{conf}, which is the thermodynamic reason for the presence of vacancies.

In the absence of interactions, all distinct configurations of n_{1V} vacancies on N' lattice sites have the same energy. The configurational entropy can be expressed through the equation of Boltzmann

$$S_{\text{conf}} = k_B \ln W_{1V}, \tag{5.5}$$

where W_{1V} is the number of distinguishable ways of distributing n_{1V} monovacancies among the N' lattice sites. Combinatoric rules tell us that

$$W_{1V} = \frac{N'!}{n_{1V}! N!}. \tag{5.6}$$

The numbers appearing in Eq. (5.6) are very large. Then, the formula of Stirling, $\ln x! \approx x \ln x$, approximates the factorial terms and we get

$$\ln W_{1V} \approx (N + n_{1V}) \ln(N + n_{1V}) - n_{1V} \ln n_{1V} - N \ln N. \tag{5.7}$$

Thermodynamic equilibrium is imposed on a system at given temperature and pressure by minimising its Gibbs free energy. In the present case, this means

$$\Delta G \Rightarrow \text{Min}. \tag{5.8}$$

The equilibrium number of monovacancies, n^{eq}_{1V}, is obtained, when the Gibbs free energy in Eq. (5.3) is minimised with respect to n_{1V}, subject to the constraint that the number of atoms, N, is fixed. Inserting Eqs. (5.5) and (5.7) into Eq. (5.3), we get from the necessary condition for thermal equilibrium, $\partial \Delta G/\partial n_{1V} = 0$:

$$H_{1V}^F - TS_{1V}^F + k_B T \ln \frac{n_{1V}^{eq}}{N + n_{1V}^{eq}} = 0 \qquad (5.9)$$

By definition we introduce the *site fraction of monovacancies*[1] via:

$$C_{1V} \equiv \frac{n_{1V}}{N + n_{1V}}. \qquad (5.10)$$

This quantity also represents the probability to find a vacancy on an arbitrary, but particular lattice site. In thermal equilibrium we have $C_{1V}^{eq} \equiv n_{1V}^{eq}/(N + n_{1V}^{eq})$. Solving Eq. (5.9) for the equilibrium site fraction yields

$$C_{1V}^{eq} = \exp\left(-\frac{G_{1V}^F}{k_B T}\right) = \exp\left(\frac{S_{1V}^F}{k_B}\right) \exp\left(-\frac{H_{1V}^F}{k_B T}\right). \qquad (5.11)$$

This equation shows that the concentration of thermal vacancies increases via a Boltzmann factor with increasing temperature. The temperature dependence of C_{1V}^{eq} is primarily due to the formation enthalpy term in Eq. (5.11). We note that the vacancy formation enthalpy is also given by

$$H_{1V}^F = -k_B \frac{\partial \ln C_{1V}^{eq}}{\partial (1/T)}. \qquad (5.12)$$

This quantity is often determined in experiments which measure relative concentrations. Such measurements are less tedious than measurements of absolute concentrations (see below). In the analysis of experiments, it is frequently assumed that formation enthalpy and entropy are independent of temperature; this is often, though not always, justified.

5.1.2 Divacancies

Divacancies (2V) are point defects that form in a crystal as the simplest complex of monovacanies (1V). This is a consequence of the mass-action equilibrium for the reaction

$$1V + 1V \rightleftharpoons 2V. \qquad (5.13)$$

The probability that a given lattice site in a monoatomic crystal is vacant equals the site fraction of monovacancies. Let us suppose that a divacancy consists of two monovacancies on nearest-neighbour lattice sites. For non-interacting monovacancies, the probability of forming a divacancy is proportional to $(C_{1V})^2$. For a coordination lattice (coordination number Z) the

---

[1] Concentrations as number densities are given by $C_{1V} N$, when N is taken as the number density of atoms.

equilibrium fraction of divacancies C_{2V}^{eq} that form simply for statistical reasons is given by $\frac{Z}{2}(C_{1V}^{eq})^2$. However, there is also a gain in enthalpy (and entropy) when two vacancies are located on adjacent lattice sites. Fewer bonds to neighbouring atoms must be broken, when a second vacancy is formed next to an already existing one. Interactions between two vacancies are accounted for by a *Gibbs free energy of binding* G_{2V}^B, which according to

$$G_{2V}^B = H_{2V}^B - TS_{2V}^B \tag{5.14}$$

can be decomposed into an enthalpy H_{2V}^B and an entropy S_{2V}^B of interaction. For $G_{2V}^B > 0$ the interaction is attractive and binding occurs, whereas for $G_{2V}^B < 0$ it is repulsive. Combining Eq. (5.14) with the mass-action law for Eq. (5.13) yields

$$C_{2V}^{eq} = \frac{Z}{2} \exp\left(\frac{G_{2V}^B}{k_B T}\right) (C_{1V}^{eq})^2. \tag{5.15}$$

Equation (5.15) shows that at thermal equilibrium the divacancy concentration rises faster with increasing temperature than the monovacancy concentration (see Fig. 5.2). With increasing G_{2V}^B the equilibrium concentration of divacancies increases as well.

The total equilibrium concentration of vacant lattice sites, C_V^{eq}, in the presence of mono- and divacancies (neglecting higher agglomerates) is then

$$C_V^{eq} = C_{1V}^{eq} + 2C_{2V}^{eq}. \tag{5.16}$$

For a typical monovacancy site fraction in metals of 10^{-4} near the melting temperature (see below), the fraction of non-interacting divacancies would be

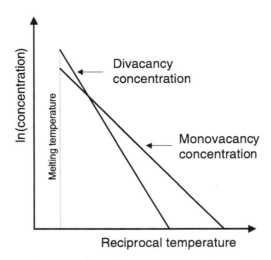

Fig. 5.2. Arrhenius diagram of equilibrium concentrations of mono- and divacancies in metals (schematic)

$\frac{Z}{2} \times 10^{-8}$. Typical interaction energies of a few 0.1 eV increase the divacancy concentration by factors of 10 to 100. Therefore, the divacancy concentration at thermal equilibrium is less or much less than that of monovacancies. Nevertheless, divacancies in close-packed metals can contribute to some extent to the diffusive transport (see Chaps. 6 and 17). The major reason is that divacancies are more effective diffusion vehicles than monovacancies, since their mobility can be considerably higher than that of monovacancies [20]. The contribution of higher agglomerates than divacancies is usually negligible.

5.1.3 Determination of Vacancy Properties

The classical method for an absolute measurement of the *total vacancy concentration*, Eq. (5.16), is **differential dilatometry (DD)**. The idea is to compare macroscopic and microscopic volume changes as functions of temperature. To understand this method, we consider a monoatomic crystal with N atoms. We denote its macroscopic volume in the defect-free state as V_0 and the volume per lattice site as Ω_0. A defect-free state can usually be realised by cooling slowly to low enough temperatures. As long as the thermal concentration of vacant lattice sites is negligible, we have $V_0 = N\Omega_0$. With increasing temperature the volume increases due to thermal expansion and due the formation of new lattice sites. Then, the macroscopic volume and the volume per lattice site take the values $V(T)$ and $\Omega(T)$, respectively. The change in the macroscopic volume is given by

$$\Delta V \equiv V(T) - V_0 = (N+n)\Omega(T) - N\Omega_0 = N\Delta\Omega + n\Omega(T), \quad (5.17)$$

where $\Delta\Omega \equiv \Omega(T) - \Omega_0$. n is the number of new lattice sites. Equation (5.17) can be rearranged to give

$$\frac{\Delta V}{V_0} = \frac{\Delta\Omega}{\Omega_0} + \frac{n}{N}\frac{\Omega(T)}{\Omega_0}. \quad (5.18)$$

This equation reflects the two major physical reasons of the macroscopic volume change: $\Delta\Omega/\Omega_0$ is the thermal expansion of the unit cell and the second term on the right-hand side stands for the additional lattice sites.

If n_V vacant sites and n_I self-interstitials are created, we have $n = n_V - n_I$ new lattice sites. The difference between the total self-interstitial fraction, C_I^{eq}, and the total site fraction of vacant lattice sites, C_V^{eq}, is given by

$$C_V^{eq} - C_I^{eq} = \frac{\Delta V}{V_0} - \frac{\Delta\Omega}{\Omega_0}. \quad (5.19)$$

In Eq. (5.19) the effect of thermal expansion in the ratio $\Omega(T)/\Omega_0$ and higher order terms in n/N have been omitted.

In metals, self-interstitials need not to be considered as equilibrium defects (see below). We then have

$$C_V^{eq} = \frac{\Delta V}{V_0} - \frac{\Delta \Omega}{\Omega_0}. \qquad (5.20)$$

For cubic crystals Eq. (5.20) can be rewritten as

$$C_V^{eq} = 3\left(\frac{\Delta l}{l_0} - \frac{\Delta a}{a_0}\right), \qquad (5.21)$$

where $\Delta l/l_0$ is the relative length change of the sample and $\Delta a/a_0$ the lattice parameter change. In deriving Eq. (5.21) from Eq. (5.20) quadratic and cubic terms in $\Delta l/l_0$ and $\Delta a/a_0$ have been neglected, because already the linear terms are of the order of a few percent or less.

Equation (5.21) shows what needs to be done in DD-experiments. The macroscopic length change and the expansion of the unit cell must be measured simultaneously[2]. The expansion of the unit cell can be measured in very precise X-ray or neutron diffraction studies. As already mentioned, near the melting temperature of metallic elements C_V^{eq} does not exceed 10^{-3} to 10^{-4} (see Table 5.1) and is much smaller at lower temperatures. Thus, precise measurements of C_V^{eq} are very ambitious. Both length and lattice parameter changes must be recorded with the extremely high accuracy of about 10^{-6}.

Differential dilatometry experiments were introduced by FEDER AND NOWICK [21] and SIMMONS AND BALLUFFI [22, 23] around 1960 and later

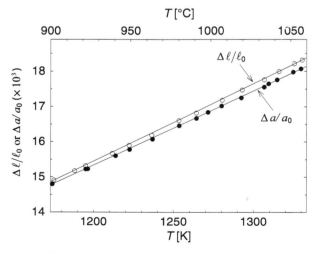

Fig. 5.3. Length and lattice parameter change *versus* temperature for Au according to SIMMONS AND BALLUFFI [23]

---

[2] For uniaxial crystals measurements in two independent directions are necessary.

used by several authors. As an example, Fig. 5.3 shows measured length and lattice parameter expansion versus temperature for gold in the interval 900 to 1060 °C according to [23]. $\Delta l/l$ is larger than $\Delta a/a$ at high temperatures due to the presence of lattice vacancies. This technique demonstrated that the dominant, thermally created defects in metals are vacancies and the exponential dependence of the vacancy concentration on temperature was also confirmed. The great advantage of DD experiments is that the total vacancy content as a function of temperature can be obtained. If monovacancies are the dominant species, both the formation enthalpy and the formation entropy can be deduced. When the divacancy contribution is not negligible, additional divacancy properties can be obtained [20].

The basic weakness of DD experiments is the unsufficient accuracy in the range below about $C_V^{eq} \approx 10^{-5}$, where the divacancy contribution would be low enough to permit a direct measurement of the formation properties of monovacancies. This is illustrated for aluminium in Fig. 5.4 according to SEEGER [24]. The thermal expansion measurements of various groups [25–27] cover, with a reasonable accuracy only the concentration range between 10^{-3} to 10^{-5}. Fortunately, there are additional techniques such as positron annihilation spectroscopy (see below) that supplement DD measurements very well. An analysis of DD measurements together with these additional data yields the line in Fig. 5.4, which corresponds to a monovacancy contribution with $H_{1V}^F = 0.66$ eV and $S_{1V}^F = 0.8 k_B$. Near the melting temperature the fraction of vacant sites associated as divacancies is about 50%.

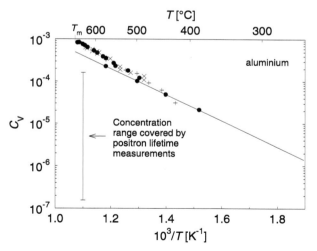

Fig. 5.4. Equilibrium concentration of vacant lattice sites in Al determined by DD measurements according to [24]. DD data: + [25], • [27], × [26]. The concentration range covered by positron lifetime measurements is also indicated

Despite the elegance of DD experiments, much information on defect properties is obtained from other ingenious experiments, which are less direct, some of which are mentioned in what follows:

Formation enthalpies can be deduced from experiments which do not involve a determination of the absolute vacancy concentration. A frequently used method is **rapid quenching (RQ)** from high temperatures, T_Q. The quenched-in vacancy population can be studied in measurements of the residual resistivity. For example, thin metal wires or foils can be rapidly quenched. Their residual resistivities before and after quenching, ρ_0 and ρ_Q, can be measured accurately at liquid He temperature. The residual resistivity after quenching increases due to the additional scattering processes of conduction electrons at 'frozen in' vacancies. The increase of the residual resistivity, $\Delta\rho$, is proportional to the frozen-in vacancy concentration $C_V^{eq}(T_Q)$:

$$\Delta\rho \equiv \rho_Q - \rho_0 = \rho_V \, C_V^{eq}(T_Q)\,. \tag{5.22}$$

ρ_V is a defect-related quantity, which accounts for the resistivity increase per vacant site. In a successful quenching experiment, the equilibrium vacancy population is completely 'frozen in'. Vacancy losses to sinks such as dislocations, grain-boundaries, or surfaces can cause problems in quenching experiments. Since the residual resistivity increase per vacant site is usually unknown, only formation enthalpies can be determined from RQ experiments when $\Delta\rho$ is measured for various quenching temperatures. Formation entropies S^F are not accessible from such experiments. Only the product $\rho_V \exp(S^F/k_B)$ can be deduced.

Transmission electron microscopy (TEM) of quenched-in vacancy agglomerates is a further possibility to determine vacancy concentrations. Upon annealing vacancies become mobile and can form agglomerates. If the agglomerates are large enough they can be studied by TEM. In addition to vacancy losses during the quenching process, the invisibility of very small agglomerates can cause problems.

A very valuable tool for the determination of vacancy formation enthalpies is **positron annihilation spectroscopy (PAS)**. The positron is the antiparticle of the electron. It is, for example, formed during the β^+ decay of radioisotopes. High-energy positrons injected in metals are thermalised within picoseconds. A thermalised positron diffuses through the lattice and ends its life by annihilation with an electron. Usually, two γ-quanta are emitted according to

$$e^+ + e \rightarrow 2\gamma\,.$$

The energy of each γ-quantum is about 511 keV. The positron lifetime depends on the total electron density. Vacancies can trap positrons. Because of the missing core electrons at the vacant lattice site, the local electron density is significantly reduced. Therefore, the lifetime of trapped positrons is enhanced as compared to that of positrons annihilating in the perfect lattice.

Positrons in a vacancy-containing crystal end their life either by annihilation as free positrons or as trapped positrons. The lifetimes of both fates are different and the trapping probability increases with the vacancy concentration. Lifetime measurements are possible using, e.g., ^{22}Na as a positron source. This nuclide emits γ-quanta simultaneously at the 'birth' of the positron. The positrons 'death' is accompanied by the emission of two 511 keV annihilation quanta.

The interpretation of positron lifetime measurements is provided by a trapping model: a thermalised positron diffusing through a metal is trapped by a vacancy with the trapping rate σ. The positron lifetime in the trapped state, τ_t, exceeds that in the free state, τ_f, when the positron is located in an interstitial position of the perfect crystal. If untrapping is disregarded two distinct lifetimes of the positron are predicted by this model:

(i) The trapped positron is annihilated with a lifetime τ_t.
(ii) A positron diffusing through the crystal may end its existence as a 'free' particle either by the annihilation rate $1/\tau_f$ or by being trapped by a vacancy with the trapping rate σC_{1V}, where σ is the trapping cross section. This results in a lifetime given by $\tau_f/(1 + \tau_f \sigma C_{1V})$. If one assumes that initially all positrons are free, one gets for their mean lifetime:

$$\bar{\tau} = \tau_f \frac{1 + \tau_t \sigma C_{1V}}{1 + \tau_f \sigma C_{1V}}. \tag{5.23}$$

Figure 5.5 shows as an example measurements of the mean lifetime of positrons in aluminium as a function of temperature [28]. The mean lifetime increases from about 160 ps near room temperature and reaches a high

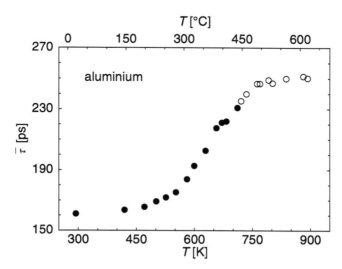

Fig. 5.5. Mean lifetime of positrons in Al according to SCHAEFER ET AL. [28]

Table 5.1. Monovacancy properties of some metals. C_{1V}^{eq} is given in site fractions

Metal	H_{1V}^F/eV	S_{1V}^F/k_B	C_{1V}^{eq} at T_m	Method(s)
Al	0.66	0.6	9.4×10^{-4}	DD + PAS
Cu	1.17	1.5	2×10^{-4}	DD + PAS
Au	0.94	1.1	7.2×10^{-4}	DD
Ag	1.09	–	1.7×10^{-4}	DD
Pb	0.49	0.7	1.7×10^{-4}	DD
Pt	1.49	1.3	–	RQ
Ni	1.7	–	–	PAS
Mo	3.0	–	–	PAS
W	4.0	2.3	1×10^{-4}	RQ + TEM

temperature value of about 250 ps. From a fit of Eq. (5.23) to the data the product σC_{1V} can be deduced. If the trapping cross section is known the vacancy concentration is accessible. If σ is constant, the vacancy formation enthalpy can be deduced from the temperature variation of σC_{1V}. At high temperature, i.e. for high vacancy concentrations, all positrons are annihilated from the trapped state. Under such conditions the method is no longer sensitive to a further increase of the vacancy concentration and the curve $\bar{\tau}$ versus T saturates. The maximum sensitivity of positron annihilation measurements occurs for vacant site fractions between about 10^{-4} and 10^{-6} (see Fig. 5.4).

A unique feature of PAS is that it is sensitive to vacancy-type defects, but insensitive to interstitials. Measurements of the mean positron lifetime is one technique of PAS. Other techniques, not described here, are measurements of the line-shape of the annihilation line and lifetime spectroscopy. Review articles on PAS applications for studies of vacancy properties in metals are provided by SEEGER [24], DOYAMA AND HASIGUTI [29], HAUTOJÄRVI [30], and SCHAEFER ET AL. [31]. Vacancy properties of metals are listed in Table 5.1 according to [6].

5.1.4 Self-Interstitials

Using statistical thermodynamics and a reasoning analogous to that for vacancies, the equilibrium fraction of self-interstitials in pure metals can be written as

$$C_I^{eq} = g_I \exp\left(-\frac{G_I^F}{k_B T}\right) = g_I \exp\left(\frac{S_I^F}{k_B}\right) \exp\left(-\frac{H_I^F}{k_B T}\right). \quad (5.24)$$

G_I^F denotes the Gibbs free energy of formation, S_I^F and H_I^F the corresponding formation entropy and enthalpy, and g_I a geometric factor. For example, in fcc metals $g_I = 3$ accounts for the fact that self-interstitials occur in the

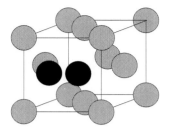

Fig. 5.6. Dumbbell configuration of a self-interstitial in an fcc lattice

so-called dumbbell configuration illustrated in Fig. 5.6, which implies three possible $\langle 100 \rangle$ orientations for a self-interstitial with its midpoint at the same lattice site.

In close-packed metals the formation enthalpy of a self-interstitial is considerably higher than that of a vacancy (see, e.g., Table 6 in the review of WOLLENBERGER [6] and the data compilation of ULLMAIER [13]):

$$H_I^F \approx (2 \text{ to } 3) \times H_{1V}^F. \tag{5.25}$$

Therefore, at thermal equilibrium

$$C_V^{eq} >>> C_I^{eq}, \tag{5.26}$$

i.e. the overwhelming thermal defect population is of the vacancy type.

Self-interstitials are produced athermally (together with an equal number of vacancies), when a metal is subject to irradiation with energetic particles. Thus, self-interstitials play a significant rôle in the radiation damage and in radiation-enhanced diffusion [9, 11]. In some ionic crystals, Frenkel disorder is established at thermal equilibrium (see Sect. 5.3 and Chap. 26). For example, in silver halides Frenkel pairs are formed, which consist of self-interstitials and vacancies in the cation sublattice of the crystal.

Semiconductors are less densely packed than metals and offer more space in their interstitial sites. Therefore, the formation enthalpies of self-interstitials and vacancies are not much different. Depending on the semiconductor, both types of defects can play a rôle under thermal equilibrium conditions. This is the case for example for Si, whereas in Ge vacancies dominate self-diffusion (see Sect. 5.5 and Chap. 23).

5.2 Substitutional Binary Alloys

A knowledge of the vacancy population in substitutional alloys is of considerable interest as well. Let us consider first dilute substituional alloys and then make a few remarks about the more complex case of concentrated alloys.

5.2.1 Vacancies in Dilute Alloys

A binary alloy of atoms B and A is denoted as dilute if the number of B atoms is not more than a few percent of the number of A atoms. Then, B is called the *solute* and A the *solvent* (or *matrix*). Depending on the solute/solvent combination interstitial and substitutional alloys are to be distinguished. Small solutes such as H, C, and N usually form interstitial alloys whereas solute atoms, which are similar in size to the solvent atoms form substitutional alloys.

In a substitutional alloy, A and B atoms and vacancies occupy sites of the same lattice. However, we have to distinguish whether a vacancy is formed on a site, where it is surrounded by A atoms only, or whether the vacancy is formed on a neighbouring site of a solute atom. In the latter case, we talk about a *solute-vacancy pair* (see Fig. 5.7). For simplicity let us suppose that the solute-vacancy interaction is restricted to nearest-neighbour sites, which is often reasonable for metals. The Gibbs free energy of vacancy formation in the undisturbed solvent, $G_{1V}^F(A)$, is different from the Gibbs free energy of vacancy formation next to a B atom, $G_{1V}^F(B)$:

$$G_{1V}^F(A) \neq G_{1V}^F(B). \tag{5.27}$$

For $G_{1V}^F(A) > G_{1V}^F(B)$ the vacancy-solute interaction is attractive, whereas for $G_{1V}^F(A) < G_{1V}^F(B)$ it is repulsive. According to LOMER [32] the total vacancy fraction in a dilute alloy, $C_V^{eq}(C_B)$, is given by

$$C_V^{eq}(C_B) = (1 - ZC_B)\exp\left[-\frac{G_{1V}^F(A)}{k_B T}\right] + ZC_B \exp\left[-\frac{G_{1V}^F(B)}{k_B T}\right], \tag{5.28}$$

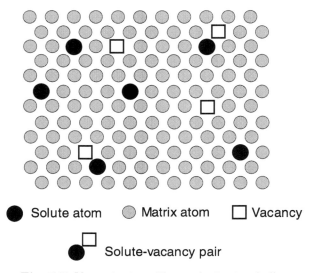

Fig. 5.7. Vacancies in a dilute substitutional alloy

where Z denotes the coordination number and C_B the solute fraction. Equation (5.28) is a good approximation for $C_B < 0.01$. We recognise that the first term corresponds to the concentration of unpaired vacancies. It is reduced by a factor $(1 - ZC_B)$ relative to that of the pure solvent. The second term is the fraction of solute-vacancy pairs. If we introduce the *Gibbs free energy of interaction* between solute and vacancy

$$G^B \equiv G_{1V}^F(A) - G_{1V}^F(B)\,, \tag{5.29}$$

Eq. (5.28) can be written as

$$C_V^{eq} = \exp\left(-\frac{G_{1V}^F(A)}{k_\mathrm{B} T}\right)\left[1 - ZC_B + ZC_B \exp\left(\frac{G^B}{k_\mathrm{B} T}\right)\right] \tag{5.30}$$

and is sometimes called the *Lomer equation*.

The first factor in Eq. (5.30) is the equilibrium vacancy fraction in the pure solvent. The factor in square brackets is larger/smaller than unity if G^B is positive/negative. For binding/repulsion between solute and vacancy the total vacancy content in the alloy is higher/lower than in the pure solvent. In dilute alloys of the noble metals with solute elements lying to their right in the periodic table, G^B is typically about 0.2 eV [13]. We note that the quantity

$$p = C_{1V}^{eq} \exp\left(\frac{G^B}{k_\mathrm{B} T}\right) \tag{5.31}$$

denotes the probability that a vacancy occupies a nearest-neighbour site of a solute, when C_{1V}^{eq} is measured in site fractions. The expressions (5.30) and (5.31) are of interest for diffusion in dilute alloys, which will be considered in Chap. 19.

5.2.2 Vacancies in Concentrated Alloys

The Lomer equation is valid for very dilute alloys ($C_B \leq 0.01$). In its derivation only associates between one solute atom and vacancy are considered. In concentrated alloys, associates between several solute atoms and vacancy and interactions between atoms of an associate become also important. To the author's knowledge robust theoretical models for the vacancy population in concentrated substitutional alloys are not available. An approximation was treated by DORN AND MITCHELL [33]. These authors attribute to each associate consisting of i solute atoms and one vacancy the (same) Gibbs free energy G_i. By standard thermodynamic reasoning, they derive the following expression for the total vacancy concentration in a concentrated alloy

$$C_V^{eq}(C_B) = \sum_{i=0}^{Z} \binom{Z}{i} C_A^{Z-i} C_B^i \exp\left(-\frac{G_i}{k_\mathrm{B} T}\right), \tag{5.32}$$

where Z denotes the coordination number. The term for $i = 0$ represents the vacancy content of free vacancies in the solvent. A limitation to the terms for $i = 0$ and $i = 1$ reproduces Lomer's equation using $G_0 \equiv G_{1V}^F(A)$ and $G_i = G_{1V}^F(B)$. In the derivation of Eq. (5.32) a random distribution of atoms has been assumed. For a generalisation of Eq. (5.32) by including an interaction between atoms we refer to [34].

5.3 Ionic Compounds

Let us consider thermal defects in ionic crystals such as the alkali halides, silver chloride and bromide. These materials crystallise in sodium chloride and cesium chloride structures. They are strongly stoichiometric and have wide band gaps so that thermally produced electrons or holes can be ignored. These materials are the *classical ion conductors*, whose conductivity arises from the presence and mobility of vacancies and/or self-interstitials.

The classical ionic conductors are to be distinguished from the *fast ion conductors*. As a general rule, fast ion conductors are materials with an open structure, which allows for the rapid motion of relatively small ions. A famous example is silver iodide, for which fast ionic conduction was reported as early as 1914 [35]. It displays a first order phase transition between a fast ion-conducting phase (α-AgI) above 147 °C and a normal conducting phase at lower temperatures. α-AgI has a body-centered cubic sublattice of practically immobile I$^-$ ions. Each unit cell displays 42 interstitial sites (6 octahedral, 12 tetrahedral, 24 trigonal) over which the two Ag$^+$ ions per unit cell are distributed (see Fig. 27.2). Since there are many more sites than Ag$^+$ ions, the latter can migrate easily. Other examples are β-alumina, some compounds with fluorite structure such as some halides such as CaF$_2$ and PbF$_2$ and oxides like doped ZrO$_2$, which are fluorine or oxygen ion conductors at elevated temperatures. These materials require a different approach, because in the sublattice of one ionic species the fraction of vacant sites is high (see Chap. 27).

To be specific, we consider here classical ionic crystals with CA stoichiometry (C=cation, A=anion). They are composed of anions and cations which carry equal charges opposite in sign. Let us further assume that all cation sites are equivalent and all anion sites likewise; in other words, there are two filled sublattices. The defect population that can develop in such a crystal has the structural constraint that the number of C atoms and of A atoms must be equal. This can also be viewed as a condition of electroneutrality by assigning ionic charges to the atoms C and A.[3] Then, only charge-preserving

---

[3] Electroneutrality must be fulfilled in the volume of ionic crystal. In the vicinity of charged dislocations, grain boundaries or surfaces, unbalanced point defect populations can develop. In compounds with additional electronic defects the requirements of structure and of electroneutrality are different (see, e.g., the textbook of MAIER [36]).

defect populations can develop. In addition, the formation of antisite defects need not to be considered due to the high Coulomb energy of an ion placed in the 'wrong' sublattice. In what follows, we consider two important cases of disorder in CA ionic crystals. For a more general treatment the reader is referred, e.g., to the textbook of ALLNATT AND LIDIARD [37].

5.3.1 Frenkel Disorder

Let us suppose that vacancies (V_C) and self-interstitials (I_C) in the C sublattice are formed from cations on cation sites (C_C) according to the quasi-chemical reaction

$$C_C \rightleftharpoons V_C + I_C. \tag{5.33}$$

This type of disorder is called *Frenkel disorder* (Fig. 5.8), as it was first suggested by the Russian scientist FRENKEL [1]. Pairs of vacancies and self-interstitials are denoted as Frenkel pairs. According to the law of mass action we may write

$$C_{V_C}^{eq} C_{I_C}^{eq} = \exp\left(\frac{S_{\text{FP}}}{k_B}\right) \exp\left(-\frac{H_{\text{FP}}}{k_B T}\right) \equiv K_{\text{FP}}. \tag{5.34}$$

Here $C_{V_C}^{eq}$ and $C_{I_C}^{eq}$ denote equilibrium site fractions of vacancies and self-interstitials in the C sublattice. K_{FP} is called the *Frenkel product*. The formation enthalpy H_{FP} and entropy S_{FP} for (non-interacting) Frenkel pairs can be split according to

$$H_{\text{FP}} = H_{V_C}^F + H_{I_C}^F \quad \text{and} \quad S_{\text{FP}} = S_{V_C}^F + S_{I_C}^F \tag{5.35}$$

into sums of formation enthalpies, $H_{V_C}^F + H_{I_C}^F$, and formation entropies, $S_{V_C}^F + S_{I_C}^F$, of vacancies and self-interstitials. Charge neutrality of undoped

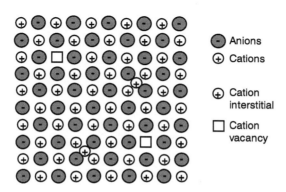

Fig. 5.8. Frenkel disorder in the cation sublattice of a CA ionic crystal

crystals requires that the numbers of vacancies and self-interstitials are equal, i.e. $C_{V_C}^{eq} = C_{I_C}^{eq}$. Then we get

$$C_{V_C}^{eq} = C_{I_C}^{eq} = \exp\left(\frac{S_{\mathrm{FP}}}{2k_{\mathrm{B}}}\right)\exp\left(-\frac{H_{\mathrm{FP}}}{2k_{\mathrm{B}}T}\right). \quad (5.36)$$

Frenkel disorder occurs in the silver sublattices of silver chloride and bromide [38, 39]. Frenkel-pair formation properties of these silver halides are listed in Table 5.2.

5.3.2 Schottky Disorder

Let us consider once more a binary ionic CA compound composed of cations on the C sublattice, C_C, and anions on the A sublattice, A_A. The constraint of electroneutrality is fulfilled, when vacancies in both sublattices, V_C and V_A, are formed according to the reaction

$$C_C + A_A + 2 \text{ new lattice sites} \rightleftharpoons V_C + V_A. \quad (5.37)$$

in equal numbers (Fig. 5.9). Applying the law of mass-action to this reaction, we get for thermal equilibrium

$$C_{V_C}^{eq} C_{V_A}^{eq} = \exp\left(\frac{S_{\mathrm{SP}}}{k_{\mathrm{B}}}\right)\exp\left(-\frac{H_{\mathrm{SP}}}{k_{\mathrm{B}}T}\right) \equiv K_{\mathrm{SP}}, \quad (5.38)$$

where C_{V_C} and C_{V_A} denote site fractions of cation and anion vacancies, respectively. H_{SP} and S_{SP} denote enthalpy and entropy for the formation of a *Schottky pair* (cation vacancy plus anion vacancy).

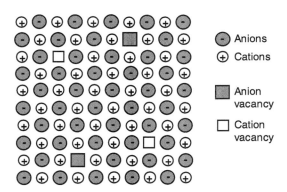

Fig. 5.9. Schottky disorder in an CA ionic crystal

Table 5.2. Formation enthalpies of Schottky- and Frenkel pairs of ionic crystals

Ionic compound	(H_{SP} or H_{FP})/eV	(S_{SP} or S_{FP})/k_{B}	Type of disorder
NaCl	2.44	9.8	Schottky
KCl	2.54	7.6	Schottky
NaI	2.00	7.6	Schottky
KBr	2.53	10.3	Schottky
LiF	2.68		Schottky
LiCl	2.12		Schottky
LiBr	1.80		Schottky
LiI	1.34		Schottky
AgCl	1.45–1.55	5.4–12.2	Frenkel
AgBr	1.13–1.28	6.6–12.2	Frenkel

This type of disorder is called *Schottky disorder* and K_{SP} is denoted as the *Schottky product*. Charge neutrality in an undoped crystal requires equal concentrations of cation and anion vacancies:

$$C_{V_C}^{eq} = C_{V_A}^{eq} = \exp\left(\frac{S_{\mathrm{SP}}}{2k_{\mathrm{B}}}\right) \exp\left(-\frac{H_{\mathrm{SP}}}{2k_{\mathrm{B}}T}\right). \tag{5.39}$$

For non-interacting Schottky pairs, the enthalpy and entropy of pair formation according to

$$H_{\mathrm{SP}} = H_{V_C}^F + H_{V_A}^F \quad \text{and} \quad S_{\mathrm{SP}} = S_{V_C}^F + S_{V_A}^F \tag{5.40}$$

can be expressed in terms of the formation enthalpies, $H_{V_C}^F$ and $H_{V_A}^F$, and entropies, $S_{V_C}^F$ and $S_{V_A}^F$, of cation and anion vacancies. Experience shows that Schottky disorder dominates the defect population in most alkali halides and in many oxides. Schottky-pair formation properties are listed in Table 5.2. Crystals doped with aliovalent ions are considered in detail in Chap. 26. In doped crystals, the Schottky product is still valid.

5.4 Intermetallics

Intermetallics are a fascinating group of materials, which attract attention from the viewpoints of fundamentals as well as applications [40, 41]. Binary intermetallics are composed of two metals or of a metal and a semimetal. Their crystal structures are different from those of the elements. This definition includes both intermetallic phases and ordered alloys. Intermetallics form a numerous and manifold group of materials and comprise a greater variety of crystal structures than metallic elements [48]. They crystallise in structures with ordered atomic distributions in which atoms are preferentially surrounded by unlike atoms. Some frequent structures are illustrated in Chap. 20. Some intermetallics are ordered up to their melting temperature, others undergo order-disorder transitions in which an almost random arrangement of atoms is favoured at high temperatures. Such transitions occur, for example, between the β' and β phases of the Cu–Zn system or in Fe–Co. There are intermetallic phases with wide phase fields and others which exist as stoichiometric compounds. Examples for both types can even be found in the same binary alloy system. For example, the Laves phase in the Co-Nb system (approximate composition Co_2Nb) exists over a composition range of about 5 at. %, whereas the phase Co_7Nb_2 is a line compound. Some intermetallics occur for certain stoichiometric compositions only. Others are observed for off-stoichiometric compositions. Some phases compensate off-stoichiometry by vacancies, others by antisite atoms.

Thermal defect populations in intermetallics can be rather complex and we shall confine ourselves to a few remarks. Intermetallic compounds are physically very different from the ionic compounds considered in the previous section. Combination of various types of disorder are conceivable: vacancies and/or antisite defects on both sublattices can form in some intermetallics. As self-interstitials play no rôle in thermal equilibrium for pure metals, it is reasonable to assume that this holds true also for intermetallics.

To be specific, let us suppose a formula A_xB_y for the stoichiometric compound and that there is a single A sublattice and a single B sublattice. This is, for example, the case in intermetallics with the B2 and $L1_2$ structure (see Fig. 20.1). The basic structural elements of disorder are listed in Table 5.3.

A first theoretical model for thermal disorder in a binary AB intermetallic with two sublattices was treated in the pioneering work of WAGNER AND SCHOTTKY [2]. Some of the more recent work on defect properties of intermetallic compounds has been reviewed by CHANG AND NEUMANN [42] and BAKKER [43].

In some binary AB intermetallics so-called *triple defect disorder* occurs. These intermetallics form V_A defects on the A sublattice on the B rich side and A_B antisites on the B sublattice on the A rich side of the stoichiometric composition. This is, for example, the case for some intermetallics with B2 structure where A = Ni, Co, Pd ... and B = Al, In, ... Some other intermetallics also with B2 structure such as CuZn, AgCd, ... can maintain high concentrations of vacancies on both sublattices.

Table 5.3. Elements of disorder in intermetallic compounds

A_A	=	A atom on A sublattice
B_B	=	B atom on B sublattice
V_A	=	vacancy on A sublattice
V_B	=	vacancy on B sublattice
B_A	=	B antisite on A sublattice
A_B	=	A antisite on B sublattice

Triple defects ($2V_A + A_B$), *bound triple defects* ($V_A A_B V_A$) and *vacancy pairs* ($V_A V_B$) have been suggested by STOLWIJK ET AL. [46]. They can form according to the reactions

$$V_A + V_B \rightleftharpoons \underbrace{2V_A + A_B}_{\text{triple defect}} \rightleftharpoons \underbrace{V_A A_B V_A}_{\text{bound triple defect}} \quad \text{and} \quad V_A + V_B \rightleftharpoons \underbrace{V_A V_B}_{\text{vacancy pair}} . \tag{5.41}$$

Very likely bound agglomerates are important in intermetallics for thermal disorder and diffusion in addition to single vacancies. In this context it is interesting to note that neither triple defects nor vacancy pairs disturb the stoichiometry of the compound.

The physical understanding of the defect structure of intermetallics is still less complete compared with metallic elements. However, considerable progress has been achieved. Differential dilatometry (DD) and positron annihilation studies (PAS) performed on intermetallics of the Fe-Al, Ni-Al and Fe-Si systems have demonstrated that the total content of vacancy-type defects can be one to two orders of magnitude higher than in pure metals [44, 45]. The defect content depends strongly on composition and its temperature dependence can show deviations from simple Arrhenius behaviour. According to SCHAEFER ET AL. [44] and HEHENKAMP [45] typical defect concentrations in these compounds near the solidus temperature can be as high as several percent.

5.5 Semiconductors

Covalent crystals such as diamond, Si, and Ge are more different from the defect point of view as one might expect from their chemical classification as group IV elements. Diamond is an electrical insulator, whose vacancies are mobile at high temperatures only. Si is a semiconductor which supports vacancies and self-interstitials as intrinsic defects. By contrast, Ge is a semiconductor in which vacancies as intrinsic defects predominate like in the metallic group IV elements Sn and Pb.

Because Si and Ge crystallise in the diamond structure with coordination number 4, the packing density is considerably lower than in metals. This holds true also for compound semiconductors. Most compound semiconductors formed by group III and group V elements like GaAs crystallise in the zinc blende structure, which is closely related to the diamond structure. Semiconductor crystals offer more space for self-interstitials than close-packed metal structures. Formation enthalpies of vacancies and self-interstitials in semiconductors are comparable. In Si, both self-interstitials and vacancies are present in thermal equilibrium and are important for self- and solute diffusion. In Ge, vacancies dominate in thermal equilibrium and appear to be the only diffusion-relevant defects (see Chap. 23 and [47, 50]).

5.5 Semiconductors

Semiconductors have in common that the thermal defect concentrations are orders of magnitude lower than in metals or ionic crystals. This is a consequence of the covalent bonding of semiconductors. Defect formation energies in semiconductors are higher than in metals with comparable melting temperatures. Neither thermal expansion measurements nor positron annihilation studies have sufficient accuracy to detect the very low thermal defect concentrations.

Point defects in semiconductors can be neutral and can occur in various electronic states. This is because point defects introduce energy levels into the band gap of a semiconductor. Whether a defect is neutral or ionised depends on the position of the Fermi level as illustrated schematically in Fig. 5.10. A wealth of detailed information about the electronic states of point defects in these materials has been obtained by a variety of spectroscopic means and has been compiled, e.g., by SCHULZ [14].

Let us consider vacancies and self-interstitials $X \in (V, I)$ and suppose that both occur in various ionised states, which we denote by $j \in (0, 1\pm, 2\pm, \dots)$. The total concentration of the defect X at thermal equilibrium can be written as

$$C_X^{eq} = C_{X^0}^{eq} + C_{X^{1+}}^{eq} + C_{X^{1-}}^{eq} + C_{X^{2+}}^{eq} + C_{X^{2-}}^{eq} + \dots \ . \tag{5.42}$$

Whereas the equilibrium concentration of uncharged defects depends only on temperature (and pressure), the concentration of charged defects is additionally influenced by the position of the Fermi energy and hence by the doping level. If the Fermi level changes due to, e.g., *background doping* the concentration of charged defects will change as well.

The densities of electrons, n, and of holes, p, are tied to the intrinsic carrier density, n_i, via the law of mass action relation

$$np = n_i^2 \ . \tag{5.43}$$

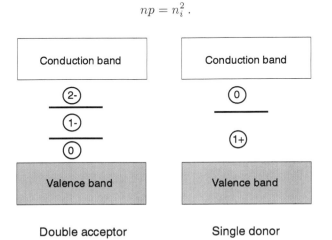

Fig. 5.10. Electronic structure of semiconductors, with a defect with double acceptor character (*left*) and donor character (*right*)

Then, Eq. (5.42) can be rewritten as

$$C_X^{eq} = C_{X^0}^{eq} + C_{X^{1+}}^{eq}(n_i)\frac{n_i}{n} + C_{X^{1-}}^{eq}(n_i)\frac{n}{n_i} + C_{X^{2+}}^{eq}(n_i)\left(\frac{n_i}{n}\right)^2$$
$$+ C_{X^{2-}}^{eq}(n_i)\left(\frac{n}{n_i}\right)^2 + \ldots, \quad (5.44)$$

where $C_{X^{j\pm}}^{eq}(n_i)$ denotes the equilibrium concentration under intrinsic conditions for defect X with charge state $j\pm$. From Eq. (5.44) it is obvious that n-doping will enhance (decrease) the equilibrium concentration of negatively (positively) charged defects. Correspondingly, p-doping will enhance (decrease) the equilibrium concentration positively (negatively) charged defects.

Furthermore, the ratio n/n_i varies with temperature because the intrinsic carrier density according to

$$n_i = \sqrt{N_{\text{eff}}^c N_{\text{eff}}^v} \exp\left(-\frac{E_g}{2k_B T}\right) \quad (5.45)$$

increases with increasing temperature. N_{eff}^c and N_{eff}^v denote the effective densities of states in the conduction and valence band, respectively. The values of n_i at different temperatures are determined mainly by the band gap energy E_g of the semiconductor. For a given background doping concentration the ratio n/n_i will be large at low temperatures and approaches unity at high temperatures. Then, the semiconductor reaches intrinsic conditions. The band gap energy is characteristic for a given semiconductor. It increases in the sequence Ge (0.67 eV), Si (1.14 eV), GaAs (1.43 eV). The intrinsic carrier density at a fixed temperature is highest for Ge and lowest for GaAs. Thus, doping effects on the concentration of charged defects are most prominent for GaAs and less pronounced for the elemental semiconductors.

Let us consider as an example a defect X which introduces a single X^{1-} and a double X^{2-} acceptor state with energy levels $E_{X^{1-}}$ and $E_{X^{2-}}$ above the valence band edge. Then, the ratios between charged and uncharged defect populations in thermal equilibrium are given by

$$\frac{C_{X^{1-}}^{eq}}{C_{X^0}^{eq}} = \frac{1}{g_{X^{1-}}} \exp\left(\frac{E_f - E_{X^{1-}}}{k_B T}\right),$$
$$\frac{C_{X^{2-}}^{eq}}{C_{X^0}^{eq}} = \frac{1}{g_{X^{2-}}} \exp\left(\frac{2E_f - E_{X^{2-}} - E_{X^{1-}}}{k_B T}\right), \quad (5.46)$$

where E_f denotes the position of the Fermi level. The degeneracy factors $g_{X^{1-}}$ and $g_{X^{2-}}$ take into account the spin degeneracy of the defect and the degeneracy of the valence band. The total concentration of point defects in thermal equilibrium for the present example is given by

$$C_X^{eq} = C_{X^0}^{eq}\left(1 + \frac{C_{X^{1-}}^{eq}}{C_{X^0}^{eq}} + \frac{C_{X^{2-}}^{eq}}{C_{X^0}^{eq}}\right). \quad (5.47)$$

Diffusion in semiconductors is affected by doping since defects in various charge states can act as diffusion-vehicles. Diffusion experiments are usually carried out at temperatures between the melting temperature T_m and about 0.6 T_m. As the intrinsic carrier density increases with increasing temperature, doping effects in diffusion are more pronounced at the low temperature end of this interval. One can distinguish two types of doping effects:

- *Background doping* is due to a homogeneous distribution of donor or acceptor atoms, that are introduced during the process of crystal growing. Background doping is relevant for diffusion experiments, when at the diffusion temperature the carrier density exceeds the intrinsic density.
- *Self-doping* is relevant for diffusion experiments of donor or acceptor elements. If the in-diffused dopant concentration exceeds either the intrinsic carrier density or the available background doping, complex diffusion profiles can arise.

References

1. J.I. Frenkel, Z. Physik **35**, 652 (1926)
2. C. Wagner, W. Schottky, Z. Physik. Chem. B **11**, 163 (1931)
3. C.P. Flynn, *Point Defects and Diffusion*, Clarendon Press, Oxford, 1972
4. A.M. Stoneham, *Theory of Defects in Solids*, Clarendon Press, Oxford, 1975
5. F. Agullo-Lopez, C.R.A. Catlow, P. Townsend, *Point Defects in Materials*, Academic Press, London, 1988
6. H.J. Wollenberger, *Point Defects*, in: *Physical Metallurgy*, R.W. Cahn, P. Haasen (Eds.), North-Holland Publishing Company, 1983, p. 1139
7. L.W. Barr, A.B. Lidiard, *Defects in Ionic Crystals*, in: *Physical Chemistry – An Advanced Treatise*, Academic Press, New York, Vol. X, 1970
8. R.G. Fuller, *Ionic Conductivity including Self-diffusion*, in: *Point Defects in Solids*, J. M. Crawford Jr., L.M. Slifkin (Eds.), Plenum Press, 1972, p. 103
9. A. Seeger, D. Schumacher, W. Schilling, J. Diehl (Eds.), *Vacancies and Interstitials in Metals*, North-Holland Publishing Company, Amsterdam, 1970
10. N.L. Peterson, R.W. Siegel (Eds.), *Properties of Atomic Defects in Metals*, North-Holland Publishing Company, Amsterdam, 1978
11. C. Abromeit, H. Wollenberger (Eds.), *Vacancies and Interstitials in Metals and Alloys*; also: Materials Science Forum **15–18**, 1987
12. O. Kanert, J.-M. Spaeth (Eds.), *Defects in Insulating Materials*, World Scientific Publ. Comp., Ltd., Singapore, 1993
13. H. Ullmaier (Vol. Ed.), *Atomic Defects in Metals*, Landolt-Börnstein, New Series, Group III: Crystal and Solid State Physics, Vol. 25, Springer-Verlag, Berlin and Heidelberg, 1991
14. M. Schulz, Landolt-Börnstein, New Series, Group III: Crystal and Solid State Physics, Vol. 22: Semiconductors, Subvolume B: *Impurities and Defects in Group IV Elements and III-V Compounds*, M. Schulz (Ed.), Springer-Verlag, 1989

15. N.A. Stolwijk, Landolt-Börnstein, New Series, Group III: Crystal and Solid State Physics, Vol. 22: Semiconductors, Subvolume B: *Impurities and Defects in Group IV Elements and III-V Compounds*, M. Schulz (Ed.), Springer-Verlag, 1989, p. 439
16. N.A. Stolwijk, H. Bracht, Landolt-Börnstein, New Series, Group III: Condensed Matter, Vol. 41: Semiconductors, Subvolume A2: *Impurities and Defects in Group IV Elements, IV-IV and III-V Compounds*, M. Schulz (Ed.), Springer-Verlag, 2002, p. 382
17. H. Bracht, N.A. Stolwijk, Landolt-Börnstein, New Series, Group III: Condensed Matter, Vol. 41: Semiconductors, Subvolume A2: *Impurities and Defects in Group IV Elements, IV-IV and III-V Compounds*, M. Schulz (Ed.), Springer-Verlag, 2002, p. 77
18. F. Benière, *Diffusion in Alkali and Alkaline Earth Halides*, in: *Diffusion in Semiconductors and Non-metallic Solids*, Landolt-Börnstein, New Series, Group III: Condensed Matter, Vol. 33, Subvolume B1, D.L. Beke (Vol.Ed.). Springer-Verlag, 1999
19. G. Erdelyi, *Diffusion in Miscelaneous Ionic Materials*, in: *Diffusion in Semiconductors and Non-metallic Solids*, Landolt-Börnstein, New Series, Group III: Condensed Matter, Vol. 33, Subvolume B1, D.L. Beke (Vol.Ed.). Springer-Verlag, 1999
20. A. Seeger, H. Mehrer, *Analysis of Self-diffusion and Equilibrium Measurements*, in: *Vacancies and Interstitials in Metals*, A. Seeger, D. Schumacher, J. Diehl and W. Schilling (Eds.), North-Holland Publishing Company, Amsterdam, 1970, p. 1
21. R. Feder, A.S. Nowick, Phys. Rev. **109**, 1959 (1958)
22. R.O. Simmons, R.W. Balluffi, Phys. Rev. **117**, 52 (1960); Phys. Rev. **119**, 600 (1960); Phys. Rev. **129**, 1533 (1963); Phys. Rev. **125**, 862 (1962)
23. R.O. Simmons, R.W. Balluffi, Phys. Rev. **125**, 862 (1962)
24. A. Seeger, J. Phys. F. Metal Phys. **3**, 248 1973)
25. R.O. Simmons, R.W. Balluffi, Phys. Rev. **117**, 52 (1960)
26. G. Bianchi, D. Mallejac, Ch. Janot, G. Champier, Compt. Rend. Acad. Science, Paris **263** 1404 (1966)
27. B. von Guerard, H. Peisl, R. Sizmann, Appl. Phys. **3**, 37 (1973)
28. H.-E. Schaefer, R. Gugelmeier, M. Schmolz, A. Seeger, J. Materials Science Forum **15–18**, 111 (1987)
29. M. Doyama, R.R. Hasiguti, Cryst. Lattice Defects **4**, 139 (1973)
30. P. Hautojärvi, Materials Science Forum **15–18**, 81 (1987)
31. H.-E. Schaefer, W. Stuck, F. Banhart, W. Bauer, Materials Science Forum **15–18**, 117 (1987)
32. W.M. Lomer, *Vacancies and other Point Defects in Metals and Alloys*, Institute of Metals, 1958
33. J. E. Dorn, J.B. Mitchell, Acta Metall. **14**, 71 (1966)
34. G. Berces, I. Kovacs, Philos. Mag. **15**, 883 (1983)
35. C. Tubandt, E. Lorenz, Z. Phys. Chem. **87**, 513, 543 (1914)
36. J. Maier, *Physical Chemistry of Ionic Solids – Ions and Electrons in Solids*, John Wiley & Sons, Ltd, 2004
37. A.R. Allnatt, A.B. Lidiard, *Atomic Transport in Solids*, Cambridge University Press, 1993
38. R. Friauf, Phys. Rev. **105**, 843 (1957)

39. R. Friauf, J. Appl. Phys. Supp. **33**, 494 (1962)
40. J.H. Westbrook, *Structural Intermetallics*, R. Dariola, J.J. Lewendowski, C.T. Liu, P.L. Martin, D.B. Miracle, M.V. Nathal (Eds.), Warrendale, PA, TMS, 1993
41. G. Sauthoff, *Intermetallics*, VCH Verlagsgesellschaft, Weinheim, 1995
42. Y.A. Chang, J.P. Neumann, Progr. Solid State Chem. **14**, 221 (1982)
43. H. Bakker, *Tracer Diffusion in Concentrated Alloys*, in: *Diffusion in Crystalline Solids*, G.E. Murch, A.S. Nowick (Eds.), Academic Press, Inc., 1984, p. 189
44. H.-E. Schaefer, K. Badura-Gergen, Defect and Diffusion Forum **143–147**, 193 (1997)
45. Th. Hehenkamp, J. Phys. Chem. Solids **55**, 907 (1994)
46. N.A. Stolwijk, M. van Gend, H. Bakker, Philos. Mag. A **42**, 783 (1980)
47. W. Frank, U. Gösele, H. Mehrer, A. Seeger, *Diffusion in Silicon and Germanium*, in: *Diffusion in Crystalline Solids*, G.E. Murch, A.S. Nowick (Eds.), Academic Press, Inc., 1984, p. 64
48. K. Girgis, *Structure of Intermetallic Compounds*, in: *Physical Metallurgy*, R.W. Cahn, P. Haasen (Eds.), North-Holland Physics Publishing, 1983, p. 219
49. M. Schulz (Vol.-Ed.), *Impurities and Defects in Group IV Elements, IV-IV and III-V Compounds*, Landolt-Börnstein, Group III, Vol. 41, Subvolume A2, Springer-Verlag, 2002
50. T.Y. Tan, U. Gösele, *Diffusion in Semiconductors*, in: *Diffusion in Condensed Matter – Methods, Materials, Models*, P. Heitjans, J. Kärger (Eds.), Springer-Verlag, 2005

6 Diffusion Mechanisms

Any theory of atom diffusion in solids should start with a discussion of diffusion mechanisms. We must answer the question: 'How does this particular atom move from here to there?' In crystalline solids, it is possible to describe diffusion mechanisms in simple terms. The crystal lattice restricts the positions and the migration paths of atoms and allows a simple description of each specific atom displacements. This contrasts with a gas, where random distribution and displacements of atoms are assumed, and with liquids and amorphous solids, which are neither really random nor really ordered.

In this chapter, we catalogue some basic atomic mechanisms which give rise to diffusion in solids. As discussed in Chap. 4, the hopping motion of atoms is an universal feature of diffusion processes in solids. Furthermore, we have seen that the diffusivity is determined by jump rates and jump distances. The detailed features of the atomic jump process depend on various factors such as crystal structure, size and chemical nature of the diffusing atom, and whether diffusion is mediated by defects or not. In some cases atomic jump processes are completely random, in others correlation between subsequent jumps is involved. Correlation effects are important whenever the atomic jump probabilities depend on the direction of the previous atom jump. If jumps are mediated by atomic defects, correlation effects always arise. The present chapter thus provides also the basis for a discussion of correlation effects in solid-state diffusion in Chap. 7.

6.1 Interstitial Mechanism

Solute atoms which are considerably smaller than the solvent atoms are incorporated on interstitial sites of the host lattice thus forming an interstitial solid solution. Interstitial sites are defined by the geometry of the host lattice. In fcc and bcc lattices, for example, interstitial solutes occupy octahedral and/or tetrahedral interstitial sites (Fig. 6.1). An interstitial solute can diffuse by jumping from one interstitial site to one of its neighbouring sites as shown in Fig. 6.2. Then the solute is said to diffuse by an interstitial mechanism.

To look at this process more closely, we consider the atomic movements during a jump. The interstitial starts from an equilibrium position, reaches the saddle-point configuration where maximum lattice straining occurs, and

96   6 Diffusion Mechanisms

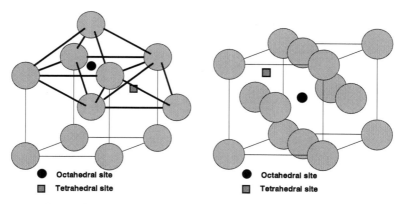

Fig. 6.1. Octahedral and tetrahedral interstitial sites in the bcc (*left*) and fcc (*right*) lattice

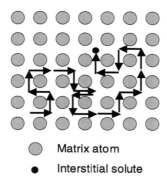

Fig. 6.2. Direct interstitial mechanism of diffusion

settles again on an adjacent interstitial site. In the saddle-point configuration neighbouring matrix atoms must move aside to let the solute atom through. When the jump is completed, no permanent displacment of the matrix atoms remains. Conceptually, this is the simplest diffusion mechanism. It is also denoted as the *direct interstitial mechanism*. It has to be distinguished from the *interstitialcy mechanism* discussed below, which is also denoted as the *indirect interstitial mechanism*. We note that no defect is necessary to mediate direct interstitial jumps, no defect concentration term enters the diffusivity and no defect formation energy contributes to the activation energy of diffusion. Since the interstitial atom does not need to 'wait' for a defect to perform a jump, diffusion coefficients for atoms migrating by the direct interstitial mechanism tend to be fairly high. This mechanism is relevant for diffusion of small foreign atoms such as H, C, N, and O in metals and other materials. Small atoms fit in interstitial sites and in jumping do not greatly displace the solvent atoms from their normal lattice sites.

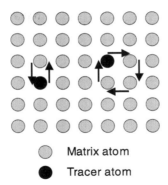

Fig. 6.3. Direct exchange and ring diffusion mechanism

6.2 Collective Mechanisms

Solute atoms similar in size to the host atoms usually form substitutional solid solutions. The diffusion of substitutional solutes and of solvent atoms themselves requires a mechanism different from interstitial diffusion. In the 1930s it was suggested that self- and substitutional solute diffusion in metals occurs by a **direct exchange** of neighbouring atoms (Fig. 6.3), in which two atoms move simultaneously. In a close-packed lattice this mechanism requires large distortions to squeeze the atoms through. This entails a high activation barrier and makes this process energetically unfavourable. Theoretical calculations of the activation enthalpy for self-diffusion of Cu performed by HUNTINGTON ET AL. in the 1940s [1, 2], which were confirmed later by more sophisticated theoretical approaches, led to the conclusion that direct exchange at least in close-packed structures was not a likely mechanism.

The so-called **ring mechanism** of diffusion was proposed for crystalline solids by the American metallurgist JEFFRIES [3] already in the 1920s and advocated by ZENER in the 1950s [4]. The ring mechanism corresponds to a rotation of 3 (or more) atoms as a group by one atom distance. The required lattice distortions are not as great as in a direct exchange. Ring versions of atomic exchanges have lower activation energies but increase the amount of collective atomic motion, which makes this more complex mechanism unlikely for most crystalline substances.

Direct exchange and ring mechanisms have in common that lattice defects are *not* involved. The observation of the so-called *Kirkendall effect* in alloys by KIRKENDALL AND COWORKERS [5, 6] during the 1940s had an important impact on the field (see also Chaps. 1 and 10). The Kirkendall effect showed that the self-diffusivities of atoms in a substitutional binary alloy diffuse at different rates. Neither the direct exchange nor the ring mechanism can explain this observation. As a consequence, the ideas of direct or ring exchanges were abandoned in the diffusion literature. It became evident that

98   6 Diffusion Mechanisms

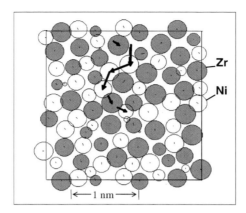

Fig. 6.4. Atom chain motion in an amorphous Ni-Zr alloy according to molecular dynamics simulations of TEICHLER [13]

vacancies are responsible for self-diffusion and diffusion of substitutional solutes in metals in practically all cases. Further historical details can be found in [7].

There is, however, some renewed interest in *non-defect mechanisms* of diffusion in connection with the enhanced diffusivity near phase transitions [8, 9]. For substitutionally dissolved boron in Cu there appears to be evidence from β-NMR experiments for a non-defect mechanism of diffusion [10].

Collective mechanisms, which involve the simultaneous motion of several atoms appear to be quite common in amorphous systems. Molecular dynamic simulations by TEICHLER [13] as well as diffusion and isotope experiments on amorphous metallic alloys reviewed by FAUPEL ET AL. [11, 12] suggest that collective mechanism operate in undercooled metallic melts and in metallic glasses. Such collective mechanisms involve the simultaneous motion of several atoms in a chain-like or caterpillar-like fashion. An example observed in molecular dynamic simulations of an amorphous Ni-Zr alloy is illustrated in Fig. 6.4.

It appears that collective jump processes play also a rôle for the motion of alkali ions in ion-conducting oxide glasses [14]. Finally, we note that *interstitialcy mechanisms* involving self-interstitials are collective in the sense that more than one atom is displaced permanently during a jump event (see Sect. 6.5).

6.3 Vacancy Mechanism

As knowledge about solids expanded, vacancies have been accepted as the most important form of thermally induced atomic defects in metals and ionic crystals (see Chaps. 5, 17, 26). It has also been recognised that the dominant

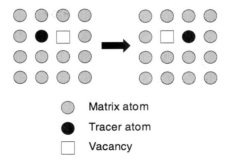

Fig. 6.5. Monovacancy mechanism of diffusion

mechanism for the diffusion of matrix atoms and of substitutional solutes in metals is the *vacancy mechanism*. An atom is said to diffuse by this mechanism, when it jumps into a neighbouring vacancy (Fig. 6.5). The constriction, which inhibits motion of an adjacent atom into a vacancy in a close-packed lattice is small, as compared to the constriction against the direct or ring exchange. Each atom moves through the crystal by making a series of exchanges with vacancies, which from time to time are in its vicinity.

In thermal equilibrium, the site fraction of vacancies in a monoatomic crystal, C_{1V}^{eq}, is given by Eq. (5.11), which we repeat for convenience:

$$C_{1V}^{eq} = \exp\left(-\frac{G_{1V}^F}{k_B T}\right) = \exp\left(\frac{S_{1V}^F}{k_B}\right) \exp\left(-\frac{H_{1V}^F}{k_B T}\right). \quad (6.1)$$

G_{1V}^F is the Gibbs free energy of vacancy formation. S_{1V}^F and H_{1V}^F denote the *formation entropy* and the *formation enthalpy* of a monovacancy, respectively. Typical values for the site fraction of vacancies near the melting temperature of metallic elements lie between 10^{-4} and 10^{-3}. From Eqs. (4.31) and (6.1) we get for the *exchange jump rate* Γ of a vacancy-mediated jump of a matrix atom to a particular neighbouring site

$$\Gamma = \omega_{1V} C_{1V}^{eq} = \nu^0 \exp\left(\frac{S_{1V}^F + S_{1V}^M}{k_B}\right) \exp\left(-\frac{H_{1V}^F + H_{1V}^M}{k_B T}\right). \quad (6.2)$$

ω_{1V} denotes the exchange rate between an atom and a vacancy and ν^0 the pertinent attempt frequency. H_{1V}^M and S_{1V}^M denote the migration enthalpy and eutropy of vacancy migration, respectively. The total jump rate of a matrix atom in a coordination lattice with Z neighbours is given by $\Gamma_{tot} = Z\Gamma$. The vacancy mechanism is the dominating mechanism of self-diffusion in metals and substitutional alloys. It is also relevant for diffusion in a number of ionic crystals, ceramic materials, and in germanium (see Parts III, IV and V of this book).

In substitutional alloys, attractive or repulsive interactions between solute atoms and vacancies play an important rôle. These interactions modify the probability, p, to find a vacancy on a nearest-neighbour site of a solute atom. For a *dilute alloy* this probability is given by the Lomer equation (see Chap. 5)

$$p = C_{1V}^{eq} \exp\left(\frac{G^B}{k_\mathrm{B}T}\right), \tag{6.3}$$

where G^B denotes the *Gibbs free energy of binding* of a solute-vacancy pair. The quantity $G_{1V}^F - G^B$ is the Gibbs free energy for the formation of a vacancy on a nearest-neighbour site of the solute. For an attractive interaction ($G^B > 0$) p is enhanced and for a repulsive interaction ($G^B < 0$) p is reduced compared to the vacancy concentration in the pure solvent. For the total jump rate of a solute atom in a coordination lattice we get

$$\Gamma_2 = Z\omega_2 p = Z\omega_2 C_{1V}^{eq} \exp\left(\frac{G^B}{k_\mathrm{B}T}\right), \tag{6.4}$$

where ω_2 is the rate of solute-vacancy exchange and Z the coordination number.

6.4 Divacancy Mechanism

When a binding energy exists, which tends to create agglomerates of vacancies (divacancies, trivacancies, ...), diffusion can also occur via aggregates of vacancies. This is illustrated for divacancies in Fig. 6.6. At thermal equilibrium divacancies are formed from monovacancies and their concentration is given by Eq. (5.15). The concentrations of mono- and divacancies at equilibrium increase with temperature. However, the concentration of divacancies rises more rapidly and may become significant at high temperatures (see Fig. 5.2). Furthermore, divacancies in fcc metals are more mobile than monovacancies. Thus, self-diffusion of fcc metals usually has some divacancy contribution in addition to the monovacancy mechanism. The latter is the dominating mechanism at temperatures below 2/3 of the melting temperature [15, 16]. Because of divacancay binding and the lower defect symmetry, diffusion via divacancies obeys slightly modified equations as compared to monovacancies (see Chap. 17). Otherwise, the two mechanisms are very similar. Diffusion by bound trivacancies is usually negligible.

6.5 Interstitialcy Mechanism

When an interstitial atom is nearly equal in size to the lattice atoms (or the lattice atoms on a given sublattice in a compound), diffusion may occur by the *interstitialcy mechanism* also called the *indirect interstitial mechanism*. Let us illustrate this for self-diffusion. Self-interstitials – extra atoms located

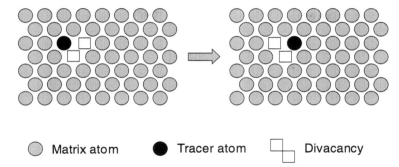

Fig. 6.6. Divacancy mechanism of diffusion in a close-packed structure

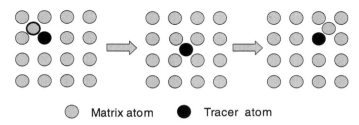

Fig. 6.7. Interstitialcy mechanism of diffusion (colinear jumps)

between lattice sites – act as diffusion vehicles. Figure 6.7 illustrates a colinear interstitialcy mechanism. Both atoms move in unison – a self-interstitial replaces an atom on a substitutional site, which then itself replaces a neighbouring lattice atom. Non-colinear versions of the interstitialcy mechanism, whereby the atoms move at an angle to one another also can occur (see Fig. 26.8). As already mentioned, interstitialcy mechanisms are collective mechanisms because at least two atoms move simultaneously.

The equilibrium configuration of a self-interstitial in metals is that of a 'dumbbell' (see Chap. 5). The dumbbell axis is $\langle 100 \rangle$ for fcc metals and $\langle 110 \rangle$ for bcc metals. In a dumbbell configuration two atoms occupy a lattice site symmetrically, and each atom is displaced by an equal amount from the regular lattice position. The motion of a dumbbell interstitial is a fairly collective process, because the simultaneous displacements of three atoms is necessary to move the center of the dumbbell from one lattice site to the next one.

In metals the interstitialcy mechanism is negligible for thermal diffusion. This is because self-interstitials have fairly high formation enthalpies compared to vacancies (see Chap. 5). The interstitialcy mechanism is, however, important for *radiation-induced diffusion*. When a crystal is irradiated with energetic particles (protons, neutrons, electrons, ...), lattice atoms are knocked out from their lattice positions. The knocked-out atom leaves behind a vacancy. The atom itself is deposited in the lattice as a self-interstitial. In this way, pairs of vacancies and self-interstitials (Frenkel pairs) are formed athermally. When these defects become mobile, they both mediate diffusion and give rise to radiation-induced diffusion, which is a topic of radiation damage in crystals.

Self-interstitials are responsible for thermal diffusion in the silver sublattice of some silver halides (see Chap. 26). In silicon, the base material of microelectronic devices, both the interstitialcy and the vacancy mechanism contribute to self-diffusion. Self-interstitials also play a prominent rôle in the diffusion of some solute atoms including important doping elements in silicon (see Chap. 23). This is not surprising, since the diamond lattice of silicon is a relatively open structure with sufficient space for interstitial species.

6.6 Interstitial-substitutional Exchange Mechanisms

Some solute atoms (B) may be dissolved on both interstitial (B_i) and substitutional sites (B_s) of a solvent crystal (A). Then, they can diffuse via one of the interstitial-substitutional exchange mechanisms ilustrated in Fig. 6.8. Such foreign atoms are denoted as *'hybrid solutes'*. The diffusivity of hybrid solutes in the interstitial configuration, D_i, is usually much higher than their diffusivity in the substitutional configuration, D_s. In contrast, the solubility in the interstitial state, C_i^{eq}, is often less or much less than the solubility in the substitutional state, C_s^{eq}:

$$D_i \gg D_s \quad \text{but} \quad C_s^{eq} > C_i^{eq}. \tag{6.5}$$

Under such conditions the incorporation of B atoms can occur by fast diffusion of B_i and subsequent change-over to B_s. Two types of interstitial-substitutional exchange mechanisms can be distinguished (see also Chap. 25):

When the change-over involves vacancies (V) according to

$$B_i + V \rightleftharpoons B_s, \tag{6.6}$$

the mechanism is denoted as the *dissociative mechanism*. This mechanism was proposed by FRANK AND TURNBULL [17] for the rapid diffusion of copper in germanium. Later on, diffusion of some foreign metallic elements in polyvalent metals such as lead, titanium, and zirconium was also attributed to this mechanism.

When the change-over involves self-interstitials (A_i) according to

$$B_i \rightleftharpoons B_s + A_i, \tag{6.7}$$

the mechanism is denoted as the *kick-out mechanism*. This mechanism was proposed by GÖSELE ET AL. [18, 19] for the fast diffusion of Au in silicon. Nowadays, the diffusion of several hybrid foreign elements, e.g., Au, Pt, Zn in silicon and Zn in gallium arsenide is also attributed to this mechanism.

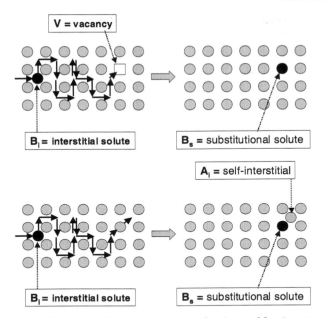

Fig. 6.8. Interstitial-substitutional exchange mechanisms of foreign atom diffusion. *Top*: dissociative mechanism. *Bottom*: kick-out mechanism

For a description of diffusion processes which involve interstitial-substitutional exchange reactions, Fick's equations must be supplemented by reaction terms which account for either Eq. (6.6) and/or Eq. (6.7). Because several species (interstitial solute, substitutional solute, defects) are involved, sets of coupled (non-linear) diffusion-reaction equations are necessary to describe the diffusion process. Solutions of these equations – apart from a few (but interesting) special cases – can only be obtained by numerical methods [20, 21]. These solutions also explain unusual (non-Fickian) shapes of concentration-distance profiles observed for hybrid diffusers. Details are discussed in Part IV of this book.

References

1. H.B. Huntington, F. Seitz, Phys. Rev. **61**, 315 (1942)
2. H.B. Huntington, Phys. Rev. **61**, 325 (1942)
3. Z. Jeffries, Trans. AIME **70**, 303 (1924)
4. C. Zener, Acta. Cryst. **3**, 346 (1950)
5. E.O. Kirkendall, Trans. AIME **147**, 104 (1942)
6. A.D. Smigelskas, E.O. Kirkendall, Trans. AIME **171**, 130 (1947)
7. C. Tuijn, Defect and Diffusion Forum **143–147**, 11 (1997)

8. A. Seeger, in: *Ultra-High-Purity Metals*, K. Abiko, K. Hirokawa, S. Takaki (Eds.), The Japan Institute of Metals, Sendai, 1995, p. 27
9. A. Seeger, Defect and Diffusion Forum **143–147**, 21 (1997)
10. B. Ittermann, H. Ackermann, H.-J. Stöckmann, K.-H. Ergezinger, M. Heemeier, F. Kroll, F. Mai, K. Marbach, D. Peters, G. Sulzer, Phys. Rev. Letters **77**, 4784 (1996)
11. F. Faupel, W. Frank, H.-P. Macht, H. Mehrer, V. Naundorf, K. Rätzke, H. Schober, S. Sharma, H. Teichler, *Diffusion in Metallic Glasses and Supercooled Melts*, Rev. of Mod. Phys. **75**, 237 (2003)
12. F. Faupel, K. Rätzke, *Diffusion in Metallic Glasses and Supercooled Melts*, Ch. 6 in: *Diffusion in Condensed Matter – Methods, Materials, Models*, P. Heitjans, J. Kärger (Eds:), Springer-Verlag, 2005
13. H. Teichler, J. Non-cryst. Solids **293**, 339 (2001)
14. S. Voss, S. Divinski, A.W. Imre, H. Mehrer, J.N. Mundy, Solid State Ionics **176**, 1383 (2005); and: A.W. Imre, S. Voss, H. Staesche, M.D. Ingram, K. Funke, H. Mehrer, J. Phys. Chem. B, **111**, 5301–5307 (2007)
15. N.L. Peterson, J. Nucl. Materials **69–70**, 3 (1978)
16. H. Mehrer, J. Nucl. Materials **69–70**, 38 (1978)
17. F.C. Frank, D. Turnbull, Phys. Rev. **104**, 617 (1956)
18. U. Gösele, W. Frank, A. Seeger, Appl. Phys. **23**, 361 (1980)
19. T.Y. Tan, U. Gösele, *Diffusion in Semiconductors*, Ch. 4 in: *Diffusion in Condensed Matter – Methods, Materials, Models*, P. Heitjans, J. Kärger (Eds.), Springer-Verlag, 2005
20. W. Frank, U. Gösele, H. Mehrer and A. Seeger, in: *Diffusion in Crystalline Solids*, G.E. Murch, A.S. Nowick (Eds.), Academic Press, 1984, p. 63
21. H. Bracht, N.A. Stolwijk, H. Mehrer, Phys. Rev. B, **52**, 16542 (1995)

7 Correlation in Solid-State Diffusion

It was not until 1951 that BARDEEN[1] AND HERRING [1, 2] drew attention to the fact that, for the vacancy mechanism, correlation exists between the directions of consecutive jumps of tracer atoms. After this pioneering work, it was soon appreciated that correlation effects play an important rôle for any solid-state diffusion process, when point defects act as diffusion vehicles.

In pure random-walk diffusion, it is assumed that the jump probabilities of atoms do not depend on the direction of the preceding jump. In real crystals, however, the jump probabilities often depend on the directions of preceding jumps. Then, successive atom jumps are correlated. Instead of following a pure random walk, each atom follows a correlated walk. This is why we have introduced the *correlation factor* in Chap. 4. We shall see below that the correlation factor depends on both the diffusion mechanism and on the lattice geometry. Clearly, an understanding of correlation effects is an important topic of solid-state diffusion. Considerable effort has gone into the study of the effects of correlation on diffusion rates in solids. In addition, methods have been devised whereby its contribution to the diffusivity can be isolated and measured experimentally (see Chap. 9).

Detailed calculations of correlation factors can be quite involved. It is the aim of the present chapter to explain the physical nature of correlation for some basic diffusion mechanisms and to describe the added understanding of diffusion that was achieved as a result of it. More comprehensive treatments of correlation effects can be found in the literature cited at the end of this chapter.

We remind the reader to the result of Chap. 4 for the correlation factor given in Eq. (4.27). For convenience, we repeat its derivation in a slightly more detailed way. Our starting point was the Einstein-Smoluchowski relation

$$\langle \bm{R}^2 \rangle = 6D^*t, \qquad (7.1)$$

---

[1] John Bardeen is one of the few scientist who received the Nobel prize twice. SCHOCKLEY, BARDEEN AND BRATTAIN were awarded for their studies of semiconductors and the development of the transistor in 1956. BARDEEN, COOPER, AND SCHRIEFER received the 1972 price for the so-called BCS theory of superconductivity.

7 Correlation in Solid-State Diffusion

which relates the diffusion coefficient of tagged atoms, D^*, and the mean square displacement of an ensemble of N such atoms, where

$$\langle R^2 \rangle = \frac{1}{N} \sum_{k=1}^{N} R_k^2 . \tag{7.2}$$

It is assumed that the net displacement of the kth atom is the result of a large number of n_k jumps with microscopic jump vectors r_i (i=1, ... k), so that

$$R_k = \left(\sum_{i=1}^{n_k} r_i \right)_k . \tag{7.3}$$

Thus

$$\langle R^2 \rangle = \frac{1}{N} \sum_{k=1}^{N} \left(\sum_{i=1}^{n_k} r_i^2 \right)_k + \frac{2}{N} \sum_{k=1}^{N} \left(\sum_{i=1}^{n_k-1} \sum_{j=1}^{n_k-i} r_i r_{i+j} \right)_k . \tag{7.4}$$

For simplicity, we restrict our discussion to cases for which all jump vectors have the same length $|r_i| = d$, i.e. to coordination lattices and nearest-neighbour jumps. We then obtain

$$\langle R^2 \rangle = d^2 \langle n \rangle \left[1 + \frac{2}{N} \sum_{k=1}^{N} \left(\sum_{i=1}^{n} r_i^2 \right)_k + \frac{2}{N} \sum_{k=1}^{N} \left(\sum_{i=1}^{n-1} \sum_{j=1}^{n-i} \cos \theta_{i,i+j} \right)_k \right]$$

$$= d^2 \langle n \rangle f . \tag{7.5}$$

Here

$$\langle n \rangle = \frac{1}{N} \sum_{k=1}^{N} n_k \tag{7.6}$$

is the average number of jumps per tracer atom during time t, $(\cos \theta_{i,i+j})_k$ the cosine of the angle between the ith and $(i+j)^{th}$ jump vectors of the kth atom and f, the quantity in square brackets, the correlation factor.

It is then argued that, when the ensemble average is performed and $\langle n \rangle \to \infty$ as $t \to \infty$, the correlation factor can be written as

$$f = 1 + \lim_{n \to \infty} \frac{2}{n} \sum_{i=1}^{n-1} \sum_{j=1}^{n-i} \langle \cos \theta_{i,i+j} \rangle . \tag{7.7}$$

We note that Eq. (7.7) is a rather complex expression, because the double summation contains correlation between an infinite number of pairs of jumps. In what follows, we consider whether and to what extent correlation effects play a rôle in some of the basic diffusion mechanisms catalogued in Chap. 6.

7.1 Interstitial Mechanism

Not all diffusion processes in solids entail correlation effects. Let us briefly address the question, why interstitial diffusion in crystals is usually *not* correlated. In a dilute interstitial solution each interstitial atom has a high probability of being surrounded by empty interstitial sites (see Fig. 6.2). All directions for jumps of an interstitial solute to neighbouring empty sites are equally probable and independent of the prior jump. Hence $\langle \cos \theta_{i,i+j} \rangle = 0$; the jump sequence of the interstitial is uncorrelated and the correlation factor equals unity:

$$f = 1 . \tag{7.8}$$

This statement is correct as long as the number of interstitial atoms is much less than the number of the available sites, which is the case for many interstitial solutions of the elements boron (B), carbon (C), or nitrogen (N) in metals. A famous example is C in iron (Fe). In fcc Fe, C is incorporated in octahedral interstitial sites and in bcc iron in tetrahedral interstitial sites. The solubility of C in bcc Fe is less than 0.02 wt. % and in fcc Fe it is less than 2 wt. %. Hence the probability that neighbouring interstitial sites to a certain C atom are unoccupied is close to unity.

In some systems, concentrated interstitial alloys form. Examples are hydrides, carbides, and nitrides of some metals. In such cases, some or even most sites in the H, C, or N sublattices are occupied by atoms and thus blocked for interstitial jumps. Then, the probability for a jump depends on the local arrangement of unoccupied neighbouring sites and

$$f < 1 . \tag{7.9}$$

For an almost filled sublattice a vacancy-type mechanism is a suitable concept.

7.2 Interstitialcy Mechanism

Interstitialcy mechanisms are important for crystals in which self-interstitials are present at thermal equilibrium. One can distinguish colinear and non-colinear versions of this mechanism. For examples of non-colinear interstitialcy mechanisms, we refer to Chap. 26 and [3, 4]. The colinear interstitialcy mechanism is illustrated in Fig. 6.7. In a colinear jump-event tracer and solvent atom move in the same direction. An atom in an interstitial position migrates by 'pushing' an atom on a regular lattice site onto an adjacent interstitial site. During a long sequence of jumps the tracer atom changes many times between substitutional (A_s) and interstitial positions (A_i) according to (see Fig. 6.7):

$$A_s \to A_i \to A_s \to A_i \to A_s \ldots . \tag{7.10}$$

108   7 Correlation in Solid-State Diffusion

Let us suppose that a tracer jumps first from an interstitial to a substitutional site and then pushes a lattice atom into an interstitial site. Immediately after this jump the tracer atom – now on a lattice site – has a self-interstitial next to it. Therefore, the second tracer jump has a greater than random probability of being the reverse of the first one. Hence $\langle \cos \theta_{1,2} \rangle$ is negative. However, the third jump of the tracer – now located at an interstitial site – takes place again in random direction. Thus, alternate pairs of consecutive tracer jumps are correlated only. In other words, jumps from an interstitial site to regular sites occur in random direction, jumps from a regular site to an interstitial site are correlated. Correlation concerns the sequence $A_i \rightarrow A_s \rightarrow A_i$ of Eq. (7.10), whereas the sequence $A_s \rightarrow A_i \rightarrow A_s$ is uncorrelated. For the sequence $A_i \rightarrow A_s \rightarrow A_i$ we have $\langle \cos \theta_{i,i+1} \rangle \equiv \langle \cos \theta \rangle \neq 0$, whereas $\langle \cos \theta \rangle = 0$ for $A_s \rightarrow A_i \rightarrow A_s$. Substituting everything in Eq. (7.7) yields

$$f = 1 + \langle \cos \theta \rangle , \qquad (7.11)$$

where $\langle \cos \theta \rangle$ is the average of cosines of the angles between pairs of correlated, consecutive tracer jumps.

7.3 Vacancy Mechanism of Self-diffusion

The vacancy mechanism is the most important diffusion mechanism in crystalline solids (see Chap. 6). In this section, we consider diffusion of self-atoms and of substitutional solutes via vacancies. To measure a self- or solute diffusion coefficients one usually studies the diffusion of very small concentrations of 'tracers' labelled by their radioactivity or their isotopic mass (see Chap. 13). This 'label' permits to distinguish of tracer from matrix atoms.

7.3.1 A 'Rule of Thumb'

Let us consider the motion of a tracer atom. The qualitative nature of correlation can easily be seen and a crude estimate of f is gained by considering Fig. 6.5. Vacancies migrate (in the absence of driving forces) in random directions. They approach a tracer atom with equal probabilty from any of its Z neighbouring sites. Thus, the initial vacancy-tracer exchange will occur at random in any of the possible Z directions. After a vacancy has exchanged its site with the tracer atom, the tracer-vacancy configuration is no longer random but depends on the direction of the initial jump. The probability for a vacancy jump to any of its Z neighbouring sites is $1/Z$, when we neglect small differences in the isotopic masses between tracer and matrix atoms. However, the probabilities for the tracer to jump to any of its Z neighbouring sites are not equal. Immediately after a first vacancy-tracer exchange, the vacancy is in a position which permits a reverse jump. The tracer thus has a higher probability to jump backward. Consecutive pairs of tracer jumps in opposite direction lead to no net displacement of the tracer. This 'backward correlation' effectively 'cancels' a pair of tracer jumps. In other words, the effective number of jumps is $n(1-2\frac{1}{Z})$ rather than n. Using this hand-waving argument, the diffusion coefficient is reduced by the factor

$$f \approx 1 - 2\frac{1}{Z}. \qquad (7.12)$$

An extreme example is a linear chain of atoms, for which $Z = 2$. Then Eq. (7.12) yields $f = 0$, which is in this case the correct value. Suppose that there is a vacancy in the chain and the tagged atom hops to a neighbouring site with the aid of this vacancy. The atom can only jump to and fro between two neighbouring positions. It cannot perform a long-range motion; thus $f = 0$. Equation (7.12) also indicates that the influence of correlation effects on self-diffusion in three-dimensional coordination lattices should amount to not more than a factor of 2. A comparison with the exact values shows that Eq. (7.12) is indeed a reasonable estimate (see below).

7.3.2 Vacancy-tracer Encounters

We now consider self-diffusion by a monovacancy mechanism in detail. To perform a jump the tracer atom must wait for the appearance of a vacancy on one of its neighbouring sites. At thermal equilibrium, the probability that a vacancy occupies a certain lattice site of an elemental crystal is given by the equilibrium fraction of vacancies, C_{1V}^{eq}. The total number of atomic (and tracer) jumps per unit time (*jump rate* Γ_{tot}), is proportional to this probability and to the rate of vacancy-atom exchanges, ω_{1V}:

$$\Gamma_{tot} = Z C_{1V}^{eq} \omega_{1V}. \qquad (7.13)$$

The explicit form of $C_{1V}^{eq}\omega_{1V}$ is given by Eq. (6.2). The *mean residence time of the tracer atom*, $\bar{\tau}$, on a particular lattice site is

$$\bar{\tau} = \frac{1}{\Gamma_{tot}} = \frac{1}{Z\omega_{1V}C_{1V}^{eq}}. \qquad (7.14)$$

On the other hand, the *mean residence time of a vacancy*, $\bar{\tau}_{1V}$, on a particular lattice site is given by

$$\bar{\tau}_{1V} = \frac{1}{Z\omega_{1V}}. \qquad (7.15)$$

In Chap. 5 we have seen that for metals the site fraction of vacancies, C_{1V}^{eq}, is a very small number. Even close to the melting temperature it never exceeds 10^{-3} to 10^{-4}; in semiconductors the site fractions of point defects in equilibrium are even much smaller. Due to its Arrhenius type temperature

Table 7.1. Return probability of a vacancy, π_0, and mean number of tracer jumps in a tracer-vacancy encounter, n_{enc}, for various lattices according to ALLNATT AND LIDIARD [30]

Lattice	π_0	n_{enc}
linear chain	1	∞
square planar	1	∞
diamond	0.4423	1.7929
imple cubic	0.3405	1.5164
bcc	0.2822	1.3932
fcc	0.2536	1.3447
bcc, octahedral interstices	0.4287	1.7504
bcc, tetrahedral interstices	0.4765	1.9102

dependence the defect fractions decrease further with decreasing temperatures. Thus, the average residence time of an atom on a given lattice site, $\bar{\tau}$, is much larger than that of a vacancy:

$$\bar{\tau} \gg \bar{\tau}_{1V} . \tag{7.16}$$

Following a first vacancy-tracer exchange, the vacancy will perform a random walk in the lattice. The longer this random walk continues, the less likely the vacancy returns to the tracer atom. The *return probability* π_0 that a random walker (vacancy) starting from a particular lattice site will return to this site depends on the dimension and the type of the lattice (see Table 7.1). On a two-dimensional square lattice the vacancy will revisit its site adjacent to a tracer atom with a probability of unity. For three dimensional lattices the return probability is smaller than unity and decreases with increasing coordination number. The results for the cubic Bravais lattices were obtained by MONTROLL [5] (see also [6]). The calculations have been extended to non-Bravais lattices by KOIWA AND COWORKERS [7, 8], who obtained the last two entries in Table 7.1.

Non-vanishing return probabilities show that correlation effects for vacancy-mediated diffusion are unavoidable, since the same vacancy may return several times to the tracer atom. The sequence of exchanges of the tracer atom with the same vacancy is called an *encounter*. A 'fresh' vacancy will approach the tracer from a random direction and terminate the encounter with the 'old' vacancy. A *complete encounter* can develop in lattices, which contain low vacancy concentrations (e.g., metallic elements). The average number of vacancy-tracer exchanges in a complete encounter, n_{enc}, is given by [30]

$$n_{enc} = \frac{1}{1 - \pi_0} . \tag{7.17}$$

Numerical values of n_{enc} are listed in Table 7.1. For three-dimensional lattices n_{enc} is a number not much larger than unity.

7.3 Vacancy Mechanism of Self-diffusion

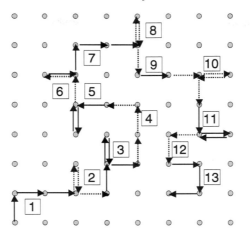

Fig. 7.1. Encounter model: tracer displacement due to encounters with several different vacancies. The *numbers* pertain to tracer jumps promoted by a particular vacancy

A macroscopic displacement of a tracer atom is the result of many encounters with different vacancies (Fig. 7.1). The displacements of the tracer that occur in different encounters are uncorrelated. Only tracer jumps within the same encounter are correlated. Each encounter gives rise to a path $\boldsymbol{R}_{enc}$ and to a mean square displacement $\langle R_{enc}^2 \rangle$. The tracer diffusion coefficient in a cubic Bravais lattice can be written as

$$D^* = \frac{1}{6} \frac{\langle R_{enc}^2 \rangle}{\tau_{enc}}, \qquad (7.18)$$

where $\tau_{enc} = \bar{\tau} n_{enc}$. In terms of the jump length d we have

$$D^* = f \frac{1}{6} \frac{d^2}{\bar{\tau}}. \qquad (7.19)$$

By comparing Eq. (7.18) and Eq. (7.19) we get for the correlation factor

$$f = \frac{\langle R_{enc}^2 \rangle}{n_{enc} d^2}. \qquad (7.20)$$

The quantities $\langle R_{enc}^2 \rangle$ and n_{enc} can be evaluated by computer simulations as functions of the number of vacancy jumps. For example, in an fcc lattice $n_{enc} = 1.3447$ and $\langle R_{enc}^2 \rangle = 1.0509 \, d^2$ has been obtained, when the number of vacancy jumps goes to infinity (see, e.g., WOLF [9]). Then, Eq. (7.20) yields $f = 0.7815$ in agreement with the value given in Table 7.2. Computer simulations also show that sequences of more than three vacancy-tracer exchanges within one encounter are rather improbable.

112   7 Correlation in Solid-State Diffusion

7.3.3 Spatial and Temporal Correlation

The correlation factor is also called the *spatial correlation factor*. The term 'spatial' refers to correlation between the jump directions of the tracer. Spatial correlations develop after a first vacancy-tracer exchange, because the vacancy retains its memory with respect to the position of the tracer atom during its excursion. Correlation effects between tracer jumps develop to its full extent only if Eq. (7.16) is fulfilled. A fresh vacancy will approach the tracer from a random direction. Its arrival destroys the chain of correlation developed in the encounter with the old vacancy. In materials with vacancy concentrations much higher than in pure metals (not considered here) encounters are no longer clearly separated and the correlation factor increases.

During an encounter between vacancy and tracer, jumps occur mainly a few multiples of $\bar{\tau}_{1V}$ after the first vacancy-tracer exchange. As recognised by EISENSTADT AND REDFIELD [10], tracer jumps experience a *bunching effect* in time illustrated in Fig. 7.2. They form small packets of jumps following the first tracer jump of an encounter. Such a packet is followed by a deadtime during which the tracer waits for a fresh vacancy. This bunching effect is equivalent to *temporal correlation* of the jump events.

Some microscopic techniques for diffusion studies such as nuclear magnetic relaxation (NMR) and Mössbauer spectroscopy (MBS) are sensitive to times between jumps. This is because these techniques have inherent time scales. In NMR experiments the Larmor frequency of the nuclear magnetic moments and in MBS the lifetime of the Mössbauer level provide such time scales. A quantitative interpretation of such experiments (see Chap. 15) must take into account temporal correlation effects in addition to the spatial ones.

7.3.4 Calculation of Correlation Factors

Various mathematical procedures for calculating correlation factors are available, to which references can be found at the end of this section. We refrain

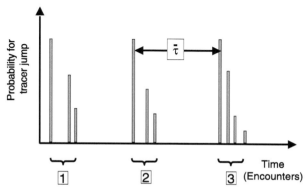

Fig. 7.2. Temporal correlation: bunching of tracer jumps within encounters

7.3 Vacancy Mechanism of Self-diffusion

from giving a comprehensive review of all methods. Instead we rather strive for a physical understanding of the underlying ideas: we consider explicitely low vacancy concentrations and cubic coordination lattices. Then, the averages in Eq. (7.7) refer to one complete encounter. Since for a given value of n there are $(n-j)$ pairs of jump vectors separated by j jumps, and since all vacancy-tracer pairs immediately after their exchange are physically equivalent, we introduce the abbreviation $\langle \cos \theta_j \rangle \equiv \langle \cos \theta_{i,j} \rangle$ and get:

$$f = 1 + \lim_{n \to \infty} \frac{2}{n} \sum_{j=1}^{n-1} (n-j) \langle \cos \theta_j \rangle. \tag{7.21}$$

Here $\langle \cos \theta_j \rangle$ is the average of the cosines of the angles between all pairs of vectors separated by j jumps in the same encounter. With increasing j the averages $\langle \cos \theta_j \rangle$ converge rapidly *versus* zero. Executing the limit $n \to \infty$, Eq. (7.21) can be written as:

$$\begin{aligned} f &= 1 + 2 \sum_{j=1}^{\infty} \langle \cos \theta_j \rangle \\ &= 1 + 2 \left(\langle \cos \theta_1 \rangle + \langle \cos \theta_2 \rangle + \ldots \right). \end{aligned} \tag{7.22}$$

To get further insight, we consider – for simplicity reasons – the x-displacements of a series of vacancy-tracer exchanges. For a suitable choice of the x-axis only two x-components of the jump vector need to be considered[2], which are equal in length and opposite in sign. Since then $\cos \theta_j = \pm 1$, we get from Eq. (7.22)

$$f = 1 + 2 \sum_{j=1}^{\infty} \left(p_j^+ - p_j^- \right), \tag{7.23}$$

where p_j^+ (p_j^-) denote the probabilities that tracer jump j occurs in the same (opposite) direction as the first jump. If we consider two consecutive tracer jumps, say jumps 1 and 2, the probabilities fulfill the following equations:

$$\begin{aligned} p_2^+ &= p_1^+ p_1^+ + p_1^- p_1^-, \\ p_2^- &= p_1^+ p_1^- + p_1^- p_1^+. \end{aligned} \tag{7.24}$$

Introducing the abbreviations $t_j \equiv p_j^+ - p_j^-$ and $t_1 \equiv t$, we get

$$t_2 = p_1^+ \underbrace{\left(p_1^+ - p_1^- \right)}_{t} - p_1^- \underbrace{\left(p_1^+ - p_1^- \right)}_{t} = t^2. \tag{7.25}$$

From this we obtain by induction the recursion formula

$$t_j = t^j. \tag{7.26}$$

---

[2] Jumps with vanishing x-components can be omitted.

The three-dimensional analogue of Eq. (7.26) was derived by COMPAAN AND HAVEN [20] and can be written as

$$\langle \cos \theta_j \rangle = (\langle \cos \theta \rangle)^j , \qquad (7.27)$$

where θ is the angle between two consecutive tracer jumps. With this recursion expression we get from Eq. (7.22)

$$f = 1 + 2\langle \cos \theta \rangle \left[\langle \cos \theta \rangle + (\langle \cos \theta \rangle)^2 + \ldots \right] . \qquad (7.28)$$

The expression in square brackets is a converging geometrical series with the sum $1/(1 - \langle \cos \theta \rangle)$. As result for the correlation factor of vacancy-mediated diffusion in a cubic coordination lattice, we get

$$f = \frac{1 + \langle \cos \theta \rangle}{1 - \langle \cos \theta \rangle}. \qquad (7.29)$$

We note that Eq. (7.29) reduces correlation between non-consecutive pairs of tracer jumps within the same encounter to the correlation between two consecutive jumps. Equation (7.29) is valid for self- and solute diffusion via a vacancy mechanism.

The remaining task is to calculate the average value $\langle \cos \theta \rangle$. At this point, it may suffice to make a few remarks: starting from Eq. (7.29), we consider the situation immediately after a first vacancy-tracer exchange (Fig. 7.3). The next jump of the tracer atom will lead to one of its Z neighbouring sites l in the lattice. Therefore, we have

$$\langle \cos \theta \rangle = \sum_{l=1}^{Z} P_l \cos \delta_l . \qquad (7.30)$$

In Eq. (7.30) δ_l denotes the angle between the first and the second tracer jump, which displaces the tracer to site l. P_l is the corresponding probability. A computation of P_l must take into account all vacancy trajectories in the lattice which start at site 1 and promote the tracer in its next jump to site l. An infinite number of such vacancy trajectories exist in the lattice. One example for a vacancy trajectory, which starts at site 1 and ends at site 4, is illustrated in Fig. 7.3. Some trajectories are short and consist of a small number of vacancy jumps, others comprise many jumps.

A crude estimate for the correlation factor can be obtained as follows: we consider the shortest vacancy trajectory, which consists of *only one* further vacancy jump after the first displacement of the tracer, i.e. we disregard the infinite number of all longer vacancy trajectories. Then, nothing else than the immediate back-jump of the tracer to site 1 in Fig. 7.3 can occur. Any other tracer jump requires vacancy trajectories with several vacancy jumps. For example, for a tracer jump to site 4 the vacancy needs at least 4 jumps (e.g.,

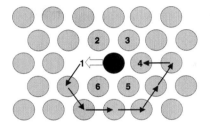

Fig. 7.3. Example of a vacancy trajectory immediately after vacancy-tracer exchange in a two-dimensional lattice

$1 \to 2 \to 3 \to 4 \to$ tracer). The probability for an immediate back-jump is $P_{back} = 1/Z$ and we have

$$\langle \cos \theta \rangle \approx P_{back} \cdot \cos 180° = -\frac{1}{Z}. \tag{7.31}$$

Inserting this estimate in Eq. (7.29), we get

$$f \approx \frac{Z-1}{Z+1} = 1 - \frac{2}{Z+1}. \tag{7.32}$$

This equation is similar to the 'rule of thumb', Eq. (7.12), discussed at the beginning of this section. Exact values of f are presented in the next section.

7.4 Correlation Factors of Self-diffusion

There are a number of publications devoted to calculations of correlation factors for defect-mediated self-diffusion. Values of f are collected in Table 7.2 for various lattices and for several diffusion mechanisms (see also the reviews by LE CLAIRE [27], ALLNATT AND LIDIARD [30], and MURCH [31]).

The correlation factor depends on the type of the lattice and on the diffusion mechanism considered. The correlation factor decreases the tracer diffusion coefficient with respect to its (hypothetical) 'random-walk value'. For self-diffusion this effect is often less than a factor of two. Nevertheless, for a complete description of the atomic diffusion process it is necessary to include f.

There are, however, additional good reasons why a study of correlations factors is of interest. The correlation factor is quite sensitive to the diffusion mechanism. For example, the correlation factor for diffusion via divacancies is smaller than that for monovacancies. An experimental determination of f could throw considerable light on the mechanism(s) of diffusion. The identification of the diffusion mechanism(s) is certainly of prime importance for the understanding of diffusion processes in solids. Unfortunately, a direct measurement of f is hardly possible. However, measurements of the isotope effect

Table 7.2. Correlation factors of self-diffusion in several lattices

Lattice	Mechanism	Correlation factor f	Reference
1d chain	vacancy	0	see text
honeycomb	vacancy	1/3	[20]
2d-square	vacancy	0.467	[20, 18]
2d hexagonal	vacancy	0.56006	[20]
diamond	vacancy	1/2	[20]
simple cubic	vacancy	0.6531	[18]
bcc cubic	vacancy	0.7272, (0.72149)	[18, 20]
fcc cubic	vacancy	0.7815	[20, 18]
fcc cubic	divacancy	0.4579	[19]
bcc cubic	divacancy	0.335 to 0.469	[17]
fcc cubic	⟨100⟩ dumb-bell interstitial	0.4395	[25]
any lattice	direct interstital	1	
diamond	colinear interstitialcy	0.727	[21]
$CaF_2(F)$	non-colinear interstitialcy	0.9855	[20]
$CaF_2(Ca)$	colinear interstitialcy	4/5	[20]
$CaF_2(Ca)$	non-colinear interstitialcy	1	[20]

of diffusion, which is closely related to f (see Chap. 9), and in some cases also measurements of the Haven ratio (see Chap. 11) can throw some light on the diffusion mechanism.

The 'rule of thumb' values from Eq. (7.12) for vacancy-mediated diffusion are listed in Table 7.3. These values are mostly within 10 % of the correct values, indicating that a large amount of correlation results from the first backward exchange of vacancy and tracer. For the diamond and honeycomb lattices and the 1d chain the 'rule of thumb' values coincide with the exact values. The exact values confirm a trend suggested already by the 'rule of thumb': correlation becomes more important, when the coordination number Z decreases.

7.5 Vacancy-mediated Solute Diffusion

Diffusion in binary alloys is more complex than self-diffusion in pure crystals. For dilute alloys there are two major aspects of diffusion: *solute diffusion* and *solvent diffusion*. In this section, we confine ourselves to correlations effects of solute diffusion in dilute alloys. Correlation effects of solvent diffusion in dilute fcc alloys have been treated by HOWARD AND MANNING [12] on the basis of the 'five-frequency-model' (see below) proposed by LIDIARD [13]. Here we concentrate on the more important case of solute diffusion. Results for correlation effects of solvent diffusion can be found in Chap. 19.

Table 7.3. Comparison between 'rule of thumb' estimates of correlation factors based on Eq. (7.12) and exact values

Lattice	Z	$1 - 2/Z$	Correlation factor
fcc	12	0.833	0.781
bcc	8	0.750	0.727
simple cubic	6	0.667	0.653
diamond	4	0.500	0.500
honeycomb	3	1/3	1/3
2d square	4	0.5	0.467
2d hexagonal	6	2/3	0.56006
1d chain	2	0	0

In a dilute alloy, isolated solute atoms are surrounded by atoms of the solvent on the host lattice. Accordingly, each solute atom can be considered to diffuse in a pure solvent. In Chap. 5 we have already discussed vacancies in dilute substitutional alloys. We remind the reader to the Lomer expression (5.31), which we repeat for convenience:

$$p = C_{1V}^{eq} \exp\left(\frac{G^B}{k_B T}\right). \tag{7.33}$$

The probability p to find a vacancy on a nearest-neighbour site of a solute atom is different from the probability C_{1V}^{eq} to find a vacancy on an arbitrary site of the solvent. This difference is due to the Gibbs free energy of binding G^B for the vacancy-solute pair. For $G^B > 0$ ($G^B < 0$) the probability p is enhanced (reduced) with respect to C_{1V}^{eq}.

The presence of the solute also influences atom-vacancy exchange rates in its surroundings. The exchange rates between vacancy and solute and between vacancy and solvent atoms near a solute atom are different from those in the pure solvent. The exchange rates enter the expression for the correlation factor. Correlation factors of solute diffusion in the fcc, bcc, and diamond lattices are considered in what follows. Note that it is common practice in the diffusion literature to use the index '2' to distinguish the solute correlation factor, f_2, from the correlation factor of self-diffusion in the pure solvent, f, and the solute diffusivity, D_2, from the self-diffusivity in the pure solvent, D.

7.5.1 Face-Centered Cubic Solvents

The influence of a solute atom on the vacancy-atom exchange rates is often described by the so-called *'five-frequency model'* proposed by LIDIARD [13,

118   7 Correlation in Solid-State Diffusion

Table 7.4. Vacancy-atom exchange rates of the 'five-frequency model'

ω_2: solute-vacancy exchange rate
ω_1: rotation rate of the solute-vacancy pair
ω_3: dissociation rate of the solute-vacancy pair
ω_4: association rate of the solute-vacancy pair
ω: vacancy-atom exchange rate in the solvent

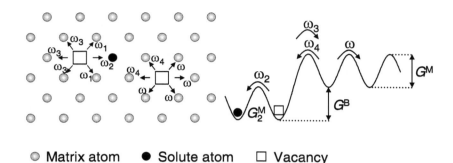

○ Matrix atom   ● Solute atom   □ Vacancy

Fig. 7.4. *Left*: 'Five-frequency model' for diffusion in dilute fcc alloys. *Right*: 'Energy landscape' for vacancy jumps in the neighbourhood of a solute atom

14]³. The five types of vacancy-atom exchange rates are illustrated in Fig. 7.4 and listed in Table 7.4. Each exchange leads to a different local configuration of solute atom, vacancy, and solvent atoms. In the framework of this model two categories of vacancies can be distinguished: vacancies located in the first coordination shell of the solute and vacancies located on lattice sites beyond this shell. The vacancy jump with rate ω_4 (ω_3) forms (dissociates) the vacancy-solute pair. Association and dissociation rate are related to the Gibbs free energy of binding of the pair via the detailed balancing equation

$$\frac{\omega_3}{\omega_4} = \exp\left(-\frac{G^B}{kT}\right). \qquad (7.34)$$

Before we discuss the correct expression for the correlation factor of solute diffusion, let us consider – as we did in the case of self-diffusion – an estimate, which may provide a better understanding of the final result. Suppose that a first solute-vacancy exchange has occurred. The crudest approximation for f_2 takes into account only that vacancy trajectory which leads to an

---

³ Lidiard uses the word 'frequency' instead of rate. Frequency and rate have the same dimension, but they are physically different. In the authors opinion, the term 'rate' is more appropriate.

immediate reversal of the first solute atom jump. The pertinent probability expressed in terms of jump rates is

$$P_{back} \approx \frac{\omega_2}{\omega_2 + 4\omega_1 + 7\omega_3}, \qquad (7.35)$$

since the vacancy from its site next to the tracer can perform one ω_2 jump, four ω_1 jumps, and seven ω_3 jumps. Herewith, we get from Eqs. (7.29) and (7.35) for the solute correlation factor

$$f_2 \approx \frac{2\omega_1 + 7\omega_3/2}{\omega_2 + 2\omega_1 + 7\omega_3/2}. \qquad (7.36)$$

This expression illustrates that a combination of vacancy jump rates enters the correlation factor.

An exact expression for f_2 was derived by MANNING [15]:

$$f_2 = \frac{\omega_1 + 7F_3\omega_3/2}{\omega_2 + \omega_1 + 7F_3\omega_3/2}. \qquad (7.37)$$

It has some similarity with the approximate relation. In Eq. (7.37) F_3 is the probability that, after a dissociation jump ω_3, the vacancy will *not* return to a neighbour site of the solute. F_3 is sometimes called the *escape probability*. It is illustrated in Fig. 7.5 as a function of the ratio $\alpha = \omega_4/\omega$. Manning derived the following numerical expression for the escape probability:

$$7F_3(\alpha) = 7 - \frac{10\alpha^4 + 180.5\alpha^3 + 927\alpha^2 + 1341\alpha}{2\alpha^4 + 40.2\alpha^3 + 254\alpha^2 + 597\alpha + 436}. \qquad (7.38)$$

When α goes to zero, no association of the vacancy-solute complex occurs and we have $F_3 = 1$. For $\omega = \omega_4$ the escape probability is $F_3 = 0.7357$. When α goes to infinity, we have $F = 2/7$ (see Fig. 7.5). The correlation factor f_2 is a function of all vacancy-atom exchange rates. We mention several special cases:

1. For self-diffusion all jump rates are equal (when isotope effects are neglected): $\omega = \omega_1 = \omega_2 = \omega_3 = \omega_4$ and we get $f_2 = 0.7814$. This value agrees (as it should) with the value listed in Table 7.2.
2. If the vacancy-solute exchange is much slower than vacancy-solvent exchanges, i.e. for $\omega_2 \ll \omega_1, \omega_3, \ldots$, the correlation factor tends towards unity. Frequent vacancy-solvent exchanges randomise the vacancy position before the next solute jump. Then, solute diffusion is practically uncorrelated.
3. If vacancy-solute exchanges occur much faster than vacancy-solvent exchanges, i.e. for $\omega_2 \gg \omega_1, \omega_3, \ldots$, the correlation factor tends to zero and the motion of the solute atom is highly correlated. The solute atom 'rattles' frequently back and forth between two adjacent lattice sites.

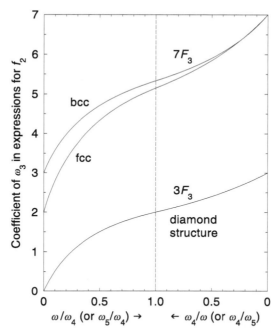

Fig. 7.5. Escape probabilities F_3 for the fcc, bcc, and diamond structure. For fcc and bcc lattices F_3 is displayed as function of ω_4/ω. For the diamond structure F_3 is a function of ω_4/ω_5. After MANNING [15]

4. Dissociation jumps ω_3 are very unlikely for a tightly bound solute-vacancy pair. Then, we get from Eq. (7.37)

$$f_2 \approx \frac{\omega_1}{\omega_2 + \omega_1}. \tag{7.39}$$

In this case the vacancy-solute pair can migrate as an entity via ω_1 and ω_2-jumps. Such pairs are sometimes called *'Johnson molecules'*.

7.5.2 Body-Centered Cubic Solvents

The simplest model for vacancy-exchange rates of solute diffusion in a bcc lattice is illustrated in Fig. 7.6. It distinguishes four jump rates: the solute-vacancy exchange rate ω_2, the dissociation rate ω_3, the association rate ω_4, and the vacancy-solvent exchange rate ω. As in the case of the fcc lattice, the rates ω_3 and ω_4 are related via the detailed balancing relation Eq. (7.34). In contrast to the fcc lattice, the bcc structure has no lattice sites which are simultaneously nearest neighbours to both the solute atom and the vacancy of a solute-vacancy pair. Hence an analogue to the rotation rate in the fcc lattice

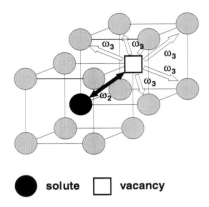

Fig. 7.6. 'Four-frequency model' of solute diffusion in the bcc lattice

does not exist in the bcc lattice. According to MANNING [15] the correlation factor for solute diffusion can be written as

$$f_2 = \frac{7\omega_3 F_3}{2\omega_2 + 7\omega_3 F_3}, \tag{7.40}$$

where the escape probability $F_3(\alpha)$ is given by

$$7F_3(\alpha) = \frac{528.4 + 779.03\alpha + 267.5\alpha^2 + 24\alpha^3}{75.50 + 146.83\alpha + 69.46\alpha^2 + 8\alpha^3} \tag{7.41}$$

illustrated in Fig. 7.5. The ratio $\alpha = \omega_4/\omega$ has the same meaning as for the fcc case.

More sophisticated models for solute diffusion in bcc metals are available in the literature (see, e.g., [16]), which we will, however, not describe here. They take into account different solute-vacancy interactions for nearest and next-nearest solute-vacancy pairs and then also several additional dissociation and association jump rates.

7.5.3 Diamond Structure Solvents

In the diamond structure one usually considers the following vacancy-jump rates (see Fig. 7.7): ω_2 for exchange with the solute, ω_3 for vacancy jumps from first to second-nearest neighbours of the solute, ω_4 for the reverse jump of ω_3, ω_5 for jumps from second- to third- or fifth-nearest neighbours of the solute, and ω for jumps originating at third-nearest neighbours or further apart from the solute. MANNING [15] derived the following expression for the correlation factor

$$f_2 = \frac{3\omega_3 F_3}{2\omega_2 + 3\omega_3 F_3} \tag{7.42}$$

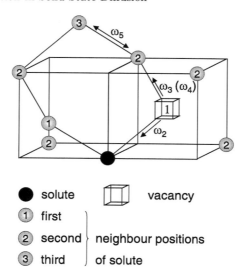

Fig. 7.7. 'Five frequency model' for jump rates in the diamond structure. ω_2: jump rate of vacancy-tracer exchange, ω_3 (ω_4): jump rate of vacancy jump from first (second) to second (first) nearest neighbour of solute atom, ω_5: jump rate of vacancy jump from second to third nearest neighbour of tracer

where the escape probability

$$F_3(\alpha) = \frac{2.76 + 4.93\alpha + 2.05\alpha^2}{2.76 + 6.33\alpha + 4.52\alpha^2 + \alpha^3} \qquad (7.43)$$

is a function of the ratio $\alpha = \omega_4/\omega_5$. For self-diffusion, all frequencies are identical, $F_3 = 2/3$ and f_2 agrees with the value $1/2$ of Table 7.2. When α goes to zero, F_3 goes to unity. In the other limit where α goes to infinity, F_3 goes to zero (see Fig.7.5).

Additional complexity can arise in semiconducting solvents, when solute and vacancy carry electrical charges. In contrast to metallic solvents, where the solute-vacancy interaction is short-ranged, the Coulomb interaction can modify the vacancy behaviour over larger distances in the surroundings of a charged solute.

7.6 Concluding Remarks

Nowadays, a number of methods for calculating correlation factors are available: COMPAAN AND HAVEN developed an analogue simulation method based on the similarity between Fick's and Ohm's law [20, 21]. A matrix method was proposed by MULLEN [22], generalised by HOWARD [23], and revisited by

BAKKER [24]. A rather elegant method based on Laplace and Fourier transforms was developed by BENOIST ET AL. [25]. KOIWA [26] designed a matrix method for return probabilities. Reviews among others were provided by LE CLAIRE [27], MEHRER [28], and in appropriate chapters of textbooks by MANNING [29], ALLNATT AND LIDIARD [30], and by MURCH [31]. Monte Carlo simulations were introduced by MURCH and coworkers [32, 33]. Meanwhile Monte Carlo methods and combinations between analytical schemes and Monte Carlo simulations play an important rôle in the calculations of correlation factors. For a recent example see, e.g., [34].

Correlation factors of self-diffusion in elements and in the sublattice of compounds are often well-defined numbers. For a given lattice they are characterstic of a certain diffusion mechanism as indicated in Table 7.2. Often the 'rule of thumb' permits already a good guess.

Correlation factors for substitutional solutes are temperature-dependent. The reason is that each vacancy-atom exchange rate is temperature dependent according to

$$\omega_i = \nu_i^0 \exp\left(-\frac{G_i^M}{k_\mathrm{B} T}\right). \tag{7.44}$$

In Eq. (7.44) the quantity G_i^M denotes the Gibbs free energy of activation for jump ω_i and ν_i^0 is the pertaining attempt frequency. At a first glance, the temperature dependence of f_2 looks rather complex. However, in a certain temperature interval it can always be approximated by an Arrhenius expression

$$f_2 \approx f_2^0 \exp\left(-\frac{C}{kT}\right), \tag{7.45}$$

where C is the 'activation enthalpy of the correlation factor' and f_2^0 some temperature-independent number. The correlation factor enters the expression for the solute diffusivity, D_2, as a multiplier. Therefore, the activation enthalpy of the correlation factor will contribute to the total activation enthalpy of solute diffusion (see Chaps. 8 and 19).

The correlation factors of solute diffusion in fcc, bcc, and diamond lattices have the following mathematical form, called 'impurity form', in common:

$$f_2 = \frac{u}{\omega_2 + u}. \tag{7.46}$$

The quantity u depends on vacancy-solvent exchange rates only and *not* on the vacancy-solute exchange rate ω_2. This should be remembered, when we discuss isotope effects of diffusion in Chap. 9.

We stress once more that Eq. (7.29) has been derived for a vacancy mechanism in a cubic coordination lattice with one type of jumps. Equation (7.29) is valid, when the defect-tracer complex contains at least a twofold rotation axis [20]. This is indeed the case for vacancy-tracer pairs in cubic coordination lattices. These condition are, however, violated for divacancy-tracer complexes in cubic crystals and for monovacancy-tracer complexes in non-cubic lattices. For generalisations of Eq. (7.29) to non-cubic lattices and to more complex diffusion mechanisms the reader should consult the literature cited at the entrance to this section.

We also remind the reader that our discussion of correlation effects has been limited to pure solids, very dilute alloys and to low defect concentrations. There are a number of solid compounds with high disorder in (at least) one of their sublattices. Examples are non-stoichiometric compounds with structural vacancies, certain concentrated interstitial alloys, and fast ion-conductors such as silver iodide. Such compounds can be viewed as solids with high apparent defect concentrations. Then the encounter model is no longer useful. In addition 'defects' may interact and correlation effects tend to be magnified and become highly temperature dependent.

In the last two decades or so intermetallic compounds have attracted considerable interest because of the substantial technological importance of some intermetallics such as aluminides and silicides. Correlation effects in intermetallic compounds are complicated by the fact that at least two sublattices are involved. In addition, several diffusion mechanisms have been suggested. Examples are the six-jump cycles of vacancies, the triple defect mechanism in B2 lattices, migration of minority atoms as antisite atoms in the majority sublattice of $L1_2$ and $D0_3$ structures, diffusion by intra-sublattice jumps of vacancies (see Chap. 20). Along with the growth in computational materials science, a number of computer simulations of atomic transport in intermetallics have been performed. The progress with analytical and random walk calculations of correlation effects for diffusion in intermetallics is reviewed by BELOVA AND MURCH [35]. Atomistic computer simulations of diffusion mechanisms using molecular dynamics methods are reviewed by MISHIN [36].

References

1. J. Bardeen, C. Herring, in: *Atom Movements* A.S.M. Cleveland, 1951, p. 87. – 288
2. J. Bardeen, C. Herring, in: *Imperfections in Nearly Perfect Solids*, W. Shockley (Ed.), Wiley, New York, 1952, p. 262
3. R. Friauf, Phys. Rev. **105**, 843 (1957)
4. R. Friauf, J. Appl. Phys. Supp. **33**, 494 (1962)
5. E.G. Montroll, Symp. Appl. Math. **16**, 193 (1964)
6. I. Stewart, Sci. Am. **280**, 111 (1999)
7. M. Koiwa, J. Phys. Soc. Japan **45**, 781 (1978)
8. M. Koiwa, S. Ishioka, Philos. Mag. **A40**, 625 (1979); J. Stat. Phys. **30**, 477 (1983); Philos. Mag. **A47**, 927 (1983)
9. D. Wolf, *Mass Transport in Solids*, Ch. 7, (1983) p. 149
10. M. Eisenstadt, A.G. Redfield, Phys. Rev. **132**, 635 (1963)
11. G.E. Murch, Solid State Ionics, **7**, 177 (1982)

12. R.E. Howard, J.R. Manning, Phys. Rev. **154**, 561 (1967)
13. A.B. Lidiard, Philos. Mag. **46**, 1218 (1955)
14. A.B. Lidiard, Philos. Mag. **5**, 1171 (1960)
15. J. Manning, Phys. Rev. **136**, A1758 (1964); see also [29]
16. J. Philibert, *Atom Movements – Diffusion and Mass Transport in Crystals*, Les Editions de Physique, Les Ulis, 1991
17. I.V. Belova, D.S. Gentle, G.E. Murch, Phil. Mag. Letters **82**, 37 (2002)
18. G. Montet, Phys. Rev. **B7**, 650 (1973)
19. G.E. Murch, J. Phys. Chem. Sol. **45**, 451 (1984)
20. K. Compaan, Y. Haven, Trans. Faraday Soc. **52**, 786 (1956)
21. K. Compaan, Y. Haven, Trans. Faraday Soc. **54**, 1498 (1958)
22. J.G. Mullen, Phys. Rev. **124**, 1723 (1961)
23. R.E. Howard, Phys. Rev. 144 650 (1966)
24. H. Bakker, Phys. Stat. Sol. **38**, 167 (1970)
25. P. Benoist, J.-L. Bocquet, P. Lafore, Acta Metall. **25**, 265 (1977)
26. M. Koiwa, Philos. Mag. **36**, 893 (1977)
27. A.D. Le Claire, *Correlation Effects in Diffusion in Solids*, in: *Physical Chemistry – an Advanced Treatise*, Vol. X, Chap. 5, Academic Press, New York and London, 1970
28. H. Mehrer, *Korrelation bei der Diffusion in kubischen Metallen*, Habilitation thesis, Universität Stuttgart, 1973
29. J.R. Manning, *Diffusion Kinetics for Atoms in Crystals*, van Norstrand Comp., Princeton 1968
30. A.R. Allnatt, A.B. Lidiard, *Atomic Transport in Solids*, Cambridge University Press, 1993
31. G.E. Murch, *Diffusion Kinetics in Solids*, Ch.3 in: *Phase Transformations in Materials*, G. Kostorz (Ed.), Wiley-Vch, 2002
32. H.J. de Bruin, G.E. Murch, Phil. Mag. **27**, 1475 (1973)
33. G.E. Murch, R.J. Thorn, J. Phys. Chem. Sol. **38**, 789 (1977)
34. S. Divinski, M. Salamon, H. Mehrer, Philos. Mag. **84**, 757 (2004)
35. I.V. Belova, G.E. Murch, Defect and Diffusion Forum **237–240**, 291 (2005)
36. Y. Mishin, *Atomistic Computer Simulation of Diffusion*, in: *Diffusion Processes in Advanced Technological Materials*, D. Gupta (Ed.), William Andrews, Inc., 2005

8 Dependence of Diffusion on Temperature and Pressure

So far, nothing has been said about the dependence of diffusion upon thermodynamic variables such as temperature, pressure, and composition. Diffusion in solids generally depends rather strongly on temperature, being low at low temperatures but appreciable at high temperatures. In an *Arrhenius diagram* the logarithm of the diffusivity is plotted *versus* the reciprocal absolute temperature. The wide range of diffusivities and activation parameters, which can occur in a solid is illustrated in the Arrhenius diagram of Fig. 8.1, where diffusion coefficients for various elements in lead are displayed. The 'spectrum' of diffusivities covers many orders of magnitude from the very fast diffusion of copper to the rather slow self-diffusion of lead. For semiconductors, an even wider spectrum of foreign atom diffusivities has been reported (see Chap. 24).

The variation of the diffusion coefficient with pressure, at least for pressures accessible in laboratory devices, is far less striking than that with temperature. Usually, the diffusivity decreases with pressure not more than a factor of ten for pressures of 1 GPa (10^4 bar). The variation of the diffusivity with composition can range from the very slight to significant. Examples of the influence of composition on diffusion can be found in Parts III to VI of this book. In this chapter, we concentrate on the dependence of the diffusivity on temperature and pressure.

8.1 Temperature Dependence

8.1.1 The Arrhenius Relation

The temperature dependence of diffusion coefficients is frequently, but by no means always, found to obey the Arrhenius formula

$$D = D^0 \exp\left(-\frac{\Delta H}{k_\mathrm{B} T}\right). \tag{8.1}$$

In Eq. (8.1) D^0 denotes the *pre-exponential factor* also called the *frequency factor*, ΔH the *activation enthalpy* of diffusion[1], T the absolute temperature,

---

[1] In the literature the symbol Q is also used instead of ΔH.

128   8 Dependence of Diffusion on Temperature and Pressure

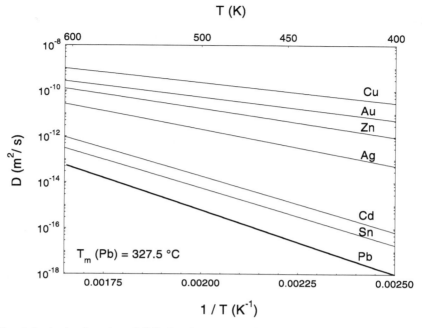

Fig. 8.1. Arrhenius plot of diffusion for various elements in Pb; activation parameters from [1]

and k_B the Boltzmann constant. Both ΔH and D^0, are called the *activation parameters* of diffusion[2]. The activation enthalpy of a diffusion process

$$\Delta H = -k_B \frac{\partial \ln D}{\partial (1/T)} \qquad (8.4)$$

corresponds to the negative slope of the Arrhenius diagram. For a temperature-independent activation enthalpy, the Arrhenius diagram is a straight line with slope $-\Delta H/k_B$. The intercept of the extrapolated Arrhenius line

---

[2] Equation (8.1) is also written as

$$D = D^0 \exp\left(-\frac{\Delta H}{RT}\right), \qquad (8.2)$$

where R denotes the gas constant. Then,

$$R = N_A k_B = 8.3145 \times 10^{-3} \, \text{kJ} \, mol^{-1} \, \text{K}^{-1} \qquad (8.3)$$

with N_A denoting the Avogadro number. The activation enthalpy ΔH is either measured in SI units kJ mol^{-1} K^{-1} or in eV per atom. Note that 1 eV per atom $= 96.472 \, \text{kJ} \, \text{mol}^{-1}$.

for $T^{-1} \Rightarrow 0$ yields the pre-exponential factor D^0. It can usually be written as

$$D^0 = gf\nu^0 a^2 \exp\left(\frac{\Delta S}{k_B}\right), \qquad (8.5)$$

where ΔS is called the *diffusion entropy*, g is a geometrical factor, f is the correlation factor, ν^0 is the attempt frequency, and a some lattice parameter. Combining Eqs. (8.1) and (8.5) we can write

$$D = gf\nu^0 a^2 \exp\left(\frac{\Delta S}{k_B}\right)\exp\left(-\frac{\Delta H}{k_B T}\right) = gf\nu^0 a^2 \exp\left(-\frac{\Delta G}{k_B T}\right). \qquad (8.6)$$

On the right-hand side of Eq. (8.6), the *Gibbs free energy of activation*

$$\Delta G = \Delta H - T\Delta S \qquad (8.7)$$

has been introduced to combine the activation enthalpy and entropy. Thermodynamics tells us that

$$\frac{\partial \Delta H}{\partial T} = T\frac{\partial \Delta S}{\partial T}. \qquad (8.8)$$

Thus, the temperature variations of enthalpy and entropy are coupled. If ΔH is temperature independent, this must hold for ΔS as well and vice versa.

Activation parameters for diffusion in metals and alloys can be found in the data collection edited by the present author [1] and for semiconductors and other non-metallic materials in a collection edited by BEKE [2].

The physical interpretation of the activation parameters ΔH and of D^0 depends on the diffusion mechanism, on the type of diffusion process, and on the lattice geometry. Simple Arrhenius behaviour should not, however, be assumed to be universal. Departures from it may arise for reasons which range from fundamental aspects of the mechanisms of atomic migration (multiple mechanisms, multiple jump vectors, ...) to effects associated with impurities and/or with microstructural features such as grain boundaries or dislocations. In this chapter, we consider lattice diffusion. Diffusion along high-diffusivity paths is the subject of Chaps. 31 and 32.

If several diffusion mechanisms with diffusion coefficients $D_I, D_{II}, \ldots$ and activation parameters $\Delta H_I, \Delta H_{II}, \ldots$ and $D_I^0, D_{II}^0, \ldots$ contribute to the total lattice diffusivity, D, we have

$$D = D_I + D_{II} + \ldots = D_I^0 \exp\left(-\frac{\Delta H_I}{k_B T}\right) + D_{II}^0 \exp\left(-\frac{\Delta H_{II}}{k_B T}\right) + \ldots . \qquad (8.9)$$

In such cases, the Arrhenius diagram will show an upward curvature. With increasing (decreasing) temperature the contribution of the process with the highest (lowest) activation enthalpy becomes more and more important. The activation enthalpy defined by Eq. (8.4) is then an effective (or apparent) value

$$\Delta H_{eff} = \Delta H_I \frac{D_I}{D_I + D_{II} + \ldots} + \Delta H_{II} \frac{D_{II}}{D_I + D_{II} + \ldots} + \ldots , \quad (8.10)$$

which represents a weighted average of the individual activation enthalpies. A well-studied example is self-diffusion in metals. Compared with divacancies (index: 2V), usually the monovacancy mechanism (index: 1V) is the dominating contribution to self-diffusion in fcc metals at temperatures below about 2/3 of the melting temperature since $\Delta H_{1V} < \Delta H_{2V}$. At higher temperatures divacancies contribute with a magnitude that varies from metal to metal (see [3, 4] and Chap. 17).

8.1.2 Activation Parameters – Examples

In what follows, we consider explicitly the physical interpretation of the activation parameters for three examples, all concerning cubic lattices: interstitial diffusion, self-diffusion via vacancies, and solute diffusion in a dilute substitutional alloy.

For **direct interstitial diffusion** (see Chap. 6) in a dilute interstitial alloy an unoccupied interstice is available next to the jumping atom. Random walk theory of Chap. 4 tells us that the diffusivity of solute interstitials can be written as

$$D = ga^2\omega = ga^2\nu^0 \exp\left(-\frac{G^M}{k_B T}\right), \quad (8.11)$$

where g is a geometrical factor, a the lattice parameter, and ω the jump rate to a neighbouring interstitial site. For octahedral interstitials in the fcc lattice we have $g = 1$ and in the bcc lattice $g = 1/6$.

As discussed in Chap. 5, the Gibbs free energy of migration G^M – the major parameter in the jump rate ω – can be separated according to

$$G^M = H^M - TS^M, \quad (8.12)$$

where H^M denotes the enthalpy and S^M the entropy of migration of the interstitial solute. Comparing Eqs. (8.1) and (8.11), the activation parameters of interstitial diffusion have the following meaning:

$$\Delta H \Rightarrow H^M \quad \text{and} \quad \Delta S \Rightarrow S^M. \quad (8.13)$$

Direct interstitial diffusion is the simplest diffusion process. The enthalpy and entropy of diffusion are identical with the pertinent migration quantities of the interstitial. Activation enthalpies of interstitial diffusers, such as H, C, N, and O in metals, tend to be fairly low since no defect formation enthalpy is required. For the same reason, interstitial diffusion is a much faster process than self- or substitutional solute diffusion.

Self-diffusion and diffusion of substitutional solutes (impurities) are defect-mediated. The diffusivity is, in essence, a product of geometrical terms ga^2, a correlation factor (f for self-diffusion or f_2 for impurity diffusion) and of the atomic jump rate (Γ for self-diffusion or Γ_2 for impurity diffusion):

$$D = fga^2\Gamma \quad \text{(self-diffusion)}, \quad D_2 = f_2 ga^2 \Gamma_2 \quad \text{(impurity diffusion)}. \quad (8.14)$$

In the case of self-diffusion, the defect availability equals the equilibrium site fraction of the defect, C_D^{eq}, discussed in Chap. 5. For a monovacancy mechanism in an elemental crystal, we have $C_D^{eq} \equiv C_{1V}^{eq}$. The *jump rate of a self-atom* can be written as

$$\Gamma = C_D^{eq}\omega = g_D \exp\left(-\frac{G_D^F}{k_B T}\right)\omega, \quad (8.15)$$

where ω denotes the *defect jump rate*. On the right-hand side of Eq. (8.15), the expression for the equilibrium fraction of defects from Chap. 5 has been inserted with the Gibbs free energy of defect formation G_D^F. g_D is a geometric factor depending on the lattice geometry and the type of the defect. For monovacancies in a monoatomic solid $g_D = 1$. For self-interstitials in $\langle 100 \rangle$-dumbbell configuration in an fcc crystal $g_D = 3$.

For solute diffusion in a very dilute substitutional alloy (often denoted as impurity diffusion) the Lomer relation Eq. (5.31) from Chap. 5 describes the defect-availability, p, on a site adjacent to a solute. We then find for the *jump rate of a substitutional impurity*:

$$\Gamma_2 = p\omega_2 = C_D^{eq} \exp\left(\frac{G^B}{k_B T}\right)\omega_2 = g_D \exp\left(-\frac{G_D^F - G^B}{k_B T}\right)\omega_2. \quad (8.16)$$

G^B denotes the Gibbs free energy of binding between defect and solute and ω_2 the defect-solute exchange rate. The quantity $G_D^F - G^B$ is the Gibbs free energy of defect formation on a site adjacent to the solute. For an attractive interaction ($G^B > 0$) the defect availability p is enhanced whereas for a repulsive interaction ($G^B < 0$) it is reduced compared to the equilibrium defect concentration in the pure host lattice. As usual, G^B can be decomposed according to $G^B = H^B - TS^B$, where H^B is the binding enthalpy and S^B the binding entropy (see Chap. 5).

The Gibbs free energies of the defect-mediated jumps can be separated into the enthalpic and entropic terms according to:

$$G^M = H^M - TS^M \quad \text{(self-atom)}, \quad G_2^M = H_2^M - TS_2^M \quad \text{(impurity)}. \quad (8.17)$$

G^M is the Gibbs free energy of motion for an exchange of the self-atom with the defect in a pure solvent. G_2^M is the barrier for an exchange-jump between impurity and defect (e.g., a vacancy). Then, the exchange jump rates read either

$$\omega = \nu^0 \exp\left(-\frac{G^M}{k_B T}\right) = \nu^0 \exp\left(\frac{S^M}{k_B}\right) \exp\left(-\frac{H^M}{k_B T}\right), \quad (8.18)$$

8 Dependence of Diffusion on Temperature and Pressure

or

$$\omega_2 = \nu_2^0 \exp\left(-\frac{G_2^M}{k_B T}\right) = \nu_2^0 \exp\left(\frac{S_2^M}{k_B}\right) \exp\left(-\frac{H_2^M}{k_B T}\right), \quad (8.19)$$

where ν^0, ν_2^0 denote the corresponding *attempt frequencies*. Usually, the Debye frequency of the lattice is an adequate approximation for the attempt frequencies. Of course, values of the Gibbs energies, enthalpies, and entropies of motion depend on the defect involved and on the material considered.

Inserting the expressions of f and Γ or f_2 and Γ_2 into Eq. (8.14), we arrive at the following activation parameters:

- For self-diffusion via one type of defect (subscript D) we get

$$\Delta H \Rightarrow H_D^F + H_D^M \quad \text{and} \quad \Delta S \Rightarrow S_D^F + S_D^M. \quad (8.20)$$

The activation enthalpy (entropy) of self-diffusion equals the sum of the formation and migration enthalpies (entropies) of the diffusion-mediating defect. For a monovacancy we have $\Delta H = H_{1V}^F + H_{1V}^M$.

- For solute diffusion in a dilute substitutional alloy the activation enthalpy is a more slightly complex quantity. Combining Eqs. (8.4), (8.14), and (8.16) we get

$$\Delta H_2 \Rightarrow H_D^F - H^B + H_2^M + C. \quad (8.21)$$

The correlation term

$$C = -k_B \frac{\partial \ln f_2}{\partial (1/T)} \quad (8.22)$$

arises from the temperature dependence of the correlation factor f_2 of impurity diffusion (see Chap. 7). The quantity $H_D^F - H^B$ is the formation enthalpy of the defect-impurity complex and H_2^M the barrier for a impurity-defect exchange. Depending on the various contributions, the activation enthalpy of substitutional impurity diffusion, ΔH_2, can be higher or lower than the activation enthalpy of self-diffusion (see Chap. 19).

8.2 Pressure Dependence

The effect of hydrostatic pressure p on diffusion can easily be recognised from the Arrhenius expression Eq. (8.6). A variation of the diffusivity with pressure is largely due to the fact that the Gibbs free energy of activation varies with pressure according to

$$\Delta G = \Delta H - T\Delta S = \Delta E - T\Delta S + p\Delta V. \quad (8.23)$$

Here ΔE denotes the *activation energy* (ΔE is the change in internal energy) and ΔV the *activation volume* of diffusion. Thermodynamic tells us that

$$\Delta V = \left(\frac{\partial \Delta G}{\partial p}\right)_T. \quad (8.24)$$

Equation (8.24) can be considered as the definition of the activation volume.

A comprehensive characterisation of a diffusion process requires information about three activation parameters, namely:

$$\Delta E, \Delta S, \text{ and } \Delta V.$$

The activation energy ΔE and the entropy ΔS are usually well appreciated in the diffusion literature, whereas the activation volume ΔV is often a 'forgotten' parameter. Activation enthalpy and activation energy are related via

$$\Delta H = \Delta E + p\Delta V. \qquad (8.25)$$

The term $p\Delta V$ can be significant at high pressures. At ambient pressure, it is almost negligible for solids. Then $\Delta E \approx \Delta H$ and activation energy and activation enthalpy are synonymous.

Equations (8.6) and (8.24) show that the activation volume can be obtained from measurements of the pressure dependence of the diffusion coefficient at constant temperature via

$$\Delta V = -k_\mathrm{B} T \cdot \left(\frac{\partial \ln D}{\partial p}\right)_T + \underbrace{k_\mathrm{B} T \cdot \frac{\partial \ln \left(f a^2 \nu^0\right)}{\partial p}}_{corr.\ term}. \qquad (8.26)$$

As an example, Fig. 8.2 shows the self-diffusion coefficient of gold as a function of pressure at constant temperature. The slope of the logarithm of D as a function of p corresponds to the first term of Eq. (8.26). The second term on the right-hand side of Eq. (8.26) is a correction term. It can be estimated from the isothermal compressibility κ_T and the Grüneisen constant γ_G [5–7]:

$$corr.\ term \approx k_\mathrm{B} T \kappa_T \gamma_G. \qquad (8.27)$$

Estimates for the correction terms on the basis of Eq. (8.27) lead to small corrections in the range of 0.01 to 0.03 Ω, where Ω denotes the atomic volume. Often the correction term can be neglected within the experimental accuracy. In the case of Fig. 8.2, the activation volume is $\Delta V = 0.76\,\Omega$.

If several mechanisms with diffusivities $D_I, D_{II}, \ldots$ operate simultaneously (see Eq. 8.9), measurements of the pressure dependence give an effective activation volume

$$\Delta V_{eff} = \Delta V_I \frac{D_I}{D_I + D_{II} + \ldots} + \Delta V_{II} \frac{D_{II}}{D_I + D_{II} + \ldots} + \ldots, \qquad (8.28)$$

which is a weighted average of the activation volumes of the individual activation volumes $\Delta V_I, \Delta V_{II}, \ldots$. Since the relative contributions of several mechanisms vary with temperature (and pressure) the effective activation volume is temperature (and pressure) dependent.

Activation volumes of ionic conduction (see Chaps. 26 and 27) can be determined from the pressure dependence of the dc conductivity, σ_{dc}, according to

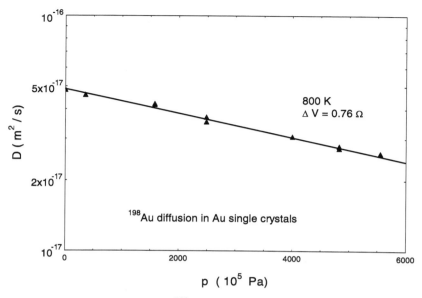

Fig. 8.2. Pressure dependence of ^{198}Au diffusion in Au single crystals at constant temperature according to WERNER AND MEHRER [11]. Ω denotes the atomic volume of Au

$$\Delta V_\sigma \approx -k_\mathrm{B} T \frac{\partial \ln \sigma_{dc}}{\partial p} . \qquad (8.29)$$

The atomistic meaning of ΔV_σ depends on the type of the ion-conducting material, on the type of disorder (Frenkel or Schottky disorder), and on the temperature region studied (intrinsic or extrinsic region). Examples are discussed below.

There are good reasons why the study of pressure effects has consumed energies of many researcher in the past. First, a thorough understanding of diffusion requires knowledge about the influence of pressure on the diffusivity. Second, in favourable cases the value of ΔV itself, its magnitude and sign, can throw light on the mechanism(s) of diffusion that is (are) operating (see below). Some selected values for activation volumes are listed in Table 8.1. For elemental crystals the unit is the atomic volume Ω; for compounds the unit is the molar volume V_m. For a comprehensive collection of activation volumes for metals and alloys available until 1990 the reader is referred to Chap. 10 in [1]. A more recent review about the effects of pressure on self- and solute diffusion in metals and semiconductors is given in [6].

The microscopic interpretation of the activation volume ΔV depends on the mechanism of diffusion as it also does for ΔH and ΔS. As discussed in Chap. 6, self-diffusion in crystalline solids is mediated by defects. In metals, monovacancies dominate self-diffusion at low and moderate temperatures,

Table 8.1. Activation volumes of diffusion and ionic conduction; in units of the atomic volume Ω for elements; in units of the molar volume V_m for compounds.

Diffusion process	$\Delta V/\Omega$ or $\Delta V/V_m$	Reference
Cu self-diffusion	+0.93 to +1.09	Beyeler and Adda [8]
Ag self-diffusion	+0.66 to +0.88	Beyeler and Adda [8], Rein and Mehrer [9]
Au self-diffusion	+0.72 to +0.75	Dickerson et al. [10], Beyeler and Adda [8] Werner and Mehrer [11], Rein and Mehrer [9]
Na self-diffusion	+0.4 to +0.75	Mundy [12]
Ge diffusion in silicon	-0.68 to -0.28	Södervall et al. [15], Aziz et al. [16]
N in α-iron	+0.05	Bosman et al. [13]
C in α-iron	-0.08 to -0.02	Bosman et al. [14]
Al self-diffusion	+1.29	Beyeler and Adda [8]
Ge in Al	+1.16 to +1.24	Thürer et al. [17]
Zn in Al	+0.74 to +1.09	Erdelyi et al. [18]
Mn in Al	+1.67	Rummel et al. [19]
Co in Al	+1.64 to 1.93	Rummel et al. [19]
Schottky pair formation: V_{SP}^F	1.63 for KBr 2.04 for NaCl 1.23 for KBr 1.37 for NaBr	Yoon and Lazarus [22]
Cation migration: $V_{V_C}^M$	0.21 for KCl 0.26 for NaCl 0.25 for KBr 0.25 for NaBr	Yoon and Lazarus [22]
Intrinsic conduction: $V_{V_C}^M + V_{SP}^F/2$	1.03 for KCl 1.28 for NaCl 0.87 for KBr 0.93 for NaBr	Yoon and Lazarus [22]
Ag ion conduction α-AgI	≈ 0	Mellander [23]

whereas divacancies contribute to some extent as temperatures approach the melting temperature. Self-interstitials are important, for example, in silicon. In what follows, we illustrate the activation volumes for various mechanisms.

8.2.1 Activation Volumes of Self-diffusion

For a defect mechanism of self-diffusion the Gibbs free energy of activation is composed of a formation (superscript F) and a migration (superscript M) term (see Sect. 8.1):

$$\Delta G = G_D^F + G_D^M. \tag{8.30}$$

Considering Eq. (8.24) it is obvious that the activation volume is also composed of a formation volume, V_D^F, and a migration volume, V_D^M, of the diffusion-mediating defect according to

$$\Delta V = V_D^F + V_D^M. \tag{8.31}$$

The **formation volume of a monovacancy** is illustrated in Fig. 8.3. The left-hand side of the figure indicates a vacancy in a 'rigid' lattice. Without relaxation the volume of the crystal would increase by one atomic volume Ω. The situation illustrated on the right-hand side corresponds to a vacancy with some relaxation of the neighbouring atoms. This picture indicates for the formation volume of a monovacancy:

$$V_{1V}^F = +\Omega - V_{rel,1V}. \tag{8.32}$$

Usually, inward relaxation is found, which implies that the formation volume of the vacancy V_{1V}^F is somewhat smaller than Ω as the relaxation volume $V_{rel,1V}$ is positive. The amount and sign of relaxation depend on the material.

The **formation volume of a divacancy** encompasses the increase of the crystal volume due to the creation of two new lattice sites minus the relaxation volume $V_{rel,2V}$:

$$V_{2V}^F = +2\Omega - V_{rel,2V}. \tag{8.33}$$

The formation volume of the divacancy is larger than that of a monovacancy, i.e. $V_{2V}^F > V_{1V}^F$. For a material in which diffusion is mediated by mono- and divacancies the total activation volume increases as the relative divacancy contribution to the total diffusivity increases with temperature (see Chap. 17).

The **formation volume of a self-interstitial** is illustrated in Fig. 8.4. Without relaxation the formation of a self-interstitial would decrease the crystal volume by one atomic volume. On the other hand, the formation of a self-interstitial causes (considerable) outward relaxation of the surrounding lattice.

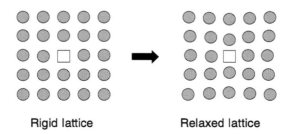

Fig. 8.3. Schematic illustration of the formation volume of a vacancy

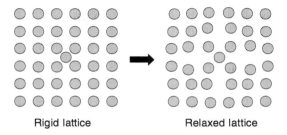

Fig. 8.4. Illustration of the formation volume of a self-interstitial

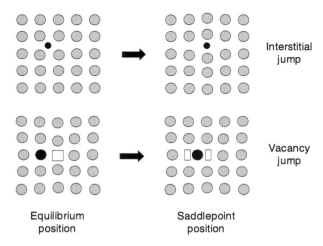

Fig. 8.5. Illustration of the migration volume. *Upper part*: interstitial migration. *Lower part*: vacancy migration

$$V_I^F = -\Omega + V_{rel,I} \,. \tag{8.34}$$

Whether V_I^F is positive or negative depends on the amount of outward relaxation. If the relaxation volume $V_{rel,I}$ is positive and larger than one atomic volume – as is the case for close-packed metals [20] – the formation volume is positive. For a less densely packed structure such as silicon, a negative formation volume of a self-interstitial can be expected.

The **migration volume** of an atom (or of a defect) refers to the volume change when the jumping atom is transferred from its equilibrium position to the saddle-point position. Its illustration is a somewhat 'dangerous' procedure. The jump event occurs in a short time interval of about 10^{-12} s. During this short period a complete relaxation of the saddle-point configuration cannot occur because atomic displacements in a solid proceed by the velocity of sound. Nevertheless, with some precaution Fig. 8.5 may serve as an illustration of the migration volume.

138   8 Dependence of Diffusion on Temperature and Pressure

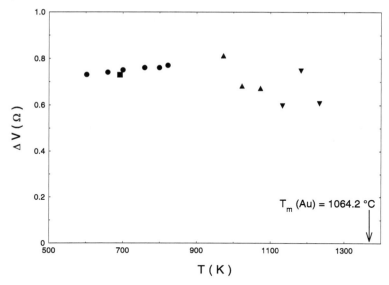

Fig. 8.6. Activation volumes for self-diffusion in Au *versus* temperature: *triangles* [8], *square* [9], *full circles* [11]

The migration volume of a vacancy in close-packed metals is fairly small. Experimental values around $V^M = 0.15\,\Omega$ have been reported for Au. These value were determined by studying the effect of hydrostatic pressure on the annealing rate of vacancies, which had been produced by quenching Au wires from high temperatures [21]. Similar numbers are reported for Pt. These values suggest that the major part of the activation volume of vacancy-mediated self-diffusion in metallic elements (see Table 8.1) must be attributed to the formation volume.

Figure 8.6 shows the activation volumes ΔV for self-diffusion of Au between about 600 K and the melting temperature T_M. ΔV is almost independent of temperature indicating that a single mechanism dominates in the whole temperature range. For Au this is the monovacancy mechanism. Values between 0.6 and 0.9 Ω are typical for vacancy-mediated diffusion in close-packed metals such as Cu, Ag, and Au (see Table 8.1). For silver, an increase of the activation volume from about 0.6 to 0.9 Ω has been reported [6, 9] and taken as evidence for the simultaneous action of mono- and divacancies (see Chap. 17).

A comparison between the activation volumes of self-diffusion of noble metals and of sodium indicates (see Table 8.1) that the relaxation around a vacancy is more pronounced for bcc metal. Negative activation volumes between about -0.6Ω and -0.3Ω have been reported for the diffusion of Ge in silicon (see Table 8.1). Solute diffusion of Ge in silicon is very similar to Si self-diffusion and Ge diffuses by the same mechanism as Si. Negative activation volumes are considered as evidence (among others) for self-interstitial mediated diffusion in silicon (see [6, 15] and Chap. 23).

8.2.2 Activation Volumes of Solute Diffusion

For diffusion of **interstitial solutes** no defect formation term is required. Then, from Eqs. (8.11) and (8.24) the activation volume is

$$\Delta V = V^M, \tag{8.35}$$

where V^M is the migration volume of the interstitial solute. As already mentioned, 'small' atoms such as C, N, and O in metals diffuse by this mechanism. The effect of pressure was studied for C and N in α-Fe, for C in Co and for C in Ni and for N and O diffusion in V (for references see Chap. 10 in [1]). Interstitial diffusion is characterised by small values of the activation volume. For example, for C and N diffusion in α-iron small values between -0.08 and +0.05 Ω were reported (see Table 8.1). This implies that interstitial diffusion is only very weakly pressure dependent.

Diffusion of **substitutional impurities** is mediated by vacancies. According to Sect. 8.1 the diffusivity can be written as

$$D_2 = ga^2 \nu_0 \exp\left(-\frac{G_{1V}^F - G^B}{k_B T}\right) \exp\left(-\frac{H_2^M}{k_B T}\right) f_2, \tag{8.36}$$

where G^B is the Gibbs ebergy of binding between solute and vacancy. H_2^M denotes the activation enthalpy for defect-impurity exchange and f_2 the correlation factor of impurity diffusion. Using Eq. (8.24), we get for the activation volume of solute diffusion:

$$\Delta V_2 = V_{1V}^F - V^B + V_2^M \underbrace{- k_B T \frac{\partial \ln f_2}{\partial p}}_{C_2}. \tag{8.37}$$

The term $V_{1V}^F - V^B$ represents the formation volume of the impurity-vacancy pair. It is different from the formation volume of the vacancy in the pure solvent due to the volume change V^B associated with pair formation. V_2^M is the migration volume of the vacancy-solute exchange, which in general is different from the migration volume V_{1V}^M of the vacancy in the pure matrix. Finally, the term C_2 arises from the pressure dependence of the solute correlation factor. $V_2^M + C_2$ can be interpreted as the migration volume of the solute-vacancy complex.

The activation volumes for various solutes in aluminium listed in Table 8.1 show a considerable variation. As we shall see in Chap. 19, transition metal solutes are slow diffusers, whereas non-transition elements are normal diffusers in Al. Self-diffusion in Al has been attributed to the simultaneous action of mono- and divacancies and a similar interpretation is tenable for the

140   8 Dependence of Diffusion on Temperature and Pressure

diffusion of non-transition elements such as Zn and Ge. On the other hand, transition elements in Al have high activation enthalpies and entropies of diffusion, which can be attributed to a repulsive interaction between vacancy and solute (see Chapt 19). The high activation volumes for the transition metals diffusers Mn and Co indicate large formation and/or migration volumes of the solute-vacancy complex [6].

8.2.3 Activation Volumes of Ionic Crystals

The pressure dependence of the ionic conductivity has been studied in several alkali halide crystals (KCl, NaCl, NaBr, KBr) with Schottky disorder by YOON AND LAZARUS [22]. These crystals consist of sublattices of cations (index C) and anions (index A). In the intrinsic region, i.e. at high temperatures, cation and anion vacancies, V_C and V_A, are simultaneously present in equal numbers (Schottky pairs). In the extrinsic region of crystals, doped with divalent cations, additional vacancies in the cation sublattice are formed to maintain charge neutrality (see Chaps. 5 and 26):

(i) In the **intrinsic region** the conductivity is due to Schottky pairs. The formation volume of Schottky pairs is

$$V_{SP}^F = V_{V_C}^F + V_{V_A}^F, \qquad (8.38)$$

where $V_{V_C}^F$ and $V_{V_A}^F$ denote the formation volumes of cation and anion vacancies, respectively. The following values for the formation volume of Schottky pairs have been reported [22] in units $cm^3\,mol^{-1}$:

V_{SP}^F: 61 ± 9 for KCl, 55 ± 9 for NaCl, 54 ± 9 for KBr, 44 for NaBr.

(ii) In the **extrinsic region** the conductivity is dominated by the motion of cation vacancies because anion vacancies are less mobile. Thus, from the pressure dependence of the conductivity one obtains the migration volume of the cation vacancy, $V_{V_C}^M$. The following values have been reported [22]:

$V_{V_C}^M$: 8 ± 1 for KCl, 11 ± 1 for NaCl, 11 ± 1 for KBr, 8 ± 1 for NaBr.

Due to the higher mobility of cation vacancies the activation volume of the ionic conductivity in the intrinsic region, ΔV_σ, is practically given by

$$\Delta V_\sigma = V_{SP}^F/2 + V_{V_C}^M. \qquad (8.39)$$

In principle, anion vacancies also contribute to the conductivity (see Chap. 26). However, as the anion component of the total conductivity is usually small this contribution has been neglected in Eq. (8.39).

A comparison between activation volumes in metals and ionic crystals with Schottky disorder may be useful. In units of the molar volumes of the crystals, V_m, the activation volumes of the ionic conductivity in the intrinsic region are:

ΔV_σ:1.03 V_m for KCl, 1.28 V_m for NaCl, 1.23 V_m for KBr, 1.37 V_m for NaBr.

The activation volumes for intrinsic ionic conduction, which is due to the motion of vacancies, are of the order of one molar volume. These values are similar to activation volumes of self-diffusion in close-packed metals, where the activation volume is also an appreciable fraction of the atomic volume of the material (see above). In contrast, the migration volumes of cation vacancies are smaller:

$V_{V_C}^M$:0.21 V_m for KCl, 0.26 V_m for NaCl, 0.25 V_m for KBr, 0.25 V_m for NaBr.

These values indicate a further similarity between metals and ionic crystals. In both cases, the migration volumes of vacancies are only a small fraction of the atomic (molar) volume.

The α-phase of silver iodide is a typical example of a *fast ion conductor* (see Chap. 27). The immobile I^- ions form a body-centered cubic sublattice, while no definite sites can be assigned to the Ag^+ ions. In the cubic unit cell 42 sites are available for only two Ag^+ ions. Because of these structural features, Ag^+ ions are easily mobile and no intrinsic defect is needed to promote their migration. The pressure dependence of the dc conductivity in α-AgI was studied up to 0.9 GPa by MELLANDER [23]. The activation volume is 0.8 to 0.9 cm^3 mol^{-1}. This very low value can be attributed to the migration of Ag^+ ions, confirming the view that migration volumes are small.

8.3 Correlations between Diffusion and Bulk Properties

Thermodynamic properties of solids such as melting points, heats of melting, and elastic moduli reflect different aspects of the lattice stability. It is thus not surprising that the diffusion behaviour correlates with thermodynamic properties. Despite these correlations, diffusion remains a kinetic property and cannot be based solely on thermodynamic considerations. In this section, we survey some correlations between self-diffusion parameters and bulk properties of the material. These relationships, which can be qualified as 'enlightened empirical guesses', have contributed significantly to the growth of the field of solid-state diffusion. The most important developments in this area were: (i) the establishment of correlations between diffusion and melting parameters and (ii) Zener's hypothesis to relate the diffusion entropy with the temperature dependence of elastic constants. These old and useful correlations have been re-examined by BROWN AND ASHBY [24] and by TIWARI ET AL. [25].

8.3.1 Melting Properties and Diffusion

Diffusivities at the Melting Point: The observation that the self-diffusivity of solids at the melting point, $D(T_m)$, roughly equals a constant is an old one, dating back to the work of VAN LIEMPT from 1935 [27]. But it was not until the mid 1950s that enough data of sufficient precision were available to recognise that $D(T_m)$ is only a constant for a given structure and for a given type of bonding: the bcc structure, the close-packed structures fcc and hdp, and the diamond structured elements all differ significantly. As data became better, additional refinements were added: the bcc metals were subdivided into two groups each with characteristic values of $D(T_m)$ [26]; also alkali halides were seen to have a characteristic value of $D(T_m)$ [31]. Figure 8.7 shows a comparison of self-diffusion coefficients extrapolated to the melting point for various classes of crystalline solids according to BROWN AND ASHBY [24]. The width of the bar is either twice the standard deviation of the geometric mean, or a factor of four, whichever is greater. Data for the solidus diffusivities of bcc and fcc alloys coincide with the range shown for pure metals. It is remarkable that $D(T_m)$ varies over about 6 orders of magnitude, being very small for semiconductors and fairly large for bcc metals.

At the melting temperature T_m according to Eq. (8.6) the self-diffusivity is given by

$$D(T_m) = D^0 \exp\left(-\frac{\Delta H}{k_B T_m}\right) = gfa^2\nu^0 \exp\left(\frac{\Delta S}{k_B}\right) \exp\left(-\frac{\Delta H}{k_B T_m}\right) . \quad (8.40)$$

The constancy of the diffusivity at the melting point reflects the fact that for a given crystal structure and bond type the quantities D^0 and $\Delta H/(k_B T_m)$ are roughly constant:

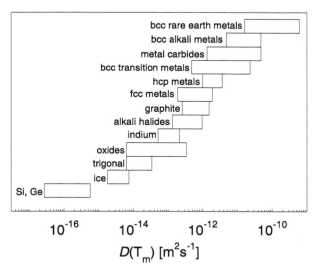

Fig. 8.7. Self-diffusivities at the melting point, $D(T_m)$, for various classes of crystalline solids according to BROWN AND ASHBY [24]

8.3 Correlations between Diffusion and Bulk Properties

The pre-exponential factor D^0 is indeed almost a constant. According to Eq. (8.5) it contains the attempt frequency ν^0, the lattice parameter a, geometric and correlation factors, and the diffusion entropy ΔS. Attempt frequencies are typically of the order of the Debye frequency, which lies in the range of 10^{12} to 10^{13} s^{-1} for practically all solids. The diffusion entropy is typically of the order of a few k_B. Correlation factors and geometric terms are not grossly different from unity.

The physical arguments for a constancy of the ratio $\Delta H/(k_\mathrm{B} T_m)$ are less clearcut. One helpful line of reasoning is to note that the formation of a vacancy, like the process of sublimation, involves breaking half the bonds that link an atom in the interior of the crystal to its neighbours; the enthalpy required to do so should scale as the heat of sublimation, H_s. The migration of a vacancy involves a temporary loss of positional order – it is somehow like local melting – and involves an energy that scales as the heat of melting (fusion), H_m. One therefore may expect

$$\frac{\Delta H}{k_\mathrm{B} T_m} \approx \alpha \frac{H_s}{k_\mathrm{B} T_m} + \beta \frac{H_m}{k_\mathrm{B} T_m}, \tag{8.41}$$

where α and β are constants. The first term on the right-hand side contains the sublimation entropy at the melting temperature, $S_s = H_s/T_m$; the second term contains the entropy of melting, $S_m = H_m/T_m$. These entropy changes are roughly constant for a given crystal structure and bond type; it follows that $\Delta H/(k_\mathrm{B} T_m)$ should be approximately constant, too.

Activation Enthalpy and Melting Properties: From practical considerations, correlations between melting and activation enthalpy are particularly useful. Figure 8.8 shows the ratio $\Delta H/(k_\mathrm{B} T_m)$ for various classes of crystalline solids. It is approximately a constant for a given structure and bond type. The constants defined in this way vary over a factor of about 3.5. The activation enthalpy was related to the melting point many years ago [27–29]. These correlations have been reconsidered for metals and alloys by BROWN AND ASHBY [24] and for pure metals recently by TIWARI ET AL. [25]. The activation enthalpy of diffusion is related via

$$\Delta H \approx K_1 T_m \tag{8.42}$$

to the melting temperature (expressed in Kelvin) of the host crystal. This relation is called the *van Liempt rule* or sometimes also the *Bugakov – van Liempt rule* [30].

One may go further by invoking the thermochemical *rule of Trouton*, which relates the melting point of materials to their (nearly) constant entropy of melting, S_m. Trouton's rule, $S_m = H_m/T_m \approx 2.3\ cal/mol = 9.63\ J/mol$, allows one to replace the melting temperatuire in Eq. (8.42) by the enthalpy of melting, H_m. Then, the van Liempt rule may be also expressed as

$$\Delta H \approx \frac{K_1}{S_m} H_m \equiv K_2 H_m. \tag{8.43}$$

144   8 Dependence of Diffusion on Temperature and Pressure

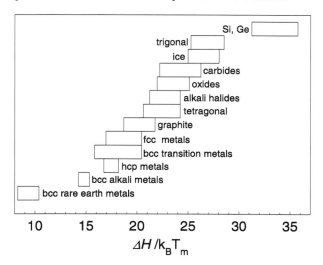

Fig. 8.8. Normalised activation enthalpies of self-diffusion, $\Delta H/(k_B T_m)$, for classes of crystalline solids according to BROWN AND ASHBY [24]

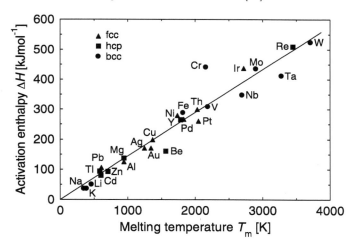

Fig. 8.9. Activation enthalpies of self-diffusion in metals, ΔH, *versus* melting temperatures, T_m, according to TIWARI ET AL. [25]

K_1 and K_2 are constants for a given class of solids. Plots of Eqs. (8.42) and (8.43) for metals are shown in Figs. 8.9 and 8.10. Values of the slopes for metals are: $K_1 = 146\ Jmol^{-1}K^{-1}$ and $K_2 = 14.8$ [25].

The validity of Eq. (8.42) has been demonstrated for alkali halides by BARR AND LIDIARD [31]. For inert gas solids and molecular organic solids,

8.3 Correlations between Diffusion and Bulk Properties   145

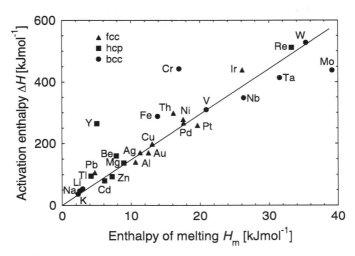

Fig. 8.10. Activation enthalpies of self-diffusion in metals, ΔH, *versus* melting enthalpies, H_m, according to TIWARI ET AL. [25]

the validity of Eqs. (8.42) and (8.43) has been established by CHADWICK AND SHERWOOD [32].

The correlations above are based on self-diffusion, which is indeed the most basic diffusion process. Diffusion of foreign elements introduces additional complexities such as the interaction between foreign atom and vacancy and temperature-dependent correlation factors (see Chaps. 7 and 19). Correlations between the activation enthalpies of self-diffusion and substitutional impurity diffusion have been proposed by BEKE ET AL. [33].

Activation Volume and Melting Point: The diffusion coefficient is pressure dependent due to the term $p\Delta V$ in the Gibbs free energy of activation. The activation volume of diffusion, ΔV, has been discussed in Sect. 8.2. NACHTRIEB ET AL. [34, 35] observed that the diffusivity at the melting point is practically independent of pressure. For example, in Pb and Sn the lattice diffusivity, as for most metals, decreases with increasing pressure in such a way that the increased melting point resulted in a constant rate of diffusion at the same homologous temperature. If one postulates that $D(T_m)$ is independent of pressure, we have

$$\frac{d\left[\ln D(T_m)\right]}{dp} = 0. \qquad (8.44)$$

Then, we get from Eq. (8.6)

$$\Delta V = \frac{\Delta H(p=0)}{T_m(p=0)} \frac{dT_m}{dp} \qquad (8.45)$$

if the small pressure dependence of the pre-exponential factor is neglected. This equation predicts that ΔV is controlled by the sign and magnitude of dT_m/dp. In fact, BROWN AND ASBY report reasonable agreement of Eq. (8.45) with experimental data [24]. In general, dT_m/dp is positive for most metals and indeed their activation volumes are positive as well. For plutonium dT_m/dp is negative and, as expected from Eq. (8.45), the activation volume of Pu is negative [36].

Later, however, also remarkable exceptions have been reported, which violate Eq. (8.45). For example, dT_m/dp is negative for Ge [37], but the activation volume of Ge self-diffusion is positive [38] (see also Chap. 23). Neither the variation of the activation volume with temperature due to varying contributions of different point defects to self-diffusion nor the differences between the activation volumes of various solute diffusers are reflected by this rule.

8.3.2 Activation Parameters and Elastic Constants

A correlation between the elastic constants and diffusion parameters was already proposed in the pioneering work of WERT AND ZENER [39, 40]. They suggested that the Gibbs free energy for migration (of interstitials), G^M, represents the elastic work to deform the lattice during an atomic jump. Thus, the temperature variation of G^M should be the same as that of an appropriate elastic modulus μ:

$$\frac{\partial (G^M/G_0^M)}{\partial T} = \frac{\partial (\mu/\mu_0)}{\partial T}. \tag{8.46}$$

The subscript 0 refers to values at absolute zero. The migration entropy S^M is obtained from the thermodynamic relation

$$S^M = -\frac{\partial G^M}{\partial T} \tag{8.47}$$

and $H^M = G^M + TS^M$ yields the migration enthalpy. In the Wert-Zener picture both H^M and S^M are independent of T if μ varies linearly with temperature. If this is not the case, both S^M and H^M are temperature dependent. Substituting the thermodynamic relation and $G_0^M \approx H^M$ in Eq. (8.46), we get:

$$S^M \approx -\frac{H^M}{T_m} \cdot \frac{\partial (\mu/\mu_0)}{\partial (T/T_m)} = \Theta \frac{H^M}{T_m}. \tag{8.48}$$

At temperatures well above the Debye temperature, elastic constants usually vary indeed linearly with temperature. The derivative $\Theta \equiv -\partial(\mu/\mu_0)/\partial(T/T_m)$ is then a constant. Its values lie between -0.25 to -0.45 for most metals. Then H^M and S^M are proportional to each other and the model of Zener predicts a positive migration entropy.

For a vacancy mechanism, Zener's idea is strictly applicable only to the migration and not to the formation property of the defect. One can, however, always deduce a diffusion entropy via $\Delta S = k_\mathrm{B} \ln[D^0/(gfa^2\nu^0)]$ from the measured value of D^0. Experimental observations led Zener to extend his relation to the activation properties of atoms on substitutional sites:

$$\Delta S = \lambda \Theta \frac{\Delta H}{T_m} . \qquad (8.49)$$

Then the pre-exponential factor can be written as

$$D^0 = fga^2\nu^0 \exp\left(\lambda \Theta \frac{\Delta H}{T_m}\right) . \qquad (8.50)$$

λ is a constant that depends on the structure and on the diffusion mechanism. For self-diffusion in fcc metals $\lambda \approx 0.55$ and for bcc metals $\lambda \approx 1$. The relation Eq. (8.49) is often surprisingly well fulfilled. We note that this relation also suggests that the diffusion entropy $\Delta S = S^M + S^F$ is positive. This conclusion is supported by the well-known fact that the formation entropy for vacancies, S^F, is positive (see Chap. 5).

8.3.3 Use of Correlations

The value of the correlations discussed above is that they allow diffusivities to be estimated for solids for which little or no data are available. For example, when diffusion experiments are planned for a new material, these rules may help in choosing the experimental technique and adequate thermal treatments. The correlations should be used with clear appreciation of the possible errors involved; in some instances, the error is small.

We emphasise that the correlations have been formulated for self-diffusion. Solute diffusion of substitutional solutes in most metals differs by not more than a factor of 100 for many solvent metals and the activation enthalpies by less than 25 % from that of the host metal (see Chap. 19).

There are, however, remarkable exceptions: examples are the very slow diffusion of transition metals solutes in Al and the very fast diffusion of noble metals in lead and other 'open metals' (see Chap. 19). Also diffusion of interstitial solutes (see Sect. 18.1), hydrogen diffusion (see Sect. 18.2), and fast diffusion of hybrid foreign elements in Si and Ge (see Chap. 25) do not follow these rules.

References

1. H. Mehrer (Vol. Ed.), *Diffusion in Solid Metals and Alloys*, Landolt-Börnstein, Numerical Data and Functional Relationships in Science and Technology, New Series, Group III: Condensed Matter, Vol. 26, Springer-Verlag, 1990
2. D.L. Beke (Vol. Ed.), *Diffusion in Semiconductors and Non-Metallic Solids*, Landolt-Börnstein, Numerical Data and Functional Relationships in Science and Technology, New Series, Group III: Condensed Matter; Vol. 33, Subvolume A: Diffusion in Semiconductors, Springer-Verlag, 1998; Subvolume B1: Diffusion in Non-Metallic Solids, Springer-Verlag, 1999

3. N.L. Peterson, J. of Nucl. Materials **69–70**, 3 (1978)
4. H. Mehrer, J. of Nucl. Materials **69–70**, 38 (1978)
5. R.N. Jeffrey, D. Lazarus, J. Appl. Phys. **41**, 3186 (1970)
6. H. Mehrer, Defect and Diffusion Forum **129–130**, 57 (1996)
7. H. Mehrer, A. Seeger, Cryst. Lattice Defects **3**, 1 (1972)
8. M. Beyeler, Y. Adda, J. Phys. (Paris) **29**, 345 (1968)
9. G. Rein, H. Mehrer, Philos. Mag. **A 45**, 767 (1982)
10. R.H. Dickerson, R.C. Lowell, C.T. Tomizuka, Phys. Rev. **137**, A 613 (1965)
11. M. Werner, H. Mehrer, in: *DIMETA-82, Diffusion in Metals and Alloys*, F.J. Kedves, D.L. Beke (Eds.), Trans Tech Publications, Diffusion and Defect Monograph Series No. 7, 392 (1983)
12. J.N. Mundy, Phys. Rev. **B3**, 2431 (1971)
13. A.J. Bosman, P.E. Brommer, G.W. Rathenau, Physica **23**, 1001 (1957)
14. A.J. Bosman, P.E. Brommer, L.C.H. Eickelboom, C. J. Schinkel, G.W. Rathenau, Physica **26**, 533 (1960)
15. U. Södervall, A. Lodding, W. Gust, Defect and Diffusion Forum **66–59**, 415 (1989)
16. E. Nygren, M.J. Aziz, D. Turnbull, J.F. Hays, J.M. Poate, D.C. Jacobson, R. Hull, Mat. Res. Soc. Symp. Proc. **86**, 77 (19?8)
17. A. Thürer, G. Rummel, Th. Zumkley, K. Freitag, H. Mehrer, Phys. Stat. Sol. (a) **143**, 535 (1995)
18. G. Erdelyi, H. Mehrer, D.L. Beke, I. Gödeny, Defect and Diffusion Forum **143–147**, 121 (1997)
19. G. Rummel, Th. Zumkley, M. Eggersmann, K. Freitag, H. Mehrer, Z. Metallkd. **86**, 132 (1995)
20. H. Wollenberger, *Point Defects*, in: *Physical Metallurgy*, R.W. Cahn, P. Haasen (Eds.), North-Holland Physics Publishing, 1983, p.1139
21. R.M. Emrick, Phys. Rev. **122**, 1720–1733 (1961)
22. D.N. Yoon, D. Lazarus, Phys. Rev. B **5**, 4935 (1972)
23. B.E. Mellander, Phys. Rev. B **26**, 5886 (1982)
24. A.M. Brown, M.F. Ashby, Acta Metall. **28**, 1085 (1980)
25. G.P. Tiwari, R.S. Mehtrota, Y. Iijima, *Solid-State Diffusion and Bulk Properties*, in: *Diffusion Processes in Advanced Technological Materials*, D. Gupta (Ed.), William Andrew, Inc., 2005, p. 69
26. N.L. Peterson, Solid State Physics **22**, 481 (1968)
27. J.A.M. van Liempt, Z. Physik **96**, 534 (1935)
28. O.D. Sherby, M.T. Simnad, Trans. A.S.M. **54**, 227 (1961)
29. A.D. Le Claire, in: *Diffusion in Body Centered Cubic Metals*, J.A. Wheeler, F.R. Winslow (Eds.), A.S.M., Metals Parks, 1965, p.3
30. B.S. Bokstein, S.Z. Bokstein, A.A. Zhukhvitskii, *Thermodynamics and Kinetics of Diffusion in Solids*, Oxonian Press, New Dehli, 1985
31. L.W. Barr, A.B. Lidiard, *Defects in Ionic Crystals*, in: *Physical Chemistry – an Advanced Treatise*, Academic Press, New York, Vol. X, 1970
32. A.V. Chadwick, J.N. Sherwood, *Self-diffusion in Molecular Solids*, in: *Diffusion Processes*, Vol. 2, Proceedings of the Thomas Graham Memorial Symposium, University of Strathclyde, J.N. Sherwood, A.V.Chadwick, W.M. Muir, F.L. Swinton (Eds.), Gordon and Breach Science Publishers, 1971, p. 472
33. D. Beke, T. Geszti, G. Erdelyi, Z. Metallkd. **68**, 444 (1977)
34. N.H. Nachtrieb, H.A. Resing, S. A. Rice, J. Chem. Phys. **31**, 135 (1959)

35. N.H. Nachtrieb, C. Coston, in: *Physics of Solids at High Pressure*, C.T. Tomizuka, R.M. Emrick (Eds.), Academic Press, 1965, p. 336
36. J.A. Cornet, J. Phys. Chem. Sol. **32**, 1489 (1971)
37. D.A. Young, *Phase Diagrams of the Elements*, University of California Press, 1991
38. M. Werner, H. Mehrer, H.D. Hochheimer, Phys. Rev. B **32**, 3930 (1985)
39. C. Wert, C. Zener, Phys. Rev. **76**, 1169 (1949)
40. C. Zener, J. Appl. Phys. **22**, 372 (1951)
41. C. Zener, in: *Imperfections in Nearly Perfect Crystals*, W. Shockley et al. (Eds.), John Wiley, 1952, p. 289

9 Isotope Effect of Diffusion

In this chapter, we consider the diffusion of two chemically identical atoms that differ in their atomic masses. Their diffusivities are different and this difference is denoted as *isotope effect*. The isotope effect, sometimes also called the mass effect, is of considerable interest. It provides an important experimental means of gaining access to correlation effects. Correlation factors of self- and solute diffusion are treated in Chap. 7 and values for correlation factors of self-diffusion in several lattices and for various diffusion mechanisms are listed in Table 7.2. Correlation factors of solute diffusion are the subject of Sect. 7.5. We shall see below that the isotope effect is closely related to the correlation factor. Since correlation factors of self-diffusion often take values characteristic for the diffusion mechanism, isotope effects experiments can throw light on the mechanism.

9.1 Single-jump Mechanisms

Let us consider two isotopes α and β of the same element labelled by their isotopic masses m_α and m_β. Because of their different masses, the two isotopes have different diffusion coefficients in the same host lattice. For self- and impurity-diffusion in coordination lattices the tracer diffusivities can be written as:

$$D^*_\alpha = A\omega_\alpha f_\alpha, \quad \text{and} \quad D^*_\beta = A\omega_\beta f_\beta. \tag{9.1}$$

The quantity A contains a geometrical factor, the lattice parameter squared, and for a defect mechanism also the equilibrium fraction of defects or the defect availability next to the solute. The atom-defect exchange rates ω_α or ω_β are factors in Eq. (9.1). The correlation factors for vacancy-mediated diffusion in fcc, bcc, and diamond lattices according to Eq. (7.46) have the same mathematical form, sometimes called the *'impurity form'*:

$$f_\alpha = \frac{u}{\omega_\alpha + u}, \quad \text{and} \quad f_\beta = \frac{u}{\omega_\beta + u}. \tag{9.2}$$

The quantity u in Eq. (9.2) depends on the exchange rates between vacancy and solvent atoms *but not* on the vacancy-tracer exchange rate (see Chap. 7). Correlation factors of self- and impurity diffusion have the 'impurity form' because tracer isotopes of the same element as the matrix itself are formally 'solutes', whose jump rates differ slightly from those of the host atoms due to the different masses. When the small differences between vacancy-tracer and vacancy-host atom exchange rates are neglected the correlation factor of self-diffusion is reduced to one of the values listed in Table 7.2.

After taking the ratio D_α^*/D_β^* and eliminating u and f_β using Eqs. (9.1) and (9.2), we find

$$\frac{D_\alpha^* - D_\beta^*}{D_\beta^*} = f_\alpha \frac{\omega_\alpha - \omega_\beta}{\omega_\beta}. \tag{9.3}$$

The tracer-defect exchange rates can be written as

$$\omega_{\alpha,\beta} = \nu_{\alpha,\beta}^0 \exp\left(-\frac{G_{\alpha,\beta}^M}{k_B T}\right) \tag{9.4}$$

where $\nu_{\alpha,\beta}^0$ denote the attempt frequencies of the isotopes α and β, and $G_{\alpha,\beta}^M$ the Gibbs free migration energies of their jumps. In the following discussion we assume

$$G_\alpha^M = G_\beta^M = G^M. \tag{9.5}$$

In other words, the activation enthalpies and entropies of the jump are independent of the isotopic masses of the tracers. This is usually well justified, since the barrier for an atomic jump is determined by the electronic interaction, which is identical for two isotopes of the same element, and not by the masses of the nuclei[1]. Because of Eq. (9.5) the ratio of the jump rates reduces to the ratio of the attempt frequencies:

$$\frac{\omega_\alpha}{\omega_\beta} = \frac{\nu_\alpha^0}{\nu_\beta^0}. \tag{9.6}$$

In what follows, we first mention a simple approximation to this frequency ratio: Einstein's model for the vibration frequencies of atoms in a solid describes a crystal as a set of independent harmonic oscillators. WERT [1] has shown in 1950 that in the classical rate theory, ν^0 is the vibration frequency of an atom in its jump direction. Harmonic oscillator theory tells us that the vibration frequencies are inversely proportional to the square-root of their isotopic masses:

$$\frac{\nu_\alpha^0}{\nu_\beta^0} \approx \sqrt{\frac{m_\beta}{m_\alpha}}. \tag{9.7}$$

---

[1] Hydrogen diffusion is an exception. For hydrogen isotopes quantum effects (see Sect. 18.2), such as zero-point vibrations and tunnelling are relevant. Both effects are mass-dependent. For atoms heavier than Li, quantum effects are usually negligible.

Inserting Eq. (9.7) into Eq. (9.3) and making use of Eq. (9.6) yields

$$\frac{(D_\alpha^* - D_\beta^*)/D_\beta^*}{\sqrt{m_\beta/m_\alpha} - 1} \approx f_\alpha. \qquad (9.8)$$

This result was derived by SCHOEN in 1958 [2]. It suggests that a measurement of the isotope effect permits a determination of the correlation factor.

Unfortunately the derivation of Eq. (9.8) is based on the Einstein approximation, which assumes that all atoms in the crystal vibrate independently. In other words, the Einstein model neglects many-body effects. Lattice dynamics shows that the coupling between atomic vibrations is important and manifests itself, among other effects, in a spectrum of phonon frequencies. Based on VINEYARD'S [3] classical statistical mechanics treatment of the atomic jump process (see Chap. 4), MULLEN [4] and LE CLAIRE [5, 6] took into account the influence of many-body effects. They obtain the relation

$$\frac{\omega_\alpha - \omega_\beta}{\omega_\beta} = \Delta K \left(\sqrt{\frac{m_\beta}{m_\alpha}} - 1 \right), \qquad (9.9)$$

where ΔK is denoted as the *kinetic energy factor*. It is a dimensionless parameter and denotes the fraction of the kinetic energy of the jumping atom at the saddle-point with respect to the total kinetic energy, associated with the motion of all atoms in the jump direction. From Eqs. (9.3) and (9.9) we find

$$\frac{(D_\alpha^* - D_\beta^*)/D_\beta^*}{\sqrt{m_\beta/m_\alpha} - 1} = f_\alpha \Delta K \equiv E_{\alpha,\beta}, \qquad (9.10)$$

which replaces the approximation of Eq. (9.8). The abbreviation $E_{\alpha,\beta}$ introduced in Eq. (9.10) is denoted as the *isotope-effect parameter*. In an analogous way, we arrive at

$$\frac{(D_\beta^* - D_\alpha^*)/D_\alpha^*}{\sqrt{m_\alpha/m_\beta} - 1} = f_\beta \Delta K \equiv E_{\beta,\alpha}. \qquad (9.11)$$

In principle, the two isotope-effect parameters, $E_{\alpha,\beta}$ and $E_{\beta,\alpha}$, are different because f_α and f_β are also slightly different. However, the relative mass differences between two isotope pairs are often small. Then, the differences between $E_{\beta,\alpha}$ and $E_{\alpha,\beta}$ are usually of the order of a few percent and often smaller than the errors in a typical isotope effect experiment. Therefore, it is common practice in the literature to use the following approximation:

$$f_\alpha \approx f_\beta \approx f. \qquad (9.12)$$

Equation (9.12) drops the distinction between f_α and f_β. f is sometimes called the *geometric correlation factor*. f refers to a 'hypothetical' tracer isotope with the same jump rate as the isotopes of the solvent. Then, we may drop the distinction between the two isotope effect parameters (i.e. $E_{\alpha,\beta} \approx E_{\beta,\alpha} \approx E$) and get

$$E = f \Delta K \,. \tag{9.13}$$

Because both f and ΔK are positive and not larger than unity we have the following limits for the isotope effect parameter

$$0 < E \leq 1 \,. \tag{9.14}$$

Equation (9.13) expresses in compact form the relation between isotope effect and correlation factor mentioned at the beginning of this chapter. If the tracer jump is completely decoupled from the motion of other atoms, we have $\Delta K = 1$. This represents the upper limit for the kinetic energy factor. Since a certain amount of coupling between the diffusing atom and the surrounding atoms always exists, we expect $\Delta K < 1$.

For the interstitial mechanism we have

$$E = \Delta K \tag{9.15}$$

since $f = 1$. If several mechanisms with tracer diffusivities $D_I^*, D_{II}^*, \ldots$ operate simultaneously (see Eq. 8.9), measurements of the isotope-effect give an effective isotope-effect parameter, which corresponds to a weighted average

$$E_{eff} = E_I \frac{D_I^*}{D_I^* + D_{II}^* + \ldots} + E_{II} \frac{D_{II}^*}{D_I^* + D_{II}^* + \ldots} + \ldots \,, \tag{9.16}$$

of the isotope effect parameters $E_I, E_{II}, \ldots$ of the individual mechanisms [7].

A measurement of the isotope effect parameter may not uniquely determine f and hence the diffusion mechanism. Nevertheless, it is definitely useful to identify mechanisms, which are consistent with an experimental value of E, and to reject ones, which are not acceptable.

We remind the reader that in the derivation of Eq. (9.13) we have made use of the mathematical form of Eq. (9.2). Chap. 7 has shown that there are indeed important mechanisms for which the correlation factor has this form. This is the case for the monovacancy mechanisms in cubic coordination lattices and also for the divacancy mechanism in an fcc lattice [8]. There are, however, mechanisms where the correlation factor does not have the impurity form (9.2). Examples are mechanisms which have several jump rates such as diffusion in non-cubic crystals and diffusion mechanisms in cubic crystals, which involve more than one tracer jump rate (e.g., for nearest-neighbour and next-nearest neighbour jumps). Sometimes it is possible to derive equations equivalent to Eq. (9.10) [6]. An example is divacancy diffusion in bcc crystals involving several configurations of a divacancy and transitions between these configurations [9, 10]. For a more detailed discussion of the validity of Eq. (9.13) we refer the reader to a review on isotope effects in diffusion by PETERSON [11].

9.2 Collective Mechanisms

Equation (9.10) has been derived for single-jump mechanisms, where in a jump event only the tracer atom changes permanently its site. A more general jump process involves the simultaneous (or collective) jumping of more than one atom. A simple example is the colinear interstitialcy mechanism (see Fig. 6.7), where two atoms jump simultaneously. For a dumbbell-interstitialcy mechanism even three atoms are displaced permanently. Other examples are direct exchange, ring mechanism an chain-like motion of several atoms. All these mechanisms involve the collective motion of several atoms (see Chap. 6).

For a mechanism in which n atoms move collectively during one jump event, the masses in Eq. (9.7) must be replaced by $(n-1)m + m_{\alpha,\beta}$, where m denotes the average mass of the host atoms [5]. Then,

$$\frac{\nu_\alpha^0}{\nu_\beta^0} = \sqrt{\frac{(n-1)m + m_\beta}{(n-1)m + m_\alpha}} \qquad (9.17)$$

and the isotope effect parameter is given by [6]

$$E = \frac{(D_\alpha^* - D_\beta^*)/D_\beta^*}{\sqrt{[m_\beta + (n-1)m])/[m_\alpha + (n-1)m]} - 1}. \qquad (9.18)$$

As a consequence, the isotope effect is reduced. For a highly collective mechanism, a very small isotope effect is plausible, due to the 'dilution' of the mass effect by the participation of many solvent atoms in the jump event. For example, in metallic glasses, collective jump events of chains of atoms dominate, which typically involve ten to twenty atoms. Indeed the isotope effect parameter is close to zero (see [15, 16] and Chap. 29).

9.3 Isotope Effect Experiments

Isotope effects in diffusion are usually small effects. An exception is diffusion of hydrogen isotopes, which is considered later in Sect. 18.2. Depending on the isotope pair (see Table 9.1), the quantity $(D_\alpha^* - D_\beta^*)/D_\beta^*$ is typically of the order of a few percent. For example, for the isotope pair ^{105}Ag and ^{110m}Ag the term in brackets of Eq. (9.9) is about 0.024. Thus, resolving the effects of a relatively small mass difference on the diffusion coefficient is a challenging task. Since the errors in tracer measurements are typically a few percent, it is not feasible to deduce the isotope effect parameter from determinations of D_α^* and D_β^* in separate experiments.

Typical experimental situations for isotope effect studies are illustrated in Fig. 9.1. Two isotopes of one element are diffused simultaneously into the same sample [11–13]. In this way, errors arising from temperature and time

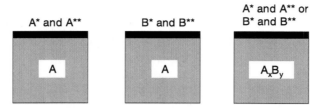

Fig. 9.1. Schematic illustration of various isotope effect experiments. *Left*: isotope pair A^*/A^{**} in a solid element A. *Middle*: isotope pair B^*/B^{**} in a pure solid A. *Right*: isotope pairs A^*/A^{**} or B^*/B^{**} in a binary A_xB_y compound

measurements and from the profiling procedure drop out, since these errors affect both isotopes in the same way. If the isotopes α and β are co-deposited in a very thin layer, the diffusion penetration curves of both isotopes (see also Chap. 13) are given by

$$C_{\alpha,\beta} = C^0_{\alpha,\beta} \exp\left(-\frac{x^2}{4D^*_{\alpha,\beta}t}\right), \tag{9.19}$$

where $C_{\alpha,\beta}$ denote their concentrations in depth x after a diffusion anneal during time t. For a given time the quantities $C^0_{\alpha,\beta}$ are constants. By taking the logarithm of the ratio C_α/C_β, we get from Eq. (9.19)

$$\begin{aligned} \ln\frac{C_\alpha}{C_\beta} &= \ln\frac{C^0_\alpha}{C^0_\beta} - \frac{x^2}{4D^*_\alpha t} + \frac{x^2}{4D^*_\beta t} \\ &= \ln\frac{C^0_\alpha}{C^0_\beta} + \frac{x^2}{4D^*_\alpha t}\left(\frac{D^*_\alpha}{D^*_\beta} - 1\right). \end{aligned} \tag{9.20}$$

Using Eq. (9.19) to eliminate x^2, we obtain

$$\ln\frac{C_\alpha}{C_\beta} = \text{const} - \left(\frac{D^*_\alpha}{D^*_\beta} - 1\right)\ln C_\alpha. \tag{9.21}$$

Equation (9.21) shows that from the slope of a plot of $\ln(C_\alpha/C_\beta)$ versus $\ln C_\alpha$ the quantity $(D^*_\alpha - D^*_\beta)/D^*_\beta$ can be deduced.

Isotope effect experiments are usually performed with radioisotope pairs. Examples of such pairs suitable for isotope effect studies are listed in Table 9.1. Suppose, for example, that the radioisotopes ^{195}Au and ^{199}Au are diffused into a single crystal of gold. Then, the isotope effect of self-diffusion in gold is studied. With the radioisotopes ^{65}Zn and ^{69}Zn diffusing in gold, the isotope effect of Zn solute diffusion is accessible. In an isotope effect experiment the specific activities (proportional to the concentrations) of both isotopes must be determined separately. Separation techniques can be based

Table 9.1. Examples for pairs of radioisotopes suitable for isotope experiments in diffusion studies

^{12}C	^{22}Na	^{55}Fe	^{57}Co	^{65}Zn	^{64}Cu	^{105}Ag	^{195}Au
^{13}C	^{24}Na	^{59}Fe	^{60}Co	^{69}Zn	^{67}Cu	^{110}Ag	^{198}Au

on the different half-lives of the isotopes, on the differences in the emitted γ- or β-radiation using γ-spectroscopy, or on a combination of γ- and β-counting. Half-life separation requires a short-lived and a long-lived isotope. Scintillation spectroscopy can be used, if the γ spectra are favourable. High-resolution intrinsic Ge or Ge(Li) detectors are recommended for separation and corrections for the Compton-scattered radiation must be made. All methods require careful monitoring of radioactive impurities by either half-life measurements or by spectroscopy. Very good counting statistics is necessary to resolve the small differences between the diffusivities of the two isotopes.

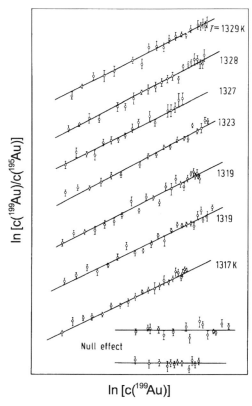

Fig. 9.2. Simultaneous diffusion of the radioisotope pair ^{199}Au and ^{195}Au in monocrystalline Au according to HERZIG ET AL. [14]

158   9 Isotope Effect of Diffusion

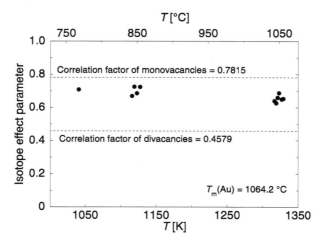

Fig. 9.3. Isotope effect parameters of self-diffusion in Au according to HERZIG ET AL. [14]

Further details about the experimental techniques in isotope effect experiments can be found in the review of ROTHMAN [12]. In a few experiments, stable isotope pairs were utilised as diffusers. Then, the separation is achieved by secondary ion mass spectroscopy (SIMS), which requires good mass resolution and careful background corrections.

Figure 9.2 shows isotope effect measurements for self-diffusion in single crystals of gold according to HERZIG ET AL. [14]. In this experiment the isotope pair ^{199}Au and ^{195}Au was used. The profile measurement was achieved by serial sectioning on a microtome. The isotope concentrations C_α and C_β in each section were separated by combining γ spectroscopy and half-life separation. The logarithm of the ratio $C(^{199}\text{Au})/C(^{195}\text{Au})$ is plotted *versus* the logarithm of $C(^{199}\text{Au})$, as suggested by Eq. (9.21). Isotope effect parameters deduced therefrom are shown in Fig. 9.3. The correlation factor for self-diffusion via monovacancies in an fcc crystal is $f_{1V} = 0.78146$ (see Table 7.2). The experimental results demonstrate that self-diffusion in gold is dominated by the monovacancy mechanism with a kinetic energy factor of $\Delta K \approx 0.9$, which is close to its upper limit 1. The slight decrease of the isotope effect parameter with increasing temperature has been attributed to a small contribution of divacancies according to Eq. (9.16). The divacancy correlation factor ($f_{2V} = 0.458$, see Table 7.2) is smaller than that of monovacanices. Near the melting temperature T_m, the divacancy contribution in Au is, however, not more than 20 % of the total diffusivity [14].

Reviews of isotope effects in diffusion were given by LE CLAIRE [6] and by PETERSON [11]. A comprehensive collection of isotope effect data, which were available until 1990, can be found in Chap. 10 of [13]. Diffusion and isotope effects in metallic glasses have been reviewed by FAUPEL ET AL. [15, 16].

In metallic glasses, isotope effects parameters are usually small and indicate a collective mechanism, which involve the simultaneous chain- or ring-like motion of several (10 to 20) atoms. As an example, isotope effect measurements involving the isotopes ^{57}Co and ^{60}Co were carried out in the deeply supercooled liquid state of bulk metallic glasses. The isotope effect parameter is very small (around 0.1) over the whole temperature range and exhibits no significant temperature dependence (see Chap. 29) . The magnitude of the isotope effect is similar to that of the isotope effect found in the glassy state of conventional metallic glasses [15, 16]. This lends support to the view that the diffusion mechanism in metallic glasses does not change at the calorimetric glass transition temperature. Highly collective hopping processes occurring in the glassy state still determine long-range diffusion in a deeply undercooled melt.

References

1. C. Wert, Phys. Rev. **79**, 601 (1950)
2. A.H. Schoen, Phys. Rev. Letters **1**, 138 (1958)
3. G.H. Vineyard, J. Phys. Chem. Sol. **3**, 121 (1957)
4. J.G. Mullen, Phys. Rev. **121**, 1649 (1961)
5. A.D. Le Claire, Philos. Mag. **14**, 1271 (1966)
6. A.D. Le Claire, *Correlation Effects in Diffusion in Solids*, in: *Physical Chemistry – an Advanced Treatise*, Vol. X, Chap. 5, Academic Press, 1970
7. A. Seeger, H. Mehrer, in: *Vacancies and Interstitials in Metals*, A. Seeger, D. Schumacher, W. Schilling, J. Diehl (Eds.), North-Holland Publishing Company, The Netherlands, 1970, p. 1
8. H. Mehrer, J. Phys. F: Metal Phys. **2**, L11 (1972):
9. H. Mehrer, *Korrelation bei der Diffusion in kubischen Metallen*, Habilitation thesis, Universität Stuttgart, 1973
10. I.V. Belova, D.S. Gentle, G.E. Murch, Phil. Mag. Letters **82**, 37 (2002)
11. N.L. Peterson, *Isotope Effects in Diffusion*, in: *Diffusion in Solids – Recent Developments*, A.S. Nowick, J.J. Burton (Eds.), Academic Press 1975, p.116
12. S.J. Rothman, *The Measurement of Tracer Diffusion Coefficients in Solids*, in: *Diffusion in Crystalline Solids*, G.E. Murch, A.S. Nowick (Eds.), Academic Press, (1984), p. 1
13. H. Mehrer, N. Stolica, *Mass and Pressure Dependence of Diffusion in Solid Metals and Alloys*, Chap. 10, in: *Diffusion in Solid Metals and Alloys*, H. Mehrer (Vol. Ed.), Landolt-Börnstein, Numerical Data and Functional Relationships in Science and Technology, New Series, Group III: Crystal and Solid State Physics, Vol.26, Springer-Verlag, 199, p.574
14. Chr. Herzig, H. Eckseler, W. Bussmann, D. Cardis, J. Nucl. Mater. **69/70**, 61 (1978)
15. F. Faupel, W. Frank, H.-P. Macht, H. Mehrer, V. Naundorf, K. Rätzke, H. Schober, S. Sharma, H. Teichler, *Diffusion in Metallic Glasses and Supercooled Melts*, Review of Modern Physics **75**, 237 (2003)
16. F. Faupel, K. Rätzke, *Diffusion in Metallic Glasses and Supercooled Melts*, in: *Diffusion in Condensed Matter – Methods, Materials, Models*, P. Heitjans, J. Kärger (Eds.), Springer-Verlag, 2005

10 Interdiffusion and Kirkendall Effect

Diffusion processes in alloys with composition gradients are of great practical interest. In the preceding chapters, we have assumed that the concentration gradient is the only cause for flow of matter. Such situations can be studied using small amounts of trace elements in otherwise homogeneous materials. We will discuss the experimental procedure for tracer experiments in Chap. 13. From a general viewpoint, the diffusion flux is proportional to the gradient of the *chemical potential*. The chemical potential is proportional to the concentration gradient only for dilute systems or ideal solid solutions. The gradient of the chemical potential gives rise to an 'internal' driving force and the intermixing of a binary A-B system can be described by a concentration-dependent *chemical* or *interdiffusion coefficient*. In a binary alloy there is a single interdiffusion coefficient that characterises interdiffusion. The interdiffusion coefficient is usually a composition-dependent quantity. On the other hand, interdiffusion is due to the diffusive motion of A and B atoms, which in general have different *intrinsic diffusion coefficients*. This difference manifests itself in the *Kirkendall effect*, a shift of the diffusion zone with respect to the ends of the diffusion couple. We consider first interdiffusion and the Boltzmann-Matano and Sauer-Freise methods for the determination of the interdiffusion coefficient. Further sections are devoted to intrinsic diffusion, Kirkendall effect, and to the Darken relations. The Darken-Manning relations, the so-called vacancy-wind effect, and the stability or instability of Kirkendall planes are described. A discussion of interdiffusion in ionic systems and the Nernst-Planck equation and its relation to the Darken equation is postponed to the end of the next chapter.

10.1 Interdiffusion

Let us consider a binary diffusion couple, in which the chemical composition varies in the diffusion zone over a certain range. Diffusing atoms then experience different chemical environments and hence have different diffusion coefficients. As already mentioned in Chap. 2, this situation is called *interdiffusion* or *chemical diffusion*. We use the symbol $\tilde{D}$ to indicate that the diffusion coefficient is concentration-dependent and call it the *interdiffusion or chemical diffusion coefficient*. Fick's second law Eq. (2.5) then reads

$$\frac{\partial C}{\partial t} = \frac{\partial}{\partial x}\left[\tilde{D}(C)\frac{\partial C}{\partial x}\right] = \tilde{D}(C)\frac{\partial^2 C}{\partial x^2} + \frac{\mathrm{d}\tilde{D}(C)}{\mathrm{d}C}\left(\frac{\partial C}{\partial x}\right)^2. \quad (10.1)$$

The second term on the right-hand side represents an *'internal driving force'* (see also Chap. 11). Mathematically, Eq. (10.1) is a non-linear partial differential equation. For an arbitrary concentration dependence of $\tilde{D}(C)$ it can usually be not solved analytically. In addition, theoretical models which permit the calculation of the composition-dependent diffusivity from deeper principles are at present not broadly available.

The strategy illustrated in Chaps. 2 and 3 for calculating the concentration field for given initial and boundary conditions is not applicable to interdiffusion. We shall see, however, that it is possible to determine the concentration-dependent diffusivity, $\tilde{D}$, from a measured concentration field by using Eq. (10.1). Two methods for extracting diffusivities from concentration-depth profiles – the classical *Boltzmann-Matano method* and related approaches proposed by *Sauer and Freise* – are considered below. Boltzmann's transformation of Fick's second law is fundamental for both methods and is discussed first.

10.1.1 Boltzmann Transformation

In 1894 the famous LUDWIG BOLTZMANN [1] showed that the nonlinear partial differential equation (10.1) can be transformed to a nonlinear but ordinary differential equation if $\tilde{D}$ is a function of $C(x)$ alone. He introduced the variable

$$\eta \equiv \frac{x - x_M}{2\sqrt{t}}, \quad (10.2)$$

which is a combination of the space and time variables x and t, respectively. x_M refers to a special reference plane – the so-called *Matano plane* – to be defined below. Applying chain-rule differentiation to Eq. (10.1), we get the following identity:

$$\frac{\partial}{\partial x} \equiv \frac{\mathrm{d}}{\mathrm{d}\eta}\frac{\partial \eta}{\partial x} = \frac{1}{2\sqrt{t}}\frac{\mathrm{d}}{\mathrm{d}\eta}. \quad (10.3)$$

The operator on the left-hand side of Eq. (10.1) is

$$\frac{\partial}{\partial t} \equiv \frac{\mathrm{d}}{\mathrm{d}\eta}\frac{\partial \eta}{\partial t} = -\frac{x - x_M}{4t^{3/2}}\frac{\mathrm{d}}{\mathrm{d}\eta} = -\frac{\eta}{2t}\frac{\mathrm{d}}{\mathrm{d}\eta}. \quad (10.4)$$

The right-hand side of Eq. (10.1) can also be written in terms of η as

$$\frac{\partial}{\partial x}\left[\tilde{D}(C)\frac{\partial C}{\partial x}\right] = \frac{\mathrm{d}}{\mathrm{d}\eta}\frac{\partial \eta}{\partial x}\left[\frac{\tilde{D}(C)}{2\sqrt{t}}\frac{\mathrm{d}C}{\mathrm{d}\eta}\right] = \frac{1}{4t}\frac{\mathrm{d}}{\mathrm{d}\eta}\left[\tilde{D}(C)\frac{\mathrm{d}C}{\mathrm{d}\eta}\right]. \quad (10.5)$$

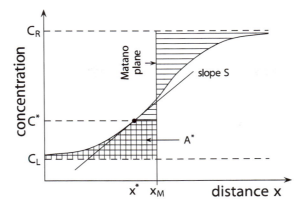

Fig. 10.1. Schematic illustration of the Boltzmann-Matano method for a binary diffusion couple with starting compositions C_L and C_R

By recombining left- and right-hand sides and using the Boltzmann variable we get Fick's second law as an ordinary differential equation for $C(\eta)$:

$$-2\eta \frac{dC}{d\eta} = \frac{d}{d\eta}\left[\tilde{D}(C)\frac{dC}{d\eta}\right]. \qquad (10.6)$$

Some authors omit the factor 2 in the definition Eq. (10.2) of η. Then, a factor of 1/2 instead of 2 appears in the equation corresponding to Eq. (10.6). However, when finally transformed in ordinary time and space coordinates, the solutions obtained are identical.

10.1.2 Boltzmann-Matano Method

The Boltzmann-transformed version of Fick's second law Eq. (10.6) is a non-linear ordinary differential equation. This equation allows us to deduce the concentration-dependent interdiffusion coefficient from an experimental concentration-depth profile, $C(x)$. Appropriate boundary conditions for an interdiffusion experiment have been suggested by the Japanese scientist MATANO in 1933 [2]. He considered a binary diffusion couple, which consists of two semi-infinite bars joined at time $t = 0$. The initial conditions are

$$\begin{aligned} C &= C_L \quad \text{for} \quad (x < 0,\ t = 0), \\ C &= C_R \quad \text{for} \quad (x > 0,\ t = 0). \end{aligned} \qquad (10.7)$$

During a diffusion anneal of time t, a concentration profile $C(x)$ develops. This profile can be measured on a cross section of the diffusion zone, e.g., by electron microprobe analysis (see Chap. 13). Such a profile is schematically illustrated in Fig. 10.1.

Carrying out an integration between C_L and a fixed concentration C^*, we get from Eq. (10.6)

$$-2\int_{C_L}^{C^*} \eta\, dC = \tilde{D}\left(\frac{dC}{d\eta}\right)_{C^*} - \tilde{D}\left(\frac{dC}{d\eta}\right)_{C_L}. \qquad (10.8)$$

Matano's geometry guarantees vanishing gradients $dC/d\eta$ as C^* approaches C_L (or C_R). Using $(dC/d\eta)_{C_L} = 0$ and solving Eq. (10.8) for $\tilde{D}$ yields

$$\tilde{D}(C^*) = -2\frac{\int_{C_L}^{C^*} \eta\, dC}{(dC/d\eta)_{C=C^*}}. \qquad (10.9)$$

We transform Eq. (10.9) back to space and time coordinates using the Boltzmann variable Eq. (10.2) and get

$$\tilde{D}(C^*) = -\frac{1}{2t}\frac{\int_{C_L}^{C^*} (x - x_M)\, dC}{(dC/dx)_{C^*}}. \qquad (10.10)$$

This relation is called the *Boltzmann-Matano equation*. It permits us to determine $\tilde{D}$ for any C^* from an experimental concentration-distance profile. For the analysis, the position of the *Matano plane*, x_M, must be known. Carrying out the integration between the limits C_L and C_R, we get from Eq. (10.6)

$$\int_{C_L}^{C_R} \eta\, dC = 0. \qquad (10.11)$$

Equation (10.11) can be considered as the definition of the Matano plane. x_M must be chosen in such a way that Eq. (10.11) is fulfilled.

In order to determine the Matano plane, we have to remember that at the beginning of the experiment the concentration of the diffusing species was C_L (C_R) on the left-hand (right-hand) side. Let us suppose, for example, $C_L < C_R$. Then, at the end of the experiment, the surplus of the diffusing species found on the left-hand side must have arrived by diffusion from the right-hand side. The location of the Matano plane can be determined from the conservation condition

$$\underbrace{\int_{-\infty}^{x_M} [C(x) - C_L]\, dx}_{\text{gain}} = \underbrace{\int_{x_M}^{\infty} [C_R - C(x)]\, dx}_{\text{loss}}. \qquad (10.12)$$

When integrated by parts, the integrals in Eq. (10.12) transform to integrals with C as the running variable instead of x. If we apply the Matano boundary conditions Eq. (10.7), we get

$$(C_L - C_R)x_M + \int_{C_L}^{C_M} x\, dC + \int_{C_M}^{C_R} x\, dC = 0, \qquad (10.13)$$

where C_M denotes the concentration at the Matano plane. If we choose the Matano plane as origin of the x-axis ($x_M = 0$), the first term in Eq. (10.13) vanishes. Then, the following integrals balance across the Matano plane:

$$\int_{C_R}^{C_M} x\mathrm{d}C + \int_{C_M}^{C_L} x\mathrm{d}C = 0. \tag{10.14}$$

Although the location of the Matano plane is not known a priori, it can be found from experimental concentration-distance data by balancing the horizontally hatched areas in Fig. 10.1.

In summary, the determination of interdiffusion coefficients from an experimental concentration-distance profile via the Boltzmann-Matano method requires the following steps:

1. Determine the position of the Matano plane from Eq. (10.11) and use this position as the origin of the x-axis.
2. Choose C^* and determine the integral $\int_{C_L}^{C^*} x\mathrm{d}C$ from the experimental concentration-distance data. The integral corresponds to the double-hatched area A^* in Fig. 10.1.
3. Determine the concentration gradient $S = (\mathrm{d}C/\mathrm{d}x)_{C^*}$. S corresponds to the slope of the concentration-distance curve at the position x^*.
4. Determine the interdiffusion coefficient $\tilde{D}$ for $C = C^*$ from the Boltzmann-Metano equation (10.10) as: $\tilde{D}(C^*) = -A^*/(2tS)$.

We draw the readers attention to the following points:

(i) The Boltzmann-Matano equation (10.10) refers to an 'infinite' system. Its application to an experiment requires that the concentration changes must not have reached the boundaries of the system.
(ii) Close to the end-member compositions, the Boltzmann-Matano procedure may incur relatively large errors in $\tilde{D}$ because far away from the Matano plane both the integral A^* and the slope S become very small.
(iii) The initial interface of a diffusion couple can be tagged by inert diffusion markers (e.g., ThO$_2$ particles, thin Mo or W wires). The plane of the markers in the diffusion couple is denoted as the *Kirkendall plane*. Usually, for $t \neq 0$ the positions of the Matano plane and of the Kirkendall plane will be different. This is called the *Kirkendall effect* and is discussed in Sect. 10.2.
(iv) This method is applicable when the volume of the diffusion couple does not change during interdiffusion.

The Boltzmann-Matano method has been modified by SAUER AND FREISE [3] and later by DEN BROEDER [4]. These authors introduce a normalised concentration variable Y defined by:

$$Y = \frac{C - C_R}{C_L - C_R}. \tag{10.15}$$

If no volume change occurs upon interdiffusion, the Sauer-Freise solution can be written in the following way:

$$\tilde{D}(C^*) = \frac{1}{2t(dC/dx)_{x^*}} \left[(1-Y) \int_{x^*}^{\infty} (C^* - C_R) dx + Y \int_{-\infty}^{x^*} (C_L - C^*) dx \right]. \tag{10.16}$$

Here C^* is the concentration at the position x^*. The Sauer-Freise approach circumvents the need to locate the Matano plane. In this way, errors associated with finding its position are eliminated. On the other hand, application of Eq. (10.16) to the analysis of an experimental interdiffusion profile, like the Boltzmann-Matano method, requires the computation of two integrals and of one slope.

10.1.3 Sauer-Freise Method

When the volume of a diffusion couple changes during interdiffusion neither the Boltzmann-Matano equation (10.10) nor Eq. (10.16) can be used. Fick's law then needs a correction term [5, 6]. Volume changes in a binary diffusion couple occur whenever the total molar volume V_m of an A-B alloy deviates from *Vegard's rule*, which states that the total molar volume of the alloy is obtained from $V_m = V_A N_A + V_B N_B$, where V_A, V_B denote the molar volumes of the pure components and N_A, N_B the molar fractions of A and B in the alloy. Vegard's rule is illustrated by the dashed line in Fig. 10.2.

Non-ideal solid solution alloys exhibit deviations from Vegard's rule, as indicated by the solid line in Fig. 10.2. Diffusion couples of such alloys change their volume during interdiffusion. Couples with positive deviations from Vegard's rule swell, couples with negative deviations shrink. The partial molar volumes of the components A and B, $\tilde{V}_A \equiv \partial V_m / \partial N_A$ and $\tilde{V}_B \equiv \partial V_m / \partial N_B$, are related to the total molar volume via:

$$V_m = \tilde{V}_A N_A + \tilde{V}_B N_B. \tag{10.17}$$

As indicated in Fig. 10.2, the partial molar volumes can be obtained graphically as intersections of the relevant tangent with the ordinate.

SAUER AND FREISE [3] deduced a solution for interdiffusion with volume changes. Instead of Eq. (10.15), they introduced the ratio of the mole fractions

$$Y = \frac{N_i - N_i^R}{N_i^L - N_i^R}, \tag{10.18}$$

with N_i^L and N_i^R being the unreacted mole fractions of component i at the left-hand or right-hand side of the diffusion couple. The interdiffusion coefficient $\tilde{D}$ is then obtained from

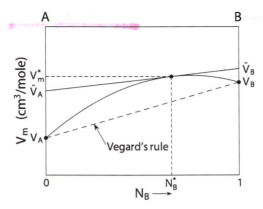

Fig. 10.2. Molar volume of an A-B solid solution alloy (*solid line*) versus composition. The *dashed line* repesents the *Vegard rule*. The partial molar volumes, $\tilde{V}_A$ and $\tilde{V}_B$, and the molar volumes of the pure components, V_A and V_B, are also indicated

$$\tilde{D}(Y^*) = \frac{V_m}{2t(dY/dx)_{x^*}} \left[(1-Y^*) \int_{-\infty}^{x^*} \frac{Y}{V_m} dx + Y^* \int_{x^*}^{+\infty} \frac{1-Y}{V_m} dx \right]. \quad (10.19)$$

In order to evaluate Eq. (10.19), it is convenient to construct from the experimental composition-distance profile and from the V_m data two graphs, namely the integrands Y/V_m and $(1-Y)/V_m$ versus x, as illustrated in Fig. 10.3. The two integrals in Eq. (10.19) correspond to the hatched areas. Equations (10.19) and (10.16) contain two infinite integrals in the running variable. Their application to the analysis of an experimental concentration-depth profile requires accurate computation of a gradient and of two integrals.

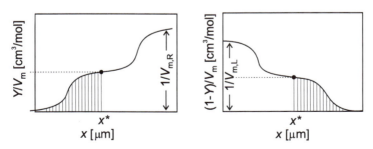

Fig. 10.3. Composition profiles constructed according to the Sauer-Freise method. $V_{m,L}$ and $V_{m,R}$ are the molar volumes of the left-hand and right-hand end-members of the diffusion couple

10.2 Intrinsic Diffusion and Kirkendall Effect

So far, we have described diffusion of a two-component system by a single interdiffusion coefficient, which depends on composition. In general, the rate of transfer of A atoms is greater/smaller than that of B atoms. Thus, there are two diffusion coefficients, D_A^I and D_B^I, which are denoted as the *intrinsic diffusion coefficients* of the components. They are concentration dependent as well. On the other hand, there is only one diffusion process, namely the intermixing of A and B. These two apparently contradictory facts are closely related to the question of how we specify the reference frame for the diffusion process. We know that the atoms in a crystalline solid are held in a lattice structure and we shall therefore retain the form of Fick's first law for the diffusion fluxes relative to a frame fixed in the local crystal lattice (intrinsic diffusion fluxes):

$$j_A = -D_A^I \frac{\partial C_A}{\partial x}, \quad j_B = -D_B^I \frac{\partial C_B}{\partial x}. \tag{10.20}$$

The inequality of these fluxes leads to a net mass flow accompanying the interdiffusion process, which causes the diffusion couple to shrink on one side and to swell on the other side. This observation is called the *Kirkendall effect*. It was discovered by KIRKENDALL AND COWORKERS in a copper-brass diffusion couple in the 1940s [7, 8]. The Kirkendall shift can be observed by incorporating inert inclusions, called markers (e.g., Mo or W wires, ThO_2 particles), at the interface where the diffusion couple is initially joined. The original Kirkendall experiment is illustrated in Fig. 10.4. It showed that Zn atoms diffused faster outwards than Cu atoms inward ($D_{Zn}^I > D_{Cu}^I$) causing the inner brass core to shrink. This in turn resulted in the movement of the inert Mo wires. In the period since, it has been demonstrated that the Kirkendall effect is a widespread phenomenon of interdiffusion in substitutional alloys.

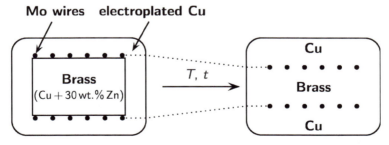

Fig. 10.4. Schematic illustration of a cross section of a diffusion couple composed of pure Cu and brass (Cu-Zn) prepared by SMIGELSKAS AND KIRKENDALL [8] before and after heat treatment. The Mo markers placed at the original contact surface moved towards each other. It was concluded that Zn atoms diffused faster outwards than Cu atoms move inwards ($D_{Zn}^I > D_{Cu}^I$)

10.2 Intrinsic Diffusion and Kirkendall Effect

The Kirkendall effect was received by contemporary scientists with much surprise. Before the 1940s it was commonly believed that diffusion in solids takes place via direct exchange or ring mechanism (see Chap. 6), which imply that the diffusivities of both components of a binary alloy are equal. The fact that in a solid-state diffusion process the species diffuse at different rates changed the existing atomistic models on solid-state diffusion completely. The Kirkendall effect lended much support to the vacancy mechanism of diffusion[1].

The position of the Kirkendall plane, x_K, moves parabolically in time with respect to the laboratory-fixed frame:

$$x_K = K\sqrt{t}. \tag{10.21}$$

Here K is a (temperature-dependent) constant. The parabolic shift indicates that we are dealing with a diffusion-controlled process. We also note that the Kirkendall plane is the only marker plane that starts moving from the beginning. The *Kirkendall velocity* v_K is given by

$$v_K \equiv \frac{dx_K}{dt} = \frac{x_K}{2t} \tag{10.22}$$

From the position of the Kirkendall plane one can deduce information about the intrinsic diffusivities. VAN LOO showed that their ratio is given by [10]:

$$\frac{D_A^I}{D_B^I} = \frac{\tilde{V}_A}{\tilde{V}_B} \left[\frac{N_A^R \int_{-\infty}^{x_K} \frac{1}{V_m}(N_A - N_A^L)dx - N_A^L \int_{x_K}^{\infty} \frac{1}{V_m}(N_A^R - N_A)dx}{-(N_B^R) \int_{-\infty}^{x_K} \frac{1}{V_m}(N_A - N_A^L)dx + N_B^L \int_{x_K}^{\infty} \frac{1}{V_m}(N_A^R - N_A)dx} \right]. \tag{10.23}$$

N_i is the mole fraction of component i, with N_i^L and N_i^R the unreacted left-hand ($x \to -\infty$) and right-hand ($x \to -\infty$) ends of the couple, respectively.

Since the discovery of the Kirkendall effect by SMIGELSKAS AND KIRKENDALL [8] and its analysis by DARKEN [9], this effect assumed a prominent rôle in the diffusion theory of metals. It was considered as evidence for vacancy-mediated diffusion in solids. There are also technological fields in which the Kirkendall effect is of great interest. Examples are composite materials, coating technologies, microelectronic devices, etc. The interactions accompanying the Kirkendall effect between constituents of such structures can, for example, induce stress and even deformation on a macroscopic scale. It can also cause migration of microscopic inclusions inside a reaction zone and Kirkendall porosity.

---

[1] Nowadays, we know that the Kirkendall effect can manifest itself in many phenomena such as the development of diffusional porosity (Kirkendall voids), generation of internal stresses [13, 14], and even by deformation of the material on a macroscopic scale [15]. These diffusion-induced processes are of concern in a wide variety of structures including composite materials, coatings, welded components, and thin-film electronic devices.

10.3 Darken Equations

The first theoretical desciption of interdiffusion and Kirkendall effect was attempted by DARKEN in 1948 [9]. For a binary substitutional alloy he used the two intrinsic diffusivities introduced above to describe the interdiffusion process. The Kirkendall velocity v_K can be expressed in terms of the intrinsic fluxes, j_A and j_B, and partial molar volumes, $\tilde{V}_A$ and $\tilde{V}_B$, as

$$v_K = -(\tilde{V}_A j_A + \tilde{V}_B j_B). \tag{10.24}$$

Given the fact that $dC_A = -(\tilde{V}_B/\tilde{V}_A)dC_B$, we can write for the Kirkendall velocity

$$v_K = \tilde{V}_B (D_B^I - D_A^I)\frac{\partial C_B}{\partial x}, \tag{10.25}$$

where $\partial C_B/\partial x$ denotes the concentration gradient at the Kirkendall plane. Following Darken's approach, the laboratory-fixed interdiffusion flux J (at the Kirkendall plane) can be written as the sum of an intrinsic diffusion flux of one of the components i plus (or minus) a Kirkendall drift term $v_K C_i$:

$$J = -D_i^I \frac{\partial C_i}{\partial x} \pm v_K C_i \quad i = A, B. \tag{10.26}$$

Substituting Eq. (10.25) in Eq. (10.26) one arrives at a general expression for the interdiffusion coefficient:

$$\tilde{D} = C_B \tilde{V}_B D_A^I + C_A \tilde{V}_A D_B^I. \tag{10.27}$$

Equations (10.25) and (10.27) provide a description of isothermal diffusion in a binary substitutional alloy. They also provide a possibility to determine the intrinsic diffusivities from measurements of the interdiffusion coefficient and the Kirkendall velocity.

From a fundamental point of view, the assumption that the concentration gradients are the driving forces of diffusion as given by Fick's laws is not correct. Instead, as already stated at the beginning of this chapter, the gradient of the chemical potential μ_i of component i is the real driving force. The flux of component i ($i = A, B$) in a binary alloy can be written as [16, 17]

$$j_i = -B_i C_i \frac{\partial \mu_i}{\partial x}, \tag{10.28}$$

where B_i denotes the mobility of component i. The chemical potential can be expressed in terms of the thermodynamic activity, a_i, via

$$\mu_i = \mu_i^0 + RT \ln a_i, \tag{10.29}$$

where μ_i^0 is the standard chemical potential (at 298 K and 1 bar) and R is the ideal gas constant ($R = 8.3143$ J mol^{-1}K^{-1}). The atomic mobility B_i

is connected to the tracer diffusion coefficient D_i^* of component i via the Nernst-Einstein relation (see Chap. 11):

$$D_i^* = B_i RT. \tag{10.30}$$

Substituting Eqs. (10.30) and (10.28) in Eq. (10.20) and knowing that $C_A = N_A/V_m$ and $dN_A = (V_m^2/\tilde{V}_A)dC_A$ one obtains relations between the intrinsic and the tracer diffusion coefficients:

$$D_A^I = D_A^* \frac{V_m}{\tilde{V}_B} \frac{\partial \ln a_A}{\partial \ln N_A} \quad \text{and} \quad D_B^I = D_B^* \frac{V_m}{\tilde{V}_A} \frac{\partial \ln a_B}{\partial \ln N_B}. \tag{10.31}$$

The quantity $\Phi \equiv \partial \ln a_i / \partial \ln N_i$ is denoted as the *thermodynamic factor*.

The thermodynamics of binary systems tells us that the thermodynamic factor can also be expressed as follows [21]:

$$\Phi = \frac{N_A N_B}{RT} \frac{d^2 G}{dN_i^2} = \frac{\partial \ln a_i}{\partial \ln N_i} = 1 + \frac{\partial \ln \gamma_i}{\partial \ln N_i}, \tag{10.32}$$

Here G denotes the Gibbs free energy and $\gamma_i \equiv a_i/N_i$ the coefficient of thermodynamic activity of species i (= A or B). In addition, as a consequence of the Gibbs-Duhem relation there is only one thermodynamic factor for a binary alloy:

$$\Phi = \frac{\partial \ln a_A}{\partial \ln N_A} = \frac{\partial \ln a_B}{\partial \ln N_B} \tag{10.33}$$

Substiting Eq. (10.31) in Eq. (10.27) and knowing the relation $C_i \equiv N_i(C_A + C_B) = N_i/V_m$ between concentrations and mole fractions, we obtain for the interdiffusion coefficient

$$\tilde{D}_{Darken} = (N_A D_B^* + N_B D_A^*)\, \Phi. \tag{10.34}$$

Equations (10.31) and (10.34) are called the *Darken equations*. Sometimes the name *Darken-Dehlinger equations* is used. These relations are widely used in practice for substitutional binary alloys. Their simplicity provides an obvious convenience. We shall assess its accuracy below and also in Chap. 12.

If thermodynamic data are available, either from activity measurements or from theoretical models, Eqs. (10.31) or (10.34) allow to relate the intrinsic diffusivities and the interdiffusion coefficient to the tracer diffusivities. For an ideal solid solution alloy we have $\gamma_i = 1$ and $a_i = N_i$ and hence $\Phi = 1$ (Raoult's law). For non-ideal solutions Φ deviates from unity. It is larger than unity for phases with negative deviations of G from ideality and smaller than unity in the opposite case. Negative deviations are expected for systems with order. Therefore, thermodynamic factors of intermetallic compounds are often larger, sometimes even considerably larger than unity due to the attractive interaction between the constituents of the intermetallic phase. As a consequence, interdiffusion coefficients are often larger than the term in paranthesis of Eq. (10.34).

10.4 Darken-Manning Equations

Soon after the detection of the Kirkendall effect and its phenomenological description via the Darken equations, SEITZ in 1948 [11] and BARDEEN in 1949 [12] recognised that the original Darken equations are approximations. From an atomistic point of view, interdiffusion in substitutional alloys is mediated by vacancies. It was shown that the Darken equations are obtained if the concentration of vacancies is in thermal equilibrium during the interdiffusion process. For the Kirkendall effect to occur, vacancies must be created at one site and annihilated on the other side of the interdiffusion zone so that a vacancy flux is created to maintain local equilibrium. This implies that sources and sinks for vacancies are abundant in the diffusion couple. The vacancy flux causes the so-called vacancy-wind effect – a correction term that must be added to the original Darken equations to obtain the Darken-Manning relations.

The relations between tracer diffusivities and intrinsic diffusivities and the interdiffusion coefficient discussed in the previous section are incomplete for a vacancy mechanism, because of correlation effects. The exact expressions are similar to those discussed above but with *vacancy-wind factors* (see, e.g., [17, 21, 22]). The intrinsic diffusion coefficients Eq. (10.31) with vacancy-wind corrections, r_A and r_B, can be written as

$$D_A^I = D_A^* \frac{V_m}{\tilde{V}_B} \Phi r_A \quad \text{and} \quad D_B^I = D_B^* \frac{V_m}{\tilde{V}_A} \Phi r_B. \tag{10.35}$$

The vacancy-wind factors can be expressed in terms of the tracer and collective correlation factors:

$$r_A = \frac{f_{AA} - N_A f_{AB}^{(A)}/N_B}{f_A} \quad \text{and} \quad r_B = \frac{f_{BB} - N_A f_{AB}^{(B)}/N_A}{f_A}. \tag{10.36}$$

The f_i are the tracer correlation factors and f_{ij} the collective correlation factors sometimes also called correlations functions [22] (see also Chap. 12).

Perhaps the best-known vacancy-wind factor is the total vacancy-wind factor, S, occuring in the generalised Darken equation:

$$\tilde{D} = (N_A D_B^* + N_B D_A^*) \Phi S = \tilde{D}_{Darken} S. \tag{10.37}$$

Equation (10.37) is also called the *Darken-Manning equation* and S is also denoted as the *Manning factor*. It can be expressed as

$$S = \frac{N_A D_B^* r_B + N_B D_A^* r_A}{N_A D_B^* + N_B D_A^*}. \tag{10.38}$$

MANNING [18, 19] developed approximate expressions for vacancy-wind factors in the framework of a model called the *random alloy model*. The term random alloy implies that vacancies and A and B atoms are distributed at random on the same lattice, although the rates at which atoms exchange with vacancies are allowed to be different. For a random alloy, the individual vacancy-wind factors are

$$r_A = 1 + \frac{(1-f)}{f} \frac{N_A(D_A^* - D_B^*)}{(N_A D_A^* + N_B D_B^*)} \quad \text{and}$$

$$r_B = 1 + \frac{(1-f)}{f} \frac{N_A(D_B^* - D_A^*)}{(N_A D_A^* + N_B D_B^*)}, \quad (10.39)$$

where f is the tracer correlation factor for self-diffusion. A transparent derivation of Eq. (10.39) can be found in [20]. For convenience let us assume that $D_A^* \geq D_B^*$. Then, from these expressions we see that the factors r_A and r_B take the limits

$$1.0 \leq r_A \leq \frac{1}{f} \quad \text{and} \quad 0.0 \leq r_B \leq 1.0. \quad (10.40)$$

There is also a 'forbidden region' $N_A \leq 1 - f$, where r_B can take negative values (unphysical for this model) if $D_A^*/D_B^* > N_B/(N_B - f)$. In other words, there is a concentration-dependent upper limit for the ratio of the tracer diffusivities in this region. Manning also provides an expression for the total vacancy-wind factor:

$$S = 1 + \frac{(1-f)}{f} \frac{N_A N_B (D_A^* - D_B^*)^2}{(N_A D_A^* + N_B D_B^*)(N_A D_B^* + N_B D_A^*)}. \quad (10.41)$$

From Eq. (10.41) it is seen that S varies within narrow limits:

$$1 \leq S \leq 1/f. \quad (10.42)$$

Thus, in the framework of the random alloy model the total vacancy-wind factor S is not much different from unity. The Manning expressions for the vacancy-wind factors have been used for some 30 years. Extensive computer simulations studies in simple cubic, fcc, and bcc random alloys by BELOVA AND MURCH [23] have shown that the Manning formalism is not as accurate as commonly thought. It is, however, a reasonable approximation when the ratio of the atom vacancy exchange rates are not too far from unity.

Vacancy-wind corrections for chemical diffusion in intermetallic compounds depend on the structure, the type of disorder and on the diffusion mechanism. BELOVA AND MURCH have also contributed significantly to chemical diffusion in ordered alloys by considering among others L1$_2$ structured compounds [24], D0$_3$ and A15 structured alloys [26], and B2 structured compounds [25].

10.5 Microstructural Stability of the Kirkendall Plane

Kirkendall effect induced migration of inert markers inside the diffusion zone and the uniqueness of the Kirkendall plane have not been questioned for quite a long time. In recent years, the elucidation of the Kirkendall effect accompanying interdiffusion has taken an additional direction. CORNET AND CALAIS [29] were the first to describe hypothetical diffusion couples in which more than one 'Kirkendall marker plane' can emerge. Experimental discoveries also revealed a more complex behaviour of inert markers situated at the original interface of a diffusion couple in both spatial and temporal domains. Systematic studies of the microstructural stability of the Kirkendall plane were undertaken by VAN LOO AND COWORKERS [30–35]. Clear evidence for the ideas of CORNET AND CALAIS was found and led to further developments in the understanding of the Kirkendall effect. In particular, it was found that the Kirkendall plane, under predictable circumstances, can be multiple, stable, or unstable.

The diffusion process in a binary A-B alloy can best be visualised by considering the intrinsic fluxes, j_A and j_B, of the components in Eq. (10.20) with respect to an array of inert markers positioned prior to annealing along the anticipated diffusion zone. According to Eq. (10.25) the sum of the oppositely directed fluxes of the components is equal to the velocity of the inert markers, v, with respect to the laboratory-fixed frame of reference:

$$v = \tilde{V}_B(D_B^I - D_A^I)\frac{\partial C_B}{\partial x}, \qquad (10.43)$$

with $\tilde{V}_B$ being the partial molar volume of component B. Multifoil experiments, in which a diffusion couple is composed of many thin foils with markers at each interface, permit a determination of v at many positions inside a diffusion couple. Thus a v versus x curve (marker-velocity curve) can be determined experimentally.

In a diffusion-controlled intermixing process, those inert markers placed at the interface where the concentration step is located in the diffusion couple is the *Kirkendall plane*. The markers in the Kirkendall plane are the only ones that stay at a constant composition and move parabolically with a velocity given by Eq. (10.22), which we repeat for convenience:

$$v_K = \frac{dx}{dt} = \frac{x_K}{2t}. \qquad (10.44)$$

x_K is the position of the Kirkendall plane at time t.

The location of the Kirkendall plane(s) in the diffusion zone can be found graphically at the point(s) of intersection(s) between the marker-velocity curve and the straight line given by Eq. (10.44) (see Fig. 10.5). In order to draw the line $v_K = x_K/2t$, one needs to know the position in the diffusion zone where the 'Kirkendall markers' were located at time $t = 0$. If the total volume does not change during interdiffusion this position can be determined via the usual Boltzmann-Matano analysis. If the partial molar volumes are composition dependent, the Sauer-Freise method should be used.

The nature of the Kirkendall plane(s) in a diffusion couple depends on the gradient of the marker-velocity curve at the point of intersection with the

10.5 Microstructural Stability of the Kirkendall Plane

straight line $x_K/2t$. For illustration, let us consider a hypothetical diffusion couple of A-B alloys with the end-members A_yB_{1-y} and A_zB_{1-z} where $y > z$. Let us suppose that on the A-rich side of the diffusion zone A is the faster diffusing species, whereas on the B-rich side B is the faster diffusing species. Figure 10.5 shows schematic representations of the marker-velocity curves in different situations. For some diffusion couples the straight line, $v_K = x/2t$, may intersect the marker-velocity curve in the diffusion zone once at a point with a negative gradient (upper part). At this point of intersection one can expect *one stable Kirkendall plane*. Markers, which by some perturbation

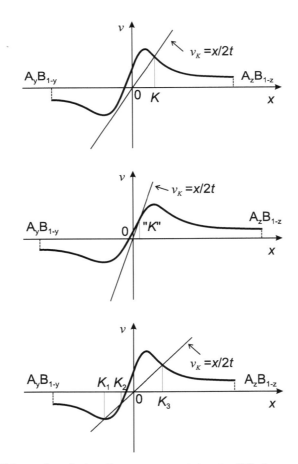

Fig. 10.5. Schematic velocity diagrams, pertaining to diffusion couples between the end-members A_yB_{1-y} and A_zB_{1-z} for $y > z$. On the A-rich side A diffuses faster and on the B-rich side B diffuses faster. Different situations are shown, which pertain to one stable Kirkendall plane (*upper part*), to an unstable plane (*middle part*), and to two stable Kirkendall planes, K_1 and K_3, and an unstable plane K_2

move ahead of the Kirkendall plane, will slow down, because of the lower velocity; markers behind this plane will move faster. The stable Kirkendall plane acts as an 'attractor for inert markers'. By changing the end-member compositions the straight line, $v_K = x/2t$, may intersect the marker-velocity curve at a point with a positive velocity gradient (middle part). Markers slightly ahead of this plan will move faster, whereas markers behind this plane will move slower. This will result in scatter of the markers and there will be no unique plane acting as the Kirkendall plane (*unstable Kirkendall plane*). The lower part of Fig. 10.5 illustrates a situation where the straight line intersects the marker-velocity curve three times at K_1, K_2, and K_3. In this case one might expect that three Kirkendall planes will be present in the sample. In reality, one finds *two stable Kirkendall planes*, K_1 and K_3. An unstable plane, K_2, is located between two stable Kirkendall planes and the stable planes will accumulate the markers during the initial stage of interdiffusion.

The presence of stable and unstable Kirkendall planes has been verified, for example, in Ni-Pd and Fe-Pd diffusion couples [31]. The marker-velocity curves over the whole homogeneity range have been determined in multifoil experiments. It was indeed found that for Ni-Pd a stable Kirkendall plane is present and the straight line, $v_K = x/2t$, intersects the marker-velocity curve at a point with a negative gradient. An unstable Kirkendall plane is found in Fe-Pd and the gradient of the marker-velocity curve is positive at the intersection point.

References

1. L. Boltzmann, Wiedemanns Ann. Phys. **53** (1894) 959
2. C. Matano, Japan. J. Phys. **8** (1933) 109
3. F. Sauer and V. Freise, Z. Elektrochem. **66** (1962) 353
4. F.J.A. den Broeder, Scr. Metall. **3** (1969) 321
5. K. Wagner, Acta Metall. **17** (1969) 99
6. F.F.J. van Loo, Acta Metall. **18** (1970) 1107
7. E.O. Kirkendall, Trans. AIME **147**, 104 (1942)
8. A.D. Smigelskas, E.O. Kirkendall, Trans. AIME **171**, 130 (1947)
9. L.S. Darken, Transactions AIME **175**, 184 (1948)
10. F.J.J. van Loo, Progr. Solid State Chemistry **20**, 47 (1990)
11. F. Seitz, Phys. Rev. **74**, 1513 (1948)
12. J. Bardeen, Phys. Rev. **76**, 1403 (1949)
13. G.B. Stephenson, Acta Metall. **36**, 2663 (1988)
14. D.L. Beke, I.A. Szabo (Eds.), Proc. of Int. Workshop on *Diffusion and Stresses*, Balatonfüred, Hungary, May 1995; also: Defect and Diffusion Forum **129–130** (1996)
15. I. Daruka, I.A. Szabo, D.L. Beke, C. Cerhati, A.A. Kodentsov, F.J.J. van Loo, Acta Mater. **44**, 4981 (1996)
16. S. Prager, J. Chem. Phys. **21** 1344–1347 (1953)

References 177

17. A.R. Allnatt, A.B. Lidiard, *Atomic Transport in Solids*, Cambridge Univerity Press, 1993
18. J.R. Manning, *Diffusion Kinetics for Atoms in Crystals*, van Norstrand, Princeton, 1968
19. J.R. Manning, Acta Metall. **15**, 817 (1967)
20. R.E. Howard, A.B. Lidiard, *Matter Transport in Solids*, in: Reports on Progress in Physics, Vol. XXVII, The Institute of Physics and the Physical Society, London, 1964, pp. 161–240
21. J. Philibert, *Atom Movements – Diffusion and Mass Transport in Solids*, Les Editions de Physique, Les Ulis, 1991
22. G.E. Murch, Z. Qin, Defect and Diffusion Forum **109**, 1 (1994)
23. I.V. Belova, G.E. Murch, Philos. Mag. **A 80**, 1469–1479 (2000); Philos. Mag. **A 81**, 1749–1758 (2001)
24. I.V. Belova, G.E. Murch, Philos. Mag, **A 78**, 1085–1092 (1998)
25. I.V. Belova, G.E. Murch, Philos. Mag, **A 79**, 193–202 (1999)
26. I.V. Belova, G.E. Murch, J. Phys. Chem. Sol. **60**, 2023–2029 (1998)
27. I.V. Belova, G.E. Murch, Philos. Mag, **A 81**, 83–94 (2001)
28. I.V. Belova, G.E. Murch, Defect and Diffusion Forum **194–199**, 533–540 (2001)
29. J.-F. Cornet, D. Calais, J. Phys. Chem. Sol. **33**, 1675 (1972)
30. F.J.J. van Loo, B. Pieragi, R.A. Rapp, Acta Metall. Mater. **38**, 1769 (1990)
31. M.J.H. van Dal, M.C.I.P. Pleumeckers, A. A. Kodentsov, F.J.J. van Loo, Acta Mater. **48**, 385 (2000)
32. M.J.H. van Dal, A.M. Gusak, C. Cerhati, A. A. Kodentsov, F.J.J. van Loo, Phys. Rev. Lett. **86**, 3352 (2001)
33. M.J.H. van Dal, A.M. Gusak, C. Cerhati, A. A. Kodentsov, F.J.J. van Loo, Philos. Mag. A **82**, 943 (2002)
34. A. Paul, M.J.H. van Dal, A. A. Kodentsov, F.J.J. van Loo, Acta Mater. **52**, 623 (2004)
35. A. Paul, A.A. Kodentsov, F.J.J. van Loo, Defect and Diffusion Forum **237–240**, 813 (2005)

11 Diffusion and External Driving Forces

11.1 Overview

Diffusing particles experience a drift motion in addition to random diffusion, when an external driving force is applied. Table 11.1 lists examples of driving forces. An electric field is the most common example of an external force and is treated in detail below. Another example is the gradient of the non-ideal part of the chemical potential, which we have considered already in the previous chapter on interdiffusion. The nature of and analytic expressions for the driving forces can be deduced from the thermodynamics of irreversible processes (see Chap. 12 and [1, 2]).

For many ionic solids (see Chaps. 26 and 27) the electrical conductivity results from the transport of ions rather than electrons as is the case in metals and semiconductors. When the charge carriers are ions in an electronically insulating crystal or glass, the ionic motion under the influence of an electric field is described by the *ionic conductivity*. The dc conductivity, σ_{dc}, relates the electrical current density, j_e, via Ohm's law to the applied electric field E:

$$j_e = \sigma_{dc} E. \tag{11.1}$$

If the ions in the material are labelled i ($= 1, 2, \ldots$), the conductivity can be written in terms of mobilities u_i and charges q_i as

$$\sigma_{dc} = \sum_i C_i \mid q_i \mid u_i, \tag{11.2}$$

where C_i is the number density of ions of kind i. An appropriate comparison between mobility and diffusivity of ions is obtained via the Nernst-Einstein relation discussed below.

The interplay of electron currents and atomic fluxes is necessary in the consideration of *electromigration* in metals [5, 6]. In metals ions are screened by the conduction electrons. Then, the effective charge of an ion can be very different from the charge of the ionic cores. An electronic current also exerts a force on the atomic species. The origin of this force is the scattering of electrons at the ion cores and the associated momentum transfer. The coupling between electronic and atomic current in metals at elevated temperatures is of considerable technical relevance, because it is the origin of electromigration.

Table 11.1. Examples of driving forces for drift motion of atoms

Force	Expression	Remarks
Gradient of electrical potential $\boldsymbol{E} = -\boldsymbol{\nabla} U$	$q^*\boldsymbol{E}$	q^*: effective electric charge
Gradient of chemical potential (non-ideal part)	$-\boldsymbol{\nabla}\mu$	μ: chemical potential
Temperature gradient $\boldsymbol{\nabla} T$	$-(Q^*/T)\boldsymbol{\nabla} T$ or $-S\boldsymbol{\nabla} T$	Q^*: heat of transport S: Soret coefficient
Stress gradient	$-\boldsymbol{\nabla} U_{el}$	U_{el}: elastic interaction energy due to stress field
Gravitational force	mgz	m: particle mass g: acceleration due to gravity
Centrifugal force	$m^*\omega^2 r$	m^*: effective atomic mass ω: angular velocity r: distance from rotation axis

The latter is a major reason for the degradation of metallic interconnects in microelectronic devices.

Temperature gradients in a material can also act as driving force on diffusing atoms. The resulting effect is called *thermotransport* (also *thermomigration*). If there are simultaneous gradients of temperature and of concentration, we can combine Fick's first law and thermotransport to give

$$j = -D\frac{\partial C}{\partial x} - S\frac{\partial T}{\partial x}. \tag{11.3}$$

A steady state (index ss) can be established for $j = 0$. We then have

$$\left(\frac{\partial C}{\partial x}\right)_{ss} = -\frac{S}{D}\left(\frac{\partial T}{\partial x}\right)_{ss}. \tag{11.4}$$

This equation describes the concentration gradient established by thermotransport. This effect is called the *Ludwig-Soret effect* or often just the *Soret effect*. The Soret coefficient, S, and the heat of transport, $Q^* \equiv S/T$, may be of either sign, whereas D is always positive. Thermotransport is a relatively complex phenomenon since the system is, *per se*, not isothermal. For a more detailed discussion the reader is referred to [3–5].

In the case of *non-ideal alloys* the gradient of the chemical potential gives rise to a driving force. The gradient of the chemical potential can be expressed as a sum of two terms: the first term contains the concentration gradient and the second term the gradient of the activity coefficient. This case is discussed in detail in Chaps. 10 and 12.

A uniform stress field cannot generate a particle flux. However, a driving force can arise from the *stress gradient*. Its effects must be considered whenever the interaction energy of the particle with the stress field is large enough. Examples are the *Gorski effect*, which can be observed for hydrogen diffusion in metals (see Chap. 14), and the formation of *Cottrell atmospheres* in the stress field of dislocations [7].

Gravitational forces are weak and accordingly play no rôle in solid-state diffusion. In gases and liquids, however, gravitational forces can cause sedimentation effects. The experimental observation of effects of *centrifugal forces* requires rapid rotation. Equilibrium sedimentation of Au in solid sodium and lead has been studied in centrifuge experiments [8]. The results have been discussed to demonstrate the interstitial nature of fast diffusers.

11.2 Fick's Equations with Drift

Suppose that an *external driving force* F acts on diffusing particles. After a short transition period, a steady state particle flux develops. The *drift velocity* $\bar{v}$ of the particles under the action of the driving force is

$$\bar{v} = uF. \tag{11.5}$$

u is called the *mobility*. The mobility is the drift velocity for a unit driving force, i.e. for $F = 1$. The particle flux is $C\bar{v} = CuF$. The total flux due to diffusion plus the action of the driving force is

$$j = -D\frac{\partial C}{\partial x} + \bar{v}C. \tag{11.6}$$

The first term is the well-known Fickian term and the second term is the drift term. Thermodynamics of irreversible processes shows that the distinction between these two terms is more practical than fundamental (see Chap. 12). Formally, the second term in Eq. (11.6) is also a flux which is proportional to a gradient. In the case of an electrical driving force, this is the electrical potential gradient. If we combine Eq. (11.6) with the continuity equation (2.4), we get

$$\frac{\partial C}{\partial t} = \frac{\partial}{\partial x}\left(D\frac{\partial C}{\partial x}\right) - \frac{\partial}{\partial x}(\bar{v}C). \tag{11.7}$$

If the driving force and, consequently, the drift velocity $\bar{v}$ and the diffusion coefficient are independent of x, Eq. (11.7) reduces to

$$\frac{\partial C}{\partial t} = D\frac{\partial^2 C}{\partial x^2} - \bar{v}\frac{\partial C}{\partial x}. \tag{11.8}$$

Then, it is possible to reduce the problem of solving the differential equation (11.8) by applying the following transformation [9, 10]

$$C = C^* \exp\left(\frac{\bar{v}}{2D}x - \frac{\bar{v}^2 t}{4D}\right). \tag{11.9}$$

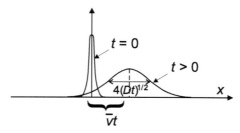

Fig. 11.1. Schematic illustration of diffusion and drift

Substitution of Eq. (11.9) in Eq. (11.8) yields a linear differential equation for C^*

$$\frac{\partial C^*}{\partial t} = D \Delta C^* . \tag{11.10}$$

Hence solutions for Eq. (11.8) are available, if Eq. (11.10) can be solved for the given boundary condition.

The concentration field $C(x,t)$ originating from a thin-film diffusion source is modified by a constant driving force. The appropriate diffusion equation is Eq. (11.8). Its solution can be obtained by inserting the thin-film solution without drift, Eq. (3.9), into Eq. (11.9). We get

$$C(x,t) = \frac{N}{2\sqrt{\pi D t}} \exp\left[-\frac{(x - \bar{v}t)^2}{4Dt}\right] . \tag{11.11}$$

This corresponds to a Gaussian distribution, the center of which shifts with the velocity $\bar{v}$. Equation (11.11) somehow justifies the distinction between the Fickian term and the drift term in Eq. (11.6). The first term leads to a broadening of the distribution of diffusing particles and the second term causes a translation of the whole distribution (Fig.11.1).

The fact that diffusion leads to a Gaussian distribution of the diffusing particles reveals their underlying random movement, whereas under the influence of an external force an additional directed motion occurs. On the other hand, both processes involve the same elementary jumps of the particles. This indicates that D and $\bar{v}$ must be related to each other. This relation is known as the *Nernst-Einstein relation*. Its derivation and some of its implications are considered in the following.

11.3 Nernst-Einstein Relation

We consider a system with one mobile component, where the flux resulting from an external driving force exactly counterbalances the diffusion flux. The combined effect of a concentration gradient and of a driving force can lead

11.3 Nernst-Einstein Relation

to a steady state, if the corresponding fluxes are equal and opposite in sign, i.e. if the total flux vanishes. Then, we get from Eq. (11.6)

$$0 = -\tilde{D}\frac{\partial C}{\partial x} + \bar{v}C. \tag{11.12}$$

The diffusion coefficient $\tilde{D}$ in Eq. (11.12) refers to a chemical composition gradient as will become evident below. It is definitely conceived as a *chemical diffusion coefficient* not as a tracer diffusion coefficient. Lack of appreciation of this fact leads to misunderstandings and inconsistencies.

Let the diffusing substance be contained in a cylinder and let us suppose that $\bar{v} = uF$ is the stationary velocity in negative x-direction due to an external field. Then, the solution of Eq. (11.12) is

$$C = C_0 \exp\left(-\frac{\bar{v}}{\tilde{D}}x\right), \tag{11.13}$$

where C_0 denotes the stationary concentration at $x = 0$. Let us further assume that the external force is the derivative of a potential U:

$$F = -\frac{\partial U}{\partial x}. \tag{11.14}$$

At thermodynamic equilibrium, the distribution of non-interacting particles must also follow the *Boltzmann distribution*[1]

$$C(x) = \alpha \exp\left(-\frac{U}{k_\mathrm{B}T}\right), \tag{11.16}$$

where k_B denotes the Boltzmann constant, T absolute temperature, and α a constant. Differentiation with respect to x yields

$$\frac{\partial C}{\partial x} = -\frac{C}{k_\mathrm{B}T}\frac{\partial U}{\partial x} = \frac{CF}{k_\mathrm{B}T}. \tag{11.17}$$

Substituting this equation in Eq. (11.12), we get

$$\tilde{D} = \frac{\bar{v}}{F}k_\mathrm{B}T = uk_\mathrm{B}T = u\frac{RT}{N_A}. \tag{11.18}$$

$R = k_\mathrm{B}N_A$ denotes the gas constant and N_A the Avogadro number. Equation (11.18) relates the chemical coefficient $\tilde{D}$ and the mobility u of the diffusing particles. This relation is called the *Nernst-Einstein* relation.

---

[1] For a gas in the gravitational field of the earth the potential acting on gas molecules is $U = mgz$, where m is their mass, g the acceleration in the gravitational field, and z the height. Then we get

$$C = C_0 \exp\left(-\frac{mgz}{k_\mathrm{B}T}\right). \tag{11.15}$$

This is the well-known formula for the decrease of gas concentration in the atmosphere with increasing height.

11.4 Nernst-Einstein Relation for Ionic Conductors and Haven Ratio

Let us now suppose that the external driving force is the result of an electric field E acting on ions with charge q. Then we have

$$F = qE. \tag{11.19}$$

Using Eq. (11.18) the flux of ions in the electric field can be written as

$$j = \bar{v}C = \frac{qC\tilde{D}}{k_B T} E. \tag{11.20}$$

The associated electric current density is

$$j_e = qj = \frac{q^2 C\tilde{D}}{k_B T} E. \tag{11.21}$$

This equation is Ohm's law, $j_e = \sigma_{dc} E$, with the ionic d.c. conductivity

$$\sigma_{dc} = \frac{q^2 C\tilde{D}}{k_B T}. \tag{11.22}$$

caused by the mobile ions. In the solid-state diffusion literature this relation is often called in a rather misleading way the Nernst-Einstein relation. However, we remind the reader that we used in its derivation Eq. (11.16) – an equation which holds for non-interacting particles only.

More generally, interactions are present between the particles. For this case, MURCH [11] showed that the general form of the Nernst-Einstein relation is

$$\sigma_{dc} = \frac{q^2 C\tilde{D}}{k_B T} \left(\frac{\partial \ln N}{\partial \mu} \right), \tag{11.23}$$

where μ is the chemical potential of the particles and N their site fraction.

When the distribution of particles is completely ideal (no interaction, not even site blocking effects), the thermodynamic factor is unity and Eq. (11.23) reduces to Eq. (11.22). In this special case, the diffusion coefficient $\tilde{D}$ equals the tracer diffusion coefficient D^*, so that

$$\sigma_{dc} = \frac{q^2 C D^*}{k_B T}. \tag{11.24}$$

When the particles are ideally distributed but subject to site blocking effects, Eq. (11.23) can be written as [11]

$$\sigma_{dc} = \frac{q^2 C D^*}{f k_B T}, \tag{11.25}$$

11.4 Nernst-Einstein Relation for Ionic Conductors and Haven Ratio

where f is the tracer correlation factor (see Chap. 4). This equation is appropriate to very dilute solutions. Ionic crystals having a virtually perfect lattice of ions fall into this category.

Charge diffusion coefficient: In the solid-state diffusion literature Eq. (11.22) is often used to calculate another 'diffusion coefficient' called the *charge diffusion coefficient* (sometimes denoted also as the *conductivity diffusion coefficient* or the *electrical diffusion coefficient*) via:

$$D_\sigma = \frac{k_\mathrm{B} T \sigma_{dc}}{C q^2} = \frac{RT \sigma_{dc}}{N_A C q^2}. \tag{11.26}$$

R denotes the gas constant and N_A the Avogadro number. D_σ has indeed the dimensions of a diffusion coefficient. However, it does not correspond to any diffusion coefficient that can be measured by way of Fick's laws. The identification of D_σ with a diffusion coefficient is only adequate for non-interacting particles.

What can be done is to use Eq. (11.26) as a definition of D_σ, recognising that at the same time D_σ has no Fickian meaning. It is, however, misleading in these circumstances to call Eq. (11.26) the Nernst-Einstein equation. Equation (11.26) is then used purely as a means of changing the d.c. conductivity σ_{dc} to a quantity that has the dimensions of a diffusion coefficient.

Haven ratio: It is common practice in solid-state ionics to define the so-called *Haven ratio* H_R, which is simply the ratio of the tracer diffusion coefficient, D^*, and the charge diffusion coefficient, D_σ:

$$H_R \equiv \frac{D^*}{D_\sigma}. \tag{11.27}$$

In view of the remarks about D_σ, it is appropriate to ask whether the Haven ratio has a straightforward physical meaning. Apart from some simple models (mentioned below) a general theory of H_R is not yet available.

For a *hopping model of ionic conduction*, provided that the ionic jumps are mediated by low concentrations of vacancies like in NaCl (see Chap. 26), one gets [11, 13]

$$H_R = \frac{f}{f_{AA}}. \tag{11.28}$$

Here f is the tracer correlation factor of the atoms and f_{AA} the physical or collective correlation factor (see Chap. 12). If one can neglect collective correlations, which is justified for an almost full or empty lattice of charge carriers, the interpretation of the Haven ratio is straightforward. This is the case for diffusion mechanisms such as the vacancy mechanism. Then, the Haven ratio simply equals the tracer correlation factor:

$$H_R \approx f. \tag{11.29}$$

186   11 Diffusion and External Driving Forces

In such a case, from a measurement of the Haven ratio the correlation factor and then the mechanism of diffusion can be exposed (see Chap. 7).

For a collective mechanism, such as the *interstitialcy mechanism*, the displacement of the charge is different from the displacement of the tracer atom. For a colinear interstitialcy mechanism the charge moves two jump distances whereas the tracer moves only one (see Chap. 6). The Haven ratio then becomes

$$H_R = \frac{fd^2}{f_{AA}d_q^2},\qquad(11.30)$$

where d and d_q are the jump distances of the tracer and of the charge, respectively.

A classic example for identifying the mechanism using experimental Haven ratios is the work of FRIAUF [12] on the motion of Ag in AgBr. It was observed that the Haven ratio varies from 0.5 at low temperatures to 0.65 at high temperatures. This behaviour of the Haven ratio was attributed to the simultaneous action of colinear and non-colinear interstitialcy jumps (see Chap. 26).

Much less satisfactorily interpreted are Haven ratios for fast ion conductors with highly disordered sublattices and for ion-conducting glasses. Haven ratios in such materials have been reviewed by MURCH [13] and by JAIN AND KANERT [14]. Further comments on the subject of correlations effects and ionic diffusion can be found in the review by MURCH [11] and the literature cited therein, as well as in a paper on Monte Carlo simulations of the Haven ratio of alkali ions in oxide glasses [15]. Pressure-dependent diffusivities and Haven ratios in alkali borate glasses are considered by IMRE ET AL. [18].

11.5 Nernst-Planck Equation – Interdiffusion in Ionic Crystals

Let us now consider two ionic system AX and BX with two ionic species A and B, which share the same sublattice of a crystal. Let us assume that only the species A and B are mobile, the X sublattice being fixed. During interdiffusion the two mobile species compete for vacancies on the common sublattice of A and B. Let us first consider the case of an ideal solution and the absence of an electric driving force. Then fluxes j_A and j_B are connected to the concentration gradients via

$$j_A = -D_A^* \frac{\partial C_A}{\partial x} \quad \text{and} \quad j_B = -D_B^* \frac{\partial C_B}{\partial x} \qquad(11.31)$$

with

$$\frac{\partial C_A}{\partial x} = -\frac{\partial C_B}{\partial x}.\qquad(11.32)$$

11.5 Nernst-Planck Equation – Interdiffusion in Ionic Crystals

Since D_A^* and D_B^* are not equal, a net flux of electric charge $j_A \neq j_B$ would develop. However, the condition of electrical neutrality requires that this net flux must be zero.

Any electrical field E gives rise to drift fluxes. More general, the flux equations should then be written as

$$j_A = -D_A^* \frac{\partial C_A}{\partial x} + \frac{qC_A D_A^*}{k_B T} E \quad \text{and} \quad j_B = -D_B^* \frac{\partial C_B}{\partial x} + \frac{qC_B D_B^*}{k_B T} E. \quad (11.33)$$

Here we have introduced the drift term from the Nernst-Einstein equation and q denotes the (identical) charges of ions A and B. The condition of electrical neutrality requires

$$j_A + j_B = 0. \quad (11.34)$$

It follows that

$$E = \frac{k_B T}{q} \frac{D_A^* - D_B^*}{C_A D_A^* + C_B D_B^*} \frac{\partial C_A}{\partial x}. \quad (11.35)$$

We now recall the intrinsic diffusivities of Eq. (10.20), which we repeat for convenience:

$$j_A = -D_A^I \frac{\partial C_A}{\partial x} \quad \text{and} \quad j_B = -D_B^I \frac{\partial C_A}{\partial x}. \quad (11.36)$$

Inserting Eqs. (11.33) and (11.35), we arrive at:

$$D_A^I = D_B^I = \frac{D_A^* D_B^*}{N_A D_A^* + N_B D_B^*}. \quad (11.37)$$

This equation is called the *Nernst-Planck equation* for ideal solutions. For interdiffusion between ionic crystals AX and BX, there is no Kirkendall effect because there is no separation of charge and therefore no net vacancy flow. Equation (11.37) can also be written as

$$\frac{1}{D_A^I} + \frac{1}{D_B^I} = \frac{N_A}{D_B^*} + \frac{N_B}{D_A^*}. \quad (11.38)$$

For a non-ideal solution Eq. (11.37) must be multiplied by the thermodynamic factor Φ:

$$D_A^I = D_B^I = \frac{D_A^* D_B^*}{N_A D_A^* + N_B D_B^*} \Phi \equiv \tilde{D}_{Nernst-Planck}. \quad (11.39)$$

We mention without prove that in terms of the transport coefficients discussed in Chap. 12 one gets:

$$\tilde{D}_{Nernst-Planck} = \frac{2f}{1-f} \frac{L_{AB}}{N_A N_B}. \quad (11.40)$$

This equation shows that interdiffusion in ionic crystals is completely due to the off-diagonal term L_{AB} of the Onsager matrix.

11.6 Nernst-Planck Equation *versus* Darken Equation

The Nernst-Planck equation (11.39) and the Darken equation (10.34) differ significantly if the ratio of the tracer diffusivities is far from unity:

$$\tilde{D}_{Nernst-Planck} \neq \tilde{D}_{Darken}. \tag{11.41}$$

We have seen above that the interdiffusion coefficient described by the Nernst-Planck equation corresponds to a consecutive connection of the two diffusivities; it is controlled by the slower diffusing component. By contrast, the Darken expression is analogous to the paralled connection of the two diffusivities, which is controlled by the faster diffusing component.

In addition, we note that the Nernst-Planck and the Darken equations represent two limiting cases of interdiffusion in a binary metallic alloy:

We recall that the Darken equation has been derived under the assumption that vacancies are practically at thermal equilibrium during the interdiffusion process and that the vacancy flux compensates the unequal fluxes of the atomic species A and B. Usually, this is justified for a macroscopic diffusion sample, when sufficient sinks and sources for vacancies (e.g., dislocations) keep the vacancy concentration close to equilibrium in spite of non-vanishing vacancy fluxes.

Non-equilibrium vacancy distributions can be generated by the difference of the intrinsic fluxes of atoms, if the density and/or efficiency of vacancy sources/sinks are insufficient. NAZAROV AND GUROV have performed an analysis of interdiffusion in binary alloys taking into account non-equilibrium vacancies [16]. They have shown that the evolution of an interdiffusion profile is then governed by

$$\tilde{D}_{Nazarov-Gurov} = \frac{D_A^* D_B^*}{N_A D_A^* + N_B D_B^*} \Phi \equiv \tilde{D}_{Nernst-Planck}. \tag{11.42}$$

The Nernst-Planck expression for interdiffusion in ionic crystals and the Nazarov-Gurov expression for interdiffusion in a binary alloy with non-equilibrium vacancies are identical.

Furthermore, the interdiffusion coefficients of Darken and of Nazarov-Gurov correspond to different space and time regimes. This has been pointed out by GUSAK AND COWORKERS (see, e.g., [17]):

– The Darken expression, Eq. (10.34), governs the interdiffusion process for long diffusion times

$$t \gg \frac{\tau_V}{C_V}. \tag{11.43}$$

Here τ_V is the mean life-time of vacancies between their creation at vacancy sources and their annihilation at vacancy sinks. C_V denotes the vacancy concetration. This condition is fulfilled when the average distance between vacancy sources and sinks, $\sqrt{D_V \tau_V}$, is small as compared to the width of the interdiffusion zone, i.e. for

$$\tilde{D}_{Darken}\, t \gg D_V \tau_V\,, \tag{11.44}$$

D_V denotes the vacancy diffusivity.

- The Nazarov-Gurov equation governs the interdiffusion process for short diffusion times

$$t \ll \frac{\tau_V}{C_V}\,. \tag{11.45}$$

This corresponds to

$$\tilde{D}_{Nazarov-Gurov}\, t \ll D_V \tau_V\,, \tag{11.46}$$

Then, the width of the interdiffusion zone is much smaller than the average distance between vacancy sources and sinks.

Loosely speaking, non-equilibrium vacancies and the Nazarov-Gurov equation are important on 'nanoscopic scales', whereas the Darken equation is relevant for interdiffusion on 'macroscopic scales'.

References

1. S.R. de Groot, P. Mazur, *Thermodynamics of Irreversible Processes*, North-Holland Publ. Comp., 1952
2. R.E. Howard, A.B. Lidiard, *Matter Transport in Solids*, in: Reports on Progress in Physics Vol. XXVII, The Institute of Physics and the Physical Society, London, 1964, pp. 161–240
3. J. Philibert, *Atom Movements – Diffusion and Mass Transport in Solids*, Les Editions de Physique, Les Ulis, 1991
4. A.R. Allnatt. A.B. Lidiard, *Atomic Transport in Solids*, Cambridge University Press, 1993
5. H. Wever, *Elektro- und Thermotransport in Metallen*, Barth-Verlag, Leipzig, 1973
6. H.B. Huntington, *Electromigration in Metals*, in: *Diffusion in Solids – Recent Developments*, A. S. Nowick, J.J. Burton (Eds.), Academic Press, 1975, p. 303
7. J. Völkl, G. Alefeld, *Hydrogen Diffusion in Metals*, in: *Diffusion in Solids – Recent Developments*, A. S. Nowick, J.J. Burton (Eds.), Academic Press, 1975, p. 303
8. L.W. Barr, A.D. Le Claire, Philos. Mag, **20**, 1289–1291 (1969)
9. R. Fürth, *Diffusion*, in: *Handbuch physikalisch, technische Mechanik*, F. Auerbach, W. Hort (Eds.), Vol. 7, Leipzig, 1932
10. W. Jost, *Diffusion in Solids, Liquids, Gases*, Academic Press, Inc., New York, 1960
11. G.E. Murch, *Diffusion Kinetics in Solids*, Chap. 3 in: *Phase Transformations in Materials*, G. Kostorz (Ed.), Wiley-VCh Verlag GmbH, Weinheim, Germany, 2001

12. R.J. Friauf, Phys. Rev. **105**, 843 (1957)
13. G.E. Murch, Solid State Ionics **7**, 177 (1982)
14. H. Jain, O. Kanert, in: *Defects in Insulating Materials*, O. Kanert, J.-M. Spaeth (Eds.), World Scientific Publishing Comp., Ltd., Singapore, 1993
15. S. Voss, S.V. Divinski, A.W. Imre, H. Mehrer, J.N. Mundy, Solid State Ionics **176**, 1383 (2005)
16. A.V. Nazarov, K.P. Gurov, Fizika Metallov Metallovedenie **37**, 496 (1974); **38**, 689 (1974); **45**, 855 (1978)
17. A.M. Gusak, S.V. Kornienko, G.V. Lutsenko, presented at Int. Workshop on *Diffusion and Stresses* (DS 2006), Lillafüred, Hungary, 2006
18. A.W. Imre, S. Voss, H. Staesche, M.D. Ingram, K. Funke, H. Mehrer, J. Phys. Chem. B, **111**, 5301–5307 (2007)

12 Irreversible Thermodynamics and Diffusion

12.1 General Remarks

So far, we have discussed the usual descriptions of atomic transport in solids such as diffusion and ionic conduction. Fick's first law was introduced as a postulate describing a linear relationship between the flux of a diffusing species i and its concentration gradient. Ohm's law describes a linear relation between the flux of charged species and the gradient of the electric potential. Fourier's law is a linear relation between the flux of heat and the temperature gradient. However, these simple laws may not be sufficient even within the stated limitation of linear effects. For example, Fick's first law is sometimes insufficient for attaining equilibrium of species i because it does not recognise all driving forces, direct or indirect, acting on i. It may be necessary to allow that a concentration gradient in one species gives rise to a flux of another.

Diffusion, electrical conduction, and heat flow are examples of irreversible processes. Non-equilibrium thermodynamics provides a general phenomenological theory of such processes. In this chapter, we briefly introduce the macroscopic equations of this theory suggested for the first time by LARS ONSAGER (1903–1976), the Norwegian Nobel laureate of 1968. Wide-ranging treatments, especially of macroscopic formulations of this theory can be found in textbooks of DE GROOT AND MAZUR [1] and HAASE [2]. A treatment with emphasis on the foundations of irreversible thermodynamics by statistical mechanics has been provided by KREUZER [3].

Non-equilibrium thermodynamics is based on three major assumptions:

1. The Onsager transport equations of atoms, heat, and electrons are linear relations between the fluxes $\boldsymbol{J}_i$ and the so-called generalised thermodynamic forces $\boldsymbol{X}_i$. The phenomenological response for the complete set of n fluxes assumes a linear form

$$\boldsymbol{J}_i = \sum_{j=1}^{n} L_{ij} \boldsymbol{X}_j , \qquad (12.1)$$

where L_{ij} are called the *phenomenological coefficients* or *transport coefficients*. The matrix of coefficients is also called the *Onsager matrix* or simply the **L** matrix. The great importance of the phenomenological coefficients stems from their independence of driving forces. For example, in solid-state diffusion problems the coefficients L_{ij} are functions of temperature and pressure, but they do not depend on the gradient of the chemical potential.

2. The Onsager matrix is composed in part of diagonal terms, L_{ii}, connecting each generalised force with its conjugate flux. For example, a gradient of the chemical potential causes a generalised diffusion 'force', and the associated diffusion response is determined by the material's diffusivity. Similarly, an applied temperature gradient creates a generalised force associated with heat flow. In this case, the amount of heat flow is determined by the thermal conductivity.

The Onsager matrix also contains off-diagonal coefficients, L_{ij}. Each off-diagonal coefficient determines the influence of a generalised force on a non-conjugate flux. For example, a concentration gradient of one species can give rise to a flux of another species. The electric field, which exerts a force on electrons in metals to produce an electric current has a cross-influence on the flow of heat, known as the *Peltier effect*. Conversely, the thermal force (temperature gradient) that normally causes heat flow, also has a cross-influence on the distribution of electrons – known as the Thomson effect. The Thomson and Peltier effects combine and provide the basis for *thermoelectric devices*: thermopiles can be used to convert heat flow into electric current; in thermocouples a voltage is produced by a temperature difference. Another example is that an electronic current and the associated 'electron wind' causes a flow of matter called *electromigration* (see also Chap. 11). Electromigration can be a major cause for the failure of interconnects in microelectronic devices.

The Onsager matrix is symmetric, provided that no magnetic field is present. The relationship
$$L_{ij} = L_{ji} \tag{12.2}$$
is known as the *Onsager reciprocity theorem*.

3. The central idea of non-equilibrium thermodynamics is that each of the thermodynamic forces acting with its flux response dissipates free energy and produces entropy. The characteristic feature of an irreversible process is the generation of entropy. The rate of entropy production, σ, is basic to the theory. It can be written as:
$$T\sigma = \sum_{i}^{n} \boldsymbol{J}_i \boldsymbol{X}_i + \boldsymbol{J}_q \boldsymbol{X}_q. \tag{12.3}$$

$\boldsymbol{J}_i$ denotes the flux of atoms i and $\boldsymbol{J}_q$ the flux of heat.
The thermodynamic forces require some explanation: $\boldsymbol{X}_i$ and $\boldsymbol{X}_q$ are measures for the imbalance generating the pertinent fluxes. The thermal force $\boldsymbol{X}_q$
$$\boldsymbol{X}_q = -\frac{1}{T}\boldsymbol{\nabla} T \tag{12.4}$$

is determined by the temperature gradient ∇T. When only external forces are acting, the $\boldsymbol{X}_i$ are identical with these forces. If, for example, an ionic system with ions of charge q_i is subject to an electric field $\boldsymbol{E}$, each ion of type i experiences a mechanical force $\boldsymbol{F}_i = q_i \boldsymbol{E}$. In the presence of a composition gradient the appropriate force is related to the gradient of chemical potential $\nabla \mu_i$. Then, the thermodynamic force X_i is the sum of the external force exerted by the electric field and the gradient of the chemical potential of species i:

$$\boldsymbol{X}_i = \boldsymbol{F}_i - T\nabla\left(\frac{\mu_i}{T}\right) = \boldsymbol{F}_i - \nabla \mu_i \,. \tag{12.5}$$

Here the gradient of the chemical potential is that part due to gradients in concentration, but not to temperature.

Thermodynamic equilibrium is achieved when the entropy production vanishes:

$$\sigma = 0\,. \tag{12.6}$$

Then, there are no irreversible processes any longer and the thermodynamic forces and the fluxes vanish.

12.2 Phenomenological Equations of Isothermal Diffusion

In this section, we apply the phenomenological transport equations to solid-state diffusion problems. We give a brief account of some major aspects relevant for transport of matter, which are treated in more detail in [4–6]. The phenomenological equations are on the one hand very powerful. On the other hand, they lead quickly to cumbersome expressions. Therefore, only a few examples will be given. Detailed expressions for the phenomenological coefficients in terms of the elementary jump characteristics must be provided by atomistic models.

Here, we consider the consequences of phenomenological equations for isothermal diffusion. In a binary system we have 3 transport coefficients – two diagonal ones and one off-diagonal coefficient. For a ternary system six transport coefficients must be taken into account. One of the crucial questions is, whether the off-diagonal terms are sufficiently different from zero to be important for data analysis. If they are negligible, the analysis can be largely simplified. This assumption in made in some models for diffusion, e.g., in the derivation of the Darken equations for a binary system (see Chap. 10). We shall see below, however, that neglecting off-diagonal terms is not always justified.

12.2.1 Tracer Self-Diffusion in Element Crystals

Fundamental mobilities of atoms in solids can be obtained by monitoring radioactive isotopes ('tracers') (see Chap. 13). Let us consider the diffusion

12 Irreversible Thermodynamics and Diffusion

of an isotope A^* in a solid A, where A^* and A are chemically identical. Let us further suppose that diffusion occurs via vacancies (index V). Taking into account the reciprocity relations, the Onsager flux equations can be written as[1]:

$$J_{A^*} = L_{A^*A^*}X_{A^*} + L_{A^*A}X_A + L_{A^*V}X_V,$$
$$J_A = L_{A^*A}X_{A^*} + L_{AA}X_A + L_{AV}X_V,$$
$$J_V = L_{A^*V}X_{A^*} + L_{AV}X_A + L_{VV}X_V. \quad (12.7)$$

Let us now suppose that in an isothermal experiment vacancies are always maintained at thermal equilibrium. This is possible if sources and sinks of vacancies such as dislocations are sufficiently numerous and active during the diffusion process. Under such conditions, the chemical potential of vacancies is constant everywhere and hence $X_V = 0$. Then, the flux equations for the components apply directly in the laboratory reference frame:

$$J_{A^*} = L_{A^*A^*}X_{A^*} + L_{A^*A}X_A$$
$$J_A = L_{A^*A}X_{A^*} + L_{AA}X_A. \quad (12.8)$$

Tracer atoms A^* and matrix atoms A form an ideal (isotopic) solution. Let C_{A^*} and C_A denote their concentrations and μ_{A^*} and μ_A their chemical potentials, respectively. These are:

$$\mu_{A^*} = \mu^0_{A^*}(p,T) + k_B T \ln C_{A^*},$$
$$\mu_A = \mu^0_A(p,T) + k_B T \ln C_A. \quad (12.9)$$

The reference potentials $\mu^0_{A^*}(p,T)$ and $\mu^0_A(p,T)$ depend on pressure and temperature but not on concentration. The corresponding forces are

$$X_{A^*} = -k_B T \frac{1}{C_{A^*}} \frac{\partial C_{A^*}}{\partial x},$$
$$X_A = -k_B T \frac{1}{C_A} \frac{\partial C_A}{\partial x}. \quad (12.10)$$

For tracer self-diffusion in an element crystal we have

$$J_{A^*} + J_A = 0, \quad (12.11)$$

i.e. the fluxes are equal in magnitude and opposite in sign. Furthermore, since $C_{A^*} + C_A$ is constant, the concentration gradients are equal in magnitude and opposite in sign, i.e. $\partial C_{A^*}/\partial x = -\partial C_A/\partial x$. We then have

$$J_{A^*} = -\underbrace{\left(\frac{L_{A^*A^*}}{C_{A^*}} - \frac{L_{A^*A}}{C_A}\right)}_{D_A^{A^*}} k_B T \frac{\partial C_{A^*}}{\partial x}, \quad (12.12)$$

$$J_A = -\left(\frac{L_{AA}}{C_A} - \frac{L_{A^*A}}{C_{A^*}}\right) k_B T \frac{\partial C_A}{\partial x}. \quad (12.13)$$

---

[1] For simplicity reasons we consider unidirectional flow (in x direction) and omit the vector notation.

Since $J_{A^*} = -J_A$ we get:

$$\frac{L_{A^*A^*}}{C_{A^*}} - \frac{L_{A^*A}}{C_A} = \frac{L_{AA}}{C_A} - \frac{L_{A^*A}}{C_A^*}. \tag{12.14}$$

Recalling that the diffusivity of a tracer is defined through Fick's law

$$J_{A^*} = -D_A^{A^*} \frac{\partial C_{A^*}}{\partial x}, \tag{12.15}$$

we obtain by comparison with Eqs. (12.12):

$$D_A^{A^*} = \left(\frac{L_{A^*A^*}}{C_{A^*}} - \frac{L_{A^*A}}{C_A}\right) k_B T = \left(\frac{L_{AA}}{C_A} - \frac{L_{A^*A}}{C_{A^*}}\right) k_B T \tag{12.16}$$

Since for tracer diffusion always $C_{A^*} \ll C_A$, we also have

$$D_A^{A^*} = \frac{L_{A^*A^*}}{C_{A^*}} k_B T = \left(\frac{L_{AA}}{C_A} - \frac{L_{A^*A}}{C_{A^*}}\right) k_B T. \tag{12.17}$$

The first term on the right-hand side of Eq. (12.17) is sometimes denoted as the 'true' self-diffusion coefficient, $D_A^A \equiv k_B T L_{AA}/C_A$. The quantity D_A^A denotes self-diffusion in the absence of a tracer – a quantity that is difficult to measure directly[2]. We then get

$$D_A^{A^*} = \underbrace{\left[1 - \frac{L_{A^*A}}{L_{AA}} \frac{C_A}{C_{A^*}}\right]}_{f} D_A^A, \tag{12.18}$$

where f is the correlation factor of tracer self-diffusion (see Chap. 7). This equation shows that the coupling between the fluxes is the origin of correlation effects. For $L_{A^*A} = 0$ the correlation factor would be unity. We also note that in addition to the defect two atomic species (here A^* and A) must be present to obtain correlation. This is a rule that we have already stated in Chap. 4.

12.2.2 Diffusion in Binary Alloys

In this section, we discuss the Onsager equations for solid-state diffusion via vacancies in a substitutional binary alloy and the structure and physical meaning of the pertinent phenomenological coefficients. We suppose that the system is isothermal and that external forces are absent, i.e. $X_q = 0$ and $F_i = 0$. We then have a system of two atomic components A and B and vacancies (index V) on a single lattice. Taking into account the reciprocity relations the Onsager flux equations can be written as:

---

[2] In favourable cases, PFG-NMR (see Chap. 13) can be applied to measure this quantity.

$$\begin{aligned}
J_A &= L_{AA}X_A + L_{AB}X_B + L_{AV}X_V, \\
J_B &= L_{AB}X_A + L_{BB}X_B + L_{BV}X_V, \\
J_V &= L_{AV}X_A + L_{BV}X_B + L_{VV}X_V.
\end{aligned} \qquad (12.19)$$

To promote atomic diffusion, either A or B atoms exchange with vacancies on the same lattice. Therefore, the fluxes defined relative to that lattice (in regions outside the diffusion zone) necessarily must obey

$$J_V = -(J_A + J_B). \qquad (12.20)$$

In other words, the flux of vacancies is equal and opposite in sign to the total flux of atoms[3]. In view of the constraint imposed by Eq. (12.20), each term on the right-hand side of Eq. (12.19) must vanish column by column. We get:

$$\begin{aligned}
(L_{AA} + L_{AB} + L_{AV})X_A &= 0, \\
(L_{AB} + L_{BB} + L_{BV})X_B &= 0, \\
(L_{AV} + L_{BV} + L_{VV})X_V &= 0.
\end{aligned} \qquad (12.21)$$

These equations must hold for arbitrary values of the forces X_i. Hence each bracket term must vanish separately:

$$\begin{aligned}
-L_{AV} &= L_{AA} + L_{AB}, \\
-L_{BV} &= L_{AB} + L_{BB}, \\
-L_{VV} &= L_{AV} + L_{BV}.
\end{aligned} \qquad (12.22)$$

These equations show that the kinetic coefficients of the vacancy flux are related to those of the atomic species. If Eq. (12.22) is combined with the flux equations (12.19), the following expressions for the fluxes of the atomic species are obtained:

$$\begin{aligned}
J_A &= L_{AA}(X_A - X_V) + L_{AB}(X_B - X_V), \\
J_B &= L_{AB}(X_A - X_V) + L_{BB}(X_B - X_V).
\end{aligned} \qquad (12.23)$$

The vacancy flux may be written as

$$J_V = L_{AV}(X_A - X_V) + L_{BV}(X_B - X_V). \qquad (12.24)$$

Let us now assume that vacancies are always maintained close to their thermal equilibrium concentration. Then, the chemical potential of vacancies is constant everywhere and $X_V = 0$. Equations (12.23) then reduce to

$$\begin{aligned}
J_A &= L_{AA}X_A + L_{AB}X_B, \\
J_B &= L_{AB}X_A + L_{BB}X_B.
\end{aligned} \qquad (12.25)$$

---

[3] In this context the reader should also see the discussion of the 'vacancy wind' in Chap. 10.

12.2 Phenomenological Equations of Isothermal Diffusion

The chemical potential of a real solid solution is given by

$$\mu_i = \mu_i^0(p,T) + k_B T \ln(N_i \gamma_i), \tag{12.26}$$

where γ_i is the activity coefficient and N_i mole fractions of species i. Using $N_A + N_B \approx 1$, i.e. $C_V \ll C_A + C_B$, we obtain:

$$J_A = -\underbrace{\left(\frac{L_{AA}}{N_A} - \frac{L_{AB}}{N_B}\right) k_B T \Phi}_{D_A^I} \frac{\partial N_A}{\partial x},$$

$$J_B = -\underbrace{\left(\frac{L_{BB}}{N_B} - \frac{L_{AB}}{N_A}\right) k_B T \Phi}_{D_B^I} \frac{\partial N_B}{\partial x}. \tag{12.27}$$

Here we have used that the Gibbs-Duhem relation of thermodynamics provides an additional constraint: the factors $(1 + \partial \ln \gamma_A / \partial \ln N_A)$ and $(1 + \partial \ln \gamma_B / \partial \ln N_B)$ are equal to each other. This common factor is abbreviated by Φ and denoted as the *thermodynamic factor* (see Chap. 10). We note that one common thermodynamic factor exists for binary alloys. For ternary or higher order systems several thermodynamic factors are necessary.

Equations (12.27) have the form of Fick's first law with diffusion coefficients, D_A^I and D_B^I, denoted as the *intrinsic diffusion coefficients*:

$$D_A^I = \frac{L_{AA}}{N_A}\left(1 - \frac{L_{AB} N_A}{L_{AA} N_B}\right) k_B T \Phi,$$

$$D_B^I = \frac{L_{BB}}{N_B}\left(1 - \frac{L_{AB} N_B}{L_{BB} N_A}\right) k_B T \Phi. \tag{12.28}$$

The intrinsic diffusivities are in general different and can be determined separately. The two quantities that are usually measured to obtain the intrinsic diffusivities are (i) the chemical interdiffusion coefficient $\tilde{D}$, and (ii) the Kirkendall velocity v_K (see Chap. 10):

In interdiffusion plus Kirkendall experiments, diffusion couples (either A-B or $A_x B_{1-x}$-$A_y B_{1-y}$) are studied and the initial interfaces contain inert markers. The fluxes J_A and J_B relative to the local crystal lattice are generally such that their sum is non-zero, which according to Eq. (12.20) implies that the vacancy flux is also non-zero. Since the fluxes vary with position one also has:

$$\text{div} J_V = \text{div}(-J_A - J_B) \neq 0. \tag{12.29}$$

This condition requires vacancies to disappear or to be created at inner sources or sinks (e.g., dislocations) in the diffusion zone of the sample. When this occurs, regions where diffusion fluxes are large will move relative to regions where fluxes are small. The concentration distribution after an interdiffusion experiment, analysed with respect to the unaffected ends of the couple, links the concentration gradients to the fluxes of atoms relative to the fixed parts of the sample. In the simplest case these fluxes are

$$J'_A = J_A - N_A(J_A + J_B),$$
$$J'_B = J_B - N_B(J_A + J_B). \quad (12.30)$$

Here we have assumed that the velocity v of the local lattice relative to the non-diffusing parts of the sample is $-v(J_A + J_B)$. This implies that there is no change in the cross-section or shape of the sample, i.e. that the vacancies condense on lattice planes perpendicular to the diffusion flow. It also implies that the volume per lattice site remains constant. From Eqs. (12.27), (12.30), and $N_A + N_B = 1$, because the vacancy site fraction is very small, it also follows

$$J'_A = -(N_B D^I_A + N_A D^I_B)\frac{\partial N_A}{\partial x},$$

$$J'_B = -(N_B D^I_A + N_A D^I_B)\frac{\partial N_B}{\partial x}. \quad (12.31)$$

The quantity
$$\tilde{D} = N_B D^I_A + N_A D^I_B. \quad (12.32)$$

is the *interdiffusion coefficient* introduced already in Chap. 10. It is the same for both components. The Kirkendall velocity v_K characterises the motion of the diffusion zone relative to the fixed end of the sample and can be observed by inserting inert markers (see Chap. 10). Since $\partial N_A/\partial x = -\partial N_B/\partial x$ it is given by:

$$v_K = (D^I_A - D^I_B)\frac{\partial N_A}{\partial x}. \quad (12.33)$$

The Kirkendall effect is a consequence of unequal intrinsic diffusivities and results from the non-zero vacancy flux in the diffusion zone.

We now have convinced ourselves that the equations of Darken and Manning discussed in Chap. 10 have a basis in the phenomenological equations. We note, however, that although measurements of interdiffusion and the Kirkendall shift have been made on a number of alloy systems it is clear that only two quantities, D^I_A and D^I_B, can be obtained. This is insufficient to deduce the three independent phenomenological coefficients L_{AA}, L_{AB}, and L_{BB} for a binary alloy system.

Such a situation, in which the number of experimentally accessible quantities is less than the number of the independent Onsager coefficients, is not uncommon. This is one reason for the theoretical interest in these coefficients. It is also the reason for the interest in relations between the phenomenological coefficients. Such relations are called 'sum rules'. Sum rules have been identified in several cases [6] and some of them are discussed in Sect. 12.3.2.

12.3 The Phenomenological Coefficients

The physical meaning of diffusion coefficients is well appreciated. However, this is less so for the phenomenological coefficients. The purpose of this section is to provide some insight into the phenomenological coefficients, their structure, relations between phenomenological coefficients and diffusion coefficients, and relations among phenomenological coefficients. An introduction for non-specialists in this area, which we follow in parts below, has been given by MURCH AND BELOVA [6].

Consider, for example, diffusion in a binary alloy. We have eliminated the vacancies as a third component because we have made the assumption that vacancies are maintained at their equilibrium concentration. Then, it suffices to introduce the phenomenological coefficients L_{AA}, L_{AB}, and L_{BB} as we have done in the previous section. According to the Onsager reciprocity theorem the **L** matrix is symmetric, i.e. $L_{AB} = L_{BA}$. We simply have three independent coefficients, two diagonal ones and one off-diagonal coefficient.

Let us now consider a 'thought experiment': suppose that A atoms in a binary AB alloy can respond to an external electric field E but the B atoms cannot. This is expressed by writing the driving force on A with charge q_A as $X_A = q_A E$, whereas the driving force on B is $X_B = 0$. One might first expect that the A atoms would simply drift in the field and the B atoms would be unaffected. The Onsager flux equations show indeed that the flux of A atoms is given by $J_A = L_{AA} q_A E$. But the Onsager flux equations also show that the flux of B atoms is not zero but given by $J_B = L_{AB} q_A E$. This equation says that the B atoms should also drift in the field, although the B atoms do not feel the field directly. The drifting A atoms appear to drag the B atoms along with them, thereby giving rise to a flux of B atoms. In principle, the off-diagonal coefficient L_{AB} can be either positive or negative depending on the type of interaction. If L_{AB} were to be negative, it would mean that the B atoms would drift up-field whilst the A atoms drift down-field. This example may suffice to illustrate that the off-diagonal coefficient can be responsible not only for an atomic flux but also that it can change the magnitude and even the direction of an atom flux.

As discussed in Chap. 4, the tracer diffusion coefficient is related to the mean square of the displacement $\boldsymbol{R}$ of a particle during time t as [8]:

$$D^* = \frac{\langle \boldsymbol{R}^2 \rangle}{6t}, \qquad (12.34)$$

where the brackets $\langle \rangle$ indicate an average over an ensemble of a large number of particles. This relation is called the *Einstein relation* or sometimes also the *Einstein-Smoluchowski relation*. The diffusion coefficient is understood to be a tracer diffusion coefficient indicated by the superscript * placed on D. The implication is that in principle we can follow each particle explicitly.

The Einstein expression for tracer diffusion coefficient is usually expanded for solid-state diffusion according to 'hopping models', in which atoms jump from one site to another in a lattice with a long residence time at lattice sites between the jumps. For the simplest case, diffusion on a cubic Bravais lattice with jump distance d and jump rate Γ, we have (see Chap. 6)

$$D^* = \frac{1}{6}d^2\Gamma f \tag{12.35}$$

where f is the tracer correlation factor. This equation shows that the tracer diffusion coefficient is the product of two parts: a *correlated part* embodied in the tracer correlation factor and an *uncorrelated part* that contains the jump distance squared and the jump rate. We remember that according to Eq. (7.22) the tracer correlation factor can be expanded as

$$f = 1 + 2\sum_{j=1}^{\infty}\langle\cos\theta^{(j)}\rangle. \tag{12.36}$$

$\langle\cos\theta^{(j)}\rangle$ is the average of the cosine of the angle between the first and the j's succeeding tracer jump. In vacancy-mediated solid-state diffusion $\langle\cos\theta^{(1)}\rangle$ is invariably negative because the first jump is more likely to be reversed, either as the result of the vacancy being still present at the nearest-neighbour site to the tracer, or perhaps as a result of re-ordering jump immediately following a disordering one, or a combination of both. For a vacancy mechanism the values of $\langle\cos\theta^{(j)}\rangle$ also alternate in sign. The phenomenon of tracer correlation has been the subject of an extensive literature over several decades and is discussed in Chap. 7.

In 1982 ALLNATT [7] showed on the basis of a linear response theory that the phenomenological coefficients for isothermal diffusion in solids can be expressed via generalised Einstein formulae, similar in character to the Einstein-Smoluchowski relation. These relations can be written as:

$$L_{ii} = \frac{\langle \boldsymbol{R}_i \cdot \boldsymbol{R}_i \rangle}{6Vk_\mathrm{B}Tt},$$
$$L_{ij} = \frac{\langle \boldsymbol{R}_i \cdot \boldsymbol{R}_j \rangle}{6Vk_\mathrm{B}Tt}, \tag{12.37}$$

where $\boldsymbol{R}_i$ and $\boldsymbol{R}_j$ are the collective displacements of atoms i and j, V is the volume of the system. The collective displacement of a species in each case can be thought of as the displacement of the centre of mass of that species. Imagine in a 'thought experiment' a volume V containing N lattice sites on which two species A and B are randomly distributed. This might represent a binary alloy. Let diffusion occur for some time t. Then, we calculate the displacements of the centres of mass of A atoms, $\boldsymbol{R}_A$, and of B atoms, $\boldsymbol{R}_B$, and repeat the experiment a large number of times in order to produce the ensemble average. In this way, we would be able to calculate the phenomenological coefficients L_{AA}, L_{BB}, and L_{AB} from Eqs. (12.37). This provides indeed a convenient route for the evaluation of the phenomenological coefficients in Monte Carlo computations (see, e.g., [9, 10]).

Similar to the tracer diffusion coefficient the phenomenological coefficients can be decomposed into correlated and into non-correlated parts

$$L_{ii} = \frac{f_{ii} d^2 C_i n_i}{6 k_B T t}, \qquad (12.38)$$

$$L_{ij} = \frac{f_{ij}^{(i)} d^2 C_i n_i}{6 k_B T t}, \quad \text{or alternatively} \qquad (12.39)$$

$$L_{ij} = \frac{f_{ij}^{(j)} d^2 C_j n_j}{6 k_B T t}. \qquad (12.40)$$

n_i denotes the number of jumps of species i during the time t. $C_i = N_i N/V$ with N_i denoting the fraction of species i and N the total number of sites in volume V. The correlated parts of the phenomenological coefficients, the f_{ij}, are denoted as the *correlation functions or collective correlation factors*. In very much the same way that the tracer correlation factors can be expressed in terms of the average cosines of the angles between a given jump of the tracer and its succeeding jumps via Eq. (12.36), the diagonal and off-diagonal collective correlation factors can also be expressed in terms of the average of the cosines of the angle between a given jump of a species and the subsequent jump of the same (diagonal) or another (off-diagonal) species. The diagonal correlation factors are given by

$$f_{ii} = 1 + 2 \sum_{k=1}^{\infty} \langle \cos \theta_{ii}^{(k)} \rangle, \qquad (12.41)$$

where $\langle \cos \theta_{ii}^{(k)} \rangle$ is the average of the cosine of the angle between some jump of an atom of species i and the k'th succeeding jump of the same or another atom of species i. The expressions for the off-diagonal collective correlation factors are a bit more complicated in notation, but they are structurally related to Eq. (12.41). For simplicity, the following expression are given only for a binary system:

$$f_{AB}^{(A)} = \sum_{k=1}^{\infty} \langle \cos \theta_{AB}^{(k)} \rangle + \frac{C_B n_B}{C_A n_A} \sum_{k=1}^{\infty} \langle \cos \theta_{BA}^{(k)} \rangle, \qquad (12.42)$$

$$f_{AB}^{(B)} = \frac{C_A n_A}{C_B n_B} \sum_{k=1}^{\infty} \langle \cos \theta_{AB}^{(k)} \rangle + \sum_{k=1}^{\infty} \langle \cos \theta_{BA}^{(k)} \rangle, \qquad (12.43)$$

where $\langle \cos \theta_{ij}^{(k)} \rangle$ is the average cosine of the angle between any given jump of the i species and the k'th succeeding jump of the j species.

12.3.1 Phenomenological Coefficients, Tracer Diffusivities, and Jump Models

Tracer diffusion coefficients are directly accessible from experiments (see Chap. 13). On the other hand, direct measurements of phenomenological coefficients are difficult. Accordingly, there is interest in relations between the tracer diffusivities and phenomenological coefficients. There are also expressions for phenomenological coefficients in terms of atomistic hopping models. In what follows, we consider examples of such relations and expressions.

Relations for Self-diffusion: In the previous section, we obtained equations which related kinetic coefficients for an element crystal to the tracer self-diffusion coefficient. After some algebra we get

$$L_{A^*A^*} \approx \frac{C_{A^*} D_A^{A^*}}{k_B T} \tag{12.44}$$

and

$$L_{AA^*} = L_{A^*A} = \frac{C_{A^*} D_A^{A^*}}{k_B T} \frac{(1-f)}{f}. \tag{12.45}$$

The off-diagonal transport coefficient is related to tracer correlation. For uncorrelated diffusion, i.e. for $f = 1$, it vanishes.

Relations for Dilute Binary Alloys: For a dilute alloy of solute B in solvent A, i.e. for $N_B \to 0$ we have $\Phi \to 1$. It can be shown that L_{BB}/N_B approaches a finite value, whereas L_{AB}/N_A goes to zero. Then, we get from the second equation (12.28):

$$D_B^I(0) = k_B T \frac{L_{BB}}{N_B} = D_A^{B^*}(0), \tag{12.46}$$

where $D_A^{B^*}(0)$ is the solute diffusion coefficient at infinite dilution. It is measurable, e.g. by using tracer B^*. From the first equation (12.28) we have

$$D_A^I(0) = \frac{L_{AA}}{C_A}\left(1 - \frac{L_{AB} N_A}{L_{BB} N_B}\right) k_B T \neq \frac{L_{AA}}{N_A} k_B T. \tag{12.47}$$

In this case the off-diagonal term L_{AB} cannot be neglected.

It may be of some interest to consider in a very dilute alloy the phenomenological coefficients in the framework of the five-frequency model of diffusion suggested by LIDIARD. This model is described in Chap. 7. It is very useful for describing solute and solvent diffusion in a very dilute binary alloy. We remind the reader that in this model five exchange jump-rates between vacancies and A or B atoms are specified, namely: ω_2 for solute-vacancy exchange, ω_1 for rotation of the vacancy-solute complex, ω_3 (ω_4) for dissociation (association) of the vacancy-solute complex, and ω for vacancy jumps in the solvent. According to [4] the phenomenological coefficients are:

12.3 The Phenomenological Coefficients

$$L_{AA} = \frac{d^2 \omega C_V}{k_B T}\left(1 - 12 N_B \frac{\omega_4}{\omega_3}\right)\omega$$

$$- \frac{d^2 C_V N_B}{k_B T}\frac{\omega_4}{\omega_3}\left[\frac{40\omega_1\omega_3 + 40\omega_3^2 + 14\omega_2\omega_3 + 4\omega_1\omega_2}{\omega_1 + \omega_2 + 7\omega_3/2} + \frac{7\omega_3\omega}{\omega_4}\right]$$

$$L_{BB} = \frac{d^2 C_V N_B}{k_B T}\frac{\omega_4}{\omega_3}\frac{\omega_2(\omega_1 + 7\omega_3/2)}{\omega_1 + \omega_2 + 7\omega_3/2}$$

$$L_{AB} = \frac{d^2 C_V N_B}{k_B T}\frac{\omega_4}{\omega_3}\frac{\omega_2(-2\omega_1 + 7\omega_3)}{\omega_1 + \omega_2 + 7\omega_3/2} \tag{12.48}$$

The so-called *Heumann relation* [18] was also derived on the basis of the five-frequency model. It can be shown that in the limit $C_B \to 0$, the ratio of $L_{AB}(0)/L_{BB}(0)$ is given by:

$$\frac{L_{AB}(0)}{L_{BB}(0)} = \frac{D_A^{A^*}(0)}{D_A^{B^*}(0)}\left[\frac{1}{f} - \frac{D_A^I(0)}{D_A^{A^*}(0)}\right], \tag{12.49}$$

where $D_A^{A^*}(0), D_A^{B^*}(0)$ are the tracer diffusivities of A and B and $D_A^I(0)$ the intrinsic diffusion coefficient of A in the limit $C_B \to 0$.

Relations for Concentrated Binary Alloys: Let us first briefly consider relations based on the original DARKEN equations [11]. In essence, Darken neglects the off-diagonal coefficients entirely. If this is assumed, the diagonal phenomenological coefficients can be related to the corresponding tracer diffusion coefficient in a binary alloy:

$$L_{AA} \approx \frac{N_A D_A^{A^*}}{k_B T}, \quad L_{BB} \approx \frac{N_B D_B^{B^*}}{k_B T}, \quad L_{AB} = 0. \tag{12.50}$$

The Darken equations neglect all correlation information as embodied in tracer correlation factors, collective correlation factors, and vacancy-wind factors [24]. The neglect of the off-diagonal phenomenological coefficients can be dangerous. However, in most cases, it is reasonable as a first approximation.

The *random alloy model* with the vacancy mechanism introduced by MANNING is probably the most important model for dealing with diffusion in concentrated alloys that are disordered [12, 13]. The atomic species A and B exchange sites with vacancies with the jump rates ω_A or ω_B, respectively. Then, the Darken relation is replaced by the *Darken-Manning relation* which includes the *vacancy wind corrections* (see Chap. 10). In the random alloy model, the phenomenological coefficients are directly related to the tracer diffusion coefficients via the *Manning relations*:

$$L_{ii} = \frac{N_i D_i^*}{k_B T}\left[1 + \frac{(1-f)}{f}\frac{N_i D_i^*}{N_A D_A^{A^*} + N_B D_B^{B^*}}\right],$$

$$L_{AB} = \frac{(1-f)}{f}\frac{N_A D_A^{A^*} N_B D_B^{B^*}}{k_B T(N_A D_A^{A^*} + N_B D_B^{B^*})}. \tag{12.51}$$

f is the tracer correlation factor: e.g., $f = 0.781$ for the fcc lattice and $f = 0.732$ for the bcc lattice.

The random alloy model seems to have more general validity. Computer simulations by MURCH AND COWORKERS have shown that the Manning relations are quite good approximations, even for ordered alloys, at least at low levels of order [14–16]. The Manning relations have also been re-derived for ordered alloys [17].

12.3.2 Sum Rules – Relations between Phenomenological Coefficients

Various relations usually called *sum rules* have been identified between the phenomenological coefficients in randomly mixed systems. Sum rules reduce the number of independent phenomenological coefficients.

As an example, we consider the sum rules between the phenomenological coefficients in the random alloy model with a vacancy mechanism. The atom-vacancy exchange rates ω_i can be considered in two rather different ways. In the first way, one can consider them as explicit jump rates that depend only on the species i and not on the surroundings. For example, ω_A in the binary random alloy then simpls represents the average jump rate of a given A atom at all compositions and environments. In the second more general way, one considers that the ω_i represent an average jump rate of species i as it migrates through the lattice sampling the various environments. Since the average environment of an atom will change with composition, the ω_i can also be expected to change with composition.

MOLEKO AND ALLNATT [19] identified the following sum rules for an n-component random alloy for diffusion via a vacancy mechanism:

$$\sum_{i=1}^{n} L_{ij} \frac{\omega_j}{\omega_i} = \frac{Zd^2}{6k_\mathrm{B}T} C_V \omega_j N_j, \quad j = 1, \ldots, n. \tag{12.52}$$

Here Z is the coordination number and d the jump distance. Equations (12.52) relate the phenomenological coefficients to the vacancy-atom exchange rates and reduce the number of independent phenomenological coefficients.

In addition, we mention that the sum rules can be restated in terms of collective correlation factors as:

$$\sum_{i=1}^{n} f_{ij}^{(j)} \frac{\omega_j}{\omega_i} = 1, \quad j = 1, \ldots n. \tag{12.53}$$

For a *binary random alloy* the sum rule relations (12.52) reduce to two equations:

$$L_{AA} + L_{AB}\frac{\omega_A}{\omega_B} = \frac{Zd^2}{6k_\mathrm{B}T}C_V\omega_A N_A, \tag{12.54}$$

$$L_{BB} + L_{AB}\frac{\omega_B}{\omega_A} = \frac{Zd^2}{6k_\mathrm{B}T}C_V\omega_B N_B. \tag{12.55}$$

Hence there is only one independent coefficient and not three. In a ternary random alloy, the number of independent coefficients is reduced from six to three. On the other hand, the reader should keep in mind that the random alloy model introduces two vacancy jump rates for a binary alloy and three jump rates for a ternary alloy.

For the derivation of sum rules we refer the reader to the original papers. The sum rules introduced by MOLEKO AND ALLNATT for the random alloy model were the first ones that were discovered. In the meantime, various sum rules have been identified for a number of other mechanisms and situations by MURCH AND COWORKERS. Such situations include the dumb-bell interstitial mechanism in the binary random alloy [20], the divacancy mechanism in the fcc random alloy [21], the vacancy-pair mechanism in ionic materials with randomly mixed cations [22], and for an intermetallic compound with randomly mixed sublattices [23]. Interdiffusion data in multicomponent alloys as a source of quantitative fundamental diffusion information are summarised in [24].

References

1. S.R. de Groot, P. Mazur, *Non-equilibrium Thermodynamics*, North-Holland, Amsterdam, 1962
2. R. Haase, *Thermodynamics of Irreversible Processes*, Addison Wesley, Reading, 1969
3. J.H. Kreuzer, *Non-equilibrium Thermodynamics and its Statistical Foundations*, Clarendon Press, Oxford, 1981
4. J. Philibert, *Atom Movements – Diffusion and Mass Transport in Solids*, Les Editions de Physique, Les Ulis, 1991
5. A.R. Allnatt, A.B. Lidiard, *Atomic Transport in Solids*, Cambridge University Press, 1993
6. G.E. Murch, I.V. Belova, *Phenomenological Coefficients in Solid-State Diffusion: an Introduction*, in: *Diffusion Fundamentals*, p. 105; J. Kärger, F. Grinberg, P. Heitjans (Eds.), Universitätsverlag Leipzig, 2005
7. A.R. Allnatt, J. Phys. C: Solid State Phys. **15**, 5605 (1982)
8. A. Einstein, Annalen der Physik **17**, 549 (1905)
9. A.R. Allnatt, E.L. Allnatt, Philos. Mag. A **49**, 625 (1984)
10. I.V. Belova, G.E. Murch, Philos. Mag. A **49**, 625 (2000)
11. L.S. Darken, Trans-Am. Inst. Min. Metall. Engrs. **175**, 184 (1948)
12. J.R. Manning, *Diffusion Kinetics for Atoms in Crystals*, van Norstrand, Princeton, 1968
13. J.R. Manning, Phys. Rev. **B4**, 1111 (1971)
14. L. Zhang, W.A. Oates, G.E. Murch, Philos. Mag. **60**, 277 (1989)

15. I.V. Belova, G.E. Murch, J. Phys. Chem. Solids **60**, 2023 (1999):
16. I.V. Belova, G.E. Murch, Philos. Mag. **A 78**, 1085 (1998)
17. I.V. Belova, G.E. Murch, Philos. Mag. **A 75**, 1715 (1997)
18. Th. Heumann, J. Phys. F: Metal Physics **9**, 1997 (1979)
19. L.K. Moleko, A.R. Allnatt, Philos. Mag. **A 59**, 141 (1989)
20. S. Sharma, D.K. Chatuvedi, I.V. Belova, G. E. Murch, Philos. Mag. **A 81**, 431 (2001)
21. I.V. Belova, G.E. Murch, Philos. Mag. Lett. **81**, 101 (2001)
22. I.V. Belova, G.E. Murch, Philos. Mag. Lett. **84**, 3637 (2004)
23. I.V. Belova, G.E. Murch, Defect and Diffusion Forum **194–199**, 547 (2001)
24. I.V. Belova, G.E. Murch, Defect and Diffusion Forum **263**, 1 (2007)

Part II

Experimental Methods

13 Direct Diffusion Studies

13.1 Direct *versus* Indirect Methods

There are numerous experimental methods for studying diffusion in solids. They can be grouped roughly into two major categories (see Table 13.1).

Direct methods are based on the laws of Fick and the phenomenological definition of the diffusion coefficient therein. They are sensitive to long-range diffusion and in this sense they are macroscopic.

The radiotracer method is the standard technique for the study of self- and solute diffusion, if radioisotopes with suitable half-lives are available. The tracer method is element-selective and due to the use of nuclear counting facilities highly sensitive. It can cover a large range of diffusivities provided that both mechanical and sputter sectioning techniques are used for depth profiling. Further profiling techniques for diffusion studies are secondary ion mass

Table 13.1. Survey of experimental methods for direct and indirect diffusion studies in solids

Direct methods	Indirect methods
Tracer diffusion plus depth profiling	**Mechanical spectroscopy** (after effect, internal friction, Gorski effect)
Chemical diffusion plus profiling *Profiling techniques*: - Mechanical and sputter profiling	**Magnetic relaxation** (for ferromagnetc materials) **Nuclear magnetic relaxation** (NMR):
- Secondary ion mass spectrometry (SIMS)	- Line-shape spectroscopy
- Electron microprobe analysis (EMPA)	- Spin lattice relaxation spectroscopy
- Auger electron spectroscopy (AES)	- Spin alignment experiments (SAE)
Spreading resistance profiling (SRP) for semiconductors	**Impedance spectroscopy** (IS) for ion conductors
Rutherford backscattering (RBS)	**Mössbauer spectroscopy** (MBS)
Nuclear reaction analysis (NRA)	**Quasielastic neutron scattering** (QENS)
Field gradient NMR (FG-NMR) **Pulsed fieldgradient NMR** (PFG-NMR)	

spectrometry (SIMS), electron microprobe analysis (EMPA), Auger electron spectroscopy (AES). SIMS and AES both utilise sputter profiling and are appropriate for small diffusion distances and low diffusivities. AES is applicable for diffusion of foreign atoms, since it discriminates between different elements but not between isotopes of the same element. EMPA is the major tool for the study of chemical diffusion (interdiffusion) and suitable for relatively large diffusion coefficients, since the size of the specimen volume excited by the electron beam limits the depth resolution. Rutherford back scattering (RBS) or nuclear reaction analysis (NRA) are both nuclear techniques, which use ion beams of several MeV energy for profile analysis. RBS is particular suitable for heavy solutes in a light solvent whereas NRA is appropriate for some light solutes including hydrogen. A prerequisite for NRA studies is a nuclear reaction with a narrow resonance. The penetration depth and energy straggling of the ion beam limit RBS and NRA to small diffusivities. Spreading resistance profiling (SRP) of dopant diffusion profiles in semiconductors is direct in the sense that it provides a depth profile of the spreading resistance. However, some transformation is needed to convert spreading resistance to dopant concentration. Usually, NMR techniques are indirect (see below). However, field-gradient NMR, either with static field gradients (FG-NMR) or with pulsed field gradients (PFG-NMR), are methods that permit diffusivity measurements without referring to a microscopic model.

Indirect methods are not directly based on the laws of Fick. Indirect methods usually study phenomena which are influenced by the diffusion jumps of atoms.

Some of these methods are sensitive to one or a few atomic jumps only. Quantities such as relaxation times, relaxation rates, or line-widths are measured and the mean residence time of the diffusing atoms, $\bar{\tau}$, is deduced therefrom. A microscopic model of the atomic jump process is needed to deduce the diffusivity via the Einstein-Smoluchowski relation (see Chap. 4). In simple cases, the (uncorrelated) diffusivity D_E is given by

$$D_E = \frac{d^2}{6\bar{\tau}}, \qquad (13.1)$$

where d denotes the length of an atomic jump.

The numbers of atomic jumps performed by the diffusing species during anelastic or magnetic after-effect measurements (e.g., Snoek or Zener effect) are typically of the order of one. Internal friction studies are particularly sensitive to diffusion processes, when the atomic jump rate, $1/\bar{\tau}$, is comparable with the vibration frequency of the internal friction device. When applicable, these techniques can monitor very small to small diffusion coefficients. The Gorski effect is an anelastic after-effect, which can be observed in hydrogen-metal systems. Its origin is the hydrogen redistribution in a strained sample. The associated after-effect can be monitored because hydrogen diffusion is a very fast process.

13.1 Direct versus Indirect Methods

Among the nuclear methods, nuclear magnetic relaxation (NMR) covers the widest range of diffusivities. Spin-alignment experiments (SAE), line-shape spectroscopy, and spin-lattice relaxation spectroscopy can be used. Favourable are materials with large gyromagnetic ratios and small non-diffusive contributions to line-width or relaxation rates. Mössbauer spectroscopy (MBS) requires a suitable Mössbauer isotope. The usual workhorse of MBS is ^{57}Fe, which permits studies of Fe diffusion. There is a short list of further Mössbauer probes such as ^{119}Sn, ^{151}Eu, and ^{161}Dy. Quasielastic neutron scattering (QENS) is applicable to isotopes with large enough quasi-elastic scattering cross sections. Both techniques are limited to relatively fast diffusion processes. The main virtues of MBS and QENS are that these techniques can unravel microscopic information such as jump length and jump direction of the diffusing atoms.

Impedance spectroscopy (IS) measures the complex conductivity of ion-conducting materials as a function of the frequency. For materials in which only one type of ion contributes to the dc conductivity, Eq. (11.26) can be used to 'translate' the dc conductivity, σ_{dc}, into a charge diffusion coefficient of the ions, D_σ (see Chap. 11).

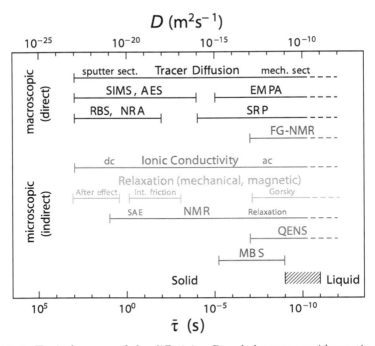

Fig. 13.1. Typical ranges of the diffusivity D and the mean residence time $\bar{\tau}$ of direct and indirect methods for diffusion studies

Figure 13.1 shows typical ranges of diffusivity (D: *upper abscissa*) and mean residence time ($\bar{\tau}$: *lower abscissa*) for direct and indirect methods, respectively. D and $\bar{\tau}$ have been converted via Eq. (13.1), adopting a typical jump length d in solids of some tenths of a nanometer. The length scale for the diffusion processes which are probed by IS and NMR varies with the applied measuring frequency. Thus a combination of various techniques and/or experimental devices may be desirable.

The present chapter is devoted to direct methods. Diffusion of atoms in a certain direction x is described by one of the following versions of Fick's second law (see Chaps. 2 and 10)

$$\frac{\partial C}{\partial t} = D \frac{\partial^2 C}{\partial x^2} \quad \text{or} \quad \frac{\partial C}{\partial t} = \frac{\partial}{\partial x}\left[\tilde{D}(C)\frac{\partial C}{\partial x}\right]. \qquad (13.2)$$

The first version corresponds to a concentration-independent diffusivity. As outlined below, in experiments with trace elements very tiny amounts of the diffusing species can be studied and the chemical composition of the sample is unchanged. The second version is applicable for a concentration-dependent diffusion coefficient, $\tilde{D}(c)$, denoted as the *interdiffusion- or chemical diffusion coefficient* (see Chap. 10). We discuss relaxation and internal friction methods based on the anelastic behaviour of materials in Chap. 14, the nuclear methods NMR, MBS, and QENS in Chap. 15, and the electrical methods IS and SRP in Chap. 16. For further details the reader may consult textbooks [1–3], reviews [4, 5], and conference proceedings [6–9].

13.2 The Various Diffusion Coefficients

Before discussing experimental methods in detail, we describe situations which entail various types of diffusion coefficients. In this section, we distinguish the various diffusion coefficients by lower and upper indices. We drop the indices in the following text again, whenever it is clear which diffusion coefficient is meant. We concentrate on lattice (bulk) diffusion in unary and binary systems. Diffusion in ternary systems produces complexities, which are not treated in this book. We focus on lattice diffusion since diffusion along grain boundaries and dislocations is considered separately in Chaps. 32 and 33.

13.2.1 Tracer Diffusion Coefficients

In diffusion studies with trace elements (labelled by their radioactivity or by their isotopic mass) tiny amounts of the diffusing species (in the ppm range or even less) can be used. Although in a diffusion experiment a concentration gradient of the trace element is necessary, the total tracer concentration can be kept so small that the overall composition of the sample during the investigation practically does not change. From an atomistic viewpoint this implies that a tracer atom is not influenced by other tracer atoms. The analysis of such a diffusion experiment yields a *tracer diffusion coefficient*, which is independent of tracer concentration. Tracers are appropriate to study self-diffusion of matrix atoms. They can also be used to study diffusion of foreign atoms under very dilute conditions. The latter phenomenon is called *impurity diffusion*. The expressions *foreign diffusion* or *solute diffusion* are also used.

Self-diffusion: The diffusion of A atoms in a solid element A is called *self-diffusion*. Studies of self-diffusion with tracers utilise an isotope A^* of the same element. A typical initial configuration for a tracer self-diffusion experiment is illustrated in Fig. 13.2a. The *tracer self-diffusion coefficient* $D_A^{A^*}$ is obtained from the diffusion broadening of a narrow initial distribution.

The connection between the macroscopically defined tracer self-diffusion coefficient and the atomistic picture of diffusion is the *Einstein-Smoluchowski relation* discussed in Chap. 4. In simple cases, it reads

$$D_A^{A^*} = f D_E \quad \text{with} \quad D_E = \frac{d^2}{6\bar{\tau}}, \tag{13.3}$$

where d denotes the jump length and $\bar{\tau}$ the mean residence time of an atom on a particular but arbitrary site of the crystal. Equation (13.3) is applicable for cubic structures when only jumps to nearest-neighbour sites occur. f is the tracer correlation factor discussed in Chap. 7. For self-diffusion in cubic crystals f is usually a known numeric factor, which depends on the lattice geometry and the diffusion mechanism. In some textbooks the quantity D_E is denoted as the *Einstein diffusion coefficient*. In the author's opinion, the notation Einstein diffusion coefficient is misleading, since the original Einstein-Smoluchowski equation relates the total macroscopic mean square displacement of atoms to the diffusion coefficient (see Chap. 4), in which correlation effects are included.

In a homogeneous binary $A_X B_{1-X}$ alloy or compound two tracer self-diffusion coefficients for A^* and B^* tracer atoms can be measured using the initial configuration displayed in Fig. 13.2b. We denote the corresponding tracer diffusion coefficients by $D_{A_X B_{1-X}}^{A^*}$ and $D_{A_X B_{1-X}}^{B^*}$. In general, the two tracer diffusivities are not equal:

$$D_{A_X B_{1-X}}^{A^*} \neq D_{A_X B_{1-X}}^{B^*}.$$

Depending on the specific alloy system one component will be more mobile than the other. The difference depends on the crystal structure of the material, on the atomic mechanisms which mediate diffusion, and on the constituents themselves. For example, in B2 structured intermetallic compounds the tracer diffusivities of the constituents are usually similar, whereas in $L1_2$ or DO_3 structured compounds the component diffusivities can be very different (see [10–12] and Chap. 20). In ionic crystals and ceramics the diffusivities of the components also can differ significantly (see Chaps. 26 and 27). Of

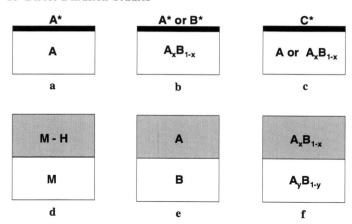

Fig. 13.2a–f Initial configurations for direct diffusion studies: **a)** Thin layer of A^* on solid A: tracer self-diffusion in pure elements. **b)** Thin layer of A^* or B^* on homogeneous A-B alloy: tracer self-diffusion of alloy components. **c)** Thin layer of C^* on solid A or on a homogeneous alloy: Impurity diffusion. **d)** Diffusion couple between metal-hydrogen alloy and a pure metal. **e)** Diffusion couple between pure end-members. **f)** Diffusion couple between two homogeneous alloys

course, both component diffusivities are also functions of the thermodynamic variables temperature and pressure and in general also depend on the composition.

Impurity Diffusion: When the diffusion of a trace solute C^* in a monoatomic solvent A or in a homogeneous binary solvent $A_X B_{1-X}$ (Fig. 13.2c) is measured, the tracer diffusion coefficients

$$D_A^{C^*} \text{ and } D_{A_X B_{1-X}}^{C^*}$$

are obtained. These diffusion coefficients are denoted as the *impurity diffusion coefficients* or sometimes also as the *foreign diffusion coefficients*.

13.2.2 Interdiffusion and Intrinsic Diffusion Coefficients

For interdiffusion studies on binary alloys, diffusion couples are formed consisting either of two elements from a continuous solid solution alloy or of two homo-phase alloys with different compositions ($A_X B_{1-X}$ and $A_Y B_{1-Y}$) within the same phase field (Fig. 13.2 e and f). Usually the thicknesses of the couple members are chosen large as compared to the average diffusion length. Then, each couple member can be considered to be semi-infinite. Some typical examples are:

– Pure end-member diffusion couples consisting of two slices of pure elements joined together (Ni|Pd, Au|Ag, Si|Ge, ...).

- Incremental diffusion couples consisting of two slices of homogenous alloys joined together
  ($Fe_{75}Al_{25}|Fe_{60}Al_{40}$, $Ni_{50}Pd_{50}|Ni_{70}Pd_{30}$, $Ni|Ni_{70}Pd_{30}$, ...).
- Diffusion couples which involve solutions of hydrogen
  (Pd-H|Pd, $Ag_{1-X}H_X|Ag_{1-Y}H_Y$, ...).

During hydrogen interdiffusion the metal atoms are practically immobile. Then the intrinsic diffusion coefficient and the chemical diffusion coefficient of hydrogen are identical.

Interdiffusion: The phenomenon of *interdiffusion* or *chemical diffusion* has been discussed in Chap. 10. We have seen that the interdiffusion coefficient $\tilde{D}$ can be deduced either by a Boltzmann-Matano analysis for systems without volume change or for systems with volume changes from a Sauer-Freise type analysis of the experimental concentration-distance profile. Using one of these methods, an interdiffusion coefficient

$$\tilde{D} = \tilde{D}(C)$$

for each composition C in the diffusion zone is obtained (see Chap. 10). $\tilde{D}$ characterises the intermixing of A and B atoms.

Intrinsic Diffusion: The intrinsic diffusion coefficients D_A^I and D_B^I describe diffusion of the components A and B of a binary alloy relative to the lattice planes. As discussed in Chap. 10, a determination the intrinsic diffusivities requires two measurements. The *Kirkendall velocity*, v_K, and the interdiffusion coefficient, $\tilde{D}$, permit to deduce intrinsic diffusivities as described in Chap. 10. Either the Darken equations or the more precise Darken-Manning equations can be used.

13.3 Tracer Diffusion Experiments

Many of the reliable diffusion studies on solids have been performed by radiotracer techniques as evidenced in textbooks [1–3], reviews [4, 5, 10–12] and conference proceedings [6–9]. Due to the high sensitivity of nuclear counting facilities, radiotracer studies are often superior to other techniques. A very important advantage is the fact that self-diffusion – the most basic diffusion process in a solid material – can be studied in a straightforward manner using radioisotopes of matrix atoms. Then, the tracer self-diffusion coefficient is obtained. Foreign atom diffusion studies can also be performed with tiny amounts of tracer. Typical tracer concentrations are less or even much less than a ppm, if radioisotopes with high specific activity are used. In this way, diffusion in a chemically homogeneous solid can be investigated. Complications due to chemical gradients play no rôle and the thermodynamic factor equals unity. In a tracer diffusion experiment atoms are usually labelled by their radioactivity. When stable isotopes are used as tracers the 'label' is

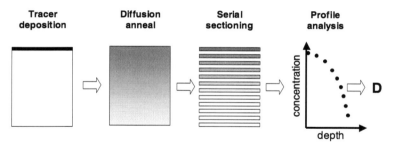

Fig. 13.3. Schematic illustration of the tracer method: The major steps – deposition of the tracer, diffusion anneal, serial sectioning, and evaluation of the penetration profile – are indicated

the isotopic mass. Sometimes, in the case of impurity diffusion atoms are labelled just by their chemical nature. The major steps of a tracer diffusion experiment are indicated in Fig. 13.3.

Preparation of a diffusion sample usually involves preparation of a flat, strain-free surface. Polishing of metals, intermetallics, semiconductors, and glasses is usually performed by standard metallographic procedures. Soft materials such as organic crystals or polymers can be cut with a microtome. Mechanical methods produce the best flatness but introduce strain. Etching or electropolishing and/or a pre-diffusion anneal should be used to remove cold-worked material.

The tracer is deposited onto the polished, flat surface of the diffusion sample. Evaporation, dripping of a liquid solution, and electrodeposition of the radiotracer onto the surface are the major deposition techniques. Complete homogeneity of the deposit is not necessary as long as iregularties in its thickness are small as compared to the mean diffusion length and as long as the same sample area is counted in each section [13]. Implantation of the radioisotope is more laborious but offers sometimes advantages [4, 14]: for example, tracer hold-up by surface oxide layers can be avoided by implantation.

Following the tracer deposition, an isothermal diffusion anneal is performed at temperature T for some diffusion time t. During the diffusion anneal the sample is usually encapsulated in a quartz ampoule under vacuum or inert atmosphere (e.g., Ar). For temperatures below 1500 K quartz ampoules and resistance furnaces are frequently used. For higher temperatures more sophisticated annealing techniques (e.g., electron-beam heating) are necessary.

Suppose that a thin layer of tracer atoms (M atoms per unit area) has been deposited at the surface $x = 0$ of a semi-infinite sample. Let us further suppose that tracer losses and tracer hold-up at the surface can be avoided. Then, the concentration distribution after a diffusion anneal is described by (see Chap. 3)

$$C(x,t) = \frac{M}{\sqrt{\pi Dt}} \exp\left(-\frac{x^2}{4Dt}\right). \qquad (13.4)$$

We recall that Eq. (13.4) is the *thin-film solution (Gaussian solution)* of Fick's second law and that the quantity $\sqrt{Dt}$ is a typical *diffusion length*.

An alternative possibility of tracer deposition is *ion-implantation* using an accelerator. Implantation can be a very suitable deposition technique for materials like Al, which readily form a thin oxide layer when exposed to air. After implantation the tracer atoms form a buried layer. For a fixed implantation energy their distribution as a function of range x is given by[1]

$$C(x,0) = \frac{M}{\sqrt{2\pi \Delta R_p}} \exp\left[-\frac{(x-R_p)^2}{4Dt}\right]. \qquad (13.5)$$

R_p denotes the *mean projected range of implantation* and ΔR_p the *standard deviation of the projected range*. Both quantities depend on the implantation energy, on the tracer, and on the matrix. Typical values of R_p lie in the range 20 to 100 nm for implantation energies of 50 keV. A layer buried in great depth broadens during diffusion annealing to

$$C(x,t) = \frac{M}{\sqrt{2\pi \Delta R_p + 4Dt}} \exp\left[-\frac{(x-R_p)^2}{2\Delta R_p + 4Dt}\right]. \qquad (13.6)$$

However, after implantation the tracer layer will usually be close to the sample surface. Then, Eq. (13.6) must be modified. If the surface acts either as a 'perfect mirror' or as a 'perfect sink' for tracer atoms the solution of Fick's second equation can be written as

$$C(x,t) = \frac{M}{\sqrt{2\pi \Delta R_p + 4Dt}} \left(\exp\left[-\frac{(x-R_p)^2}{2\Delta R_p + 4Dt}\right] \pm \exp\left[-\frac{(x+R_p)^2}{2\Delta R_p + 4Dt}\right]\right). \qquad (13.7)$$

The minus-sign stands for a perfect sink and the plus-sign for a perfect reflection at $x = 0$. Both boundary conditions are approximations and may not always hold in practical cases. If this is the case, numerical solutions of Fick's equation should be used.

13.3.1 Profile Analysis by Serial Sectioning

The major task of a diffusion experiment is to study the concentration-depth profile and to deduce the diffusion coefficient by comparison with the corresponding solution of Fick's second law. Let us assume that the experimental conditions were chosen in such a way that the deposited layer is thin compared with the diffusion length $\sqrt{Dt}$. Then, the distribution after the diffusion anneal is described by Eq. (13.4).

---

[1] For simplicity reasons we neglect channelling effects. Channeling can be neglected if the direction of the implntation beam avoids directions of high crystal symmetry,

The best way to determine the resulting concentration-depth profile is serial sectioning of the sample and subsequent determination of the amount of tracer per section. To understand sectioning the reader should think in terms of isoconcentration contours. For lattice diffusion these are parallel to the original surface, on which the thin layer is deposited, and perpendicular to the diffusion direction. The most important criterion of sectioning is the parallelness of sections to the isococentration contours. For radioactive tracers the specific activity per section, $A(x)$, is proportional to the tracer concentration:

$$A(x) = kC(x). \tag{13.8}$$

Here k is a constant, which depends on the nature and energy of the nuclear radiation and on the efficiency of the counting device. The specific activity is obtained from the section mass and the count rate. The latter can be measured in nuclear counting facilities such as γ- or β-counting devices. Usually, the count-rate must be corrected for the background count-rate of the counting device. For short-lived radioisotopes half-life corrections are also necessary. According to Eq. (13.4) a diagram of the logarithm of the specific activity *versus* the penetration distance squared is linear. From its slope, $(4Dt)^{-1}$, and the diffusion time the tracer diffusivity D is obtained.

In an ordinary thin-layer sectioning experiment, one wishes to measure diffusion over a drop of about three orders of magnitude in concentration. About twenty sections suffice to define a penetration profile. The section thickness Δx required to get a concentration decrease of three orders of magnitude over 20 sections is $\Delta x \approx \sqrt{Dt}/3.8$. Thicker sections should be avoided for the following reason: in a diffusion penetration profile the average concentrations (specific activities) per section are plotted *versus* the position of the distance of the center of each section from the surface. Errors caused by this procedure are only negligible if the sections are thin enough.

The radiotracer deposited on the front face of a sample may rapidly reach the side surfaces of a sample by surface diffusion or via transport in the vapour phase and then diffuse inward. To eliminate lateral diffusion effects, one usually removes about $6\sqrt{Dt}$ from the sample sides before sectioning. For studies of bulk diffusion, single crystalline samples rather than polycrystalline ones should be used to eliminate the effects of grain-boundary diffusion, which is discussed in Chap. 31. If no single crystals are available coarse-grained polycrystals should be used.

The following serial-sectioning techniques are frequently used for the determination of diffusion profiles:

Mechanical sectioning: For diffusion lengths, $\sqrt{Dt}$, of at least several micrometers mechanical techniques are applicable (for a review see [4]). Lathes and microtomes are appropriate for ductile samples such as some pure metals (Na, Al, Cu, Ag, Au, ...) or polymers. For brittle materials such as intermetallics, semiconductors, ionic crystals, ceramics, and inorganic glasses grinding is a suitable technique.

13.3 Tracer Diffusion Experiments   219

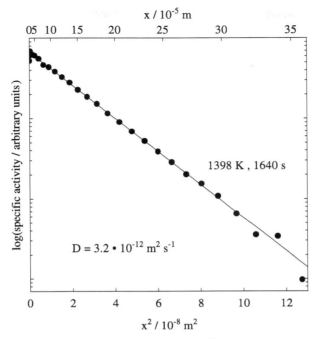

Fig. 13.4. Penetration profile of the radioisotope ^{59}Fe in Fe$_3$Si obtained by grinder sectioning [15]. The *solid line* represents a fit of the thin-film solution of Fick's second law

For extended diffusion anneals and large enough diffusivities, $D > 10^{-15}\,\mathrm{m^2\,s^{-1}}$, lathe sectioning can be used. Diffusivities $D > 10^{-17}\,\mathrm{m^2\,s^{-1}}$ are accessible via microtome sectioning. In cases where the half-life of the isotope permits diffusion anneals of several weeks, grinder sectioning can be used for diffusivities down to $10^{-18}\,\mathrm{m^2\,s^{-1}}$. Figure 13.4 shows a penetration profile of the radioisotope ^{59}Fe in the intermetallic Fe$_3$Si, obtained by grinder sectioning [15]. Gaussian behaviour as stated by Eq. (13.4) is observed over several orders of magnitude in concentration.

Ion-beam Sputter Sectioning (IBS): Diffusion studies at lower temperatures often require measurements of very small diffusivities. Measurements of diffusion profiles with diffusion lengths in the micrometer or sub-micrometer range are possible using sputtering techniques. Devices for serial sectioning of radioactive diffusion samples by ion-beam sputtering (IBS) are described in [16, 17]. Figure 13.6 shows a schematic drawing of such a device. Oblique incidence of the ion beam and low ion energies between 500 and 1000 eV are used to minimise knock-on and surface roughening effects. The sample (typically several mm in diameter) is rotated to achieve a homogeneous lateral sputtering rate. The sputter process is discussed in some detail below and

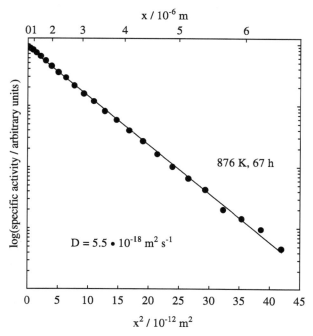

Fig. 13.5. Penetration profile of the radioisotope ^{59}Fe in Fe$_3$Al obtained by sputter sectioning [18]. The *solid line* represents a fit of the thin-film solution of Fick's second law

illustrated in Fig. 13.8, in connection with secondary ion mass spectroscopy (SIMS). An advantage of IBS devices lies in the fact that neutral atoms are collected, which comprise by far the largest amount (about 95 to 99 %) of the off-sputtered particles. In contrast, SIMS devices (see below) analyse the small percentage of secondary ions, which depends strongly on sputter- and surface conditions.

Sectioning of shallow diffusion zones, which correspond to average diffusion lengths between several ten nm and 10 μm, is possible using IBS devices. For a reasonable range of annealing times up to about 10^6 s, a diffusivity range between 10^{-23} m^2 s^{-1} and 10^{-16} m^2 s^{-1} can be examined. Depth calibration can be performed by measuring the weight loss during the sputtering process or by determining the depth of the sputter crater by interference microscopy or by profilometer techniques. The depth resolution of IBS and SIMS is limited by surface roughening and atomic mixing processes to about several nm. A penetration profile of ^{59}Fe in the intermetallic Fe$_3$Al [18], obtained with the sputtering device described in [17] is displayed in Fig. 13.5.

From diffusion profiles of the quality of Figs. 13.4 and 13.5, diffusion coefficients can be determined with an accuracy of a few percent. A determi-

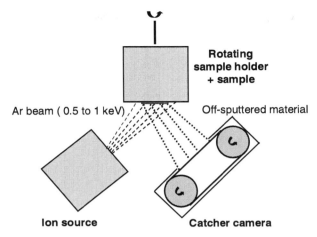

Fig. 13.6. Ion-beam sputtering device for serial sectioning of diffusion samples

nation of the absolute tracer concentration is not necessary since the diffusion coefficient is obtained from the slope, $-1/(4Dt)$, of such profiles.

Deviations from Gaussian behaviour in experimental penetration profiles (not observed in Figs. 13.4 and 13.5) may occur for several reasons:

1. *Grain-boundary diffusion*: Grain boundaries in a polycrystalline sample act as diffusion short-circuits with enhanced mobility of atoms. Grain boundaries usually cause a 'grain-boundary tail' in the deeper penetrating part of the profile (see Chap. 32 and [19]). In the 'tail' region the concentration of the diffuser is enhanced with respect to lattice diffusion. Then, one should analyse the diffusion penetration profile in terms of lattice diffusion and short-circuit diffusion terms:

$$C(x,t) = \frac{M}{\sqrt{\pi Dt}} \exp\left(-\frac{x^2}{4Dt}\right) + C_0 \exp(-A\ x^{6/5}). \qquad (13.9)$$

Here C_0 is constant, which depends on the density of grain boundaries. The quantity A is related to the grain-boundary diffusivity, the grain-boundary width, and to the lattice diffusivity. The grain-boundary tails can be used for a systematic study of grain-boundary diffusion in bi- or polycrystalline samples. Grain-boundary diffusion is discussed in Chap. 32.

2. *Evaporation losses of tracer*: A tracer with high vapour pressure will simultaneously evaporate from the surface and diffuse into the sample. Then, the thin-film solution (13.4) is no longer valid. The outward flux of the tracer will be proportional to the tracer concentration at the surface:

$$D\left(\frac{\partial C}{\partial x}\right)_{x=0} = -KC(0). \qquad (13.10)$$

K is the rate constant for evaporation. The solution for Fick's second equation for this boundary condition is [1]

$$C(x,t) = M \left[\frac{1}{\sqrt{\pi D t}} \exp\left(-\frac{x^2}{4Dt}\right) - \frac{K}{D} \exp\left(\frac{K^2}{D^2} Dt + \frac{K}{D} x\right) \right.$$
$$\left. \text{erfc}\left(\frac{x}{2\sqrt{Dt}} + \frac{K}{D}\sqrt{Dt}\right) \right]. \tag{13.11}$$

Evaporation losses of the tracer cause negative deviations from Gaussian behaviour in the near-surface region.

3. *Evaporation losses of the matrix*: For a matrix material with a high vapour pressure the surface of the sample may recede due to evaporation. A solution for continuous matrix removal at a rate v and simultaneous in-diffusion of the tracer has been given by [20]

$$C(x',t) = M \left[\frac{1}{\sqrt{\pi D t}} \exp(-\eta^2) - \frac{v}{2D} \text{erfc}(\eta) \right], \tag{13.12}$$

where x' is the distance from the surface after diffusion and $\eta = (x' + vt)/2\sqrt{Dt}$.

13.3.2 Residual Activity Method

GRUZIN has suggested a radiotracer technique, which is called the residual activity method [21]. Instead of analysing the activity in each removed section, the activity remaining in the sample after removing a section is measured. This method is applicable if the radiation being detected is absorbed exponentially. The residual activity $A(x_n)$ after removing a length x_n from the sample is then given by

$$A(x_n) = k \int_{x_n}^{\infty} C(x) \exp[-\mu(x - x_n)] dx, \tag{13.13}$$

where k is a constant and μ is the absorption coefficient. According to SEIBEL [22] the general solution of Eq. (13.13) – independent of the functional form of $C(x)$ – is given by

$$C(x_n) = kA(x_n) \left[\mu - \frac{\mathrm{d}\ln A(x_n)}{\mathrm{d}x_n} \right]. \tag{13.14}$$

If the two bracket terms in Eq. (13.14) are comparable, the absorption coefficient must be measured accurately in the same geometry in which the sample is counted. Thus, the Gruzin method is less desirable than counting the sections, except for two limiting cases:

1. *Strongly absorbed radiation*: Suppose that the radiation is so weak that it is absorbed in one section, i.e. $\mu \gg \mathrm{d}\ln A(x_n)/\mathrm{d}x_n$. Isotopes such as ^{63}Ni,

^{14}C, or ^{3}H emit weak β-radiation. Their radiation is readily absorbed and Eq. (13.14) reduces to

$$C(x_n) = \mu k A(x_n) \qquad (13.15)$$

and the residual activity $A(x_n)$ follows the same functional form as $C(x_n)$. In this case, the Gruzin technique has the advantage that it obviates the tedious preparation of sections for counting.

2. *Slightly absorbed radiation*: For $\mu \ll \mathrm{d}\ln A(x_n)/\mathrm{d}x_n$ the radiation is so energetic that absorption is negligible. Then, the activity A_n in section n is obtained by subtracting two subsequent residual activities:

$$A_n = A(x_n) - A(x_{n+1}). \qquad (13.16)$$

The Gruzin technique is useful, when the specimen can be moved to the counter repeatedly without loosing alignment in the sectioning device. In general, this method is not as reliable as sectioning and straightforward measurement of the section activity.

13.4 Isotopically Controlled Heterostructures

The use of enriched stable isotopes combined with modern epitaxial growth techniques enables the preparation of isotopically controlled heterostructures. Either chemical vapour deposition (CVD) or molecular beam epitaxy (MBE) are used to produce the desired heterostructures. After diffusion annealing, the diffusion profiles can be studied using, for example, conventional SIMS or TOF-SIMS techniques (see the next section).

We illustrate the benefits of this method with an example of Si self-diffusion. In the past, self-diffusion experiments were carried out using the radiotracer ^{31}Si with a half-life of 2.6 hours. However, this short-lived radiotracer limits such studies to a narrow high-temperature range near the melting temperature of Si. Other self-diffusion experiments utilising the stable isotope ^{30}Si (natural abundance in Si is about 3.1 %) in conjunction with neutron activation analysis, SIMS profiling and nuclear reaction analysis (NRA) overcame this difficulty (see also Chap. 23). However, these methods have the disadvantage that the ^{30}Si background concentration is high.

Figure 13.7 illustrates the technique of isotopically controlled heterostructures for Si self-diffusion studies. The sample consists of a Si-isotope heterostructure, which was grown by chemical vapour deposition on a natural floating-zone Si substrate. A 0.7 μm thick ^{28}Si layer was covered by a layer of natural Si (92.2 % ^{28}Si, 4.7 % ^{29}Si, 3.1 % ^{30}Si). The ^{28}Si profile in the as-grown state (dashed line), after a diffusion anneal (crosses), and the best fit to the data (solid line) are shown. Diffusion studies on isotopically controlled heterostructures have been used by BRACHT AND HALLER and their associates mainly for self- and dopant diffusion studies in elemental [24, 25] and compound semiconductors [26–28].

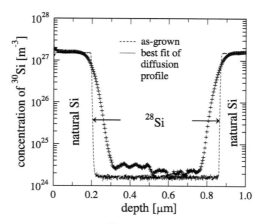

Fig. 13.7. SIMS depth profiles of ^{30}Si measured before and after annealing at 925 °C for 10 days of a ^{28}Si isotope heterostructure. The initial structure consisted of a layer of ^{28}Si embedded in natural Si

13.5 Secondary Ion Mass Spectrometry (SIMS)

Secondary ion mass spectroscopy (SIMS) is an analytical technique whereby layers of atoms are sputtered off from the surface of a solid, mainly as neutral atoms and a small fraction as ions. Only the latter can be analysed in a mass spectrometer. Several aspects of the sputtering process are illustrated in Fig. 13.8. The primary ions (typically energies of a few keV) decelerate during impact with the target by partitioning their kinetic energy through a series of collisions with target atoms. The penetration depth of the primary ions depends on their energy, on the types of projectile and target atoms and their atomic masses, and on the angle of incidence. Each primary ion initiates a 'collision cascade' of displaced target atoms, where momentum vectors can be in any direction. An atom is ejected after the sum of phonon and collisional energies focused on a target atom exceeds some threshold energy. The rest of the energy dissipates into atomic mixing and heating of the target.

The sputtering yield of atomic and molecular species from a surface depends strongly on the target atoms, on the primary ions and their energy. Typical yields vary between 0.1 to 10 atoms per primary ion. The great majority of emitted atoms are neutral. For noble gas primaries the percentage of secondary ions is below 1 %. If one uses reactive primary ions (e.g., oxygen- or alkali-ions) the percentage of secondary ions can be enhanced through the interaction of a chemically reactive species with the sputtered species by exchanging electrons.

In a SIMS instrument, schematically illustrated in Fig. 13.9, a primary ion beam hits the sample. The emitted secondary ions are extracted from the surface by imposing an electrical bias of a few kV between the sample

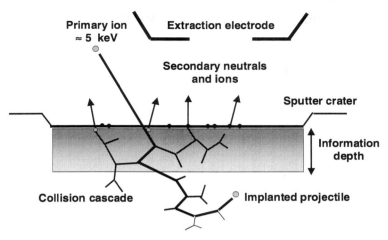

Fig. 13.8. Sputtering process at a surface of a solid

and the extraction electrode. The secondary ions are then transferred to the spectrometer via a series of electrostatic and magnetic lenses. The spectrometer filters out all but those ions with the chosen mass/charge ratios, which are then delivered to the detector for counting. The classical types of mass spectrometers are equipped either with quadrupole filters, or electric and magnetic sector fields.

Time-of-flight (TOF) spectrometers are used in TOF-SIMS instruments. The TOF-SIMS technique developed mainly by BENNINGHOVEN [35] combines high lateral resolution (< 60 nm) with high depth resolution (< 1 nm). It is nowadays acknowledged as one of the major techniques for the surface characterisation of solids. In different operational modes - surface spectrometry, surface imaging, depth profiling - this technique offers several features: the mass resolution is high; in principle all elements and isotopes can be detected and also chemical information can be obtained; detection limits in the range of ppm of a monolayer can be achieved. For details of the construction of SIMS devices we refer to [33, 34, 36, 37].

When SIMS is applied for diffusion profile measurements, the mass spectrum is scanned and the ion current for tracer and host atoms can be recorded simultaneously. In conventional SIMS, the ion beam is swept over the sample and, in effect, digs a crater. An aperture prevents ions from the crater edges from reaching the mass spectrometer. The diffusion profile is constructed from the plots of instantaneous tracer/host atom ratio *versus* sputtering time. The distance is deduced from a measurement of the total crater depth, assuming that the material is removed uniformly as a function of time. Large changes of the chemical composition along the diffusion direction can invalidate this assumption.

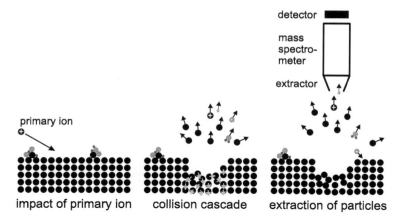

Fig. 13.9. SIMS technique (schematic illustration)

One must keep in mind that the relationship between measured secondary-ion signals and the composition of the target is complex. It involves all aspects of the sputtering process. These include the atomic properties of the sputtered ions such as ionisation potentials, electron affinities, the matrix composition of the target, the environmental conditions during the sputtering process such as the residual gas components in the vacuum chamber, and instrumental factors. Diffusion analysis by SIMS also depends on the accuracy of measuring the depth of the eroded crater and the resolution of the detected concentration profile. A discussion of problems related to quantification and standardisation of composition and distance in SIMS experiments can be found in [34, 39].

SIMS, like the IBS technique discussed above, enables the measurement of very small diffusion coefficients, which are not attainable with mechanical sectioning techniques. The very good depth resolution and the high sensitivity of mass spectrometry allows the resolution of penetration profiles of solutes in the 10 nm range and at ppm level. Several perturbing effects, inherent to the method and limiting its sensitivity are: degradation of depth resolution by surface roughening, atomic mixing, and near surface distortion of profiles by transient sputtering effects.

SIMS has mainly been applied for diffusion of foreign atoms although the high mass resolution especially of TOF-SIMS also permits separation of stable isotopes of the same element. SIMS has found particularly widespread use in studies of implantation- and diffusion profiles in semiconductors. However, SIMS is applicable to all kinds of solids. As an example, Fig. 13.10 shows diffusion profiles for both stable isotopes ^{69}Ga and ^{71}Ga of natural Ga in a ternary Al-Pd-Mn alloy (with a quasicrystalline structure) according to [38]. For metals, the relatively high impurity content of so-called 'pure metals' as compared to semiconductors can limit the dynamic range of SIMS profiles.

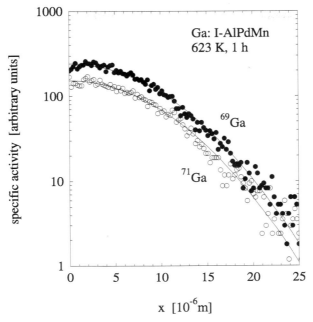

Fig. 13.10. Diffusion profiles for both stable isotopes ^{69}Ga and ^{71}Ga of natural Ga in AlPdMn (icosahedral quasicrystalline alloy) according to [38]. The *solid lines* represent fits of the thin-film solution

SIMS has in few cases also been applied to self-diffusion. This requires that highly enriched stable isotopes are available as tracers. Contrary to self-diffusion studies by radiotracer experiments, in the case of stable tracers diffused into a matrix with a natural abundance of stable isotopes the latter limits the concentration range of the diffusion profile. A fine example of this technique can be found in a study of Ni self-diffusion in the intermetallic compound Ni$_3$Al, in which the highly enriched stable ^{64}Ni isotope was used [40]. The limitation due to the natural abundance of a stable isotope in the host has been avoided in some SIMS studies of self-diffusion on amorphous Ni-containing alloys by using the radioisotope ^{63}Ni as tracer [42, 43].

An elegant possibility to overcome the limits posed by the natural abundance of stable isotopes are isotopically controlled heterostructures. This method is discussed in the previous section and illustrated in Fig. 13.7.

13.6 Electron Microprobe Analysis (EMPA)

The basic concepts of electron microprobe analysis (EMPA) can be found already in the PhD thesis of CASTAING [44]. The major components of an

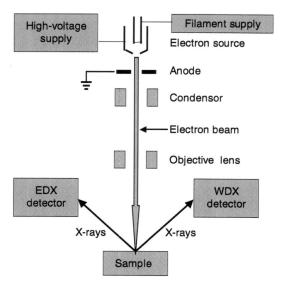

Fig. 13.11. Schematic view of an electron microprobe analyser (EMPA)

EMPA equipment are illustrated in Fig. 13.11. An electron-optical column containing an electron gun, magnetic lenses, a specimen chamber, and various detectors is maintained under high vacuum. The electron-optical column produces a finely focused electron beam, with energies ranging between 10 and 50 keV. Scanning coils and/or a mechanical scanning device for the specimen permit microanalysis at various sample positions. When the beam hits the specimen it stimulates X-rays of the elements present in the sample. The X-rays are detected and characterised either by means of an energy dispersive X-ray spectrometer (EDX) or a crystal diffraction spectrometer. The latter is also referred to as a wave-length dispersive spectrometer (WDX).

The ability to perform a chemical analysis is the result of a simple and unique relationship between the wavelength of the characteristic X-rays, λ, emitted from an element and its atomic number Z. It was first observed by MOSELEY [45] in 1913. He showed that for K radiation

$$Z \propto \frac{1}{\sqrt{\lambda}}. \tag{13.17}$$

The origin of the characteristic X-ray emission is illustrated schematically in Fig. 13.12. An incident electron with sufficient energy ejects a core electron from its parent atom leaving behind an orbital vacancy. The atom is then in an excited state. Orbital vacancies are quickly filled by electronic relaxations accompanied by the release of a discrete energy corresponding to the difference between two orbital energy levels. This energy can be emitted as an X-ray photon or it can be transferred to another orbital electron, called

13.6 Electron Microprobe Analysis (EMPA)

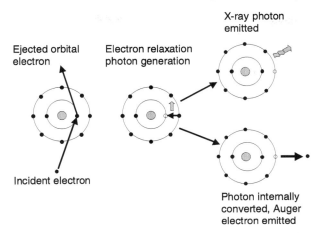

Fig. 13.12. Characteristic X-ray and Auger-electron production

an Auger electron, which is ejected from the atom. The fraction of electronic relaxations which result in X-ray emission rather than Auger emission depends strongly on the atomic number. It is low for small atomic numbers and high for large atomic numbers. The characteristic radiation is superimposed to the continuous radiation also denoted as 'Bremsstrahlung'. The continuum is the major source of the background and the principal factor limiting the X-ray sensitivity. For details about EMPA, the reader may consult, e.g., the reviews of HUNGER [46] and LIFSHIN [47].

A diffusion profile is obtained by examining on a polished cross-section of a diffusion sample the intensity of the characteristic radiation of the element(s) involved in the diffusion process along the diffusion direction. The detection limit in terms of atomic fractions is about 10^{-3} to 10^{-4}, depending on the selected element. It decreases with decreasing atomic number. Light elements such as C or N are difficult to study because their fluorescence yield is low. The diameter of the electron beam is typically 1 μm or larger depending on the instrument's operating conditions. Accordingly, the volume of X-ray generation is of the order of several μm^3. This limits the spatial resolution to above 1 to 2 μm. Thus, only relatively large diffusion coefficients $D > 10^{-15}\,\mathrm{m^2\,s^{-1}}$ can be measured (Fig. 13.1). Because of its detection limit, EMPA is mainly appropriate for interdiffusion- and multiphase-diffusion studies. An example of a single-phase interdiffusion profile for an $Al_{50}Fe_{50}$–$Al_{30}Fe_{70}$ couple is shown in Fig. 13.13 [23].

The Boltzmann-Matano method [29, 30] is usually employed to evaluate interdiffusion coefficients $\tilde{D}$ from an experimental profile. Related procedures for non-constant volume have been developed by SAUER AND FREISE and DEN BROEDER [31, 32]. These methods for deducing the interdiffusion coefficient, $\tilde{D}(c)$, from experimental concentration-depth profiles are described in Chap. 10.

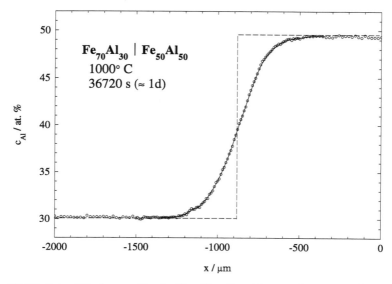

Fig. 13.13. Interdiffusion profile of a $Fe_{70}Al_{30}$–$Fe_{50}Al_{50}$ couple measured by EMPA according to SALAMON ET AL. [23]. *Dashed line*: composition distribution before the diffusion anneal

13.7 Auger-Electron Spectroscopy (AES)

Auger-electron spectroscopy (AES) is named after PIERRE AUGER, who discovered and explained the Auger effect in experiments with cloud chambers in the mid 1920s (see [48]). An Auger electron is generated by transitions within the electron orbitals of an atom following an excitation an electron from one of the inner levels (see Fig. 13.12). Auger-electron spectroscopy (AES) was introduced in the 1960s. In AES instruments the excitation is performed by a primary electron beam.

The kinetic energy of the Auger electron is independent of the primary beam but is characteristic of the atom and electronic shells involved in its production. The probability that an Auger electron escapes from the surface region decreases with decreasing kinetic energy. The range of analytical depth in AES is typically between 1 and 5 nm. AES is one of the major techniques for surface analysis.

When a primary electron beam strikes a surface, Auger electrons are only a fraction of the total electron yield. Most of the electrons emitted from the surface are either secondary electrons or backward scattered electrons. These and the inelastically scattered Auger electrons constitute the background in an Auger spectrum. Auger-electron emission and X-ray fluorescence after creation of a core hole are competing processes and the emission probability depends on the atomic number. The probability for Auger-electron emission

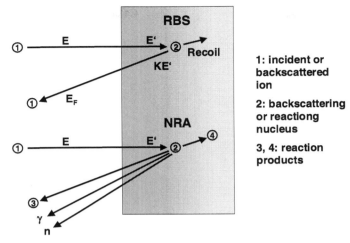

Fig. 13.14. Schematic representation of Rutherford backscattering (RBS) and of nuclear reaction analysis (NRA)

decreases with increasing atomic number whereas the probability for X-ray fluorescence increases with atomic number. AES is thus particularly well suited for light elements.

The combined operation of an AES spectrometer for chemical surface analysis and an ion sputtering device can be used for depth profiling. Information with regard to the quantification and to factors affecting their resolution can be found, e.g., in [49, 50]. AES is applicable to diffusion of foreign atoms, since AES only discriminates between different elements. It has, for example, been used to measure Au and Ag diffusion in amorphous Cu-Zr [41] and Cu and Al diffusion in amorphous $Zr_{61}Ni_{39}$-alloys [51].

13.8 Ion-beam Analysis: RBS and NRA

High-energy ion-beam analysis has several desirable features for depth profiling of diffusion samples. The technique is largely non-destructive, it offers good depth resolution, and measurements of both concentration and depth can be achieved. The depth resolution is in the range from about 0.01 to 1 µm. This is inferior to the depth resolution achieved in IBS or SIMS devices but substantially better than the resolution of mechanical sectioning techniques.

Atomic species are identified in ion-beam analysis by detecting the products of nuclear interactions, which are created by the incident MeV ions. There are several different techniques. The two more important ones are *Rutherford backscattering* (RBS) and *nuclear reaction analysis* (NRA). These two are depicted schematically in Fig. 13.14.

13 Direct Diffusion Studies

Rutherford Backscattering (RBS): The first scattering experiment was performed by RUTHERFORD in 1911 [53] and his students GEIGER AND MARSDEN [54] for verifications of the atomic model. A radioactive source of α-particles was used to provide energetic probing ions and the particles scattered from a gold foil were observed with a zinc blende scintillation screen. Nowadays, elastic backscattering analysis also denoted as Rutherford backscattering (RBS) is probably the most frequently used ion-beam analytical technique among the surface analysis tools.

In RBS experiments a high-energy beam of monoenergetic ions (usually α-particles) with energies of some MeV is used for depth profiling. The sample is bombarded along the diffusion direction with ions and one studies the number of elastically backscattered ions as a function of their energy. The particles of the analysing beam are scattered by the nuclei in the sample and the energy spectrum of scattered particles is used to determine the concentration profile of scattering nuclei. The signals from different nuclei can be separated in the energy spectrum, because of the different kinematic factors K of the scattering process. K is related to the masses of analysing particles and scattering nuclei. It is a monotonically decreasing function of the mass of the target nuclei. The backscattered particles re-emerge unchanged except for a reduction in energy. The depth information comes from the continuous energy loss of the ions in the sample. The yield of the backscattered ions is proportional to the concentration of the scattering nuclei.

RBS is illustrated schematically in Fig. 13.15 for a layer of heavy atoms (mass M) deposited on a substrate of light atoms (Mass m). Yield and energy of the backscattered ions are monitored by an energy-sensitive particle detector and a multichannel analyser. The high energy end of the spectrum (M-signal) corresponds to ions backscattered from heavy atoms at the sample surface. The low energy end of the M-signal corresponds to ions backscattered from the heavy atoms near the interface. The signals from the heavy and light nuclei are separated in the spectrum due to the different kinematic factors for heavy and light nuclei.

Although widely applicable, RBS has two inherent limitations for diffusion studies: First, the element of interest must differ in mass sufficiently – at least several atomic masses – from other constituents of the sample. Second, adequate sensitivity is achieved only when the solutes are heavier than the majority constituents of the matrix. Then, the backscattering yield from the diffuser appears at higher energies than the yield from the majority nuclei. Therefore, RBS is particular suitable for detecting heavy elements in a matrix of substantially lower atomic weight. Because of the limited penetration range of ions (several micrometers) and the associated energy straggling in a solid, only relatively small diffusion coefficients are accessible.

Nuclear Reaction Analysis (NRA): In a NRA profiling experiment monoenergetic high-energy particles (protons, α-particles, ...) are used as in RBS. NRA is applicable if the analysing particles undergo a suitable nu-

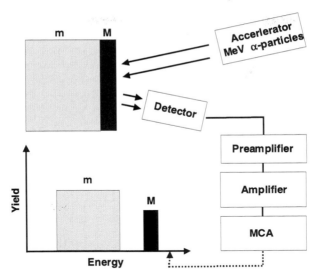

Fig. 13.15. Rutherford backscattering spectrometry: high-energy ion beam, electronics for particle detection and a schematic example of a RBS spectrum. The technique is illustrated for a thin layer of atoms with mass M deposited on a substrate of lower mass m

clear reaction with narrow resonance with the atoms of interest. The yield of out-going reaction products is measured as a function of the energy of the incident beam. From the yield *versus* energy curve the concentration profile can be deduced.

As shown schematically in Fig. 13.14, the analysis-beam particles undergo an inelastic, exothermic nuclear reaction with the target nuclei thus producing two or more new particles. Depending on the conditions it may be preferable to detect either charged reaction products, neutrons or γ-rays from the reaction. This method distinguishes specific isotopes and is therefore free from the mass-related restrictions of RBS. Suitable resonant nuclear reactions occur for at least one readily available isotope of all elements from hydrogen to fluorine and for beam energies below 2 MeV. NRA can mainly be used to investigate the diffusion of light solutes in a heavier matrix.

Concluding Remarks: Depth profiling is possible in RBS and NRA because the charged particles continuously loose energy as they traverse the specimen. Usually, this loss is almost entirely due to electronic excitations, although there is some additional contribution from small-angle nuclear scattering. The consequences may be appreciated by considering the RBS experiment illustrated in Fig. 13.15. In RBS the energy of the analysis-beam particle decreases during both inward and outward passages. When the particle is detected, the accumulated energy loss is superimposed on the recoil loss via the kinematic factor. Hence the measured energy decreases monotonically with the depth of the scattering nucleus. In NRA the situations are analogous but more varied. For example, the relevant energy loss may occur only during the inward or outward passage. Nevertheless, depth resolution is always a consequence of the charged-particle energy loss in the sample. For example, the diffusion of ion-implanted boron in amorphous $Ni_{59.5}Nb_{40.5}$ was measured by irradiating the amorphous alloy with high energy protons and detecting α-particles emitted from the nuclear reaction $^{11}B + p \to {}^8B + \alpha$, and determining the concentration profile of ^{11}B from the number and energy of emitted α-particles as a function of the incident proton energy [52].

In NRA and in RBS the penetration range of ions is not more than several micrometer. This limits the diffusion depth. Diffusion coefficients between about 10^{-17} and $10^{-23}\,m^2\,s^{-1}$ are accessible (see also Fig. 13.1). Both RBS and NRA methods need a depth calibration, which is based on not always very accurate data of the stopping power in the matrix for the relevant particles. Also the depth resolution is usually inferior to that achievable in careful IBS radiotracer and SIMS profiling studies. For a comprehensive discussion of ion-beam techniques the reader may consult reviews by MYERS [55], LANFORD ET AL. [56], and CHU ET AL. [57].

References

1. J. Crank, *The Mathematics of Diffusion*, Oxford University Press, 2nd ed., 1975
2. J. Philibert, *Atom Movements – Diffusion and Mass Transport in Solids*, Les Editions de Physique, Les Ulis, 1991
3. Th. Heumann, *Diffusion in Metallen*, Springer-Verlag, Berlin, 1992
4. S.J. Rothman, *The Measurement of Tracer Diffusion Coefficients in Solids*, in: *Diffusion in Crystalline Solids*, G.E. Murch, A.S. Nowick (Eds.), Academic Press, 1984, p. 1
5. H. Mehrer (Vol. Ed.), Sect. 1.6 in: *Diffusion in Solid Metals and Alloys*, Landolt-Börnstein, Numerical Data and Functional Relationships in Science and Technology, New Series, Group III: Crystal and Solid State Physics, Vol. 26, Springer-Verlag, 1990
6. Proc. Int. Conf. on *Diffusion in Materials – DIMAT-92*, Kyoto, Japan, 1992, M. Koiwa, H. Nakajima, K.-I. Hirano (Eds.); also: Defect and Diffusion Forum **95–98** (1993)
7. Proc. Int. Conf. on *Diffusion in Materials – DIMAT-96*, Nordkirchen, Germany, 1996, H. Mehrer, Chr. Herzig, N.A. Stolwijk, H. Bracht (Eds.); also: Defect and Diffusion Forum **143–147** (1997)
8. Proc. Int. Conf. on *Diffusion in Materials – DIMAT-2000*, Paris, France, 2000, Y. Limoge, J.L. Bocquet (Eds.); also: Defect and Diffusion Forum **194–199** (2001)
9. Proc. Int. Conf. on *Diffusion in Materials – DIMAT-2004*, Cracow, Poland, 2004, M. Danielewski, R. Filipek, R. Kozubski, W. Kucza, P. Zieba, Z. Zurek (Eds.); also: Defect and Diffusion Forum **237–240** (2005)
10. H. Mehrer, Materials Transactions, JIM, **37**, 1259 (1996)

11. H. Mehrer, F. Wenwer, *Diffusion in Metals*, in: *Diffusion in Condensed Matter*, J. Kärger, R. Haberlandt, P. Heitjans (Eds.), Vieweg Verlag, 1998
12. H. Mehrer, *Diffusion: Introduction and Case Studies in Metals and Binary Alloys*, in: *Diffusion in Condensed Matter – Methods, Materials, Models*, P. Heitjans, J. Kärger (Eds.), Springer-Verlag, 2005
13. D. Tannhauser, J. Appl. Phys. **27**, 662 (1956)
14. H. Mehrer, Phys. Stat. Sol. (a) **104**, 247 (1987)
15. A. Gude, H. Mehrer, Philos. Mag. A **76**, 1 (1996)
16. F. Faupel, P.W. Hüppe, K. Rätzke, R. Willecke. T. Hehenkamp, J. Vac. Sci. Technol. a **10**, 92 (1992)
17. F. Wenwer, A. Gude, G. Rummel, M. Eggersmann, Th. Zumkley, N.A. Stolwijk, H. Mehrer, Meas. Sci. Technol. **7**, 632 (1996)
18. M. Eggersmann, B. Sepiol, G. Vogl, H. Mehrer, Defect and Diffusion Forum **143–147**, 339 (1997)
19. I. Kaur, Y. Mishin, W. Gust, *Fundamentals of Grain and Interphase Boundary Diffusion*, John Wiley & Sons Ltd., 1995
20. R.N. Ghoshtagore, Phys. Stat. Sol. **19**, 123 (1967)
21. P.L. Gruzin, Dokl. Akad. Nauk. SSSR **86**, 289 (1952)
22. G. Seibel, Int. J. Appl. Radiat. Isot. **15**, 679 (1964)
23. M. Salamon, S. Dorfman, D. Fuks, G. Inden, H. Mehrer, Defect and Diffusion Forum **194–199**, 553 (2001)
24. H.D. Fuchs, W. Walukiewicz, E.E. Haller, W. Dondl, R. Schorer, G. Abstreiter, A.I. Rudnev, A.V. Tikomirov, V.I. Ozhogin, Phys. Rev. **B 51**, 16817 (1995)
25. H. Bracht, E.E. Haller, R. Clark-Phelps, Phys. Rev. Lett. **81** 393 (1998)
26. L. Wang, L. Hsu, E.E. Haller, J.W. Erickson, A. Fisher, K. Eberl, M. Cardona, Phys. Rev. Lett. **76**, 2342 (1996)
27. L. Wang, J.A. Wolk, L. Hsu, E.E. Haller, J. W. Erickson, M. Cardona, T. Ruf, J.P. Silveira, F. Briones, Appl. Phys. Lett. **70**, 1831 (1997)
28. H. Bracht, E.E. Haller, K. Eberl, M. Cardona, R. Clark-Phelps, Mat. Res. Soc. Symp. **527**, 335 (1998)
29. L. Boltzmann, Wiedemanns Ann. Physik **53**, 959 (1894)
30. C. Matano, Jap. J. Phys. **8**, 109–113 (1933)
31. F. Sauer, V. Freise, Z. Elektrochem. **66**, 353 (1962)
32. F.J.A. den Broeder, Scr. Metall. **3**, 321 (1969)
33. C.-E. Richter, *Sekundärionen-Massenspektroskopie und Ionenstrahl-Mikroanalyse*, in: *Ausgewählte Untersuchungsverfahren der Metallkunde*, H.-J. Hunger et al. (Eds.), VEB Verlag, 1983, p. 197
34. W.T. Petuskey, *Diffusion Analysis using Secondary Ion Mass Spectroscopy*, in: *Nontraditional Methods in Diffusion*, G.E. Murch, H.K. Birnbaum, J.R. Cost (Eds.), The Metallurgical Society of AIME, Warrendale, 1984, p. 179
35. A. Benninghoven, *The History of Static SIMS: a Personal Perspective*, in: *TOF-SIMS – Surface Analysis by Mass Spectrometry*, J.C. Vickerman, D. Briggs (Eds.), IM Publications and Surface Spectra Limited, 2001
36. A. Benninghoven, F.G. Rüdenauer, H.W. Werner, *Secondary Ion Mass Spectrometry – Basic Concepts, Instrumental Aspects, Applications and Trends*, John Wiley and Sons, Inc., 1987
37. J.C. Vickerman, D. Briggs (Eds.), *TOF-SIMS – Surface Analysis by Mass Spectrometry*, IM Publications and Surface Spectra Limited, 2001

38. H. Mehrer, R. Galler, W. Frank, R. Blüher, T. Strohm, *Diffusion in Quasicrystals*, in: *Quasicrystals: Structure and Physical Properties*, H.-R. Trebin (Ed.), J. Wiley VCH, 2003
39. M.-P. Macht, V. Naundorf, J. Appl. Phys. **53**, 7551 (1982)
40. S. Frank, U. Södervall, Chr. Herzig, Phys. Stat. Sol. (b) **191**, 45 (1995)
41. E.C. Stelter, D. Lazarus, Phys. Rev. B **36**, 9545 (1987)
42. A.K. Tyagi, M.-P. Macht, V. Naundorf, Scripta Metall. et Mater. **24**, 2369 (1999)
43. A.K. Tyagi, M.-P. Macht, V. Naundorf, Acta Metall. et Mater. **39**, 609 (1991)
44. R. Castaing, Ph.D. thesis, Univ. of Paris, 1951
45. H.G.J. Moseley, Philos. Mag. **26**, 1024 (1913)
46. H.-J. Hunger, *Elektronenstrahl-Mikroanalyse und Rasterelektronen-Mikroskopie*, in: *Ausgewählte Untersuchungsverfahren der Metallkunde*, H.-J. Hunger et al. (Eds.), VEB Deutscher Verlag für Grundstoffindustrie, Leipzig, 1983, p.175
47. E. Lifshin, *Electron Microprobe Analysis*, in: *Materials Science and Technology*, R.W. Cahn, P. Haasen, E.J. Kramer (Eds.), Vol. 2B: Characterisation of Materials, VCH, 1994, p. 351
48. P. Auger, Surf. Sci **1**, 48 (1975)
49. S. Hofmann, Surf. Interface Anal. **9**, 3 (1986)
50. A. Zalar, S. Hofmann, Surf. Interface Anal. **12**, 83 (1988)
51. S.K. Sharma, P. Mukhopadhyay, Acta Metall. **38**, 129 (1990)
52. M.M. Kijek, D.W. Palmer, B. Cantor, Acta Metall. **34**, 1455 (1986)
53. E. Rutherford, Philos. Mag. **21**, 669 (1911)
54. H. Geiger, E. Marsden, Philos. Mag. **25**, 206 (1913)
55. S.M. Myers, *Ion-beam Analysis and Ion Implantation in the Study of Diffusion*, in: *Nontraditional Methods in Diffusion*, G.E. Murch, H.K. Birnbaum, J.R. Cost (Eds.), The Metallurgical Society of AIME, Warrendale, 1984, p. 137
56. W.A. Lanford, R. Benenson, C. Burman, L. Wielunski, *Nuclear Reaction Analysis for Diffusion Studies*, in: *Nontraditional Methods in Diffusion*, G.E. Murch, H.K. Birnbaum, J.R. Cost (Eds.), The Metallurgical Society of AIME, Warrendale, 1984, p. 155
57. W.K. Chu, J. Liu, Z. Zhang, K.B. Ma, *High Energy Ion Beam Analysis Techniques*, in: *Materials Science and Technology*, R.W. Cahn, P. Haasen, E.J. Cramer (Eds.), Vol. 2B: Characterisation of Materials, VCH Weinheim, 1994, p. 423

14 Mechanical Spectroscopy

14.1 General Remarks

The discoveries of thermally-activated anelastic relaxation processes in solids by SNOEK [1], ZENER [2, 3] and GORSKI [4] were made more than half a century ago. Since then, anelastic measurements have become an established tool for the study of atomic movements in solids. *Relaxation methods* and the closely related *internal friction (or damping)* methods make use of the fact that atomic motion in a solid can be induced by the application of constant or oscillating mechanical stress. Nowadays, anelastic measurements are also denoted by the title *mechanical spectroscopy*.

Under the influence of an applied stress or strain, an instantaneous elastic effect (Hooke's law) is observed, followed by strain or stress which varies with time. The latter effect is called *anelasticity* or *anelastic relaxation*. Anelastic behaviour is reversible. If stress (strain) is removed the sample will return – after some time – to its initial shape. This distinguishes anelastic from plastic behaviour.

Light interstitials, such as H, C, N, and O as well as substitutional solutes and solute-defect complexes are accompanied by local straining of the surrounding lattice. The presence of microstrains surrounding a diffusing atom allows interaction between a macroscopic stress field arising from external forces applied to the material. This interaction generates a rich variety of stress-assisted diffusion effects. Stress-mediated motion can cause time-dependent anelastic (recoverable) strains that result in several types of internal friction processes encountered in many materials.

Sometimes, anelastic relaxation involves the reorientation of point defects which act as elastic dipoles as illustrated in Fig. 14.1. Reorientation relaxations are short-range processes, which in some cases involve only one or few atomic jump(s). However, only in some special cases, exemplified by Snoek relaxation, the same jump produces both reorientation and diffusion. Only then, a simple relationship exists between the relaxation time and the long-range diffusion coefficient. Long-range diffusion controls the so-called Gorski relaxation illustrated in Fig. 14.2. Gorsky relaxation can be produced by bending a sample containing defects, which act as dilatation centers. In practice, the only experimentally known example of Gorski relaxation is due to hydrogen diffusion metals. It can be observed because hydrogen diffusion is very fast.

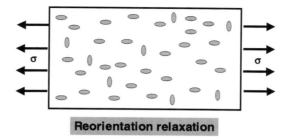

Fig. 14.1. Schematic illustration of anelastic relaxation caused by reorientation of elastic dipoles (represented by *grey ellipses*)

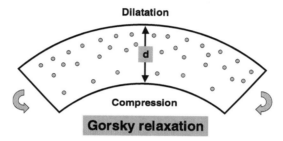

Fig. 14.2. Schematic illustration of Gorski-effect

One should, however, keep in mind that mechanical relaxation and internal friction may arise from various sources. These can range from point-defect reorientations, long-range diffusion, dislocation effects, grain-boundary processes, and phase transformations to visco-elastic behaviour and plastic deformation. Some point-defect relaxations are diffusion-related, some are not. For point-defect relaxations of trapped and paired defects, the nature and the activation enthalpy of the reorientation jump can be significantly different from those associated with long-range diffusion. A review of the substantial body of work that has been accumulated on the study of atomic movement by anelastic methods is beyond the scope of this chapter.

Several textbooks, e.g., those of ZENER [3] and NOWICK AND BERRY [5] and reviews by BERRY AND PRITCHET [6, 7] are available for the interested reader. A review about the potential of mechanical loss spectroscopy for inorganic glasses and glass ceramics has been given by ROLING [8]. A comprehensive treatment of magnetic relaxation effects can be found in a textbook of KRONMÜLLER [9].

In the present chapter, we first mention the basic concepts of mechanical loss spectroscopy, i.e. of anelastic behaviour and internal friction. Then, we describe some examples of diffusion-related anelasticity such as the *Snoek effect*, the *Zener effect*, the *Gorski effect*, and give an example of a mechanical loss spectrum of glasses.

14.2 Anelasticity and Internal Friction

From the viewpoint of mechanical stress-strain behaviour, we may regard an ideal solid as one which obeys Hooke's law and thus behaves in an ideally elastic manner. Such a solid would always recover completely and instantaneously on removal of an applied stress. If set into vibration, the solid would vibrate forever with undiminished amplitude if totally isolated from its surroundings. The mechanical behaviour of real solids at low stress levels (below the yield stress) is modified by the appearance of anelasticity, which develops at a rate controlled by the atomic movements. It can often be traced back to the presence of mobile atoms or point defects.

A quantitative description of the anelastic behaviour of materials can be found by analysing a model having the name *standard linear solid*, which was originally proposed by VOIGT [10] and by POYNTING AND THOMSON [11]. In this model, stress σ, strain ϵ, and their respective time derivatives, $\dot{\sigma}$ and $\dot{\epsilon}$, are related through a linear response equation:

$$\sigma + \tau_\epsilon \dot{\sigma} = M_R(\epsilon + \tau_\sigma \dot{\epsilon}) \,. \tag{14.1}$$

This anelastic equation of state is a generalisation of Hooke's law of linear elasticity. Equation (14.1) contains three material parameters: the *strain relaxation time* τ_ϵ, the *stress relaxation time* τ_σ (sometimes also denoted as the stress retardation time), and the *relaxed elastic modulus* M_R. Figure 14.3 illustrates in its left part the strain response of a standard linear solid induced by an instantaneous application and subsequent removal of a constant stress. The continued relaxation of the strain after removal of the stress is also termed the *elastic aftereffect*. The stress response induced by instantaneous application and removal of strain is illustrated in the right part. Note that τ_σ and τ_ϵ are different. It is obvious from Eq. (14.1) that for vanishing time derivatives Eq. (14.1) reduces to Hooke's law. Under uniaxial stress M_R is termed the Young modulus, whereas under applied shear M_R is termed the shear modulus.

Periodic Stress and Strain: Let us now suppose that a uniaxial, periodic stress-time function of frequency ω and amplitude σ_0 of the form

$$\sigma = \sigma_0 \exp\left[i\omega t\right] \tag{14.2}$$

is imposed on the material. The time-dependent strain response of an anelastic solid then is

$$\epsilon = \epsilon_0 \exp\left[i(\omega t - \delta)\right] \,, \tag{14.3}$$

where δ is the phase shift between σ and ϵ. For a completely elastic material, σ and ϵ are in phase and the phase shift is zero for all frequencies. The stress-strain behaviour for an anelastic material under periodic stress is illustrated in Fig. 14.4. For an anelastic material a hysteresis loop is obtained. The area

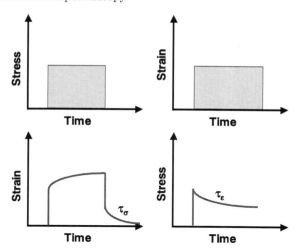

Fig. 14.3. Schematic illustration of anelastic behaviour. The strain response for an instantaneous stress-time function is shown in the *left half*. The stress response for an instantaneous strain-time function corresponds to the *right half*

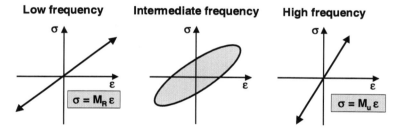

Fig. 14.4. Stress-strain relations for a periodically driven anelastic material at three different frequencies

inside the hysteresis represents the dissipated energy per unit volume and per cycle (see below).

It is convenient to introduce a *complex elastic modulus* $\hat{M}$ via

$$\sigma = \hat{M}\epsilon, \tag{14.4}$$

which can be split up according to

$$\hat{M} = M' + iM'', \tag{14.5}$$

i.e. into real and imaginary parts M' and M'', respectively. Assuming periodic strain with a frequency ω and substituting Eqs. (14.4) and (14.5) into Eq. (14.1) yields after a few steps of algebra

$$\hat{M} = M_R \frac{1 + \tau_\sigma i\omega}{1 + \tau_\epsilon i\omega}. \tag{14.6}$$

14.2 Anelasticity and Internal Friction

After separation into real and imaginary parts we get

$$M'(\omega) = M_R \frac{1 + \tau_\epsilon \tau_\sigma \omega^2}{1 + \omega^2 \tau_\epsilon^2} = M_R + \Delta M \frac{\omega^2 \tau_\epsilon^2}{1 + \omega^2 \tau_\epsilon^2} \qquad (14.7)$$

and

$$M''(\omega) = M_R \frac{(\tau_\sigma - \tau_\epsilon)\omega}{1 + \omega^2 \tau_\epsilon^2} = \Delta M \frac{\omega \tau_\epsilon}{1 + \omega^2 \tau_\epsilon^2}, \qquad (14.8)$$

where the abbreviations

$$\Delta M \equiv M_U - M_R \quad \text{and} \quad \Delta \equiv \Delta M / M_R \qquad (14.9)$$

have been introduced. At high frequencies, the time scale for stress and strain removals becomes small compared to the relaxation times. Then M' approaches an *unrelaxed elastic modulus*

$$M_U = \frac{M_R \tau_\sigma}{\tau_\epsilon}, \qquad (14.10)$$

which denotes the stress increment per unit strain at high frequency. Note that M_U and M_R are different because τ_σ and τ_ϵ are different. The tangent of the *loss angle* δ is given by

$$\tan \delta \equiv M''/M' = \Delta M \frac{\omega \tau_\epsilon}{M_R + M_U \omega^2 \tau_\epsilon^2} \equiv \Delta \frac{\omega(\tau_\sigma - \tau_\epsilon)}{1 + \tau_\sigma \tau_\epsilon \omega^2}. \qquad (14.11)$$

Internal Friction: Internal friction is the dissipation of mechanical energy caused by anelastic processes occurring in a strained solid. The internal friction, usually called Q^{-1}, in a cyclically driven anelastic solid is defined as

$$Q^{-1} \equiv \frac{\Delta E_{dissipated}}{E_{stored}}, \qquad (14.12)$$

where $\Delta E_{dissipated}$ is the energy dissipated as heat per unit volume of the material over one cycle. E_{stored} denotes the peak elastic energy stored per unit volume. For a periodically strained solid subject to sinusoidal stress, the internal friction is given by the following ratio of energy integrals:

$$Q^{-1} = \frac{\int_0^{2\pi} \sigma(\omega t) \dot{\epsilon}(\omega t)_{out-of-phase} d(\omega t)}{\int_0^{2\pi} \sigma(\omega t) \dot{\epsilon}(\omega t)_{in-phase} d(\omega t)}. \qquad (14.13)$$

Substituting the out-of-phase and in-phase components of the strain rate $\dot{\epsilon}$ yields after some algebra the following relation between internal friction and the tangent of the loss angle:

$$Q^{-1} = \pi \tan \delta. \qquad (14.14)$$

It is convenient to combine the stress and strain relaxation times to a *mean relaxation time* τ, which is defined as the geometric mean of the two fundamental times:

$$\tau \equiv \sqrt{\tau_\sigma \tau_\epsilon}. \tag{14.15}$$

We will see later that τ sometimes can be associated with atomic jump processes occurring in the strained solid, having a well-defined activation enthalpy. It is also convenient to combine the relaxed and the unrelaxed moduli to a *mean modulus* M via

$$M \equiv \sqrt{M_R M_U} = \sqrt{\frac{\tau_\sigma}{\tau_\epsilon}} M_R = \sqrt{\frac{\tau_\epsilon}{\tau_\sigma}} M_U. \tag{14.16}$$

Using the definitions of the mean modulus Eq. (14.16), the mean relaxation time Eq. (14.15) and Eq. (14.11), yields a basic expression for internal friction:

$$Q^{-1} = \pi \tan \delta = \pi \frac{\Delta M}{M} \frac{\omega \tau}{1 + \omega^2 \tau^2}. \tag{14.17}$$

The term $\pi \Delta M / M$ is called the *relaxation strength*. The second term describes the frequency dependence of internal friction. Figure 14.5 shows a diagram of Q^{-1} *versus* the logarithm of $\omega \tau$. The frequency-dependent modulus M' is also shown, which varies between the relaxed modulus M_R at low frequencies and the unrelaxed modulus M_U at high frequencies. The maximum of internal friction occurs when

$$\omega \tau = 1 \tag{14.18}$$

is fulfilled. This relation is an important condition for the analysis of anelasticity. If an anelastic solid is strained periodically with a frequency ω the maximum energy loss occurs, when the imposed frequency and relaxation time of the process match.

14.3 Techniques of Mechanical Spectroscopy

Usually, the relaxation time τ is thermally activated according to

$$\tau \propto \exp\left(\frac{\Delta H}{k_B T}\right), \tag{14.19}$$

where ΔH denotes some activation enthalpy. Thus, by varying the temperature at constant frequency ω a maximum of internal friction occurs on the temperature scale. This is the usual way of measuring internal friction peaks, as temperature is easier to vary than frequency. The latter is often more or less fixed by the internal friction device.

By using different experimental techniques, the mechanical loss can be determined at frequencies roughly between 10^{-5} and 5×10^{10} Hz. It is convenient to perform temperature-dependent measurements at fixed frequencies.

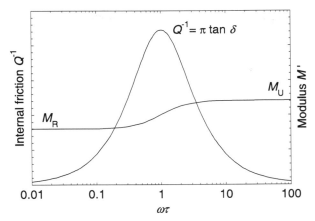

Fig. 14.5. Internal friction, $Q^{-1} = \pi \tan \delta$, and frequency dependent modulus, M', as functions of $\omega\tau$

In this case, a thermally activated process manifests itself in a loss peak, which shifts to higher temperatures as the frequency is increased. Information on the activation enthalpy is then obtained from the peak temperatures, T_{peak}, shifting with frequencies ω by using the equation:

$$\Delta H = -k_B \frac{d \ln \omega}{d(1/T_{peak})} . \qquad (14.20)$$

In the Hz regime *torsional pendulums* operating at their natural frequencies can be used. A major disadvantage of this technique is that the range of available frequencies is very narrow, often less than half a decade. This makes it difficult to determine accurate values of the activation enthalpies and to analyse frequency-temperature relations in detail. In order to overcome this limitation devices with *forced oscillations* are in use. The frequency window of this technique ranges approximately from 30 Hz up to 10^5 Hz.

At higher frequencies, the mechanical loss of solids can be studied by resonance methods [14, 15]. At even higher frequencies, in the MHz and GHz regimes, ultrasonic absorption and Brillouin light scattering can be used. However, most mechanical loss studies have been done and are still done with the help of low-frequency methods.

Starting in the 1990s, there have been efforts to make use of commercially available instrumentation for dynamic mechanical thermal analysis (DMTA) These devices usually operate in the three-point-bending mode. Among other systems, this technique has been applied to study relaxation processes in oxide glasses [16–18].

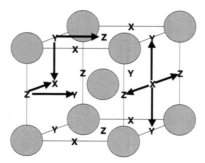

Fig. 14.6. Octahedral interstitial sites in the bcc lattice

14.4 Examples of Diffusion-related Anelasticty

14.4.1 Snoek Effect (Snoek Relaxation)

The Snoek effect is the stress-induced migration of interstitials such as C, N, or O in bcc metals. Although effects of internal friction in bcc iron were reported as early as the late 19th century, this phenomenon was first carefully studied and analysed by the Dutch scientist SNOEK [1]. Interstitial solutes in bcc crystals usually occupy octahedral interstitial sites illustrated in Fig. 14.6. Octahedral sites in the bcc lattice have tetragonal symmetry inasmuch the distance from an interstitial site to neighbouring lattice atoms is shorter along $\langle 100 \rangle$ than along $\langle 110 \rangle$ directions. The microstrains surrounding interstitial solutes have tetragonal symmetry as well, which is lower than the cubic symmetry of the matrix. Another way of expressing this is to say that interstitial solutes give rise to permanent elastic dipoles.

Figure 14.6 illustrates the three possible orientations of octahedral sites denoted as X-, Y-, and Z-sites. Without external stress all sites are energetically equivalent, i.e. $E_X = E_Y = E_Z$, and the population n_j^0 of interstitial sites by solutes is $n_X^0 = n_Y^0 = n_Z^0 = n^0/3$. n^0 denotes the total number of interstitials. If an external stress is applied this degeneracy is partly or completely removed, depending on the orientation of the external stress. For example, with uniaxial stress in the Z-direction Z-sites are energetically slightly different from X- and Y-sites, i.e. $E_Z \neq E_X = E_Y$. In contrast, uniaxial stress in $\langle 111 \rangle$ direction does not not remove the energetic degeneracy, because all sites are energetically equivalent. In thermodynamic equilibrium the distribution of interstitial solutes on the X-, Y-, and Z-sites is given by

$$n_i^{eq} = n^0 \frac{\exp(-E_i/k_B T)}{\sum_{j=X,Y,Z} \exp(-E_j/k_B T)} . \tag{14.21}$$

In general, under the influence of a suitable oriented external stress the 'solute dipoles' reorient, if the interstitial atoms have enough mobility. This redistribution gives rise to a strain relaxation and/or to an internal friction peak.

14.4 Examples of Diffusion-related Anelasticty

The relaxation time or the frequency/temperature position of the internal friction peak can be used to deduce information about the mean residence time of a solute on a certain site.

In order to deduce this information, we consider the temporal development of interstitial subpopulations n_X, n_Y, n_Z on X-, Y-, and Z-sites. Suppose that uniaxial stress is suddenly applied in Z-direction. This stress disturbs the initial equipartition of interstitials on the various types of sites and redistribution will start. Fig 14.6 shows that every X-site interstitial that performs a single jump ends either on a Y- or on a Z-site. Interstitials on Y- and Z-sites jump with equal probabilities to X-sites. The rate of change of the interstitial subpopulations can be expressed in terms of the interstitial jump rate, Γ_{int}, as follows:

$$\frac{dn_X}{dt} = -2\Gamma_{int} n_X + \Gamma_{int}(n_Y + n_Z). \tag{14.22}$$

The first term on the right-hand side in Eq. (14.22) represents the loss of interstitials located at X-sites due to hops to either Y- or Z-sites. The second term on the right-hand side represents the gain of interstitials at X-sites from other interstitials jumping from either Y- or Z-sites. Corresponding equations are obtained for n_Y and n_Z by cyclic permutation of the indices. Since the total number of interstitials, n^0, is conserved, we have

$$n^0 = n_X + n_Y + n_Z. \tag{14.23}$$

Substitution of Eq. (14.23) into Eq. (14.22) yields

$$\frac{dn_X}{dt} = -\Gamma_{int} n_X + \frac{\Gamma_{int}}{2}(n^0 - n_X^{eq}) = -\frac{3}{2}\Gamma_{int}\left(n_X - n^0/3\right). \tag{14.24}$$

Equation (14.24) is an ordinary differential equation for the population dynamics of interstitial solutes. Its solution can be written in the form

$$n_X(t) = n_X^{eq} + \left(n_X^0 - n_X^{eq}\right)\exp\left(-\frac{t}{\tau_R}\right), \tag{14.25}$$

where the relaxation time τ_R is given by

$$\tau_R = \frac{2}{3\Gamma_{int}}. \tag{14.26}$$

The relaxation time is closely related to the mean residence time, $\bar{\tau}$, of an interstitial solute on a given site. Because an interstitial solute on an octahedral site can leave its site in four directions with jump rate Γ_{int}, we have

$$\bar{\tau} = \frac{1}{4\Gamma_{int}}. \tag{14.27}$$

The solute jump rate can be written in the form

$$\Gamma_{int} = \nu^0 \exp\left(-\frac{H^M_{int}}{k_B T}\right), \tag{14.28}$$

where ν^0 and H^M_{int} denote attempt frequency and activation enthalpy of a solute jump. Then, the relaxation time of the Snoek effect is

$$\tau_R = \frac{4}{3}\bar{\tau} = \frac{1}{6\nu^0} \exp\left(\frac{H^M_{int}}{k_B T}\right). \tag{14.29}$$

The jump of an interstitial solute which causes Snoek relaxation and the elementary diffusion step (jump length $d = a/2$, a = lattice parameter) are identical. The diffusion coefficient developed from random walk theory for octahedral interstitials in the bcc lattice is given by

$$D = \frac{1}{6}\Gamma_{int} d^2 = \frac{1}{24}\Gamma_{int} a^2. \tag{14.30}$$

Substituting Eqs. (14.27) and (14.29) into Eq. (14.30) yields

$$D = \frac{1}{36}\frac{a^2}{\tau_R}. \tag{14.31}$$

This equation shows that Snoek relaxation can be used to study diffusion of interstitial solutes in bcc metals by measuring the relaxation time. It is also applicable to interstitial solutes in hcp metals since the non-ideality of the c/a-ratio gives rise to an asymmetry in the octahedral sites. Very pure and very dilute interstitial alloys must be used, if the Snoek effect of isolated interstitials is in focus. Otherwise, solute-solute or solute-impurity interactions could cause complications such as broadening or shifts of the internal friction peak.

Figure 14.7 shows an Arrhenius diagram of carbon diffusion in α-iron. For references the reader may consult LE CLAIRE'S collection of data for interstitial diffusion [12] and/or a paper by DA SILVA AND MCLELLAN [13]. The data above about 700 K have been obtained with various direct methods including diffusion-couple methods, in- and out-diffusion, or thin layer techniques. The data below about 450 K were determined with indirect methods, including internal friction, elastic after-effect, or magnetic after-effect measurements. The data cover an impressive range of about 14 orders of magnitude in the carbon diffusivity. Extremely small diffusivities around 10^{-24} m^2 s^{-1} are accessible with the indirect methods, illustrating the potential of these techniques. The Arrhenius plot of C diffusion is linear over a wide range at lower temperatures. There is some small positive curvature at higher temperatures. One possible origin of this curvature could be an influence the magnetic transition, which takes place at the Curie temperature T_C. In the case of self-diffusion of iron this influence is well-studied (see Chap. 17).

14.4 Examples of Diffusion-related Anelasticty   247

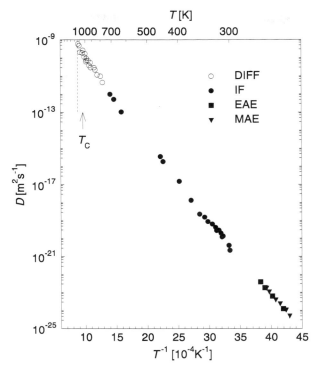

Fig. 14.7. Diffusion coefficient for C diffusion in α-Fe obtained by direct and indirect methods: DIFF = in- and out-diffusion or diffusion-couple methods; IF = internal friction; EAE = elastic after effect, MAE = magnetic after effect

It is interesting to note that the Snoek effect cannot be used to study interstitial solutes in fcc metals. Interstitial solutes in fcc metals are also incorporated in octahedral sites. In contrast to octahedral sites in the bcc lattice, which have tetragonal symmetry, octahedral sites in the fcc lattice and the microstrains associated with an interstitial solute in such sites have cubic symmetry. Interstitial solutes produce some lattice dilation but no elastic dipoles. Therefore, an external stress will not result in changes of the interstitial populations in an fcc matrix.

14.4.2 Zener Effect (Zener Relaxation)

The Zener effect, like the Snoek effect, is a stress-induced reorientation of elastic dipoles by atomic jumps. Atom pairs in substitutional alloys, pairs of interstitial atoms, solute-vacancy pairs possessing lower symmetry than the lattice can form dipoles responsible for Zener relaxation. For example, in strain-free dilute substitutional fcc alloys solute atoms are distributed randomly and isotropically. Solute-solute pairs on nearest-neighbour sites are uniformly distributed among the six $\langle 110 \rangle$ directions. The size difference between solute and solvent atoms causes pairs to create microstrains with strain fields of lower symmetry than that of the cubic host crystal.

A well-studied example of solute-solute pair reorientation in fcc materials was reported already by ZENER [2]. He observed a strong internal friction peak in Cu-Zn alloys (α-brass) around 570 K. The stress-mediated reorientation of random Zn-Zn pairs along $\langle 110 \rangle$ in fcc crystals is somewhat analogous to the Snoek effect. LE CLAIRE AND LOMER interpreted this relaxation on the basis of changing directional short-range order under the influence of external stress. In reality, the Zener effect in dilute substitutional fcc alloys depends on several exchange jump frequencies between solute atoms and vacancies. Therefore, it is difficult to relate the effect to the diffusion of solute atoms in a quantitative manner. A satisfactory model, such as is available for the Snoek effect of dilute interstitial bcc alloys, is not straightforward. The activation enthalpy of the process can be determined. However, in a pair model for low solute concentrations the activation energy is more characteristic of the rotation of the dipoles than of long-range diffusion.

14.4.3 Gorski Effect (Gorski Relaxation)

In contrast to reorientation relaxations discussed above, the *Gorski effect* is due to the long-range diffusion of solutes B which produce a lattice dilatation in a solvent A. This effect is named after the Russian scientist GORSKI [4]. Relaxation is initiated, for example, by bending a sample to introduce a macroscopic strain gradient. This gradient induces a gradient in the chemical potential of the solute, which involves the size-factor of the solute and the gradient of the dilatational component of the stress. Solutes redistribute by 'up-hill' diffusion and develop a concentration gradient, as indicated in Fig. 14.2. This transport produces a relaxation of elastic stresses, by the migration of solutes from the regions in compression to those in dilatation. The associated anelastic relaxation is finished when the concentration gradient equalises with the chemical potential gradient across the sample. For a strip of thickness d, the Gorski relaxation time, τ_G, is given by

$$\tau_G = \frac{d^2}{\pi^2 \Phi D_B}, \tag{14.32}$$

where D_B is the diffusion coefficient of solute B and Φ is the thermodynamic factor. A thermodynamic factor is involved, because Gorski relaxation establishes a chemical gardient.

Equation (14.32) shows that with the Gorski effect one measures the time required for diffusion of B atoms across the sample. The Gorski relaxation time is a macroscopic one, in contrast to the relaxation time of the Snoek relaxation. If the sample dimensions are known, an absolute value of the

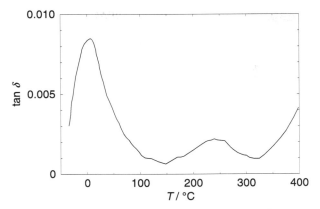

Fig. 14.8. Mechanical loss spectrum of a Na_2O4SiO_4 at a frequency of 1 Hz according to ROLING AND INGRAM [18, 19]

diffusivity is obtained. For a derivation of Eq. (14.32) we refer the reader to the review by VÖLKL [20]. The Gorski effect is detectable if the diffusion coefficient of the solute is high enough. Gorski effect measurements have been widely used for studies of hydrogen diffusion in metals [6, 20–22].

14.4.4 Mechanical Loss in Ion-conducting Glasses

Diffusion and ionic conduction in ion-conducting glasses is the subject of Chap. 30. Mechanical loss spectroscopy is also applicable for the characterisation of dynamic processes in glasses and glass ceramics. This method can provide information on the motion of mobile charge carriers, such as ions and polarons, as well as on the motion of network forming entities. Mixed mobile ion effects in different types of mixed-alkali glasses, mixed alkali-alkaline earth glasses, mixed alkaline earth glasses, and mixed cation anion glasses. For references see, e.g., a review of ROLING [8].

Let us consider an example: Fig. 14.8 shows the loss spectrum of a sodium silicate glass according to ROLING AND INGRAM [18, 19]. Such a spectrum is typical for ion conducting glasses. The low-temperature peak near 0 °C is attributed to the hopping motion of sodium ions, which can be studied by conductivity measurements in impedance spectroscopy and by tracer diffusion techniques as well (for examples see Chap. 30). The activation enthalpy of the loss peak is practically identical to the activation enthalpy of the dc conductivity, which is due to the long-range motion of sodium ions [19]. The intermediate-temperature peak at 235 °C is attributed to the presence of water in the glass. The increase of $\tan \delta$ near 350 °C is caused by the onset of the glass transition.

14.5 Magnetic Relaxation

In ferromagnetic materials, the interaction between the magnetic moment and local order can give rise to various relaxation phenomena similar to those observed in anelasticity. Their origin lies in the induced magnetic anisotropy energy, the theory of which was developed by the French Nobel laureate NEEL [24].

An example, which is closely related to the Snoek effect, was reported for the first time in 1937 by RICHTER [23] for α-Fe containing carbon. The direction of easy magnetisation in α-iron within a ferromagnetic domain is one of the three $\langle 100 \rangle$ directions. Therefore, the octahedral X-, Y-, and Z-positions for carbon interstitials are energetically not equivalent. A repopulation among these sites takes place when the magnetisation direction changes. This can happen when a magnetic field is applied. Suppose that the magnetic susceptibility χ is measured by applying a weak alternating magnetic field. Beginning with a uniform population of the interstitials, after demagnetisation a redistribution into the energetically favoured sites will occur. This stabilises the magnetic domain structure and reduces the mobility of the Bloch walls. As a consequence, a temporal decrease of the susceptibility χ is observed, which can be described by

$$\chi(t) = \chi_0 - \Delta\chi_s \left[1 - \exp\left(-\frac{t}{\tau_R}\right) \right], \tag{14.33}$$

where $\Delta\chi_s = \chi_0 - \chi(\infty)$ is denoted as the stabilisation susceptibility, t is the time elapsed since demagnetisation, and τ_R is the relaxation time. The relationship between jump frequency, relaxation time, and diffusion coefficient is the same as for anelastic Snoek relaxation.

The magnetic analogue to the Zener effect is directional ordering of ferromagnetic alloys in a magnetic field, which produces an *induced magnetic anisotropy*. The kinetics of the establishment of magnetic anisotropy after a thermomagnetic treatment can yield information about the activation energy of the associated diffusion process. The link between the relaxation time and diffusion coefficient is as difficult to establish as in the case of the Zener effect.

A magnetic analogue to the Gorski effect is also known. In a magnetic domain wall, the interaction between magnetostrictive stresses and the strain field of a defect (such as interstitials in octahedral sites of the bcc lattice, divacancies, etc.) can be minimised by diffusional redistribution in the wall. This diffusion gives rise to a magnetic after-effect. The relaxation time is larger by a factor δ_B/a (δ_B = thickness of the Bloch wall, a = lattice parameter) than for magnetic Snoek relaxation. The variation of the susceptibility with time is more complex than in Eq. (14.33). A comprehensive treatment of magnetic relaxation effects can be found in the textbook of KRONMÜLLER [9]. Obviously, magnetic methods are applicable to ferromagnetic materials at temperatures below the Curie point only.

References

1. J.L. Snoek, Physica **8**, 711 (1941)
2. C. Zener, Trans. AIME **152**, 122 (1943)
3. C. Zener, *Elasticity and Anelasticicty of Metals*, University of Chicago Press, Chicago, 1948
4. W.S. Gorski, Z. Phys. Sowjetunion **8**, 457 (1935)
5. A.S. Nowick, B.S. Berry, *Anelastic Relaxation in Crystalline Solids*, Academic Press, New York, 1972
6. B.S. Berry, W.C. Pritchet, *Anelasticity and Diffusion of Hydrogen in Glassy and Crystalline Metals*, in: *Nontraditional Methods in Diffusion*, G.E. Murch, H.K. Birnbaum, J.R. Cost (Eds.), The Metallurgical Society of AIME, Warrendale, 1984, p.83
7. R.D. Batist, *Mechanical Spectroscopy*, in: *Materials Science and Technology*, Vol. 2B: Characterisation of Materials, R.W. Cahn, P. Haasen, E.J. Cramer (Eds.), VCH, Weinheim, 1994. p. 159
8. B. Roling, *Mechanical Loss Spectroscopy on Inorganic Glasses and Glass Ceramics*, Current Opinion in Solid State Materials Science **5**, 203–210 (2001)
9. H. Kronmüller, *Nachwirkung in Ferromagnetika*, Springer Tracts in Natural Philosophy, Springer-Verlag, 1968
10. W. Voigt, Ann. Phys. **67**, 671 (1882)
11. J.H. Poynting, W. Thomson, *Properties of Matter*, C. Griffin & Co., London, 1902
12. A.D. Le Claire, *Diffusion of C, N, and O in Metals*, Chap. 8 in: *Diffusion in Solid Metals and Alloys*, H. Mehrer (Vol.Ed,), Landolt-Börnstein, Numerical Data and Functional Relationships in Science and Technology, New Series, Group III: Crystal and Solid State Physics, Vol. 26, Springer-Verlag, 1990
13. J.R.G. da Silva. R.B. McLellan, Materials Science and Engineering **26**, 83 (1976)
14. J. Woirgard, Y. Sarrazin, H. Chaumet, Rev. Sci. Instrum. **48**, 1322 (1977)
15. S. Etienne, J.Y. Cavaille, J. Perez, R. Point, M. Salvia, Rev. Sci. Instrum. **53**, 1261 (1982)
16. P.F. Green, D.L. Sidebottom, R.K. Brown, J. Non-cryst. Solids **172–174**, 1353 (1994)
17. P.F. Green, D.L. Sidebottom, R.K. Brown, J.H. Hudgens, J. Non-cryst. Solids **231**, 89 (1998)
18. B. Roling, M.D. Ingram, Phys. Rev. **B 57**, 14192 (1998)
19. B. Roling, M.D. Ingram, Solid State Ionics **105**, 47 (1998)
20. J. Völkl, Ber. Bunsengesellschaft **76**, 797 (1972)
21. J. Völkl, G. Alefeld, in: *Hydrogen in Metals I*, G. Alefeld, J. Völkl (Eds.), Topics in Applied Physics **28**, 321 (1978)
22. H. Wipf, *Diffusion of Hydrogen in Metals*, in: *Hydrogen in Metals III*, H. Wipf (Ed.), Topics in Applied Physics **73**, 51 (1995)
23. G. Richter, Ann. d. Physik **29**, 605 (1937)
24. L. Neel, J. Phys. Rad. **12**, 339 (1951); J. Phys. Rad. **13**, 249 (1952); J. Phys. Rad. **14**, 225 (1954)

15 Nuclear Methods

15.1 General Remarks

Several nuclear methods are important for diffusion studies in solids. They are listed in Table 13.1 and their potentials are illustrated in Fig. 13.1. The first of these methods is nuclear magnetic resonance or nuclear magnetic relaxation (NMR). NMR methods are mainly appropriate for self-diffusion measurements on solid or liquid metals. In favourable cases self-diffusion coefficients between about 10^{-20} and 10^{-10} m^2 s^{-1} are accessible. In the case of foreign atom diffusion, NMR studies suffer from the fact that a signal from nuclear spins of the minority component must be detected.

Mössbauer spectroscopy (MBS) and quasielastic neutron scattering (QENS) are techniques, which have considerable potential for understanding diffusion processes on a microscopic level. The linewidths $\Delta\Gamma$ in MBS and in QENS have contributions which are due to the diffusive motion of atoms. This diffusion broadening is observed only in systems with fairly high diffusivities since $\Delta\Gamma$ must be comparable with or larger than the natural linewidth in MBS experiments or with the energy resolution of the neutron spectrometer in QENS experiments. Usually, the workhorse of MBS is the isotope ^{57}Fe although there are a few other, less favourable Mössbauer isotopes such as ^{119}Sn, ^{115}Eu, and ^{161}Dy. QENS experiments are suitable for fast diffusing elements with a large incoherent scattering cross section for neutrons. Examples are Na self-diffusion in sodium metal, Na diffusion in ion-conducting rotor phases, and hydrogen diffusion in metals.

Neither MBS nor QENS are routine methods for diffusion measurements. The most interesting aspect is that these methods can provide *microscopic information* about the elementary jump process of atoms. The linewidth for single crystals depends on the atomic jump frequency and on the crystal orientation. This orientation dependence allows the deduction of the *jump direction* and the *jump length* of atoms, information which is not accessible to conventional diffusion studies.

15.2 Nuclear Magnetic Relaxation (NMR)

The technique of nuclear magnetic relaxation has been widely used for many years to give detailed information about condensed matter, especially about its atomic and electronic structure. It was recognised in 1948 by BLOEMBERGEN, PURCELL AND POUND [1] that NMR measurements can provide information on diffusion through the influence of atomic movement on the width of nuclear resonance lines and on relaxation times. Atomic diffusion causes fluctuations of the local fields, which arise from the interaction of nuclear magnetic moments with their local environment. The fluctuating fields either can be due to magnetic dipole interactions of the magnetic moments or due to the interaction of nuclear electric quadrupole moments (for nuclei with spins $I > 1/2$) with internal electrical field gradients. In addition, external magnetic field gradients can be used for a direct determination of diffusion coefficients.

We consider below some basic principles of NMR. Our prime aim is an understanding of how diffusion influences NMR. Solid state NMR is a very broad field. For a comprehensive treatment the reader is referred to textbooks of ABRAGAM [2], SLICHTER[3], MEHRING [4] and to chapters in monographs and textbooks [5–9]. In addition, detailed descriptions of NMR relaxation techniques are available, e.g., in [10]). Corresponding pulse programs are nowadays implemented in commercial NMR spectrometers.

15.2.1 Fundamentals of NMR

NMR methods are applicable to atoms with non-vanishing nuclear spin moment, $\hbar I$, and an associated magnetic moment

$$\mu = \gamma \hbar I, \tag{15.1}$$

where γ is the gyromagnetic ratio, I the nuclear spin, and $\hbar$ the Planck constant divided by 2π. In a static magnetic field $\boldsymbol{B_0}$ in z-direction, a nuclear magnetic moment $\boldsymbol{\mu}$ performs a precession motion around the z-axis governed by the equation

$$\frac{d\boldsymbol{\mu}}{dt} = \boldsymbol{\mu} \otimes \boldsymbol{B_0}. \tag{15.2}$$

The precession frequency is the Larmor frequency

$$\omega_0 = \gamma B_0. \tag{15.3}$$

The degeneracy of the $2I+1$ energy levels is raised due to the nuclear Zeeman effect. The energies of the nuclear magnetic dipoles are quantised according to

$$U_m = -m\gamma \hbar B_0, \tag{15.4}$$

where the allowed values correspond to $m = -I, -I+1, \ldots, I-1, I$. For example, for nuclei with $I = 1/2$ there are only two energy levels with the energy difference $\hbar \omega_0$.

At thermal equilibrium, the spins are distributed according to the Boltzmann statistics on the various levels. Since the energy difference between

15.2 Nuclear Magnetic Relaxation (NMR)

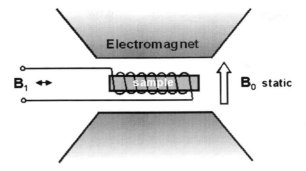

Fig. 15.1. Set-up for a NMR experiment (schematic)

levels for typical magnetic fields (0.1 to 1 Tesla) is very small, the population difference of the levels is also small. A macroscopic sample in a static magnetic field $\boldsymbol{B}_0$ in the z-direction displays a magnetisation $\boldsymbol{M}^{eq}$ along the z-direction and a transverse magnetisation $M_\perp = 0$. The equilibrium magnetisation of an ensemble of nuclei (number density N) is given by

$$\boldsymbol{M}^{eq} = N \frac{\gamma^2 \hbar^2 I(I+1)}{3k_B T} \boldsymbol{B}_0. \tag{15.5}$$

A typical experimental set-up for NMR experiments (Fig. 15.1) consists of a sample placed in a strong, homogeneous magnetic field $\boldsymbol{B}_0$ of the order of a few Tesla. A coil wound around the sample permits the application of an alternating magnetic field $\boldsymbol{B}_1$ perpendicular to the z-direction with frequency ω. Typically, these fields are radio-frequency (r.f.) fields. If the frequency ω of the transverse r.f. field $\boldsymbol{B}_1$ is close to the Larmor frequency, this field will induce transitions between the Zeeman levels of the nuclear spins. In NMR-spectrometers the coil around the sample is used for several steps of the experiment, such as irradiation of r.f. pulses and detection of the free induction decay of the ensemble of nuclei (see below).

The analysis of NMR experiments proceeds via a consideration of detailed interactions among nuclear moments and between them and other components of the solid such as electrons, point defects, and paramagnetic impurities. This theory has been developed over the past decades and can be found, e.g., in the textbooks of ABRAGAM [2] and SLICHTER [3]. Although this demands the use of quantum mechanics, much can be represented by semi-classical equations proposed originally by BLOCH. The effect of rf-pulse sequences on the time evolution of the total magnetisation $\boldsymbol{M}$ in an external field

$$\boldsymbol{B} = \boldsymbol{B}_0 + \boldsymbol{B}_1 \tag{15.6}$$

is given by the Bloch equation [2, 3]:

$$\frac{\mathrm{d}\boldsymbol{M}}{\mathrm{d}t} = \gamma \boldsymbol{M} \otimes \boldsymbol{B} - \frac{M_\perp}{T_2} - \frac{M_z - M_z^{eq}}{T_1} + \nabla\left[D\nabla(\boldsymbol{M} - \boldsymbol{M}^{eq})\right] . \quad (15.7)$$

The first term in Eq. (15.7) describes the precession of the spins around the magnetic field $\boldsymbol{B}$. The second and third terms give the rate of relaxation of the magnetisation and define two phenomenological constants, T_1 and T_2, denoted as relaxation times. They pertain to the longitudinal and transverse components of the magnetisation. In the absence of any transverse field, T_1 determines the rate at which M_z returns to its equilibrium value M_z^{eq}. This relaxation corresponds to an energy transfer between the spin-system and the so-called 'lattice', where the 'lattice' represents all degrees of freedom of the material with the exception of those of the spin-system. Therefore, T_1 is denoted as the *spin-lattice relaxation time*. T_2 refers to the transverse part of the nuclear magnetisation and is called the *spin-spin relaxation time*. Nuclear spins can be brought to a state of quasi-thermal equilibrium among themselves without being in thermal equilibrium with the lattice. T_2 describes relaxation to such a state. It follows that $T_2 \leq T_1$. T_2 is closely related to the width of the NMR signal.

The last term in Eq. (15.7) was introduced by TORREY [11] and describes the time evolution of the magnetisation $\boldsymbol{M}$, when the sample is also put into a magnetic field gradient. $\boldsymbol{M}^{eq}$ is the equilibrium value of the magnetic moment in field B_0 and D the diffusion coefficient. Equation (15.7) shows that various NMR techniques can be used to deduce information about atomic diffusion.

Elegant pulse techniques of radiofrequency spectroscopy permit the direct determination of D and of the relaxation times T_1 and T_2 (see, e.g., GERSTEIN AND DYBOWSKI [10]).

15.2.2 Direct Diffusion Measurement by Field-Gradient NMR

When a sample is placed deliberately in a magnetic field gradient, $G = \partial B_z/\partial z$, in addition to a static magnetic field, a direct determination of diffusion coefficients is possible. The basis of such NMR experiments in an inhomogeneous magnetic field is the last term of the Bloch equation. In a magnetic field gradient the Larmor frequency of a nuclear moment depends on its positions. Field-gradient NMR (FG-NMR) utilises the fact that nuclear spins that diffuse in a magnetic field-gradient experience an irreversible phase shift, which leads to a decrease in transversal magnetisation. This decay can be observed in so-called spin-echo experiments [12, 13]. The amplitude of the spin-echo is given by

$$M_G(t_{echo}) = M_0(t_{echo}) \exp\left[-\gamma^2 D \int_0^{t_{echo}} \left(\int_0^{t'} G(t'') \, \mathrm{d}t''\right)^2 \mathrm{d}t'\right] , \quad (15.8)$$

15.2 Nuclear Magnetic Relaxation (NMR)

where

$$M_0(t_{echo}) = M_0(0) \exp\left(-\frac{t_{echo}}{T_2}\right). \tag{15.9}$$

t_{echo} denotes the time of the spin echo. $M_G(t_{echo})$ and $M_0(t_{echo})$ are the echo amplitudes with and without field-gradient $G(t)$. $M_0(0)$ is the equilibrium magnetisation of the spin system.

For a 90-τ-180-τ spin-echo pulse sequence we have $t_{echo} = 2\tau$. In a constant magnetic field gradient G_0 the solution of Eq. (15.8) is proportional to the transversal magnetisation $M_\perp$, which is given by

$$M_G(2\tau) = M_0(0) \exp\left(-\frac{2\tau}{T_2}\right) \exp\left(-\frac{2}{3}\gamma^2 D G_0^2 \tau^3\right). \tag{15.10}$$

By varying τ or G_0 the diffusion coefficient can be determined from the measured spin-echo amplitude. The diffusion of spins is followed directly by FG-NMR. Thus, FG-NMR is comparable to tracer diffusion. For a known G_0 value a measurement of the diffusion-related part of the spin echo *versus* time can provide the diffusion coefficient without any further hypothesis. In contrast to tracer diffusion, the FG-NMR technique permits diffusion measurements in isotopically pure systems.

Equation (15.10) shows that the FG-NMR technique is applicable when the spin-spin relaxation time T_2 of the sample is large enough. A significant diffusion-related decay of the spin-echo amplitude must occur within T_2. For fixed values of T_2 and G_0 this requires D-values that are large enough. The measurement of small D-values requires high field-gradients. This can be achieved by using pulsed magnetic field-gradients (PFG) as suggested by MCCALL [14]. The first experiments with PFG-NMR were performed by STEJSKAL AND TANNER [13] for diffusion studies in aqueous solutions. For a comprehensive review of PFG-NMR spectroscopy the reader is referred, for example, to the reviews of STILBS [15], KÄRGER ET AL. [16], and MAJER [7]. PFG-NMR has been widely applied to study diffusion of hydrogen in metals and intermetallic compounds [7]. Applications to anomalous diffusion processes such as diffusion in porous materials and polymeric matrices can be found in [16]. Diffusion of hydrogen in solids is a relatively fast process and the proton is particularly suited for NMR studies due to its high gyromagnetic ratio. Diffusivities of hydrogen between 10^{-10} and 10^{-13} have been studied by PFG-NMR [7].

A fine example for the application of PFG-NMR are measurements of self-diffusion of liquid lithium and sodium [17]. Figure 15.2 displays self-diffusivities in liquid and solid Li obtained by PFG-NMR according to FEINAUER AND MAJER [18]. At the melting point, the diffusivity in liquid Li is almost three orders of magnitude faster than in the solid state. Also visible is the isotope effect of Li diffusion. The diffusivity of ^{6}Li is slightly faster than that of ^{7}Li.

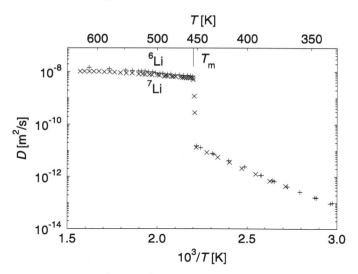

Fig. 15.2. Self-diffusion of ^{6}Li and ^{7}Li in liquid and solid Li studied by PFG-NMR according to FEINAUER AND MAJER [18]

15.2.3 NMR Relaxation Methods

Indirect NMR methods for diffusion studies measure either the relaxation times T_1 and T_2, or the linewidth of the absorption line. In addition, other relaxation times not contained in the Bloch equation can be operationally defined. The best known of these is the spin-lattice relaxation time in the rotating frame, $T_{1\rho}$. This relaxation time characterises the decay of the magnetisation when it is 'locked' parallel to B_1 in a frame of reference rotating around B_0 with the Larmor frequency $\omega_0 = \gamma B_0$. In such an experiment, M starts from M^{eq} and decays to $B_1 M^{eq}/B_0$. Since $T_{1\rho}$ is shorter than T_1, measurements of $T_{1\rho}$ permit the detection of slower atomic motion than T_1.

Let us consider a measurement of the spin-lattice relaxation time T_1. If a magnetic field is applied in the z-direction, T_1 describes the evolution of the magnetisation M_z towards equilibrium according to

$$\frac{dM_z}{dt} = \frac{M_z^{eq} - M_z}{T_1}. \tag{15.11}$$

A measurement of T_1 proceeds in two steps. (i) At first, the nuclear magnetisation is inverted by the application of an 'inversion pulse'. (ii) Then, the magnetisation $M_z(t)$ is observed by a 'detection pulse' as it relaxes back to the equilibrium magnetisation.

The effect of r.f. pulses can be discussed on the basis of the Bloch equation (15.7). If the resonance condition, $\omega_0 = \gamma B_0$, is fulfilled for the alternating B_1 field, the magnetisation will precess in the y-z plane with a precession

15.2 Nuclear Magnetic Relaxation (NMR)   259

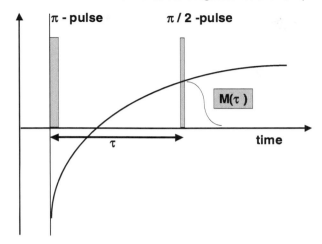

Fig. 15.3. Schematic iluustration of a T_1 measurement with an inversion-recovery (π-τ-$\pi/2$) pulse sequence

frequency γB_1. The application of a pulse of the r.f. field B_1 with a duration t_p will result in the precession of the magnetisation to the angle $\Theta_p = \gamma B_1 t_p$. By suitable choice of the pulse length the magnetisation can be inverted ($\Theta_p = \pi$) or tilted into the x-y plane ($\Theta_p = \pi/2$). During precession in the x-y plane the magnetisation will induce a voltage in the coil (Fig. 15.1). This signal is called the *free induction decay* (FID). If, for example, an initial π-pulse is applied, $M_z(t)$ can be monitored by the amplitude of FID after a $\pi/2$-reading pulse at the evolution time t, which is varied in the experiment[1]. This widely used pulse sequence for the measurement of T_1 is illustrated in Fig. 15.3.

NMR is sensitive to interactions of nuclear moments with fields produced by their local environment. The relaxation times and the linewidth are determined by the interaction between nuclear moments either directly or via electrons. Apart from coupling to the spins of conduction electrons in metals or of paramagnetic impurities in non-metals, two basic mechanisms of interaction must be considered in relation to atomic movements. The first interaction is dipole-dipole coupling among the nuclear magnetic moments. The second interaction is due to nuclear electric quadrupole moments with internal electric field gradients. Nonzero quadrupolar moments are present for nuclei with nuclear spins $I > 1/2$. The diffusion of nuclear moments causes variations in both of these interactions. Therefore, the width of the resonance line and the relaxation times have contributions which are due to the thermally activated jumps of atoms.

---

[1] Without discussing further details, we mention that more complex pulse sequences have been tailored to overcome limitations of the simple sequence, which suffers from the dead-time of the detection system after the strong r.f. pulse.

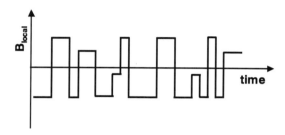

Fig. 15.4. Temporal fluctuations of the local field – the origin of motional narrowing

Spin-Spin Relaxation and Motional Narrowing: Let us suppose for the moment that we need to consider only magnetic dipole interactions, which is indeed the case for nuclei with $I = 1/2$. Each nuclear spin precesses, in fact, in a magnetic field $\boldsymbol{B} = \boldsymbol{B}_0 + \boldsymbol{B}_{local}$, where $\boldsymbol{B}_{local}$ is the local field created by the magnetic moments of neighbouring nuclei. The local field experienced by a particular nucleus is dominated by the dipole fields created by the nuclei in its immediate neighbourhood, because dipolar fields vary as $1/r^3$ with the distance r between the nuclei. Since the nuclear moments are randomly oriented, the local field varies from one nucleus to another. This leads to a dispersion of the Larmor frequency and to a broadening of the resonance line according to

$$\Delta \omega_0 = \frac{1}{T_2} \propto \gamma \Delta B_{local}. \tag{15.12}$$

ΔB_{local} is an average of the local fields in the sample. In solids without internal motion, local fields are often quite large and give rise to rather short T_2 values. Typical values without motion of the nuclei are the following: $\Delta B_{local} \approx 2 \times 10^{-4}$ Tesla, $T_2 \approx 100 \mu s$ and $\Delta \omega_0 \approx 10^4$ rad s^{-1}. Such values are characteristic of a 'rigid lattice' regime. The pertaining spin-spin relaxation time is denoted as T_2 (*rigid lattice*).

Let us now consider how diffusion affects the spin-spin relaxation time and the linewidth of the resonance line. Diffusion comes about by jumps of individual atoms from one site to another. The mean residence time of an atom, $\bar{\tau}$, is temperature dependent via

$$\bar{\tau} = \tau_0 \exp\left(\frac{\Delta H}{k_B T}\right) \tag{15.13}$$

with an activation enthalpy ΔH and a pre-factor τ_0. Each time when an atom jumps into a new site, its nuclear moment will find itself in another local field. As a consequence, the local field sensed by a nucleus will fluctuate between $\pm B_{local}$ on a time-scale characterised by the mean residence time (Fig. 15.4). If the mean residence time of an atom is much shorter than the spin-spin relaxation time of the rigid lattice, i.e. for $\bar{\tau} \ll T_2$ (*rigid lattice*),

15.2 Nuclear Magnetic Relaxation (NMR)

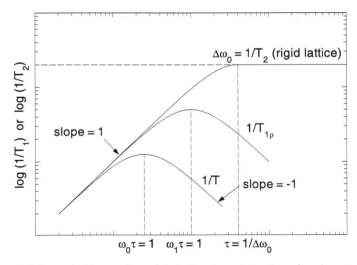

Fig. 15.5. Schematic illustration of diffusional contributions (random jumps) to spin-lattice relaxation rates, $1/T_1$ and $1/T_{1\rho}$, and to the spin-spin relaxation rate $1/T_2$

a nuclear moment will sample many different local fields. The nuclear moment will behave as though it were in some new effective local field, which is given by the average of all the local fields sampled. If the sampled local fields vary randomly in direction and magnitude this average will be quite small, depending on how many are sampled. The dephasing between the spins grows more slowly with time than in a fixed local field. The effective local fields of all the nuclear moments will be small, and the nuclear moments will precess at nearly the same frequency. Thus, the nuclear moments will not lose their coherence as rapidly during a FID, and T_2 will be longer. A longer FID is equivalent to a narrower resonance line.

If the diffusion rate is increased, it can be shown by statistical considerations that the width of the resonance line becomes

$$\Delta\omega' = \frac{1}{T_2'} = \Delta\omega_0^2 \bar{\tau}. \tag{15.14}$$

This phenomenon is called *motional narrowing*. A schematic illustration of the temperature dependence the spin-spin relaxation rate $1/T_2$ is displayed in Fig. 15.5: at low temperatures the relaxation rate of the rigid lattice is observed, since diffusion is so slow that an atom does not even jump once during the FID; as $\bar{\tau}$ gets shorter with increasing temperature $1/T_2'$ decreases and the width of the resonance line gets narrower.

Spin-Lattice Relaxation: When discussing the Bloch equations we have seen that the spin-lattice relaxation time T_1 is the characteristic time during which the nuclear magnetisation returns to its equilibrium value. We could also say the nuclear spin system comes to equilibrium with its environment, called 'lattice'. In contrast to spin-spin relaxation, this process requires an exchange of energy with the 'lattice'. Spin-lattice relaxation either takes place by the absorption or emission of phonons or by coupling of the spins to conduction electrons (via hyperfine interaction) in metals. The relaxation rate due to the coupling of nuclear spins with conduction electrons is approximately given by the Koringa relation

$$\left(\frac{1}{T_1}\right)_e = \text{const} \times T, \qquad (15.15)$$

where T denotes the absolute temperature. The relaxation rate due to dipolar interactions, $(1/T_1)_{dip}$, and due to quadrupolar interactions, $(1/T_1)_Q$, is added to that of electrons, so that the total spin-lattice relaxation rate is

$$\frac{1}{T_1} = \left(\frac{1}{T_1}\right)_e + \left(\frac{1}{T_1}\right)_{dip} + \left(\frac{1}{T_1}\right)_Q. \qquad (15.16)$$

For systems with nuclear spins $I = 1/2$, quadrupolar contributions are absent.

The fluctuating fields can be described by a correlation function $G(t)$, which contains the temporal information about the atomic diffusion process [2, 3]. Let us assume as in the original paper by BLOEMBERGEN, PURCELL AND POUND [1] that the correlation function decays exponentially with the correlation time τ_c, i.e. as

$$G(t) = G(0) \exp\left(-\frac{|t|}{\tau_c}\right). \qquad (15.17)$$

This behaviour is characteristic of jump diffusion in a three dimensional system and τ_c is closely related to the mean residence time between successive jumps. The Fourier transform of Eq. (15.17), which is called the *spectral density function* $J(\omega)$, is a Lorentzian given by

$$J(\omega) = G(0) \frac{2\tau_c}{1 + \omega^2 \tau_c^2}. \qquad (15.18)$$

Transitions between the energy levels of the spin-system can be induced, i.e. spin-lattice relaxation becomes effective, when $J(\omega)$ has components at the transition frequency. The spin-lattice relaxation rate is then approximately given by

$$\left(\frac{1}{T_1}\right)_{dip} \approx \frac{3}{2} \gamma^4 \hbar^2 I(I+1) J(\omega_0). \qquad (15.19)$$

Detailed expressions for the relaxation rates $\frac{1}{T_1}$, $\frac{1}{T_2}$ and $\frac{1}{T_{1\rho}}$ can be found, e.g., in [2, 6].

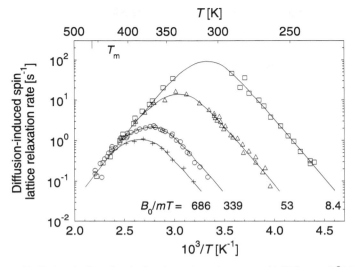

Fig. 15.6. Diffusion-induced spin-lattice relaxation rate, $(1/T_1)_{dip}$, of ^{8}Li in solid Li as a function of temperature according to HEITJANS ET AL. [8]. The B_0 values correspond to Larmor frequencies $\omega_0/2\pi$ of 4.32 MHz, 2.14 MHz, 334 kHz, and 53 kHz

The correlation time τ_c, like the mean residence time $\bar{\tau}$, will usually obey an Arrhenius relation

$$\tau_c = \tau_c^0 \exp\left(\frac{\Delta H}{k_B T}\right), \qquad (15.20)$$

where ΔH is the activation enthalpy of the diffusion process. Since the movement of either atom of a pair will change the correlation function we may identify τ_c with one half of the mean residence time $\bar{\tau}$ of an atom at a lattice site.

The diffusion-induced spin-lattice relaxation rate, $(1/T_1)_{dip}$, is shown in Fig. 15.6 for self-diffusion of ^{8}Li in lithium according to HEITJANS ET AL. [8]. In a representation of the logarithm of the relaxation rate as function of the reciprocal temperature, a symmetric peak is observed with a maximum at $\omega_0\tau_c \approx 1$. At temperatures well above or below the maximum, which correspond to the cases $\omega_0\tau_c \ll 1$ or $\omega_0\tau_c \gg 1$, the slopes yield $\Delta H/k_B$ or $-\Delta H/k_B$.

The work of BLOEMBERGEN, PURCELL AND POUND [1] is based on the assumption of the exponential correlation function of Eq. (15.17), which is appropriate for diffusion in liquids. Later on, the theory was extended to random walk diffusion in lattices by TORREY [19]. Based on the *encounter model* (see Chap. 7) the influence of defect mechanisms of diffusion and the associated correlation effects have been included into the theory by WOLF [20] and MACGILLIVRAY AND SHOLL [21]. These refinements lead to results that

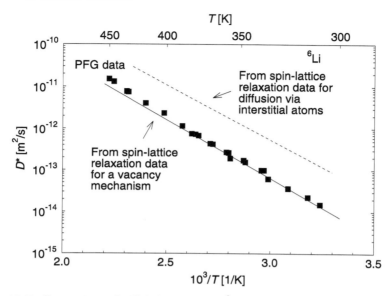

Fig. 15.7. Comparison of self-diffusivities for ^{6}Li in solid Li determined by PFG-NMR with spin-lattice relaxation results assuming a vacancy mechanism (*solid line*) and an interstitial mechanism (*dashed line*) according to MAJER [22]

are broadly similar to those of [1]. However, the refinements are relevant for a quantitative interpretation of NMR results in terms of diffusion coefficients. We illustrate this by an example:

Figure 15.7 shows a comparison of diffusion data of ^{6}Li in solid lithium obtained with PFG-NMR and data deduced from relaxation measurements. PFG-NMR yields directly ^{6}Li self-diffusion coefficients in solid lithium. No assumption about the elementary diffusion steps is needed for these data. The dashed and solid lines are deduced from $(1/T_1)_{dip}$ data, assuming two different atomic mechanisms. Good coincidence of diffusivities from spin-lattice relaxation and the PFG-NMR data is obtained with the assumption that Li diffusion is mediated by vacancies. Direct interstitial diffusion clearly can be excluded [22].

15.3 Mössbauer Spectroscopy (MBS)

The Mössbauer effect has been detected by the 1961 Nobel laureate in physics R. MÖSSBAUER [23]. The Mössbauer effect is the recoil-free emission and absorption of γ-radiation by atomic nuclei. Among many other applications, Mössbauer spectroscopy can be used to deduce information about the movements of atoms for which suitable Mössbauer isotopes exist. There are only

15.3 Mössbauer Spectroscopy (MBS)

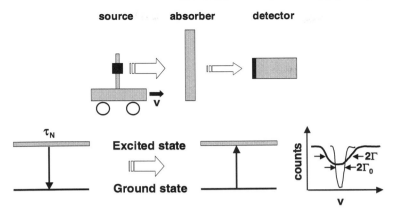

Fig. 15.8. Mössbauer spectroscopy. *Top*: moving source experiment; *bottom*: principles

a few nuclei, ^{57}Fe, ^{119}Sn, ^{151}Eu, and ^{161}Dy, for which Mössbauer spectroscopy can be used. ^{57}Fe is the major 'workhorse' of this technique

Information about atomic motion is obtained from the broadening of the otherwise very narrow γ-line. Thermally activated diffusion of Mössbauer atoms contributes to the linewidth in a way first recognised by SINGWI AND SJÖLANDER in 1960 [24] soon after the detection of the Mössbauer effect.

Mössbauer spectroscopy uses two samples, one playing the rôle of the source, the other one the rôle of an absorber of γ-radiation as indicated in Fig. 15.8. In the source the nuclei emit γ-rays, some of which are absorbed without atomic recoil in the absorber. The radioisotope ^{57}Co is frequently used in the source. It decays with a half-life time of 271 days into an excited state of the Mössbauer isotope ^{57}Fe. The Mössbauer level is an excited level of ^{57}Fe with lifetime $\tau_N = 98$ ns. It decays by emission of γ-radiation of the energy $E_\gamma = 14.4$ keV to the ground state of ^{57}Fe, which is a stable isotope with a 2.2% natural abundance. If the Mössbauer isotope is incorporated in a crystal, the recoil energy of the decay is transferred to the whole crystal. Then, the width of the emitted γ-line becomes extremely narrow. This is the effect for which Mössbauer received the Nobel price. The absorber also contains the Mössbauer isotope. A fraction f of the emitted γ-rays is absorbed without atomic recoil in the absorber. In the experiment, the source is usually moved relative to the absorber with a velocity v. Experimantal set-ups with static source and a moving absorber are also possible. This motion causes a Doppler shift

$$\Delta E = \frac{v}{c} E_\gamma \qquad (15.21)$$

of the source radiation, where c denotes the velocity of light. The linewidth in the absorber is then recorded as a function of the relative velocity or as a function of the Doppler shift ΔE.

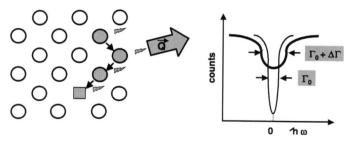

Fig. 15.9. Simplified, semi-classical explanation of the diffusional line-broadening of a Mössbauer spectrum. Q denotes the wave vector of the γ-rays

Diffusion in a solid, if fast enough, leads to a diffusional broadening of the Mössbauer spectrum. This can be understood in a simplified picture as illustrated in Fig. 15.9 [25]: at low temperatures, the Mössbauer nuclei stay on their lattice sites during the emission process. Without diffusion the natural linewidth Γ_0 is observed, which is related to the lifetime of the excited Mössbauer level, τ_N, via the Heisenberg uncertainty relation:

$$\Gamma_0 \tau_N \geq \hbar. \tag{15.22}$$

At elevated temperatures, the atoms become mobile. A diffusing atom resides on one lattice site only for a time $\bar{\tau}$ between two successive jumps. If $\bar{\tau}$ is of the same order or smaller than τ_N, the Mössbauer atom changes its position during the emission process. When an atom is jumping the wave packet emitted by the atom is 'cut' into several shorter wave packets. This leads to a broadening of the linewidth Γ, in addition to its natural width Γ_o. If $\bar{\tau} \ll \tau_N$, the broadening, $\Delta\Gamma = \Gamma - \Gamma_0$, is of the order of

$$\Delta\Gamma \approx \hbar/\bar{\tau}. \tag{15.23}$$

Neglecting correlation effects (see, however, below) and considering diffusion on a Bravais lattice with a jump length d the diffusion coefficient is related to the diffusional broadening via

$$D \approx \frac{d^2}{12} \frac{\Delta\Gamma}{\hbar}. \tag{15.24}$$

Experimental examples for Mössbauer spectra of ^{57}Fe in iron are shown in Fig. 15.10 according to VOGL AND PETRY [27]. The Mössbauer source was ^{57}Co. The linewidth increases with increasing temperature due to the diffusional motion of Fe atoms. Figure 15.11 shows an Arrhenius diagram of self-diffusion for γ- and δ-iron, in which the Mössbauer data are compared with tracer results [27]. The jump length d in Eq. (15.24) was assumed to be the nearest neighbour distance of Fe. It can be seen that

15.3 Mössbauer Spectroscopy (MBS)   267

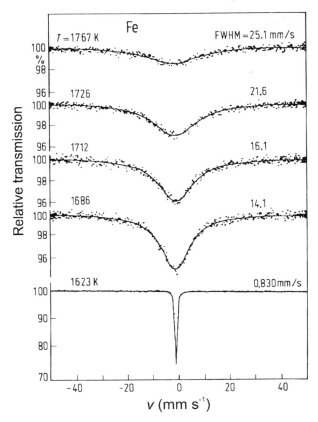

Fig. 15.10. Mössbauer spectra for self-diffusion in polycrystalline Fe from a review of VOGL AND PETRY [27]. FWHM denotes the full-width of half maximum of the Mössbauer line. The spectrum at 1623 K pertains to γ-iron and the spectra at higher temperatures to δ-iron

the diffusivities determined from the Mössbauer study agree within error bars with diffusivities from tracer studies. Equation (15.24) is an approximation and follows from the more general Eq. (15.27). For this aim Eq. (15.27) is specified to polycrystalline samples and considered for $Q \gg 1/d$. For 14.4 keV γ-radiation we have $Q = 73$ nm^{-1}, which is indeed much larger than $1/d$. The broadening is more pronounced in the high-temperature δ-phase of iron with the bcc structure as compared to the fcc γ-phase of iron. This is in accordance with the fact that self-diffusion increases by about one order of magnitude, when γ-iron transforms to δ-iron (see Chap. 17).

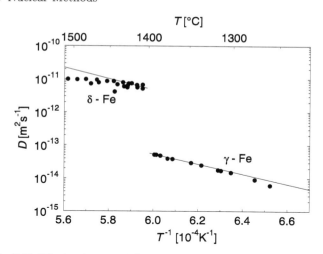

Fig. 15.11. Self-diffusion in γ- and δ-iron: comparison of Mössbauer (*symbols*) and tracer results (*solid lines*) according to VOGL AND PETRY [27]

Diffusional Broadening of MBS Signals: A quantitative analysis of diffusional line-broadening uses the fact that according to VAN HOVE [28] the displacement of atoms in space and time can be described by the self-correlation function $G_s(\bm{r}, t)$. This is the probability density to find an atom displaced by the vector $\bm{r}$ within a time interval t. We are interested in the self-correlation function because the Mössbauer absorption spectrum, $\sigma(\bm{Q}, \omega)$, is related to the double Fourier transform of G_s in space and time via

$$\sigma(\bm{Q},\omega) \propto Re\left[\int\int G_s(\bm{r},t)\exp\left[i(\bm{Q}\cdot\bm{r}-\omega t)-\Gamma_0\mid t\mid/2\hbar\right]\mathrm{d}\bm{r}\mathrm{d}t\right],$$

(15.25)

where Γ_0 is the natural linewidth of the Mössbauer transition.

The self-correlation function contains both diffusional motion as well as lattice vibrations. Usually, these two contributions can be separated. The vibrational part leads to the so-called Debye-Waller factor, f_{DW}, which governs the intensity of the resonantly absorbed radiation. The diffusional part determines the shape of the Mössbauer spectrum. As the wave packets are emitted by the same nucleus, they are coherent. The interference between these packets depends on the orientation between the jump vector of the atom and the wave vector (see Fig. 15.9). If a single-crystal specimen is used, in certain crystal directions the linewidth is small and in other directions it is larger.

To exploit Eq. (15.25) a diffusion model is necessary to calculate $\sigma(\bm{Q},\omega)$. For random jumps on a Bravais lattice (Markov process) the shape of the resulting Mössbauer spectrum is a Lorentzian [25, 26]

$$\sigma(\boldsymbol{Q},\omega) \propto f_{DW} \frac{\Delta\Gamma(\boldsymbol{Q})/2}{[\Delta\Gamma(\boldsymbol{Q})/2]^2 + (\hbar\omega)^2}, \quad (15.26)$$

where $\Delta\Gamma(\boldsymbol{Q})$ is the full peak-width at half maximum. This diffusional broadening depends on the relative orientation between radiation and crystal:

$$\Delta\Gamma(\boldsymbol{Q}) = \frac{2\hbar}{\bar{\tau}}\left(1 - \sum_j W_j E_j\right) \quad \text{where} \quad E_j = \frac{1}{N_j}\sum_{k=1}^{N_j} \exp(i\boldsymbol{Q}\cdot\boldsymbol{r}_k). \quad (15.27)$$

W_j is the probability for a displacement to coordination shell j, E_j the corresponding structure factor, N_j denotes the number of sites in the coordination shell j, and $\boldsymbol{r}_k$ are the displacement vectors to sites in shell j.

For diffusion mediated by vacancies, successive jumps of an atom are correlated. An extension of Eq. (15.27) for correlated diffusion has been developed by WOLF [20] on the basis of the so-called encounter model (see Chap. 7). The mean time between encounters is

$$\tau_{enc} = \bar{\tau} Z_{enc}, \quad (15.28)$$

where Z_{enc} is the average number of jumps performed by a Mössbauer atom in one encounter. Each complete encounter is treated as an effective displacement not correlated to the previous or following encounter. Wolf showed that the line broadening can be expressed as

$$\Delta\Gamma(\boldsymbol{Q}) = \frac{2\hbar}{\bar{\tau} Z_{enc}}\left(1 - \sum_j W_j^{enc} E_j\right), \quad (15.29)$$

where W_j^{enc} is the probability for a displacement of $\boldsymbol{r}_j$ by an encounter with a defect. For further details and for an extension to non-Bravais lattices the reader is referred to [25, 29].

An important consequence of Eq. (15.26) and of Eq. (15.29) is that both $\sigma(\boldsymbol{Q},\omega)$ and $\Delta\Gamma$ depend on the relative orientation between $\boldsymbol{Q}$ and the jump vector $\boldsymbol{r}$ and hence on the orientation of the crystal lattice. This can be exploited by measurements on monocrystals. By varying the crystal orientation, information about the length and direction of the jump vector is obtained. In that respect MBS and QENS are analogous. Examples for the deduction of elementary diffusion jumps will be given in the next section.

15.4 Quasielastic Neutron Scattering (QENS)

The scattering of beams of slow neutrons obtained from nuclear reactors or other high-intensity neutron sources can be used to study structural and dynamic properties of condensed matter. Why neutron scattering is a tool with

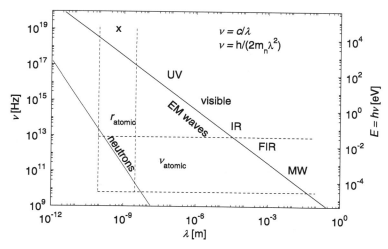

Fig. 15.12. Comparison between the dispersion relations of electromagnetic waves (EM waves) and neutrons

unique properties can be seen from Fig. 15.12, which shows a comparison of the dispersion relations of electromagnetic waves (EM waves) and neutrons. For EM waves the frequency ν and the wavelength λ are related via $\nu = c/\lambda$, where c is the velocity of light. For (non-relativistic) neutrons of mass m_n the dispersion relation is $\nu = h/(2m_n\lambda^2)$. Typical atomic vibration frequencies in a solid, ν_{atomic}, match with far infrared and microwave frequencies of EM waves. On the other hand, typical interatomic distances, r_{atomic}, match with wavelengths of X-rays. Slow and thermal neutrons have the unique feature that their wavelengths and frequencies match atomic frequencies and interatomic distances simultaneously.

Neutrons are uncharged probes and interact with nuclei. In contrast to photons, neutrons have only a weak interaction with matter. This means that neutron probes permit easy access to bulk properties. Since neutrons can penetrate suitable sample containers easily. One can also use sophisticated sample environments, such as wide temperature ranges and high magnetic fields.

The scattering cross section for neutrons is determined by the sample nuclei. The distribution of scattering cross sections in the periodic table is somehow irregular. For example, protons have very high scattering cross sections and are mainly incoherent scatterers. For deuterons the coherent cross section is larger than the incoherent scattering cross section. Carbon, nitrogen, and oxygen have very small incoherent scattering cross sections and are mainly coherent scatterers. For sodium, coherent and incoherent scattering cross sections are similar in magnitude.

15.4 Quasielastic Neutron Scattering (QENS)

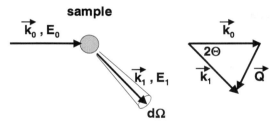

Fig. 15.13. Neutron scattering geometry: in real space (*left*); in momentum space (*right*)

Neutron scattering leads to a spectrum of energy and momentum transfers (Fig. 15.13). The energy transfer is

$$\hbar\omega = E_1 - E_0 ,\qquad(15.30)$$

where E_1 and E_0 denote the neutron energies after and before the scattering process, respectively. The corresponding momentum transfer is $\hbar\boldsymbol{Q}$. The scattering vector is

$$\boldsymbol{Q} = \boldsymbol{k}_1 - \boldsymbol{k}_0 ,\qquad(15.31)$$

where $\boldsymbol{k}_0$ and $\boldsymbol{k}_1$ are the neutron wave vectors before and after the scattering event. The corresponding neutron wavelenghts are $\lambda_1 = 2\pi/k_1$ and $\lambda_0 = 2\pi/k_0$. The values of

$$Q = \frac{4\pi}{\lambda_0}\sin(\Theta/2)\qquad(15.32)$$

(Θ = scattering angle Q = modulus of the scattering vector) vary typically between 1 and 50 nm^{-1}. Therefore, $1/Q$ can match interatomic distances. The scattered intensity in such an experiment is proportional to the so-called scattering function or dynamic structure factor, $S(\boldsymbol{Q},\omega)$, which can be calculated for diffusion processes (see below).

A schematic energy spectrum for neutron scattering with elastic, quasielastic, and inelastic contributions is illustrated in Fig. 15.14. Inelastic peaks are observed, due to the absorption and emission of phonons.

Quasielastic Scattering: Quasielastic scattering must be distinguished from the study of periodic modes such as phonons or magnons by inelastic scattering, which usually occurs at higher energy transfers.

For samples with suitable scattering cross sections, diffusion of atoms in solids can be studied by quasielastic neutron scattering (QENS), if a high-resolution neutron spectrometer is used. QENS, like MBS, is a technique which has considerable potential for elucidating diffusion steps on a microscopic level. Both techniques are applicable to relatively fast diffusion processes only (see Fig. 13.1). QENS explores the diffusive motion in space for a range comparable to the neutron wavelength. Typical jump distances and

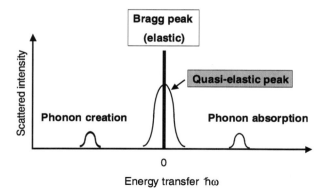

Fig. 15.14. Energy spectrum of neutron scattering (schematic)

diffusion paths between 10^{-8} and 10^{-10} m can be studied. Let us briefly anticipate major virtues of QENS. The full peak-width at half maximum of a Lorentzian shaped quasielastic line is given for small values of Q by

$$\Delta \Gamma = 2DQ^2 \,, \qquad (15.33)$$

where D is the self-diffusion coefficient [31, 32]. Quasielastic line broadening is due to the diffusive motion of atoms. The pertinent energy transfers $\hbar\omega$ typically range from 10^{-3} to 10^{-7} eV. For larger scattering vectors, $\Delta\Gamma$ is periodic in reciprocal space and hence depends on the atomic jump vector like in MBS. For a particle at rest, we have $\Delta\Gamma = 0$ and a sharp line at $\hbar\omega = 0$ is observed. This elastic line (Bragg peak) results from a scattering process in which the neutron transmits the momentum $\hbar\boldsymbol{Q}$ to the sample as a whole, without energy transfer. For resonance absorption of γ-rays this corresponds to the well-known Mössbauer line (see above). We have already seen that in MBS a diffusing particle produces a line broadening. QENS is described by similar theoretical concepts as used in MBS [25, 30–32].

Figure 15.15 shows an example of a quasielastic neutron spectrum measured on a monocrystal of sodium according to GÖLTZ ET AL. [33]. The number of scattered neutrons N is plotted as a function of the energy transfer $\hbar\omega$ for a fixed scattering vector with $Q = 1.3 \times 10^{-10}$ m^{-1}. The dashed line represents the resolution function of the neutron spectrometer. The observed line is broadened due to the diffusive motion of Na atoms. The quasielastic linewidth depends on the orientation of the momentum transfer and hence of the crystallographic orientation of the crystal (see below).

The Dynamic Structure Factor (Scattering Functions): Let us now recall some theoretical aspects of QENS. The quantity measured in neutron scattering experiments is the intensity of neutrons, ΔI_s, scattered from a collimated mono-energetic neutron beam with a current density I_0. The intensity

15.4 Quasielastic Neutron Scattering (QENS) 273

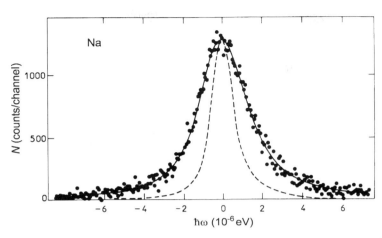

Fig. 15.15. QENS spectrum of a Na monocrystal at 367.5 K according to GÖLTZ ET AL. [33]. *Dashed line*: resolution functionof the neutron spectrometer

of neutrons scattered into a solid angle element, $\Delta\Omega$, and an interval, $\Delta\omega$, from a sample with volume V and number density of scattering atoms, N, (see Fig. 15.13) is given by [31]

$$\Delta I_s = I_0 N V \left(\frac{d^2\sigma}{d\Omega d\omega}\right) \Delta\Omega\Delta\omega\,, \qquad (15.34)$$

where the double differential scattering cross section is

$$\frac{d^2\sigma}{d\Omega d\omega} = \frac{k_1}{k_0}\frac{\sigma}{4\pi}S(\boldsymbol{Q},\omega)\,. \qquad (15.35)$$

The cross section is factorised into three components: the ratio of the wave numbers k_1/k_0; the cross section for a rigidly bound nucleus, $\sigma = 4\pi b^2$, where b is the corresponding scattering length of the nucleus; the scattering intensity is proportional to the *dynamical structure factor* $S(\boldsymbol{Q},\omega)$. The dynamical structure factor depends on the scattering vector and on the energy transfer defined in Eqs. (15.30) and (15.31). It describes structural and dynamical properties of the sample which do not depend on the interaction between neutron and nuclei.

The interaction of a neutron with a scattering nucleus depends on the chemical species, the isotope, and its nuclear spin. In a mono-isotopic sample, all nuclei have the same scattering length. Then, only coherent scattering will be observed. In general, however, several isotopes are present according to their natural abundance. Each isotope i is characterised by its scattering length b_i. The presence of different isotopes distributed randomly in the sample means that the total scattering cross section is made up of two parts, called *coherent* and *incoherent*.

The theory of neutron scattering is well developed and can be found, e.g., in reviews by ZABEL [30], SPRINGER [31, 32] and in textbooks of SQUIRES [37], LOVESEY [38], BEE [36], and HEMPELMANN [39]. Theory shows that the differential scattering cross section can be written as the sum of a coherent and an incoherent part

$$\frac{d^2\sigma}{d\Omega d\omega} = \left(\frac{d^2\sigma}{d\Omega d\omega}\right)_{coh} + \left(\frac{d^2\sigma}{d\Omega d\omega}\right)_{inc}$$
$$= \frac{k_1}{k_0}\left[\frac{\sigma_{coh}}{4\pi}S_{coh}(\bm{Q},\omega) + \frac{\sigma_{inc}}{4\pi}S_{inc}(\bm{Q},\omega)\right]. \quad (15.36)$$

Coherent (index: *coh*) and incoherent (index: *inc*) contributions depend on the composition and the scattering cross sections of the nuclei in the sample. The *coherent scattering* cross section σ_{coh} is due to the average scattering from different isotopes

$$\sigma_{coh} = 4\pi\bar{b}^2 \quad \text{with} \quad \bar{b} = \Sigma c_i b_i. \quad (15.37)$$

The *incoherent scattering* is proportional to the deviations of the individual scattering lengths from the mean value

$$\sigma_{inc} = 4\pi\left(\overline{b^2} - \bar{b}^2\right) \quad \text{with} \quad \overline{b^2} = \Sigma c_i b_i^2. \quad (15.38)$$

The bars indicate ensemble averages over the various isotopes present and their possible spin states. The c_i are the fractions of nuclei i.

Coherent scattering is due to interference of partial neutron waves originating at the positions of different nuclei. The coherent scattering function, $S_{coh}(\bm{Q},\omega)$, is proportional to the Fourier transform of the correlation function of any nuclei. Coherent scattering leads to interference effects and collective properties can be studied. Among other things, this term gives rise to Bragg diffraction peaks.

Incoherent scattering monitors the fate of individual nuclei and interference effects are absent. The incoherent scattering function, $S_{inc}(\bm{Q},\omega)$, is proportional to the Fourier transform of the correlation function of individual nuclei. Only a mono-isotopic ensemble of atoms with spin $I = 0$ would scatter neutrons in a totally coherent manner. Incoherent scattering is connected to isotopic disorder and to nuclear spin disorder.

We emphasise that it is the theory of neutron scattering that leads to the separation into coherent and incoherent terms. The direct experimental determination of two separate functions, $S_{coh}(\bm{Q},\omega)$ and $S_{inc}(\bm{Q},\omega)$, is usually not straightforward, unless samples with different isotopic composition are available. However, sometimes the two contributions can be separated without the luxury of major changes in the isotopic composition. The coherent and incoherent length b_i of nuclei are known and can be found in tables [40]. For example, the incoherent cross section of hydrogen is 40 times larger than the coherent cross section. Then, coherent scattering can be disregarded.

Incoherent Scattering and Diffusional Broadening of QENS Signals: Incoherent quasielastic neutron scattering is particularly useful for diffusion studies. The incoherent scattering function can be calculated for a given diffusion mechanism. Let us first consider the influence of diffusion on the scattered neutron wave in a simplified, semi-classical way: in an ensemble of incoherent scatterers, only the waves scattered by the same nucleus can interfere. At low temperatures the atoms stay on their sites during the scattering process; this contributes to the elastic peak. The width of the elastic peak is then determined by the energy resolution of the neutron spectrometer. At high temperatures the atoms are in motion. Then the wave packets emitted by diffusing atoms are 'cut' to several shorter 'packets', which leads to diffusional broadening of the elastic line. This is denoted as incoherent quasielastic scattering. Like in MBS the interference between wave packets emitted by the same nucleus depends on the relative orientation between the jump vector of the atom and the scattering direction. Therefore, in certain crystal direction the linewidth will be small while in other directions it will be large.

For a quantitative description of the incoherent scattering function the van Hove self-correlation function $G_s(\boldsymbol{r}, t)$ is used as a measure of diffusive motion. The incoherent scattering function is proportional to the Fourier transform of the self-correlation function

$$S_{inc}(\boldsymbol{Q}, \omega) = \frac{1}{2\pi} \int \int G_s(\boldsymbol{r}, t) \exp\left[i(\boldsymbol{Q}\boldsymbol{r} - \omega t)\right] d\boldsymbol{r} dt . \tag{15.39}$$

When atomic motion can simply be described by continuous translational diffusion in three dimensions, the self-correlation function $G_s(\boldsymbol{r}, t)$ takes the form of a Gaussian (see Chap. 3)

$$G_s(\boldsymbol{r}, t) = \frac{1}{(4\pi Dt)^{3/2}} \exp\left(-\frac{r^2}{4Dt}\right) , \tag{15.40}$$

with D denoting the self-diffusion coefficient of atoms. Its Fourier transform in space

$$S(\boldsymbol{Q}, t) = \exp\left(-Q^2 Dt\right) \tag{15.41}$$

is an exponential function of time, the time Fourier transform of which is a Lorentzian:

$$S_{inc}(\boldsymbol{Q}, \omega) = \frac{1}{\pi} \frac{DQ^2}{(DQ^2)^2 + \omega^2} . \tag{15.42}$$

This equation shows that for small Q values the linewidth of the quasielastic line is indeed given by Eq. (15.33). It is thus possible to determine the diffusion coefficient from a measurement of the linewidth as a function of small scattering vectors.

In the derivation of Eq. (15.42) the continuum theory of diffusion was used for the self-correlation function. This assumption is only valid for small scattering vectors $|Q| \ll 1/d$, where d is the length of the jump vectors in the

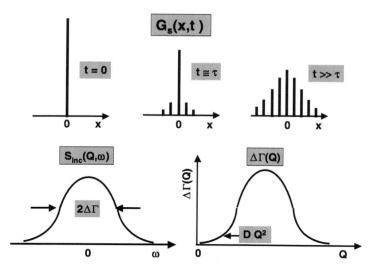

Fig. 15.16. *Top*: Self-correlation function G_s for a one-dimensional lattice. *Top*: The height of the *solid lines* represents the probability of occupancy per site. Asymptotically, the envelope approaches a Gaussian. *Bottom*: Incoherent contribution $S_{inc}(Q,\omega)$ to the dynamical structure factor and quasi-elastic linewidth $\Delta\Gamma$ *versus* scattering vector Q. According to [32]

lattice. For jump diffusion of atoms on a Bravais lattice the self-correlation function G_s can be obtained according to CHUDLEY AND ELLIOT [41]. The probability $P(\boldsymbol{r}_n, t)$ to find a diffusing atom on a site $\boldsymbol{r}_n$ at time t is calculated using the *master equation* for $P(\boldsymbol{r}_n, t)$:

$$\frac{\partial P(\boldsymbol{r}_n, t)}{\partial t} = -\frac{1}{\bar{\tau}} P(\boldsymbol{r}_n, t) + \frac{1}{Z\bar{\tau}} \sum_{i=1}^{Z} P(\boldsymbol{r}_n + \boldsymbol{l}_i, t). \quad (15.43)$$

$\boldsymbol{l}_i$ (i =1, 2, ... Z) is a set of jump vectors connecting a certain site with its Z neighbours. $\bar{\tau}$ denotes the mean residence time. The two terms in Eq. (15.43) correspond to loss and gain rates due to jumps to and from adjacent sites respectively. With the initial condition $P(\boldsymbol{r}_n, 0) = \delta(\boldsymbol{r}_n)$, the probability $P(\boldsymbol{r}, t)$ becomes equivalent to the self-correlation function $G_s(\boldsymbol{r}_n, t)$. A detailed theory of the master equation can be found in [42, 43]. If $P(\boldsymbol{r}_n, t)$ is known the incoherent scattering function is obtained by Fourier transformation in space and time according to Eq. (15.39). For a one-dimensional lattice G_s is illustrated in Fig. 15.16.

The classical model for random jump diffusion on Bravais lattices via nearest-neighbour jumps was derived by CHUDLEY AND ELLIOT in 1961 [35] (see also [36]). The incoherent scattering function for random jump motion on a Bravais lattice is given by

15.4 Quasielastic Neutron Scattering (QENS)

$$S_{inc}(\boldsymbol{Q},\omega) = \frac{2}{\pi} \frac{\Delta\Gamma(\boldsymbol{Q})}{\Delta\Gamma(\boldsymbol{Q})^2 + \omega^2} . \tag{15.44}$$

The function $\Delta\Gamma(\boldsymbol{Q})$ is determined by the lattice structure, the jumps which are possible, and the jump rate with which they occur. For the scattering of neutrons in a particular direction $\boldsymbol{Q}$, the variation with change in energy $\hbar\omega$ is Lorentzian in shape with a linewidth given by $\Delta\Gamma$.

In the case of polycrystalline samples, the scattering depends on the modulus $Q = |\boldsymbol{Q}|$ only, but still consists of a single Lorentian line with linewidth

$$\Delta\Gamma = \frac{2}{\bar{\tau}}\left[1 - \frac{\sin(Qd)}{Qd}\right] . \tag{15.45}$$

Here $\bar{\tau}$ is the mean residence time for an atom on a lattice site and d the length of the jump vector.

For a monocrystal with a simple cubic Bravais lattice one gets for the orientation dependent linewidth

$$\Delta\Gamma(\boldsymbol{Q}) = \frac{2}{3\bar{\tau}}\left[3 - \cos(Q_x d) - \cos(Q_y d) - \cos(Q_z d)\right] , \tag{15.46}$$

where Q_x, Q_y, Q_z are the components of $\boldsymbol{Q}$ and d is the length of the jump vector. The linewidth is a periodic function in reciprocal space. It has a maximum at the boundary of the Brillouin zone and it is zero if a reciprocal lattice point $\boldsymbol{G}$ is reached. This line narrowing is a remarkable consequence of quantum mechanics.

For vacancy-mediated diffusion successive jumps of atoms are correlated. Like in the case of MBS, the so-called encounter model can be used for low vacancy concentrations (see Chap. 7). A vacancy can initiate several correlated jumps of the same atom, such that one encounter comprises Z_{enc} atomic jumps. As we have seen in Chap. 7, the time intervals between subsequent atomic jumps within the same encounter are very short as compared to the time between encounters. As a consequence, the quasielastic spectrum can be calculated within the framework of the Chudley and Elliot model, where the rapid jumps within the encounters are treated as instantaneous. The linewidth of the quasielastic spectrum is described by [33]

$$\Delta\Gamma = \frac{2}{\bar{\tau}Z_{enc}}\left[1 - \sum_{\boldsymbol{r}_m} W_{enc}(\boldsymbol{r}_m)\cos(\boldsymbol{Q}\boldsymbol{r}_m)\right] , \tag{15.47}$$

where $W_{enc}(\boldsymbol{r}_m)$ denotes the probability that, during an encounter, an atom originally at $\boldsymbol{r}_m = 0$ has been displaced to lattice site $\boldsymbol{r}_m$ by one or several jumps. The probabilities can be obtained, e.g., by computer simulations. A detailed treatment based on the encounter model can be found in a paper by WOLF [46]. Equation (15.47) is equivalent to Eq. (15.29) already discussed in the section about Mössbauer spectroscopy.

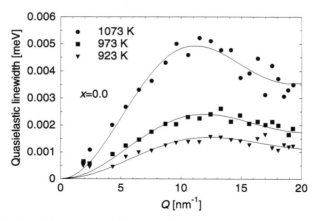

Fig. 15.17. Quasielastic linewidth as a function of the modulus $Q = |Q|$ for polycrystalline Na_2PO_4 according to WILMER AND COMBET [47]. *Solid lines*: fits of the Chudley-Elliot model

15.4.1 Examples of QENS studies

Let us now consider examples of QENS studies, which illustrate the potential of the technique for polycrystalline material and for monocrystals.

Na self-diffusion in ion-conducting rotor phases: Sodium diffusion in solid solutions of sodium orthophosphate and sodium sulfate, xNa_2SO_4 $(1-x)Na_3PO_4$, has been studied by WILMER AND COMBET [47]. These materials belong to a group of high-temperature modifications with both fast cation conductivity and anion rotational disorder and are thus termed as fast ion-conducting rotor phases. The quasielastic linewidth of polycrystalline samples has been measured as a function of the momentum transfer. In the case of polycrystalline samples, the scattering depends on the modulus of $Q = |Q|$ only, but still consist of a single Lorentzian line with linewidth Eq. (15.45). The Q-dependent linebroadening is shown for Na_2PO_4 at various temperatures in Fig. 15.17. The linewidth parameters $\bar{\tau}$ and d have been deduced. Obviously, the jump rates $\bar{\tau}^{-1}$ incresase with increasing temperature. Much more interesting is that the jump distance could be determined. It turned out that sodium diffusion is dominated by jumps between neighbouring tetrahedrally coordinated sites on an fcc lattice, the jump distance being half of the lattice constant.

At very low values of Q, quasielastic broadening does no longer depend on details of the jump geometry since the linewidth is dominated by the long-range diffusion via Eq. (15.33). The linebroadening at the two lowest accessible Q values (1.9 and 2.9 nm^{-1}) was used to determine the sodium self-diffusivites [47]. An Arrhenius plot of the sodium diffusivities is shown in

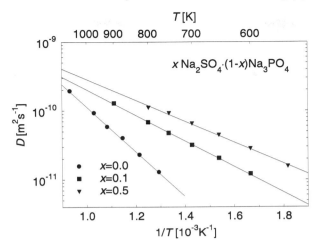

Fig. 15.18. Self-diffusion of Na in three xNa$_2$SO$_4$(1-x)Na$_3$PO$_4$ rotor phases according to WILMER AND COMBET [47]

Fig. 15.18. The activation enthalpies decrease from 0.64 eV for pure Na$_3$PO$_4$ to 0.3 eV for a sulphate content of 50%:

Na self-diffusion in Na single-crystals: Quasielastic scattering of sodium single crystals has been investigated by GÖLTZ ET AL. [33] and AIT-SALEM ET AL. [34] and analysed in terms of Eq. (15.47). It was demonstrated that self-diffusion of sodium occurs by nearest-neighbour jumps in the bcc lattice. Figure 15.19 shows the linebroadening as a function of the momentrum transfer Q in directions parallel to $\langle 111 \rangle$, $\langle 110 \rangle$, and $\langle 100 \rangle$ at 362.2 K. Model calculations are also shown, assuming a monovacancy mechnaism with nearest-neighbour jumps on with a $\langle 111 \rangle$ jumps. The results show that diffusion proceeds via nearest-neighbour jumps.

H diffusion in palladium: QENS measurements have been widely used to study diffusion of H-atoms in interstitial solutions of hydrogen in palladium. Interstitial diffusion is uncorrelated (see Chap. 7). It was shown, for example, that H-atoms jump between nearest-neighbour octahedral sites of the interstitial lattice of fcc Pd [44, 45].

15.4.2 Advantages and Limitations of MBS and QENS

For MBS diffusion studies it is necessary to heat the sample to sufficiently high temperatures that the mean residence time of an atom on a lattice site, $\bar{\tau}$, is comparable to or less than the half-life of the Mössbauer level τ_N. For metals, this implies temperatures not much below the melting temperature.

Mössbauer spectroscopy is sensitive to the elementary steps of diffusion on a microscopic scale. A direct determination of *jump vectors* and *jump rates*

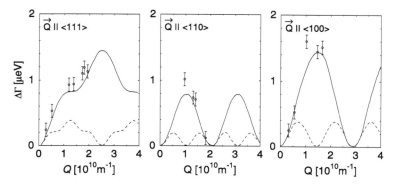

Fig. 15.19. Self-diffusion of Na metal. Dependence of the QENS line broadening in three major crystallographic directions. Theoretical curves have been calculated for a monovacancy mechanism assuming nearest-neighbour junps (*solid lines*) and $a\langle 111 \rangle$ jumps (*dotted line*). From VOGL AND PETRY [27] according to [33, 34]

is possible, when single-crystal samples are used and the line-broadening is measured as a function of crystal orientation. In addition, one can deduce the diffusion coefficient and compare it with data obtained, e.g., by tracer diffusion studies. However, this is not the main virtue of a microscopic method.

The nuclei studied in MBS must have a large value of the recoilless fraction, which limits the number of good isotopes to a few species. As already mentioned the major 'workhorse' of Mössbauer spectroscopy is ^{57}Fe. The isotopes ^{119}Sn, ^{151}Eu, and ^{161}Dy are less favourable but still useful isotopes for diffusion studies. Mössbauer diffusion studies in practice require a diffusional line-broadening that is comparable or larger than the natural linewidth of the Mössbauer transition. Only relatively large diffusion coefficients can be measured. For example, ^{57}Fe diffusion coefficients in the range 10^{-14} to 10^{-10} m^2 s^{-1} are accessible.

For diffusion studies by quasielastic neutron scattering (QENS) it is necessary to keep the sample at temperatures where the mean residence time of atoms on a lattice site, $\bar{\tau}$, is short enough to produce a diffusional broadening, which exceeds the energy resolution of the neutron spectrometer. For time-of-flight spectrometry the resolution is in the range of µeV to 0.1 meV. This allows a range of diffusion coefficients between about 10^{-12} and 10^{-8} m^2 s^{-1} to be covered. Diffusion coefficients can be determined directly from the Q^2 dependence of the linewidth.

QENS has mainly been used to study hydrogen and sodium diffusion in solids. A prerequisite of QENS is that the element of interest has a large enough incoherent scattering cross section as compared to the coherent scattering cross section. Only few elements such as hydrogen, sodium, and vanadium fulfill this condition. In these cases QENS is unique, since there are no Mössbauer isotopes for these elements. Otherwise, in luxury experiments different mixtures of isotopes can be used to separate coherent and incoherent contributions to scattering. The major merits of QENS and MBS are that both permit the investigation of the elementary steps of diffusion in solids on a scale of atomic dimensions and times. Both techniques are applicable to fast diffusion processes (see Fig. 13.1).

References

1. N. Bloembergen, E.M. Purcell, R.V. Pound, Phys. Rev. **73**, 679 (1948)
2. A. Abragam, *The Priciples of Nuclear Magnetism*, University Press, Oxford, 1973
3. C.P. Slichter, *Principles of Magnetic Resonance*, Springer-Verlag, Berlin, 1989
4. M. Mehring, *High Resolution NMR Spectroscopy in Solids*, Springer-Verlag, Berlin, 1983
5. H.T. Stokes, *Study of Diffusion in Solids by Pulsed Nuclear Magnetic Resonance*, in: *Nontraditional Methods in Diffusion*, G.E. Murch, H.K. Birnbaum, J.R. Cost (Eds.), The Metallurgical Society of AIME, Warrendale, 1984. p. 39
6. A.R. Allnatt, A.B. Lidiard, *Atomic Transport in Solids*, Cambridge University Press, 1993
7. G. Majer, *Die Methoden der Kernspinresonanz zum Studium der Diffusion von Wasserstoff in Metallen und intermetallischen Verbindungen*, Cuvillier Verlag, Göttingen, 2000
8. P. Heitjans, A. Schirmer, S. Indris, *NMR and β-NMR Studies of Diffusion in Interface-Dominated and Disordered Solids*, in: *Diffusion in Condensed Matter – Methods, Materials, Models*, P. Heitjans, J. Kärger (Eds.), Springer-Verlag, 2005
9. P. Heitjans, S. Indris, M. Wilkening, *Solid-State Diffusion and NMR*, in *Diffusion Fundamentals – Leipzig 2005*, J. Kärger, F. Grinberg, P. Heitjans (Eds.), Leipziger Universitätsverlag, 2005
10. B.C. Gerstein, C.R. Dybowski, *Transient Techniques in NMR of Solids*, Academic Press, New York, 1985
11. H.C. Torrey, Phys. Rev. **104**, 563 (1956)
12. E.L. Hahn, Phys. Rev. **80**, 580 (1950)
13. E.O. Stejskal, J.E. Tanner, J. Chem. Phys. **42** 288 (1965)
14. D.W. McCall, D.C. Douglas, E.W. Anderson, Ber. Bunsenges. **67**, 336 (1963)
15. P. Stilbs, in: *Progress in NMR Spectroscopy*, J.W. Emsley, J. Feeney, L.H. Sutcliffe (Eds.), Pergamon Press, Oxford, **19**, 1988
16. J. Kärger, G. Fleischer, U. Roland, in: *Diffusion in Condensed Matter*, J. Kärger, P. Heitjans, R. Haberlandt (Eds.), Vieweg-Verlag, 1998, p. 144
17. A. Feinauer, G. Majer, A. Seeger, Defect and Diffusion Forum **143–147**, 881 (1997)
18. A. Feinauer, G. Majer, Phys. Rev. **B 64**,134302 (2001)
19. H.C. Torrey, Phys. Rev. **96**, 690 (1954)
20. D. Wolf, *Spin Temperature and Nuclear Spin Relaxation in Matter*, Clarendon Press, Oxford, 1979
21. I.R. MacGillivray, C.A. Sholl, J. Phys. C **19**, 4771 (1986); C.A. Sholl, J. Phys. C **21**, 319 (1988)

22. G. Majer, presented at: International Max Planck Research School for Advanced Materials, April 2006
23. R.L. Mössbauer, Z. Physik **151**, 124 (1958)
24. K.S. Singwi, A. Sjölander, Phys. Rev. **120** 1093 (1960)
25. G. Vogl, R. Feldwisch, *The Elementary Diffusion Step in Metals Studied by Methods from Nuclear Solid-State Physics*, in: *Diffusion in Condensed Matter*, J. Kärger, P. Heitjans, R. Haberlandt (Eds.), Vieweg-Verlag, 1998, p. 40
26. G. Vogl, B. Sepiol, *The Elementary Diffusion Step in Metals Studied by the Interference of γ-Rays, X-Rays and Neutrons*, in: *Diffusion in Condensed Matter – Methods, Materials, Models*, P. Heitjans, J. Kärger (Eds.), Springer-Verlag, 2005
27. G. Vogl, W. Petry, *Diffusion in Metals studied by Mössbauer Spectroscopy and Quasielastic Neutron Scattering*, in: Festkörperprobleme XXV (Advances in Solid State Physics), P. Grosse (Ed.), Braunschweig, F. Vieweg und Sohn, 1985, P. 655
28. L. van Hove, Phys. Rev. **95**, 249 (1954)
29. J.G. Mullen, *Mössbauier Diffusion Studies*, in: *Nontraditional Methods in Diffusion*, G.E. Murch, H.K. Birnbaum, J.R. Cost (Eds.), The Metallurgical Society of AIME, Warrendale, 1984, p.59
30. H. Zabel, *Quasi-Elastic Neutron Scattering: a Powerful Tool for Investigating Diffusion in Solids*, in: *Nontraditional Methods in Diffusion*, G.E. Murch, H.K. Birnbaum, J.R. Cost (Eds.), The Metallurgical Society of AIME, Warrendale, 1984. p. 1
31. T. Springer, *Diffusion Studies of Solids by Quasielastic Neutron Scattering*, in: *Diffusion in Condensed Matter*, J. Kärger, P. Heitjans, R. Haberlandt (Eds.), Vieweg-Verlag, 1998., p. 59
32. T. Springer, R.E. Lechner, *Diffusion Studies of Solids by Quasielastic Neutron Scattering*, in: *Diffusion in Condensed Matter – Methods, Materials, Models*, P. Heitjans, J. Kärger (Eds.), Springer-Verlag, 2005
33. G. Göltz, A. Heidemann, H. Mehrer, A. Seeger, D. Wolf, Philos. Mag. **A 41**, 723 (1980)
34. M. Ait-Salem, T. Springer, A. Heidemann, B. Alefeld, Philos. Mag. **A 39**, 797 (1979)
35. C.T. Chudley, R.J. Elliiot. Proc. Phys. Soc. **77**, 353 (1961)
36. M. Bee, *Quasielastic Scattering*, Adam Hilger, Bristol, 1988
37. G.L. Squires, *Introduction to the Theory of Thermal Neutron Scattering*, Cambridge University Press, 1978
38. S.W. Lovesey, *Theory of Neutron Scattering from Condensed Matter*, Clarendon Press, Oxford, 1986
39. R. Hempelmann, *Quasielastic Neutron Scattering and Solid-State Diffusion*, Oxford Science Publication, 2000
40. L. Koester, H. Rauch, E. Seyman, Atomic Data and Nuclear Data Tables **49**, 65 (1991)
41. C.T. Chudley, R.J. Elliot, Proc. Phys. Soc. **77**, 353 (1961) (see [35])
42. K.W. Kehr, K. Mussawisade, Th. Wichmann, *Diffusion of Particles on Lattices*, in: *Diffusion in Condensed Matter*, J. Kärger, P. Heitjans, R. Haberlandt (Eds.), Vieweg-Verlag, 1998, p. 265
43. K.W. Kehr, K. Mussawisade, G. M. Schütz, G. Th. Wichmann, *Diffusion of Particles on Lattices*, in: *Diffusion in Condensed Matter – Methods, Materials, Models*, P. Heitjans, J. Kärger (Eds.), Springer-Verlag, 2005

44. K. Sköld, G. Nelin, J. Phys. Chem. Solids **28**, 2369 (1967)
45. J.M. Rowe, J.J. Rush, L.A. de Graaf, G.A. Ferguson, Phys. Rev. Lett **29**, 1250 (1972)
46. D. Wolf, Appl. Phys. Letters **30**, 617 (1977)
47. D. Wilmer, J. Combet, Chemical Physics **292**, 143 (2003)

16 Electrical Methods

Measurements of electrical properties of materials are of interest for diffusion experiments in several areas of materials science. *Impedance spectroscopy* plays an important rôle for ion-conducting materials. With the availability of commercially made impedance bridges covering wide frequency ranges impedance studies became popular among electrochemists and materials scientists. For an introduction to the field of impedance spectroscopy the reader may consult the textbook of MACDONALD [1]. In semiconducting materials electrically active foreign atoms have a strong influence on the electrical conductivity. For diffusion studies in semiconductors electrical measurements are useful. In particular *spreading resistance profiling*, introduced by MAZUR AND DICKEY in the 1960s [2], has become a powerful tool for measuring spatial distributions of electrically active atoms in semiconductors.

Electrical resistivity measurements on metals have sometimes also been used to study diffusion. These studies utilise the resistivity change that occurs upon in- or out-diffusion of foreign elements. This method is fairly indirect and diffusion profiles cannot be obtained. Therfore, we refrain from discussing such experiments.

16.1 Impedance Spectroscopy

In ion-conducting materials with negligible electronic conduction such as ionic crystals, ion-conducting glasses, and oxides the conductivity results from the hopping motion of ions. For such materials the measurement of the electrical conductivity is an indispensable complement to that of tracer diffusion. As discussed in Chap. 11, the dc conductivity, σ_{dc}, is related to the *charge or diffusivity*, D_σ, via

$$D_\sigma = \frac{\sigma_{dc} k_\mathrm{B} T}{N_{ion} q^2} \,. \tag{16.1}$$

N_{ion} denotes the number density of mobile ions and q the electrical charge per ion.

Measurements of the ionic conductivity are carried out with an ac bridge to avoid polarisation effects at the electrodes. An experimental set-up is illustrated in Fig. 16.1. The measurements are usually made with cells which have

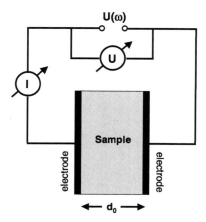

Fig. 16.1. Schematic illustration of an impedance bridge with sample and electrodes

two identical electrodes applied to the faces of a sample in the form of a circular cylinder or rectangular parallelepiped. The general approach is to apply an electrical ac stimulus of frequency ν (a known voltage or current) and to observe the resulting response (current or voltage). Usually, impedance spectroscopy can cover a frequency range from about 10^{-3} Hz to several MHz. The experimental set-up requires a variable frequency generator and a vector ammeter and volt-meter for current and voltage measurements including the phase-shift between current $\hat{I}(\nu)$ and voltage $\hat{V}(\nu)$. The complex impedance is defined as

$$\hat{Z}(\nu) \equiv \frac{\hat{V}(\nu)}{\hat{I}(\nu)}. \tag{16.2}$$

It is composed according to

$$\hat{Z}(\nu) = Z' - iZ'' \tag{16.3}$$

of real and imaginary parts, Z' and Z''. i denotes the imaginary unit. The complex conductivity

$$\hat{\sigma}(\nu) = \sigma'(\nu) + i\sigma''(\nu) \tag{16.4}$$

is also composed of a real part, σ', and an imaginary part, σ''. Conductivity and complex impedance are related via

$$\hat{\sigma}(\nu) = \frac{1}{\hat{Z}(\nu)} \frac{d_0}{A}, \tag{16.5}$$

where d_0 and A denote the thickness and the electrode area of the sample, respectively.

16.1 Impedance Spectroscopy

For a discussion of impedance measurements, it is convenient to recall that the complex impedance of a circuit formed by an ohmic resistance R and a capacitance C parallel to it is given by

$$\hat{Z} = \frac{R}{1 + i\omega CR}, \tag{16.6}$$

where $\omega = 2\pi\nu$ denotes the angular frequency. Then, the real and imaginary parts of the impedance can be written as

$$Z' = \frac{R}{1 + \omega^2 R^2 C^2} \quad \text{and} \quad Z'' = \frac{\omega C R^2}{1 + \omega^2 R^2 C^2}. \tag{16.7}$$

The representation of the impedance in the complex $Z' - Z''$ plane is denoted as the *Cole diagram*. For a RC-circuit the values of $\hat{Z}(\omega)$ plotted in the Z' versus $-Z''$ plane fall on a semicircle of diameter R, passing through the origin for $\omega \to \infty$, through $(R,0)$ for $\omega = 0$, and through $(R/2, R/2)$ for $\omega = (RC)^{-1}$.

When several circuits of this type are connected in series, the graphic representation is a series of semicircles as illustrated in Fig. 16.2. An ensemble of three RC circuits in series can represent a measurement cell. The cell may consist of a polycrstalline sample plus electrodes. Each circuit represents a conductivity process: R_V and C_V volume conduction, R_b and C_b boundary conduction, and R_e and C_e the electrode process. One should, however, keep in mind that the representation of the total impedance of an experimental set-up by RC-circuits can be oversimplified. Sometimes other equivalent circuits may better represent the actual processes. Furthermore, sometimes the centers of the arcs are below the Z'-axis.

As a simple experimental example, Fig. 16.3 shows the Cole diagram for an ion-conducting alkali-borate glass according to IMRE ET AL. [3]. The semicircles represent the volume conduction process at various temperatures. Grain boundaries are absent in a glass and the electrode process is located at lower frequencies not displayed in the figure. The dc resistance of the sample is given by the intercept of the arcs with the Z'-axis. It decreases with increasing temperature. The dc conductivity can be deduced from the ohmic resistance observed at the intersection of the 'semicircle' with the Z'-axis of Fig. 16.3. The real part of the conductivity times temperature ($\sigma' \times T$) of the same material is shown in Fig. 16.4 as a function of the frequency ν for various temperatures. The low frequency plateau in Fig. 16.4 corresponds to the dc conductivity. This plateau reflects the long-range transport of ions. The dc-conductivity increases with temperature Arrhenius activated.

Using relation Eq. (16.1), the *charge diffusivity*, D_σ, can be deduced (see Chap. 11). This quantity is displayed in the Arrhenius diagram of Fig 16.5 together with the tracer diffusivity of ^{22}Na measured on the same material according to [4]. As discussed in Chap. 11 the ratio between tracer diffusivity D^* and the conductivity diffusivity is denoted as the *Haven ratio*, $H_R \equiv$

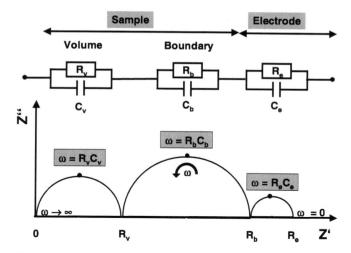

Fig. 16.2. Circuits for the complex impedance and Cole diagram

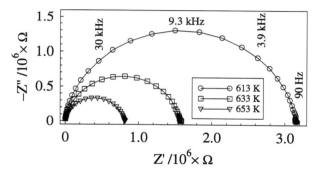

Fig. 16.3. Complex impedance (Cole diagram) for three temperatures representing volume conduction of a sodium-borate glass [3]

D^*/D_σ, which is usually less than unity. A measurement of the Haven ratio can provide useful information about the atomic mechanism of diffusion and the correlation effects involved. In the illustrated case, the Haven ratio is temperature-independent indicating that the mechanism of ionic motion does not change with temperature.

The increase of the conductivity at higher frequencies is called dispersion. The conductivity dispersion reflects the fact that ionic jumps are correlated (see, e.g., [5]). An onset frequency of dispersion, ν_{onset}, may be defined by the condition $\sigma'(\nu) = 2\sigma_{dc}$. The fact that the onset frequencies in Fig. 16.4 lie on a straight line with a slope of unity shows that $\sigma_{dc} \times T$ and ν_{onset} are thermally activated with the same activation enthalpy. This behaviour is sometimes denoted as Summerfield scaling [6]. Microscopically, it implies

16.1 Impedance Spectroscopy   289

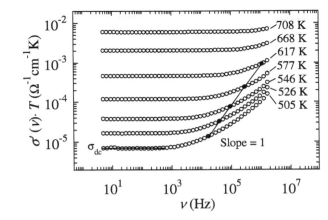

Fig. 16.4. Conductivity spectra of a sodium borate glass in a diagram of logarithm $\sigma' \times T$ *versus* logarithm of the frequency ν. The onset frequencies of dispersion for various temperatures are connected by a *straight line*

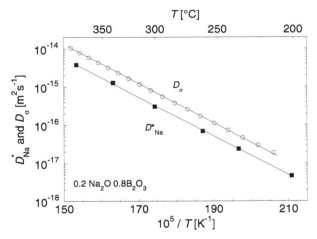

Fig. 16.5. Arrhenius diagram of the charge diffusivity D_σ and the tracer diffusivity D^* of ^{22}Na for a sodium borate glass

that the same jump processes occur at different temperatures. Of course, their jump rate is higher at higher temperatures. Summerfield scaling is observed for several materials but by no means in all ion conducting materials. For more information about conduction in disordered ionic materials the reader may consult Chap. 30 and reviews, e.g., by INGRAM [7], DIETERICH AND MAAS [8], FUNKE AND COWORKERS [9], and by BUNDE ET AL. [10].

16.2 Spreading Resistance Profiling

In semiconductors, the resistivity in the extrinsic domain is related to the concentration of electrically active foreign atoms. The concentration distribution of such atoms can be deduced by measuring the spreading resistance R_s (see below). Spreading resistance profiling (SRP) [2] has become a useful tool for measuring spatial distributions of electrically active atoms in semiconducting samples. SRP is widely used in silicon technology to monitor depth profiles of dopants after processing steps and to check the lateral uniformity in the resistivity of virgin Si wafers. In addition, SRP has been successfully applied in basic studies of diffusion processes in Si (see, e.g., [11, 12]), in Ge [13] and to a lesser extent in GaAs [14].

SRP is a two-point-probe technique, which measures the electrical resistance on semiconductor surfaces with a much higher spatial resolution than the traditional four-point-probe technique. The concentration-depth profile for foreign atoms can be established by measuring the resistance of the sample between two 'points' as indicated in Fig. 16.6. The probe tips, usually made of a tungsten-osmium alloy, are separated by a distance of typically 100 μm or less. The current between the probes spreads over a small space region near the semiconductor surface, which explains the notion 'spreading resistance'. A SRP device is commonly operated in an automatic-stepping mode with probe-tip steps varying from 5 to 25 μm. The measurements can be either performed on a cross section of the sample or for shallow profiles on a bevelled section as indicated in Fig. 16.6. An example for an experimental spreading resistance profile is displayed in the left part of Fig. 16.7. The right part shows the concentration-depth profile deduced from the spreading resistance profile. The employed procedure is described below.

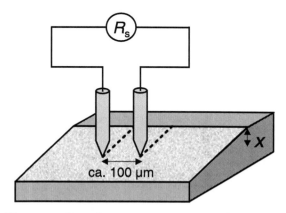

Fig. 16.6. Spreading resistance profiling (schematic)

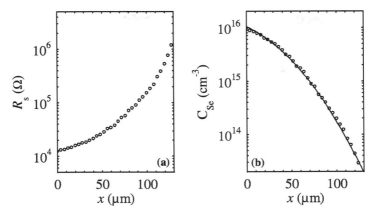

Fig. 16.7. Spreading resistance profile of Se in Si (*left*) and concentration depth profile of Se deduced therefrom (*right*)

The SRP technique relies on the fact that the ideal resistance of a small-diameter metal probe in perfect contact with the plane surface of a semi-infinite semiconductor is given by

$$R_s = \frac{\rho}{4a}, \tag{16.8}$$

where ρ denotes the resistivity of the semiconductor and a the radius of the contact area. The spreading resistance, R_s, originates from the radial flow of the current from the probe tip into the semiconductor. Due to the special configuration, about 80% of the potential drop occurs within a distance of $5a$. In practice, the SRP contact radius can be made as small as a few µm, which leads to a corresponding small sampling volume. Theoretically, the two-probe arrangement doubles R_s with regard to Eq. (16.8), provided that the probe-tip separation is much larger than the contact radius. In reality, the contacts are not planar circles but irregularly shaped microcontacts. This gives rise to substantial contact resistance, which cannot be derived with sufficient accuracy from theory. Therefore, in SRP analysis the relationship between ρ and R_s is established by calibration using homogeneously doped samples of the same material, conductivity type, and surface orientation as the test sample under consideration.

Standard semiconductor theory is used to establish the relation between the resistivity and the concentration of the electrically active foreign atoms. We illustrate this procedure for a singly ionizable donor atom X in an otherwise undoped semiconductor. For a given Fermi energy E_F, the electron concentration n is obtained as

$$n = N_C \exp\left(\frac{E_F - E_C}{k_B T}\right), \tag{16.9}$$

where E_C denotes the energy of the conduction band edge and N_C the effective density of states in the conduction band. The latter quantity is closely related to the effective electron mass m_n according to

$$N_C = 2 \left(\frac{2\pi m_n k_B T}{h^2} \right)^{3/2}, \qquad (16.10)$$

where h is the Planck constant. The hole concentration p follows from

$$np = n_i^2 = N_C N_V \exp\left[-\frac{E_G(T)}{k_B T} \right] \qquad (16.11)$$

with the intrinsic carrier density n_i, the temperature-dependent band gap energy, $E_G(T)$, and the effective density of states, N_V, in the valence band. Having obtained electron and hole concentrations, the charge neutrality condition

$$n = C_X^{ion} + p \qquad (16.12)$$

yields the concentration C_X^{ion} of ionized foreign atoms. Values of C_X^{ion} distinctly above 10^{-16}cm^{-3} lead to enhanced scattering of charge carriers, which may be taken into account by expressions for the carrier mobilities of electrons $\mu_n = \mu_n(C_X^{ion}, T)$ and holes $\mu_n = \mu_n(C_X^{ion}, T)$. Empirical expressions of this kind can be found in the literature [15]. In a subsequent step the resistivity is obtained from

$$\frac{1}{\rho} = q\mu_n n + q\mu_p p, \qquad (16.13)$$

where q denote the charge per carrier. For singly ionizable donor atoms the second term on the right hand side of Eq. (16.13) can be neglected for $C_X^{ion} \gg n_i$. Once the calculated resistivity $\rho(E_F, T)$ value has converged to the experimental one, E_F and $C_X^{ion}(E_F, T)$ are known. Then, the electrically neutral foreign atom concentration, C_X^0, results from

$$\frac{C_X^{ion}}{C_X^0} = g_X \exp\left(\frac{E_X - E_F}{k_B T} \right). \qquad (16.14)$$

This equation accounts for the electron occupation probability of a foreign atom with a donor level, E_X, and a degeneration factor, g_X. The latter equals 2 for common group-V dopants such as P in silicon. Finally, the total foreign atom concentration follows from

$$C_X = C_X^{ion} + C_X^0. \qquad (16.15)$$

Background doping can be taken into account by including additional terms in the charge neutrality Eq. (16.12). The alternative case in which the foreign atom X has acceptor character can be treated in a similar set of equations reflecting the reversed rôle of holes and electrons as majority and minority charge carriers. For further details about SRP the reader may consult a paper by VOSS ET AL. [16].

References

1. J.R. Macdonald (Ed.), *Impedance Spectroscopy – Emphasizing Solid Materials and Systems*, John Wiley and Sons, 1987
2. R.G. Mazur, D.H. Dickey, J. Electrochem. Soc. **113**, 255 (1966)
3. A.W. Imre, S. Voss, H. Mehrer, Phys. Chem. Chem. Phys. **4**, 3219 (2002)
4. H. Mehrer, A.W. Imre, S. Voss, in: *Mass and Charge Transport in Inorganic Materials II*, Proc. of CIMTEC 2002–10th Int. Ceramics Congress and 3rd Forum on New Materials, Florence, Italy, 2002;, P. Vincenzini, V. Buscaglia (Eds.), Techna, Faenza, 2003, p.127
5. K. Funke, R.D. Banhatti, S. Brückner, C. Cramer, C. Krieger, A. Mandanici, C. Martiny, I. Ross, Phys. Chem. Chem. Phys. **4**, 3155 (2002)
6. S. Summerfield, Philos. Mag. B **52**, 9 (1985)
7. M.D. Ingram, Phys. Chem of Glasses **28**, 215 (1987)
8. W. Dieterich, P. Maass, Chem. Physics **284** 439 (2002)
9. K. Funke, C. Cramer, D. Wilmer, *Concept of Mismatch and Relaxation for Self-Diffusion and Conduction in Ionic Materials with Disordered Structures*, in: *Diffusion in Condensed Matter – Methods, Materials, Models*, P. Heitjans, J. Kärger (Eds.), Springer-Verlag, 2005
10. A. Bunde, W. Dieterich, Ph. Maass, M. Mayer, *Ionic Transport in Disordered Materials*, in: *Diffusion in Condensed Matter – Methods, Materials, Models*, P. Heitjans, J. Kärger (Eds.), Springer-Verlag, 2005
11. D.A. Antoniadis, A.G. Gonzales, R.W. Dutton, J. Electrochem. Soc. **125**, 813 (1978)
12. N.A. Stolwijk, J. Hölzl, W. Frank, Appl. Phys. A: Solid surf. **39**, 37 (1966)
13. H. Bracht, N.A. Stolwijk, H. Mehrer, Phys. Rev. B **43**, 14465 (1991)
14. G. Bösker, N.A. Stolwijk, U. Södervall, W. Jäger, H. Mehrer, J. Appl. Phys. **86**, 791 (1999)
15. C. Jacobini, C. Canali, G. Ottaviani, A. Alberigi Quaranta, Solid State Electron. **20**, 77 (1977)
16. S. Voss, N.A. Stolwijk, H. Bracht, J. Applied Physics **92**, 4809 (2002)

Part III

Diffusion in Metallic Materials

17 Self-diffusion in Metals

17.1 General Remarks

Self-diffusion is the most fundamental diffusion process in a solid. This is the major reason in addition to application-oriented motives why self-diffusion studies have consumed energies of many researchers. Self-diffusion in a metallic element A is the diffusion of A atoms. In practice, in most cases tagged atoms A^* – either radioactive or stable isotopes – are used as *tracers* (see Chap. 13), which are chemically identical to the atoms of the base metal.

As already mentioned in Chap. 8, the temperature dependence of the tracer self-diffusion coefficient, D^*, is often, but by no means always, described by an Arrhenius relation[1]

$$D^* = D^0 \exp\left(-\frac{\Delta H}{k_B T}\right), \qquad (17.1)$$

with a *pre-exponential factor* D^0 and an *activation enthalpy* ΔH. The pre-exponential factor can usually be written as

$$D^0 = g f \nu^0 a^2 \exp\frac{\Delta S}{k_B}, \qquad (17.2)$$

where ΔS is called the *diffusion entropy*, g is a geometrical factor of the order of unity (e.g., $g = 1$ for the vacancy mechanism in cubic metals), f the tracer correlation factor, ν^0 an attempt frequency of the order of the Debye frequency, and a the lattice parameter. For a diffusion process with a temperature-independent activation enthalpy, the Arrhenius diagram is a straight line with slope $-\Delta H/k_B$. From its intercept – for $T^{-1} \Longrightarrow 0$ – the pre-exponential factor D^0 is obtained. The physical meaning of the activation parameters of diffusion depends on the diffusion mechanism and on the lattice geometry (see also Chap. 8).

Self-diffusion in metals is mediated by vacancy-type defects [1–6]. Strong evidence for this interpretation comes from the following observations:

1. The Kirkendall effect has shown that the diffusivities of different kinds of atoms in a substitutional metallic alloy diffuse at different rates (see also

---

[1] We use in this chapter again the upper index * to indicate tracer diffusivities.

Chaps. 1 and 10). Neither the direct exchange nor the ring mechanism can explain this observation. It became evident that vacancies are responsible for self-diffusion and diffusion of substitutional solutes in metals in practically all cases.

2. Vacant lattice sites are the dominating defect in metals at thermal equilibrium. Studies which permit the determination of vacancy properties were discussed in Chap. 5. These studies are based mainly on differential dilatometry, positron-annihilation spectroscopy, and quenching experiments.
3. Isotope-effect experiments of self-diffusion (see Chap. 9) are in accordance with correlation factors which are typical for vacancy-type mechanisms [5, 6].
4. Values and signs of activation volumes of self-diffusion deduced from high-pressure experiments (see Chap. 8) are in favour of vacancy-type mechanisms [7].
5. Formation and migration enthalpies of vacancy-type defects add up to the activation enthalpies observed for self-diffusion (see, e.g., [5, 6, 8–10]).

Self-diffusion of many metallic elements has been studied over wide temperature ranges by the techniques described in Chap. 13. As an example, Fig. 17.1 displays the tracer diffusion coefficient of the radioisotope ^{63}Ni in Ni single-crystals. A diffusivity range of about 9 orders of magnitude is covered by

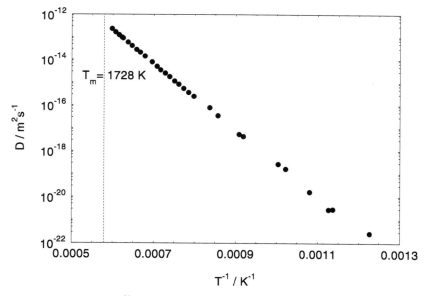

Fig. 17.1. Diffusion of ^{63}Ni in monocrystalline Ni. $T > 1200\,K$: data from grinder sectioning [11]; $T < 1200\,K$: data from sputter sectioning [12]

the combination of mechanical sectioning [11] and sputter-sectioning techniques [12]. The investigated temperature interval ranges from about $0.47\,T_m$ to temperatures close the melting temperature T_m. For some metals, data have been deduced by additional techniques. For example, nuclear magnetic relaxation proved to be very useful for aluminium and lithium, where no suitable radioisotopes for diffusion studies are available. A collection of self-diffusion data for pure metals and information about the method(s) employed can be found in [8].

A convex curvature of the Arrhenius plot – i.e. deviations from Eq. (17.1) – may arise for several reasons such as contributions of more than one diffusion mechanism (e.g., mono- and divacancies), impurity effects, grain-boundary or dislocation-pipe diffusion (see Chaps. 31–33). Impurity effects on solvent diffusion are discussed in Chap. 19. Grain-boundary influences are completely avoided, if mono-crystalline samples are used. Dislocation influences can be eliminated in careful experiments on well-annealed crystals.

17.2 Cubic Metals

Self-diffusion in metallic elements is perhaps the best studied area of solid-state diffusion. Some useful empirical correlations between diffusion and bulk properties for various classes of materials are already discussed in Chap. 8. Here we consider self-diffusivities of cubic metals and their activation parameters in greater detail.

17.2.1 FCC Metals – Empirical Facts

Self-diffusion coefficients of some fcc metals are shown in Fig. 17.2 as Arrhenius lines in a plot which is normalised to the respective melting temperatures (homologous temperature scale). The activation parameters listed in Table 17.1 were obtained from a fit of Eq. (17.1) to experimental data. The following empirical correlations are evident:

- Diffusivities near the melting temperature are similar for most fcc metals and lie between about 10^{-12} m^{2}s^{-1} and 10^{-13} m^{2}s^{-1}. An exception is self-diffusion in the group-IV metal lead, where the diffusivity is about one order of magnitude lower and the activation enthalpy higher[2].
- The diffusivities of most fcc metals, when plotted in a homologous temperature scale, lie within a relatively narrow band (again Pb provides an exception). This implies that the Arrhenius lines in the normalised plot have approximately the same slope. Since this slope equals $-\Delta H/(k_\mathrm{B} T_m)$

---

[2] The group-IV semiconductors Si and Ge have even lower diffusivities than Pb at the melting point (see Chap. 23).

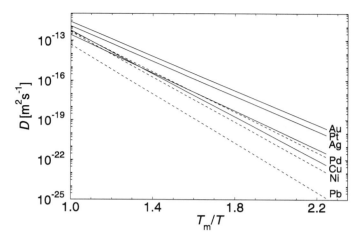

Fig. 17.2. Self-diffusion of fcc metals: noble metals Cu, Ag, Au; nickel group metals Ni, Pd, Pt; group IV metal Pb. The temperature scale is normalised to the respective melting temperature T_m

Table 17.1. Activation parameters D^0 and ΔH for self-diffusion of some fcc metals

	Cu [14]	Ag [16]	Au [18]	Ni [12]	Pd [19]	Pt [20]	Pb [23]
ΔH [kJ mol^{-1}]	211	170	165	281	266	257	107
D^0 [10^{-4} m^2 s^{-1}]	0.78	0.041	0.027	1.33	0.205	0.05	0.887
$\Delta H/k_B T_m$	18.7	16.5	14.8	19.5	17.5	15.2	21.4

a correlation between the activation enthalpy ΔH and the melting temperature T_m exists (see also Table 17.1). This correlation can be stated as follows:

$$\Delta H \approx (15 \; to \; 19) \; k_B T_m \; (T_m \; in \; K) \, . \tag{17.3}$$

Relations like Eq. (17.3) are sometimes referred to as the rule of VAN LIEMPT [13] (see also Chap. 8).
- The pre-exponential factors lie within the following interval:

$$\text{several } 10^{-6} \, \text{m}^2\text{s}^{-1} < D^0 < \text{several } 10^{-4} \, \text{m}^2\text{s}^{-1} \, . \tag{17.4}$$

The factor $gf\nu^0 a^2$ in Eq. (17.2) is typically about 10^{-6} m^2 s^{-1}. Hence the range of D^0 values corresponds to diffusion entropies ΔS between about 1 k_B and 5 k_B.
- Within one column of the periodic table, the diffusivity in homologous temperature scale is lowest for the lightest element and highest for the heaviest element. For example, in the group of noble metals Au self-

diffusion is fastest and Cu self-diffusion is slowest. In the Ni group, Pt self-diffusionm is fastest and Ni self-diffusion is slowest.

17.2.2 BCC Metals – Empirical Facts

Self-diffusion of bcc metals is shown in Fig. 17.3 on a homologous temperature scale. A comparison between fcc and bcc metals (Figs. 17.2 and 17.3) reveals the following features:

- Diffusivities for bcc metals near the melting temperature lie between about 10^{-11} m^{2}s^{-1} and 10^{-12} m^{2}s^{-1}. Diffusivities of fcc metals near their melting temperatures are about one order of magnitude lower.
- The 'spectrum' of self-diffusivities as a function of temperature is much wider for bcc than for fcc metals. For example, at 0.5 T_m the difference between the self-diffusion of Na and of Cr is about 6 orders of magnitude, whereas the difference between self-diffusion of Au and of Ni is only about 3 orders of magnitude.
- Self-diffusion is slowest for group-VI transition metals and fastest for alkali metals.
- The Arrhenius diagram of some bcc metals shows clear convex (upward) curvature (see, e.g., Na in Fig. 17.3).

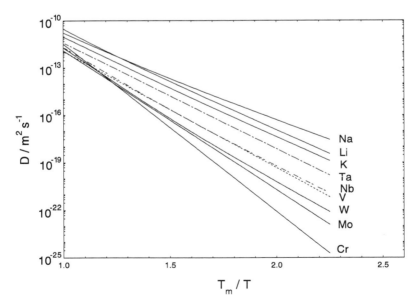

Fig. 17.3. Self-diffusion of bcc metals: alkali metals Li, Na, K (*solid lines*); group-V metals V, Nb, Ta (*dashed lines*); group-VI metals Cr, Mo, W (*solid lines*). The temperature scale is normalised to the respective melting temperature T_m

– A common feature of fcc and bcc metals is that within one group of the periodic table self-diffusion at homologous temperatures is usually slowest for the lightest and fastest for the heaviest element of the group. Potassium appears to be an exception.

Group-IV transition metals (discussed below) are not shown in Fig. 17.3, because they undergo a structural phase transition from a hcp low-temperature to a bcc high-temperature phase. The self-diffusivities in the bcc phases β-Ti, β-Zr and β-Hf are on a homologous scale even higher than those of the alkali metals. On a homologous scale self-diffusion of β-Ti – the lightest group-IV transition element – is slowest; self-diffusion of the heaviest group-IV transition element β-Hf is fastest. In addition, β-Ti and β-Zr show upward curvature in the Arrhenius diagram. β-Hf exists in a narrow temperature interval, which is too small to detect curvature.

17.2.3 Monovacancy Interpretation

Self-diffusion in most metallic elements is dominated by the monovacancy mechanism (see Fig. 6.5) at least at temperatures below $\frac{2}{3}T_m$. At higher temperatures, a certain divacancy contribution, which varies from metal to metal, may play an additional rôle (see below). Let us first consider the monovacancy contribution:

Using Eqs. (4.29) and (6.2) the diffusion coefficient of tracer atoms due to monovacancies in cubic metals can be written as

$$D^*_{1V} = g_{1V} f_{1V} a^2 C^{eq}_{1V} \omega_{1V} . \tag{17.5}$$

g_{1V} is a geometric factor ($g_{1V} = 1$ for cubic Bravais lattices), a the lattice parameter, and f_{1V} the tracer correlation factor for monovacancies. The atomic fraction of vacant lattice sites at thermal equilibrium C^{eq}_{1V} (see Chap. 5) is given by

$$C^{eq}_{1V} = \exp\left(-\frac{G^F_{1V}}{k_B T}\right) = \exp\left(\frac{S^F_{1V}}{k_B}\right) \exp\left(-\frac{H^F_{1V}}{k_B T}\right), \tag{17.6}$$

where the Gibbs free energy of vacancy formation is related via $G^F_{1V} = H^F_{1V} - TS^F_{1V}$ to the pertinent formation enthalpy and entropy. The exchange rate between vacancy and tracer atom is

$$\omega_{1V} = \nu^0_{1V} \exp\left(-\frac{G^M_{1V}}{k_B T}\right) = \nu^0_{1V} \exp\left(\frac{S^M_{1V}}{k_B}\right) \exp\left(-\frac{H^M_{1V}}{k_B T}\right), \tag{17.7}$$

where G^M_{1V}, H^M_{1V}, and S^M_{1V} denote the Gibbs free energy, the enthalpy and the entropy of vacancy migration, respectively. ν^0_{1V} is the attempt frequency of the vacancy jump.

The *standard interpretation* of tracer self-diffusion attributes the total diffusivity to monovacancies:

$$D^* \approx D_{1V}^* = D_{1V}^0 \exp\left(-\frac{H_{1V}^F + H_{1V}^M}{k_B T}\right). \qquad (17.8)$$

In the standard interpretation, the Arrhenius parameters of Eq. (17.1) have the following meaning:

$$\Delta H \approx \Delta H_{1V} = H_{1V}^F + H_{1V}^M \qquad (17.9)$$

and

$$D^0 \approx D_{1V}^0 = f_{1V} g_{1V} a^2 \nu_{1V}^0 \exp\left(\frac{S_{1V}^F + S_{1V}^M}{k_B}\right). \qquad (17.10)$$

Then, according to Eq. (17.9) the activation enthalpy ΔH_{1V} equals the sum of formation and migration enthalpies of the vacancy. The diffusion entropy

$$\Delta S \approx \Delta S_{1V} = S_{1V}^F + S_{1V}^M \qquad (17.11)$$

equals the sum of formation and migration entropies of the vacancy. Typical values for ΔS are of the order of a few k_B. As discussed in Chap. 7, the correlation factor f_{1V} accounts for the fact that for a vacancy mechanism the tracer atom experiences some 'backward correlation', whereas the vacancy performs a random walk. The tracer correlation factors are temperature-independent quantities (fcc: $f_{1V} = 0.781$; bcc: $f_{1V} = 0.727$; see Table 7.2).

17.2.4 Mono- and Divacancy Interpretation

In thermal equilibrium, the concentration of divacancies increases more rapidly than that of monovacancies (see Fig. 5.2 in Chap. 5). Even more important, individual divacancies, once formed, will avoid dissociating and thereby exhibit extended lifetimes in the crystal. In addition, divacancies in fcc metals are more effective diffusion vehicles than monovacancies since their mobility is considerably higher than that of monovacancies [2, 9]. At temperatures above about 2/3 of the melting temperature, a contribution of divacancies to self-diffusion can no longer be neglected (see, e.g., the review by SEEGER AND MEHRER [2] and the textbooks of PHILIBERT [3] and HEUMANN [4]). The total diffusivity of tracer atoms then is the sum of mono- and divacancy contributions

$$D^* = \underbrace{D_{1V}^0 \exp\left(-\frac{\Delta H_{1V}}{k_B T}\right)}_{D_{1V}^*} + \underbrace{D_{2V}^0 \exp\left(-\frac{\Delta H_{2V}}{k_B T}\right)}_{D_{2V}^*}. \qquad (17.12)$$

The activation enthalpy of the divacancy contribution can be written as

$$\Delta H_{2V} = 2H_{1V}^F - H_{2V}^B + H_{2V}^M. \qquad (17.13)$$

Here H_{2V}^M and H_{2V}^B denote the migration and binding enthalpies of the divacancy. For fcc metals the pre-exponential factor of divacancy diffusion is

$$D_{2V}^0 = g_{2V} f_{2V} a^2 \nu_{2V}^0 \exp \frac{2S_{1V}^F - S_{2V}^B + S_{2V}^M}{k_B} . \tag{17.14}$$

f_{2V} is the divacancy tracer correlation factor, g_{2V} a geometry factor, ν_{2V}^0 the attempt frequency, S_{2V}^M and S_{2V}^B denote migration and binding entropies of the divacancy.

Measurements of the temperature, mass, and pressure dependence of the tracer self-diffusion coefficient have proved to be useful to elucidate mono- and divacancy contributions in a quantitative manner. As an example of divacancy-assisted diffusion in an fcc metal, Fig. 17.4 shows the Arrhenius daigram of silver self-diffusion according to [15–17]. A fit of Eq. (17.12) to the data, performed by BACKUS ET AL. [17], resulted in the mono- and divacancy contributions displayed in Fig. 17.4. Near the melting temperature both contributions in Ag are about equal. With decreasing temperature the divacancy contribution decreases more rapidly than the monovacancy contribution. Near $2/3\ T_m$ the divacancy contribution is not more than 10 % of the total diffusivity and below about $0.5\ T_m$ it is negligible.

As a consequence, the Arrhenius diagram shows a slight upward curvature and a well-defined single value of the activation enthalpy no longer exists.

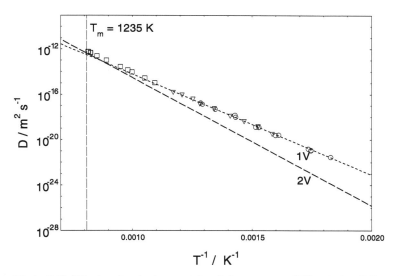

Fig. 17.4. Self-diffusion in single-crystals of Ag: *squares* [15], *circles* [16], *triangles* [17]. Mono- and divacancy contributions to the total diffusivity are shown as *dotted* and *dashed lines* with the following Arrhenius parameters: $D_{1V}^0 = 0.046 \times 10^{-4}\,\mathrm{m^2\,s^{-1}}$, $\Delta H_{1V} = 1.76\,\mathrm{eV}$ and $D_{2V}^0 = 2.24 \times 10^{-4}\,\mathrm{m^2\,s^{-1}}$, $\Delta H_{2V} = 2.24\,\mathrm{eV}$ according to an analysis of BACKUS ET AL. [17]

Instead an effective activation enthalpy, ΔH_{eff}, can be defined (see Chap. 8), which reads for the simultaneous action of mono- and divacancies

$$\Delta H_{eff} = \Delta H_{1V} \frac{D_{1V}^*}{D_{1V}^* + D_{2V}^*} + \Delta H_{2V} \frac{D_{2V}^*}{D_{1V}^* + D_{2V}^*}. \qquad (17.15)$$

Equation (17.15) is a weighted average of the individual activation enthalpies of mono- and divacancies.

Additional support for the monovacancy-divacancy interpretation comes from measurements of the pressure dependence of self-diffusion (see Chap. 8), from which an effective activation volume, ΔV_{eff}, is obtained. For the simultaneous contribution of the two mechanisms we have

$$\Delta V_{eff} = \Delta V_{1V} \frac{D_{1V}^*}{D_{1V}^* + D_{2V}^*} + \Delta V_{2V} \frac{D_{2V}^*}{D_{1V}^* + D_{2V}^*}, \qquad (17.16)$$

which is a weighted average of the activation volumes of the individual activation volumes of monovacancies, ΔV_{1V}, and divacancies, ΔV_{2V}. Since $\Delta V_{1V} < \Delta V_{2V}$ and since the divacancy contribution increases with temperature, the effective activation volume increases with temperature as well. Figure 17.5 displays effective activation volumes for Ag self-diffusion. An increase from about 0.67 Ω at 600 K to 0.88 Ω (Ω = atomic volume) near the melting temperature has been observed by BEYELER AND ADDA [21] and REIN AND MEHRER [22].

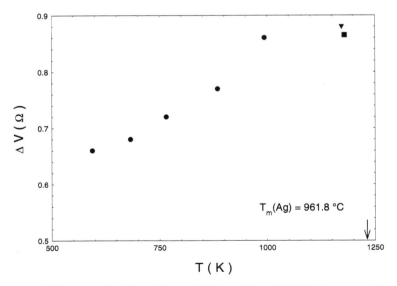

Fig. 17.5. Effective activation volumes, ΔV_{eff}, of Ag self-diffusion *versus* temperature in units of the atomic volume Ω of Ag: *triangle, square* [21], *circles* [22]

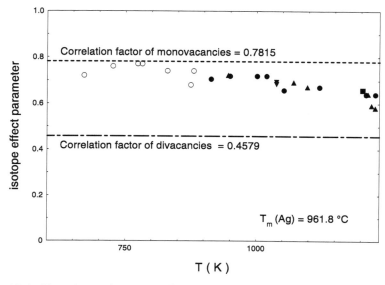

Fig. 17.6. Experimental isotope-effect parameters of Ag self-diffusion: full *circles* [15], *triangles* [25], *full square* [26], *triangles on top* [27], *open circles* [28]

Isotope-effect measurements can throw also light on the diffusion mechanism, because the isotope-effect parameter is closely related to the corresponding tracer correlation factor (see Chap. 9). If mono- and divacancies operate simultaneously, measurements of the isotope-effect yield an effective isotope-effect parameter:

$$E_{eff} = E_{1V} \frac{D^*_{1V}}{D^*_{1V} + D^*_{2V}} + E_{2V} \frac{D^*_{2V}}{D^*_{1V} + D^*_{2V}}. \qquad (17.17)$$

E_{eff} is a weighted average of the isotope-effect parameters for monovacancies, E_{1V}, and divacancies, E_{2V}. The individual isotope effect parameter are related via $E_{1V} = f_{1V}\Delta K_{1V}$ and $E_{2V} = f_{2V}\Delta K_{2V}$ to the tracer correlation factors and kinetic energy factors of mono- and divacancy diffusion (see Chap. 9). Fig 17.6 shows measurements of the isotope-effect parameter for Ag self-diffusion. According to Table 7.2, we have $f_{1V} = 0.781$ and $f_{2V} = 0.458$. The decrease of the effective isotope-effect parameter with increasing temperature has been attributed to the simultaneous contribution of mono- and divacancies in accordance with Fig. 17.4.

17.3 Hexagonal Close-Packed and Tetragonal Metals

Several metallic elements such as Zn, Cd, Mg, and Be crystallise in the hexagonal close-packed structure. A few others such as In and Sn are tetragonal.

17.3 Hexagonal Close-Packed and Tetragonal Metals

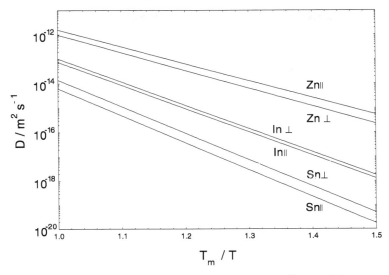

Fig. 17.7. Self-diffusion in single crystals of Zn, In, and Sn parallel and perpendicular to their unique axes

According to Chap. 2, the diffusion coefficient in a hexagonal or tetragonal single-crystal has two principal components:

$D_\perp^*$: tracer diffusivity perpendicular to the axis,

$D_\parallel^*$: tracer diffusivity parallel to the axis.

Figure 17.7 shows self-diffusion in single-crystals of Zn, In, and Sn for both principal directions. In hexagonal Zn we have $D_\parallel^* > D_\perp^*$; i.e. diffusion parallel to the hexagonal axis is slightly faster. For the tetragonal materials In and Sn $D_\parallel^* < D_\perp^*$ holds true. For all of these materials the anisotropy ratio $D_\perp^*/D_\parallel^*$ is small; it lies in the interval between about 1/2 and 2 in the temperature ranges investigated. Diffusivity values in hcp Zn, Cd, and Mg reach about 10^{-12} m^2 s^{-1} near the melting temperature. Such values are similar to those of fcc metals. This is not very surprising, since both lattices are close-packed structures.

Let us recall the atomistic expressions for self-diffusion in hcp metals. The hcp unit cell is shown in Fig. 17.8. Vacancy-mediated diffusion can be expressed in terms of two vacancy-atom exchange rates. The rate ω_a accounts for jumps within the basal plane and ω_b for jumps oblique to the basal plane. The two principal diffusion coefficients can be written as

$$D_\perp^* = \frac{a^2}{2} C_V^{eq} \left(3\omega_a f_{a\perp} + \omega_b f_{b\perp} \right) ,$$
$$D_\parallel^* = \frac{3}{4} c^2 C_V^{eq} \omega_b f_{b\parallel} . \qquad (17.18)$$

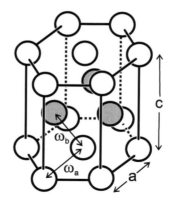

Fig. 17.8. Hexagonal close-packed unit cell with lattice paranmeters a and c. Indicated are the vacancy jump rates: ω_a is within the basal plane and ω_b oblique to it

Here a denotes the lattice parameter within the basal plane and c the lattice parameter in the hexagonal direction. $f_{a\perp}$, $f_{b\perp}$ and $f_{b\|}$ are correlation factors. The *anisotropy ratio* is then:

$$A \equiv \frac{D^*_\perp}{D^*_\|} = \frac{2}{3}\frac{a^2}{c^2}\frac{(3\omega_a f_{a\perp} + \omega_b f_{b\perp})}{\omega_b f_{b\|}}. \quad (17.19)$$

If correlation effects are negelected, i.e. for $f_{a\perp} = f_{b\perp} = f_{b\|} = 1$, we get from Eq. (17.19)

$$A \approx \frac{2}{3}\frac{a^2}{c^2}\left(3\frac{\omega_a}{\omega_b} + 1\right). \quad (17.20)$$

For the ideal ratio $c/a = \sqrt{8/3}$ and $\omega_a = \omega_b$, one finds $A = 1$; this remains correct if correlation is included. The correlation factors and A vary with the ratio ω_a/ω_b. For details the reader is referred to a paper by MULLEN [29].

17.4 Metals with Phase Transitions

Many metallic elements undergo allotropic transformations and reveal different crystalline structures in different temperature ranges. Such changes are found in about twenty metallic elements. Allotropic transitions are first-order phase transitions, which are accompanied by abrupt changes in physical properties including the diffusivity. Some metals undergo second-order phase transitions, which are accompanied by continuous changes in physical properties. A well known example is the magnetic transition from the ferromagnetic to paramagnetic state of iron. In intermetallic compounds (considered in Chap. 20) also order-disorder transitions occur, which can be second order. In what follows we consider two examples, which illustrate the effects of phase transitions on self-diffusion:

17.4 Metals with Phase Transitions

Self-diffusion in iron: Iron undergoes allotropic transitions from a bcc, to an fcc, and once more to a bcc structure, when the temperature varies according to the following scheme:

$$\alpha\text{-Fe (bcc)} \overset{1183\,K}{\Longleftrightarrow} \gamma\text{-Fe (fcc)} \overset{1653\,K}{\Longleftrightarrow} \delta\text{-Fe (bcc)} \overset{1809\,K}{\Longleftrightarrow} \text{Fe melt}.$$

Numerous heat treatments of steels benefit from these phase transformations, which are of first-order. They are associated with abrupt changes of the diffusion coefficient (see Fig. 17.9). It is interesting to note that the transition from bcc iron to close-packed fcc iron is accompanied by a decrease in the diffusivity of about one order of magnitude. This is in accordance with the observation that self-diffusion in fcc metals (Fig 17.2) is slower than self-diffusion in bcc metals (Fig. 17.3) at the same homologous temperature.

Magnetic phase transitions are prototypes of second-order phase transitions. In such transitions an order parameter passes through the transition temperature in continuos manner. Second-order transitions are associated with continuous changes of physical properties. Below the Curie temperature, $T_C = 1043$ K, iron is ferromagnetic, above T_C it is paramagnetic. Figure 17.9 shows that self-diffusion is indeed a continuous function of temperature around T_C. However, below T_C it is clearly slower that an Arrhenius extrapolation from the paramagnetic bcc region would suggest. In the literature several models have been discussed, which describe the influence of

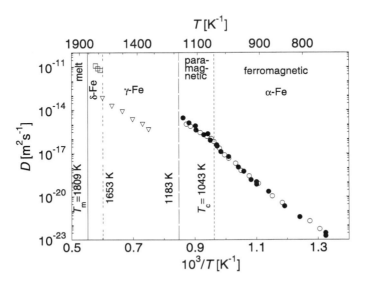

Fig. 17.9. Self-diffusion in the α-, γ- and δ-phases of Fe: *full circles* [30]; *open circles* [31]; *triagles* [32]; *squares* [33]

ferromagnetic order on diffusion. The simplest one correlates the variation of the activation enthalpy according to

$$D^* = D_p^0 \exp\left[-\frac{\Delta H_p(1+\alpha S^2)}{k_B T}\right] \qquad (17.21)$$

with some ferromagnetic order parameter S. The long-range order parameter is connected via $S \equiv M_S(T)/M_S(0)$ to the spontaneous magnetisation $M_S(T)$ at temperature T. ΔH_p and D_p^0 are the activation enthalpy and the pre-exponential factor in the paramagnetic state; α is a fitting parameter.

Self-diffusion in group-IV transition metals: These metals undergo transformations from a hcp low-temperature phase (α–phase) to a bcc high-temperature phase (β-phase). The transition temperatures, $T_{\alpha,\beta}$, are 1155 K for Ti, 1136 K for Zr, and 2013 K for Hf. As an example, Fig. 17.10 shows self-diffusion in hcp and bcc titanium according to [34–36]. We note that upon the transformation from the hcp to the bcc structure the diffusivity increases by almost three orders of magnitude. Very similar behaviour is reported for zirconium and hafnium (see, [8] for references). Self-diffusion in the close-packed hcp structure is relatively slow, whereas we have already seen in Fig. 17.3 that self-diffusion in the less densely packed bcc phases metals is relatively fast.

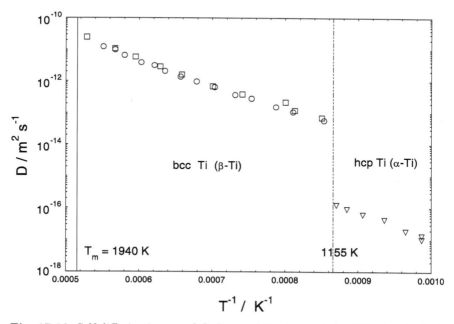

Fig. 17.10. Self-diffusion in α- und β-phases of titanium: *circles* [34]; *tiangles* [35]; *squares* [36]

The very fast self-diffusion of bcc high-temperature phases and the wide spectrum of diffusivities of bcc metals has been attributed to special features of the lattice dynamics in the bcc structure by KÖHLER AND HERZIG [37] and also by VOGL AND PETRY [38]. A nearest-neighbour jump of an atom in the bcc lattice is a jump in a $\langle 111 \rangle$ direction. For bcc metals the longitudinal phonon branch shows a minimum for $2/3 \langle 111 \rangle$ phonons. This minimum is most pronounced for group-IV metals. The associated low phonon frequencies indicate a small activation barrier for nearest-neighbour exchange jumps between atom and vacancy.

References

1. Y. Adda, J. Philibert, *La Diffusion dans les Solides*, Presses Universitaires de France, 1966
2. A. Seeger, H. Mehrer, in: *Vacancies and Interstitials in Metals*, A. Seeger, D. Schumacher, J. Diehl, W. Schilling (Eds.), North-Holland, Amsterdam, 1970, p.1
3. J. Philibert, *Atom Movements – Diffusion and Mass Transport in Solids*, Les Editions de Physique, Les Ulis, 1991
4. Th. Heumann, *Diffusion in Metallen*, Springer-Verlag, Berlin, 1992
5. N.L. Peterson, J. Nucl. Mat. **69/70**, 3 (1978)
6. H. Mehrer, J. Nucl. Mat. **69/70**, 38 (1978)
7. H. Mehrer, Defect and Diffusion Forum **129–130**, 57 (1996)
8. H. Mehrer, N. Stolica, N.A. Stolwijk, *Self-diffusion in Solid Metallic Elements*, Chap. 2 in: *Diffusion in Solid Metals and Alloys*, H. Mehrer (Vol.-Ed.), Landolt-Börnstein, Numerical Data and Functional Relationships in Science and Technology, New Series, Group III: Crystal and Solid State Physics, Vol. 26, Springer-Verlag, 1990
9. H.J. Wollenberger, *Point Defects*, in: *Physical Metallurgy*, R.W. Cahn, P. Haasen (Eds.), North-Holland Publishing Company, 1983, p. 1139
10. H. Ullmaier (Vol.Ed.), *Atomic Defects in Metals*, Landolt-Börnstein, Numerical Data and Functional Relationships in Science and Technology, New Series, Group III: Condensed Matter, Vol. 25, Springer-Verlag, 1991
11. H. Bakker, Phys. Stat. Sol. **28**, 569 (1968)
12. K. Maier, H. Mehrer, E. Lessmann, W. Schüle, Phys. Stat. Sol. (b) **78**, 689 (1976)
13. B.S. Bokstein, S.Z. Bokstein, A.A. Zhukhovitskii, *Thermodynamics and Kinetics of Diffusion in Solids*, Oxonian Press, New Dehli, 1985
14. S.J. Rothman, N.L. Peterson, Phys. Stat. Sol. **35**, 305 (1969)
15. S.J. Rothman, N.L. Peterson, J.T. Robinson, Phys. Stat. Sol. **39**, 635 (1970)
16. N.Q. Lam, S.J. Rothman, H. Mehrer, L.J. Nowicki, Phys. Stat. Sol. (b) **57**, 225 (1973)
17. J.G.E.M. Backus, H. Bakker, H. Mehrer, Phys. Stat. Sol. (b) **64**, 151 (1974)
18. M. Werner, H. Mehrer, in: *DIMETA 82 – Diffusion in Metals and Alloys*, F.J. Kedves, D.L. Beke (Eds.), Trans Tech Publications, Switzerland, 1983, p. 393
19. N.L. Peterson, Phys. Rev. A **136**, 568 (1964)
20. G. Rein, H. Mehrer, K. Maier, Phys. Stat. Sol. (a) **45**, 253 (1978)

21. M. Beyeler, Y. Adda, J. Phys. (Paris) **29**, 345 (1968)
22. G. Rein, H. Mehrer, Philos. Mag. **A 45**,767 (1982)
23. J.W. Miller, Phys. Rev. **181**, 10905 (1969)
24. R.H. Dickerson, R.C. Lowell, C.T. Tomizuka, Phys. Rev. A **137**, 613 (1965)
25. P. Reimers, D. Bartdorff, Phys. Stat. Sol. (b9 **50**, 305 (1972)
26. N.L. Peterson, L.W. Barr, A.D. Le Claire, J. Phys, **C6**, 2020 (1973)
27. Chr. Herzig, D. Wolter, Z. Metallkd. **65**, 273 (1974)
28. H. Mehrer, F. Hutter, in: *Point Defects and Defect Interactions in Metals*, J.I. Takamura, M. Doyama, M. Kiritani (Eds.), University of Tokyo Press, 1982, p. 558
29. J. Mullen, Phys. Rev. **124** (1961) 1723
30. M. Lübbehusen, H. Mehrer, Acta Metall. Mater. **38**, 283 (1990)
31. Y. Iijima, K. Kimura, K. Hirano, Acta Metall. **36**, 2811 (1988)
32. F.S. Buffington, K.I. Hirano, M. Cohen, Acta Metall. **9**, 434 (1961)
33. C.M. Walter, N.L. Peterson, Phys. Rev. **178**, 922 (1969)
34. J.F. Murdock, T.S. Lundy, E.E. Stansbury, Acta Metall. **12** 1033 (1964)
35. F. Dyment, Proc. 4th Int. Conf. on Titanium, H. Kimura, O. Izumi (Eds.), Kyoto, Japan, 1980, p. 519
36. U. Köhler, Chr. Herzig, Phys. Stat. Sol. (b) **144**, 243 (1987)
37. U. Köhler, Chr. Herzig, Philos. Mag. **A 28**, 769 (1988)
38. G. Vogl, W. Petry, Physik. Blätter **50**, 925 (1994)

18 Diffusion of Interstitial Solutes in Metals

Solute atoms which are considerably smaller than the atoms of the host lattice are incorporated in interstitial sites and form interstitial solid solutions. This is the case for hydrogen (H), carbon (C), nitrogen (N) and oxygen (O) in metals. Interstitial sites are defined by the geometry of the host lattice. For example, in fcc, bcc and hcp host metals interstitials occupy either tetrahedral or octahedral sites and diffuse by jumps from one interstitial site to neighbouring ones (see Chap. 6). No defect is necessary to mediate their diffusion jumps, no defect-formation term enters the expression for the diffusion coefficient and no defect-formation enthalpy contributes to the activation enthalpy of diffusion (see Chap. 6). In the present chapter we consider first the 'heavy' interstitial diffusers C, N and O in Sect. 18.1. Hydrogen is the smallest and lightest atom in the periodic table. Hydrogen diffusion in metals is treated separately in Sect. 18.2. Whereas non-classical isotope effects and quantum effects are usually negligible for heavier diffusers, such effects play a rôle for diffusion of hydrogen.

18.1 'Heavy' Interstitial Solutes C, N, and O

18.1.1 General Remarks

Interstitial atoms diffuse much faster than atoms of the host lattice or substitutional solutes, because the small sizes of the interstitial solutes permit rather free jumping between interstices. Typical examples are shown in Fig. 18.1, where the diffusion coefficients of C, N, and O in niobium (Nb) are displayed together with Nb self-diffusion. The corresponding activation

Table 18.1. Activation parameters of interstitial diffusants in Nb. For comparison self-diffusion paramters of Nb are also listed (for references see [1])

	C	N	O	Nb
$\Delta H / \mathrm{kJ\,mol^{-1}}$	142	161	107	395
$D^0 / \mathrm{m^2\,s^{-1}}$	1×10^{-6}	6.3×10^{-6}	4.2×10^{-7}	5.2×10^{-5}

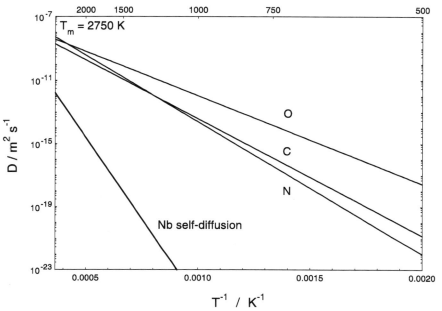

Fig. 18.1. Diffusion of interstitial solutes C, N, and O in Nb. For comparison Nb self-diffusion is also shown

enthalpies and pre-exponential factors are collected in Table 18.1. The activation enthalpies of interstitital diffusers are much smaller than those of Nb self-diffusion with the result that the diffusion coefficients for interstitial diffusion are many orders of magnitude larger than those for self-diffusion of lattice atoms. Interstitial diffusivities can be as high as diffusivities in liquids, which near the melting temperature are typically between $10^{-8}\,\mathrm{m^2\,s^{-1}}$ and $10^{-9}\,\mathrm{m^2\,s^{-1}}$ (see, e.g., Fig. 15.2).

18.1.2 Experimental Methods

The fast diffusion of C, N, and O and the fact that nitrogen and oxygen are gases, or, in the case of carbon, readily available in gas or vapour form (CO, CO_2, CH_4,...) have an impact on the choice of methods used to measure their diffusion coefficients (see also Part II of this book). In the following, we summarise the most important experimental methods:

– **Steady-state methods** are particularly appropriate for fast diffusers such as C, N, and O. The steady-state concentration gradient may be measured directly or calculated from solubility data. The flux can be

18.1 'Heavy' Interstitial Solutes C, N, and O

measured by standard gas-flow methods. With suitable electrodes supplying and removing diffusant in an electrochemical cell, which can be sometimes devised, the flux can be deduced from measurements of the electrical current.

- A suitable carbon isotope for the **radiotracer method**, *viz.* ^{14}C, is available. Because of its weak β-radiation the residual activity method (Gruzin-Seibel method, see Chap. 13) is adequate. Enriched **stable isotopes** (^{28}O, ^{12}N) in combination with SIMS profiling can be used to study O and N diffusion.

- **Diffusion-couple methods** are sometimes used, although not so commonly as with substitutional alloys. Both the simple error-function solution and the Boltzmann-Matano method of analysis are applied (see Chaps. 3 and 10).

- **Nuclear reaction analysis (NRA)** for the determination of diffusion profiles (see Chap. 15) is a natural complement for ion implantation of sample preparation. Both techniques are frequently applied together.

- **Out-diffusion methods** utilise outgassing for N or O and decarburisation for C. These processes usually entail samples containing the diffuser in primary solid solution. Often error-function type solutions apply for the analysis of the results.

- **In-diffusion experiments** often include reaction diffusion situations. Reaction of the diffuser can occur leading, in addition to the always present primary solid solution, to the growth of one or more outer layers of other phases, corresponding to the intermediate phases at the diffusion temperature. For example, carbide, oxide, or nitride phase layers may form on samples into which C, N, or O is being diffused. If the concentration-depth profile is measured, the interdiffusion coefficient can be determined, e.g., by the Boltzmann-Matano method (see Chap. 10).

- For the diffusion of C, N, and O in bcc metals by far the most commonly employed indirect methods are those based on **Snoek-effect – internal friction methods** and, at the lower temperatures, measurements of the anelastic stress or strain relaxation (for details see Chap. 14 and the reviews by NOWICK AND BERRY [2] and BATIST [3]). The latter techniques provide values of D at temperatures that may extend down to ambient and even below. With all the indirect methods, the deduced D-values are model-sensitive. For the interpretation of Snoek-effect based measurements on bcc metals it is usually assumed that the solute atoms occupy octahedral interstitial sites.

- Occasionally measurements of **magnetic relaxation** have been used in the study of C, N, and O diffusion. For details see Chap. 14 and the textbook of KRONMÜLLER [4].

18.1.3 Interstitial Diffusion in Dilute Interstial Alloys

Diffusivities of interstitial diffusers have sometimes been measured over an exceptionally wide range of temperatures by combining direct measuring techniques at high temperatures and indirect techniques, which monitor a few atomic jumps only, at low temperatures. Such measurements can provide materials for tests of the linearity of the Arrhenius relation. For example, N diffusion in bcc α-Fe and δ-Fe covers a range from 10^{-8} m^2 s^{-1} to about 10^{-24} m^2 s^{-1} without significant deviation from linearity (for references see [1]). For N and O diffusion in Ta and Nb, linear Arrhenius behaviour has been reported over wide temperature ranges as well. Diffusion coefficients, for C diffusion in α-Fe are assembled in Fig. 14.7. In this figure data from direct diffusion measurements at high temperatures and data deduced from indirect techniques such as internal friction, elastic after effect, and magnetic after effect have been included (see also Chap. 14).

Interstitial solutes migrate by the *direct interstitial mechanism* discussed already in Chap. 6. For cubic host lattices, the diffusion coefficient of interstitial solutes can be written as

$$D = ga^2\nu^0 \exp\left(\frac{S^M}{R}\right) \exp\left(-\frac{H^M}{k_B T}\right). \tag{18.1}$$

Here a is the cubic lattice parameter and g a geometric factor, which depends on the lattice geometry and on the type of interstitial sites (octahedral or tetrahedral) occupied by the solute. For jumps between neighbouring octahedral sites (see Fig. 6.1) $g = 1$ for the fcc and $g = 1/6$ for the bcc lattice. The Arrhenius parameters of Eq. (8.1) have the following simple meaning for diffusion of interstitial solutes:

$$\Delta H \equiv H^M \tag{18.2}$$

and

$$D^0 \equiv ga^2\nu^0 \exp\left(\frac{S^M}{k_B}\right), \tag{18.3}$$

where H^M and S^M denote the migration enthalpy and entropy of the intertitial. With reasonable values for the migration entropy (between zero and several k_B) and the attempt frequency ν^0 (between 10^{12} and 10^{13} s^{-1}), Eq. (18.3) yields the following limits for the pre-exponential factors of interstitial solutes:

$$10^{-7} \text{m}^2 \text{s}^{-1} \leq D^0 \leq 10^{-4} \text{m}^2 \text{s}^{-1}. \tag{18.4}$$

The experimental values of D^0 for C, N, and O from Table 18.1 indeed lie in the expected range.

Measurements of the isotope effect of ^{12}C/^{13}C diffusion in α-Fe yielded an isotope effect parameter near unity [7]. This isotope effect is well compatible with a correlation factor of $f = 1$, which is expected for a direct interstitial mechanism (see Chap. 9).

Slight upward curvatures in the Arrhenius diagram were observed, e.g., for C diffusion in Mo and in α-Fe (see Fig. 14.7) in the high-temperature region [1]. Several suggestions have been discussed in the literature to explain this curvature [5, 6]. These suggestions include formation of interstitial-vacancy pairs, formation of di-interstitials and interstitial jumps between octahedral and tetrahedral sites in addition to jumps between neighbouring octahedral sites, and in the case of iron the magnetic transition.

The simple picture developed above describes diffusion in very dilute interstitial solutions. For diffusion in interstitial alloys with higher concentration the interaction between interstitial solutes comes into play. For carbon in austenite, tracer data are available from the work of PARRIS AND MCLELLAN [8] and chemical diffusion and activity data from the work of SMITH [9]. According to BELOVA AND MURCH [10] paired carbon interstititals play a rôle in diffusion.

18.2 Hydrogen Diffusion in Metals

18.2.1 General Remarks

The field of hydrogen in metals originates from the work of THOMAS GRAHAM – one of the pioneers of diffusion in Great Britain (see Chap. 1). In 1866 he discovered the ability of Pd to absorb large amounts of hydrogen. He also observed that hydrogen can permeate Pd-membranes at an appreciable rate [11] and he introduced the concept of metal-hydrogen alloys without definite stoichiometry, in contrast to stoichiometric nonmetallic hydrides. In 1914 SIEVERTS found that the concentration of hydrogen in solution (at small concentrations) is proportional to the square root of hydrogen pressure [12]. The observation of Sieverts can be considered as direct evidence for the dissociation of the H_2 molecule, when hydrogen enters a metal-hydrogen solution. Inside the metal crystal, hydrogen atoms occupy interstices.

Since Graham and Sieverts, many scientists have devoted much of their scientific work to metal-hydrogen systems. Hydrogen in metals provides a large and fascinating topic of materials science, which has been the subject of many international conferences. Hydrogen in metals has also attracted considerable interest from the viewpoint of applications. For example, hydrogen storage in metals is based on the high solubility and fast diffusion of hydrogen in metal-hydrogen systems. Pd membranes for hydrogen purification provide an old and well-known application based on the very fast diffusion. Such membranes can be used either to extract hydrogen from a gas stream or to purify it.

Nowadays, an enormous body of knowledge about hydrogen in metals is available. In this chapter, we can only provide a brief survey of some of the striking features of H-diffusion. For more comprehensive treatments we refer the reader to books edited by ALEFELD AND VÖLKL [13, 14], WIPF [15] and to reviews of the field [16–18]. A data collection for H diffusion in metals can be found in [19] and for diffusion in metal hydrides in [20].

18.2.2 Experimental Methods

The very fast diffusion and the often high solubility of hydrogen have consequences for the experimental techniques used in hydrogen diffusion studies: In addition to concentration-profile methods, permeation methods based on Fick's first law, absorption and desorption methods, electrochemical methods, and relaxation methods are in use. Due to the favourable gyromagnetic ratio of the proton and due its large quasielastic scattering cross section for neutrons, nuclear magnetic resonance (NMR) and quasielastic neutron scattering (QENS) are specially suited for hydrogen diffusion studies. Here, we mention the major experimental methods briefly and refer for details to Part II of this book:

- The **radiotracer method** can be applied using the isotope tritium (^{3}H). Its half-life is $t_{1/2} = 12.3$ years and it decays by emitting weak β-radiation.
- **Nuclear reaction analysis (NRA)** can be used to measure concentration profiles (see Chap. 15). For example, the ^{15}N resonant nuclear reaction H(^{15}N, $\alpha\gamma$)^{12}C has been used to determine H concentration profiles.
- In **steady-state permeation** usually the stationary rate of permeation of hydrogen through a thin foil is studied. If the steady-state concentration on the entry and exit surface are maintained at fixed partial pressures of molecular hydrogen gas, p_{m1} and p_{m2}, respectively, the flux per unit are through a membrane of thickness Δx is

$$J_\infty = \frac{KD}{\Delta x}(p_{m1}^{1/2} - p_{m2}^{1/2}), \tag{18.5}$$

where K is called the Sieverts constant. The diffusivity D is then determined from the flux, provided that K is known from solubility measurements. The Sieverts constant is thermally activated via the heat of solution for hydrogen.
A fine example of permeation studies is the work of HIRSCHER AND ASSOCIATES [37], in which H permeation through Pd, Ni, and Fe membranes has been investigated.
- In **non-steady-state permeation** the rapid establishment of a fixed hydrogen concentration on the entry side of a membrane by contact with H$_2$ gas induces a transient, time-dependent flux, J_t, prior to the final steady-state flux, J_∞. It can be shown that

$$\frac{J_t}{J_\infty} \cong \frac{2\Delta x}{\sqrt{\pi D t}} \exp\left(-\frac{\Delta x^2}{4Dt}\right). \tag{18.6}$$

18.2 Hydrogen Diffusion in Metals

The diffusivity can be determined from the characteristics of the time-lag. In permeation studies, UHV gas-permeation systems are used and care must be taken to avoid problems related to surface layers.

- **Resistometric methods** make use of the resistivity change upon hydrogen incorporation in a metal. For example, the resistivity of Nb-H increases by about 0.7 $\mu\Omega$cm per percent hydrogen. Since the electric resistivity can be measured very accurately, this property can be used to monitor hydrogen redistribution in a sample.
- **Electrochemical techniques** are variations of the permeation methods. They use electrochemical techniques to measure and often, also, to produce the hydrogen flux in a sample membrane in an electrolytic cell. Hydrogen is generated on the cathode side and the entry concentration is controlled by the applied voltage. On the exit side, the potential is maintained positive so that the arriving H atoms are oxidised. The equivalent electrical current generated by the oxidising process is a sensitive measure of the flux through the membrane. Electrochemical permeation cells may be operated in several modes (step method, pulse method, oscillation method), which differ by the boundary conditions imposed experimentally on the sample. Analyses for the associated time-lag relations can be found, e.g., in a paper by ZÜCHNER AND BOES [21]. Electrochemical methods are useful over a relatively limited temperature range and may be subject to surface effects.
- **Absorption and desorption methods** are based on the continuum description of in- or out-diffusion, e.g., for cylindrical or spherical samples (see Chap. 2 for solutions of the diffusion equation). These methods are applicable at relatively high temperatures; again, care must be taken to avoid problems related to surface layers.
- **Mechanical relaxation methods** include stress and strain relaxation, internal friction and modulus change measurements:
    - The **Gorsky effect** enables studies of long-range diffusion driven by a dilation gradient. It occurs because hydrogen expands the lattice. A gradient in dilatation, produced, e.g., by bending the sample, may be used to set up a gradient in concentration. With the Gorsky technique, one measures the time required to establish the concentration gradient by hydrogen diffusion across the sample diameter. This is a macroscopic rather than a microscopic relaxation time. If the sample diameter is known, the absolute value of D can be determined (see Chap. 14).
    - Whereas the **Snoek effect** has been widely used to study the diffusion of 'heavy' interstitials in bcc metals, its application to H-diffusion has not been successful despite several attempts having been made. Hydrogen-related internal friction peaks are often reported in the literature (see, e.g., ALEFELD AND VÖLKL [13] for references), but in no case has it been established that any of these peaks results from the

Snoek effect of hydrogen alone. Combinations with impurities, lattice defects or the formation of hydrogen pairs are suggested as causes.
- **Magnetic relaxation** studies are very sensitive, but they are limited to ferromagnetic materials [22]. For example, they have been applied by HIRSCHER AND KRONMÜLLER [38] in studies of H diffusion in several ferromagnetic alloys and amorphous metals.
- The extensive use of **nuclear magnetic relaxation (NMR)** for the study of diffusion in metal-hydrogen solid solutions or in metal hydrides is favoured by the properties of the proton.
  (i) The strength of the NMR signal is proportional to the gyromagnetic ratio γ of the nucleus. The proton has the largest γ of all stable nuclei.
  (ii) The proton spin of $I = 1/2$ produces only dipolar interactions with its surroundings; the lack of quadrupole coupling (present only for nuclei with $I > 1/2$) simplifies the interpretation of NMR spectra.
  The NMR technique is a well-established tool for studying atomic motion in condensed matter. Data usually consist of measurements of longitudinal spin-lattice-relaxation time T_1, of the transverse spin-spin relaxation time T_2, or of the motional narrowing of the resonance line. A theoretical model is necessary to deduce the diffusion coefficient therefrom. For details see Chap. 15 and reviews on nuclear magnetic relaxation techniques [24–26].
- **Quasielastic neutron scattering (QENS)** is also well suited to the study of hydrogen diffusion. The neutron scattering cross-section of the proton is an order of magnitude larger than that of the deuteron and all other nuclei. The QENS method is based on the following physical phenomenon: a monoenergetic neutron beam is scattered incoherently by the protons in the metal. Because of the diffusion of the protons, the beam will be energetically broadened and the width of the line depends on the rate of diffusion. For small momentum transfer Q, i.e. for small scattering angles, the full width at half maximum in energy, ΔE, is given by

$$\Delta E = 2\hbar D Q^2, \qquad (18.7)$$

where D is the diffusion coefficient, $\hbar$ the Planck constant divided by 2π, and Q the momentum transfer. From the linewidth at large scattering angles, atomistic details of the jump process (jump length, jump direction) can be obtained as discussed in Chap. 15. Further information can be found in a textbook by HEMPELMANN [27] and several reviews of the field [28–31]. Both nuclear methods, NMR and QENS, are independent of surface related problems.

18.2.3 Examples of Hydrogen Diffusion

In this section, typical examples for the temperature dependence of hydrogen diffusion and its isotope effects are noted. Hydrogen forms an interstitial solid solution with most metallic elements. Some metals have a negative enthalpy of solution and a high solubility for hydrogen (Ti, Zr, Hf, Nb, Ta,

Pd, ...) and form hydrides at higher hydrogen concentrations. Other metals have a positive enthalpy of solution and a relatively low solubility (group-VI metals, group-VII metals, noble metals, Fe, Co, Ni, ...) [23]. We confine ourselves to fcc and bcc cubic α-hydrides, i.e. to low hydrogen concentrations. For diffusion in hydrides with higher concentrations we refer to the literature mentioned at the beginning of Sect. 18.2.

Metal-hydrogen systems are prototypes of fast solid-state diffusion. Diffusion of hydrogen can be observed still far below room temperature. The smallest activation enthalpies and the largest isotope effects have been found for metal-hydrogen systems. Below room temperature, the diffusion coefficients of hydrogen in several transition metals or alloys are the largest known for long-range diffusion in solids. The diffusivity of hydrogen in metals is even at room temperature very high. For example, H in Nb has a room-temperature diffusivity of about 10^{-9} m² s^{-1}, which corresponds to about 10^{12} jumps per second. It exceeds the diffusivity of heavy interstitial such as C, N, and O at this temperature by 10 to 15 orders of magnitude (see Sect. 18.1). Phenomenologically, this high diffusivity is a consequence of the very low activation enthalpies.

Since hydrogen atoms have a small mass, quantum effects can be expected for diffusion of hydrogen. The three isotopes of hydrogen (H, D, and T) have large mass ratios. Hence isotope effects can be studied over a wide range, which is important to shed light on possible diffusion mechanisms.

The **palladium-hydrogen** system along with nickel and iron has attracted perhaps the largest amount of attention. Most of the experimental methods were applied to Pd-H and the consistency of the data is remarkably good (for references see [19]). H-diffusion in Pd, Ni and Fe is displayed in Fig. 18.2. The Arrhenius line represents a best fit to the available data suggested by [13] with the Arrhenius parameters listed in Table 18.2. A jump model seems to be appropriate. Quasielastic neutron scattering experiments have shown that the jump distance is $a/\sqrt{2}$ (a = cubic lattice parameter). This distance corresponds to jumps between neighbouring octahedral sites and not to $a/2$ for jumps between tetrahedral sites [32].

Diffusion of **hydrogen in nickel** has been studied from about room temperature up to the melting temperature of Ni (1728 K) (for references see [19]). The data are remarkably consistent although most measurements have been performed with surface sensitive methods. Because of the low solu-

Table 18.2. Activation parameters of H diffusion in Pd, Ni, and Fe according to ALEFELD AND VÖLKL [13]

	Pd	Ni ($T < T_C$)	Ni ($T > T_C$)	Fe (above 300 K)
D^0/m² s^{-1}	2.9×10^{-7}	4.76×10^{-7}	6.87×10^{-7}	4.0×10^{-4}
ΔH/eV (or kJ mol^{-1})	0.23 (22.1)	0.41 (39.4)	0.42 (40.4)	0.047 (4.5)

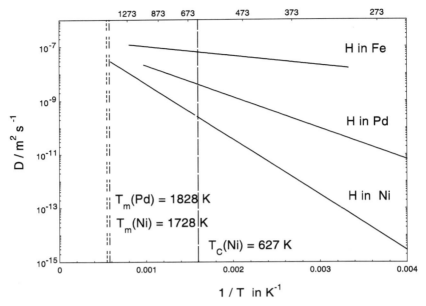

Fig. 18.2. Diffusion of H in Pd, Ni and Fe according to ALEFELD AND VÖLKL [13]

bility of H in Ni other methods such as Gorsky effect, NMR, and QENS have limited applicability for the Ni-H system. The Arrhenius parameters below the magnetic transition (Curie temperature $T_C = 627$ K) seem to be only slightly different than above (see Fig. 18.2).

Diffusion of **hydrogen in iron** has considerable importance due to the technical challenge of hydrogen embrittlement of iron and steels. The absolute values for H-diffusion in Fe are higher than those for Pd and Ni (see Fig 18.2). In spite of the technological interest and the large number of studies (for references see [19]), the scatter of the data is rather large especially below room temperature. Several reasons for this scatter have been discussed in the literature: surface effects, trapping of hydrogen by impurities, dislocations, grain boundaries, or precipitates, and the formation of molecular hydrogen in micropores, either already existing in the material or produced by excess loading (for references see [18]).

Hydrogen in niobium has been studied over a very wide temperature region (see Fig. 18.3). Because of the high diffusivities and its relatively small changes with temperature the Gorsky technique could be used for both H and D in nearly the whole temperature range. Tritium diffusion has been studied in a more limited temperature range. Isotope effects of hydrogen diffusion in Nb are very evident from Fig. 18.3. Similar isotope effect studies are available for V and Ta (see Table 18.3).

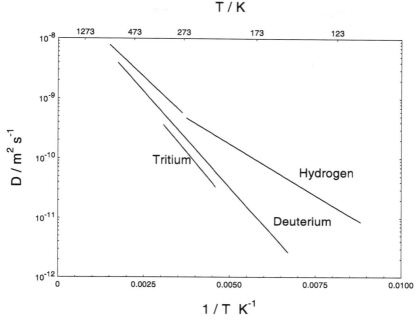

Fig. 18.3. Diffusion of H, D, and T in Nb according to [13]

18.2.4 Non-Classical Isotope Effects

Classical rate theory for the *isotope effect* in diffusion predicts, if many-body effects are neglected, for the ratio of the diffusivities [33, 34],

$$\frac{D_1}{D_2} = \sqrt{\frac{m_2}{m_1}}, \tag{18.8}$$

where m_1 and m_2 are the isotope masses. If manybody effects are taken into account, Eq. (18.8) must be modified (see Chap. 9). Nevertheless, the isotope effect is still completely attributed to the pre-exponential factors. In the classical limit, the activation enthalpies are independent of the isotope masses.

For several metal-hydrogen systems non-classical isotope effects have been reported. Figure 18.3 and Table 18.3 show that in group-V transition metals hydrogen (H) diffuses more rapidly than deuterium (D), and deuterium diffuses more rapidly than tritium (T). In addition, the activation enthalpies of hydrogen isotopes are different and hence the ratio of the diffusivities varies with temperature.

Interestingly, the characteristic features of the dependence of the diffusivity on the isotope mass are correlated with the structure of the host metal [13, 14]. In the bcc metals V, Nb and Ta, hydrogen diffuses faster than deuterium

Table 18.3. Activation parameters for diffusion of H, D, and T in Nb, Ta and V according to ALEFELD AND VÖLKL [13]

Isotope	Parameters	Nb	Ta	V
H	$D^0/\mathrm{m^2\,s^{-1}}$	5×10^{-8} ($T > 273\,\mathrm{K}$) 0.9×10^{-8} ($T < 273\,\mathrm{K}$)	4.4×10^{-8}	2.9×10^{-8}
	$\Delta H/\mathrm{eV}$	0.106 ($T > 273\,\mathrm{K}$) 0.068 ($T < 223\,\mathrm{K}$)	0.140	0.043
D	$D^0/\mathrm{m^2\,s^{-1}}$	5.4×10^{-8}	4.9×10^{-8}	3.7×10^{-8}
	$\Delta H/\mathrm{eV}$	0.129	0.163	0.08
T	$D^0/\mathrm{m^2\,s^{-1}}$	4.5×10^{-8}		
	$\Delta H/\mathrm{eV}$	0.125		

in the whole temperature regime investigated. For the activation enthalpies

$$\Delta H_H < \Delta H_D \qquad (18.9)$$

is observed (see Table 18.3). As a consequence, the ratio D_H/D_D for the group-V transition metals increases with decreasing temperature. For the fcc metals, the pre-exponential factors of H and D scale in accordance with Eq. (18.8) within the experimental errors. However, again the activation enthalpies are mass-dependent, but in contrast to bcc metals one observes

$$\Delta H_H > \Delta H_D. \qquad (18.10)$$

This fact leads to an inverse isotope effect at lower temperatures. For example, in Pd below about 773 K deuterium diffuses faster than hydrogen.

Because of the non-classical behaviour of hydrogen and the large temperature region over which diffusion coefficients were measured, it is by no means evident that the diffusion coefficient of hydrogen in metals should obey a linear Arrhenius relation over the entire temperature range. The break observed in the Arrhenius relation for H in Nb (see Fig. 18.3) was first observed with the Gorsky effect method and confirmed independently by measurements of resistivity changes [35] and by QENS [36]. The deviation of hydrogen diffusion in Nb from an Arrhenius law at lower temperatures (Fig. 18.3) has also been observed in the tantalum-hydrogen system. It has been attributed to incoherent tunneling.

References

1. A.D. Le Claire, *Diffusion of C, N, and O in Metals* Chap. 8 in: *Diffusion in Solid Metals and Alloys*, H. Mehrer (Vol.Ed.), Landolt-Börnstein, Numerical Data and Functional Relationships in Science and Technology, New Series, Group III: Crystal and Solid State Physics, Vol. 26, Springer-Verlag, 1990, p. 471

2. A.S. Nowick, B.S. Berry, *Anelastic Relaxation in Crystalline Solids*, Academic Press, New York, (1972)
3. R.D. Batist, *Mechanical Spectroscopy*, in: *Materials Science and Technology*, Vol. 2B: Characterisation of Materials, R.W. Cahn, P. Haasen, E.J. Cramer (Eds.), VCH, Weinheim, 1994, p. 159
4. H. Kronmüller, *Nachwirkung in Ferromagnetika*, Springer Tracts in Natural Philosophy, Springer-Verlag (1968)
5. J. Philibert, *Atom Movements – Diffusion and Mass Transport in Solids*, Les Editions de Physique, Les Ulis, 1991
6. Th. Heumann, *Diffusion in Metallen*, Springer-Verlag, Berlin, 1992
7. A.J. Bosman, PhD Thesis, University of Amsterdam (1960)
8. D.C. Parris, R.B. McLellan, Acta Metall. **24**, 523 (1976)
9. R.P. Smith, Acta Metall. **1**, 576 (1953)
10. I.V. Belova, G.E. Murch, Philos. Mag. **36**, 4515 (2006)
11. T. Graham, Phil. Trans. Roy. Soc. (London) **156**, 399 (1866)
12. A. Sieverts, Z. Phys. Chem. **88**, 451 (1914)
13. G. Alefeld, J. Völkl (Eds.), *Hydrogen in Metals I – Application-oriented Properties*, Topics in Applied Physics, Vol. 28, Springer-Verlag, 1978
14. G. Alefeld, J. Völkl (Eds.), *Hydrogen in Metals II – Basic Properties*, Topics in Applied Physics, Vol. 29, Springer-Verlag, 1979
15. H. Wipf (Ed.), *Hydrogen in Metals III – Properties and Applications*, Topics in Applied Physics Vol. 73, Springer-Verlag, 1997
16. H.K. Birnbaum, C.A. Wert, Ber. Bunsenges. Phys. Chem. **76**, 806 (1972)
17. R. Hempelmann, J. Less Common Metals **101**, 69 (1984)
18. J. Völkl, G. Alefeld, in *Diffusion in Solids – Recent Developments*, A.S. Nowick, J.J. Burton (Eds.) Academic Press, 1975, p. 232
19. G.V. Kidson, *The Diffusion of H, D, and T in Solid Metals*, Chap. 9 in: *Diffusion in Solid Metals and Alloys*, H. Mehrer (Vol. Ed.), Landolt-Börnstein, Numerical Data and Functional Relationships in Science and Technology, New Series, Group III: Crystal and Solid State Physics, Vol. 26, Springer-Verlag, 1990, p. 504
20. H. Matzke, V.V. Rondinella, *Diffusion in Carbides, Nitrides, Hydrides, and Borides*, in: *Diffusion in Semiconductors and Non-metallic Solids*, D.L. Beke (Vol.Ed), Landolt-Börnstein, Numerical Data and Functional Relationships in Science and Technology, Vol. 33, Subvolume B1, Springer-Verlag, 1999
21. H. Züchner, N. Boes, Ber. Bunsenges. Phys. Chem. **76**, 783 (1972)
22. H. Kronmüller, *Magnetic After-Effects of Hydrogen Isotopes in Ferromagnetic Metals and Alloys*, Ch. 11 in [13]
23. R.B. McLellan, W.A. Oates, Acta Metall. **21**, 181 (1973)
24. R.M. Cotts, Ber. Bunsenges. Phys. Chem. **76**, 760 (1972)
25. G. Majer, *Die Methoden der Kernspinresonanz zum Studium der Diffusion von Wasserstoff in Metallen und intermetallischen Verbindungen*, Cuvillier Verlag, Göttingen, 2000
26. P. Heitjans, S. Indris, M. Wilkening, *Solid-State Diffusion and NMR*, in: *Diffusion Fundamentals – Leipzig 2005*, J. Kärger, F. Grinberg, P. Heitjans (Eds.), Leipziger Universitätsverlag 2005
27. R. Hempelmann, *Quasielastic Neutron Scattering and Solid-State Diffusion*, Oxford Science Publication, 2000
28. T. Springer, Z. Phys. Chem. NF **115**, 317 (1979)

29. T. Springer, *Diffusion Studies of Solids by Quasielastic Neutron Scattering*, in: *Diffusion in Condensed Matter*, J. Kärger, P. Heitjans, R. Haberlandt (Eds.), Vieweg-Verlag, 1998, p. 59
30. T. Springer, R.E. Lechner, *Diffusion Studies of Solids by Quasielastic Neutron Scattering*, in: *Diffusion in Condensed Matter – Methods, Materials, Models*, P. Heitjans, J. Kärger (Eds.), Springer-Verlag, 2005
31. H. Zabel, *Quasielastic Neutron Scattering: a Powerful Tool for Investigating Diffusion in Solids*, in: *Nontraditional Methods in Diffusion*, G.E. Murch, H.K. Birnbaum, J.R. Cost (Eds.), The Metallurgical Society of AIME, Warrendale, 1984, p. 1
32. K. Sköld, *Quasielastic Neutron Scattering Studies of Metal Hydrides*, Ch. 11 in [13]
33. C. Wert, C. Zener, Phys. Rev. **76**, 1169 (1949)
34. C. H Vineyard, J. Phys. Chem. Sol. **3** 121 (1957)
35. H. Wipf, G. Alefeld, Phys. Stat. Sol. (a) **23**, 175 (1974)
36. D. Richter, B. Alefeld, A. Heidemann, N. Wakabayashi, J. Phys. F: Metal Phys. **7**, 569 (1977)
37. K. Yamakawa, M. Ege, B. Ludescher, M. Hirscher, H. Kronmüller, J. of Alloys and Compounds **321**, 17 (2001)
38. M. Hirscher, H. Kronmüller, J. of the Less-common Metals **172**, 658–670 (1991)

19 Diffusion in Dilute Substitutional Alloys

Diffusion in alloys is more complex than self-diffusion in pure metals. In this chapter, we consider dilute substitutional binary alloys of metals A and B with the mole fraction of B atoms much smaller than that of A atoms. Then A is denoted as the *solvent* (or *matrix*) and B is denoted as the *solute*. Diffusion in a dilute alloy has two aspects: solute diffusion and solvent diffusion. We consider at first solute diffusion at infinite dilution. This is often called *impurity diffusion*. Impurity diffusion implies concentrations of the solute less than 1 %. In practice, the sensitivity for detection of radioactive solutes enables one to study diffusion of impurities at very high dilution of less than 1 ppm. In very dilute substitutional alloys solute and solvent diffusion can be analysed in terms of vacancy-atom exchange rates.

19.1 Diffusion of Impurities

Impurity diffusion is a topic of diffusion research to which much scientific work has been devoted. We consider at first 'normal' behaviour of substitutional impurities, which is illustrated for the solvent silver. Similar behaviour is observed for the other noble metals, for hexagonal Zn and Cd, and for Ni.

There are exceptions from 'normal' impurity diffusion. A prominent example is the slow diffusion of transition elements in the trivalent solvent aluminium. Another example is the very fast diffusion of impurities in so-called 'open' metals. Lead is the most famous open metal, for which very fast impurity diffusion has been observed. The rapid diffusion of Au in Pb was discovered by the diffusion pioneer ROBERTS-AUSTEN in 1896 (see Chap. 1).

It is beyond the scope of this book to give a comprehensive overview of impurity diffusion. Instead, we refer the reader to the chapter of LE CLAIRE AND NEUMANN in the data collection edited by the present author [1] and to the review by NEUMANN AND TUIJN [2].

19.1.1 'Normal' Impurity Diffusion

Figure 19.1 shows an Arrhenius diagram for diffusion of many substitutional foreign atoms in a Ag matrix together with self-diffusion of Ag. A comparison

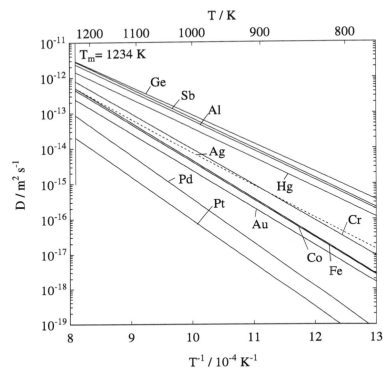

Fig. 19.1. Diffusion of substitutional impurities in Ag and self-diffusion of Ag (*dashed line*). Diffusion parameters from [1, 2]

of impurity diffusion and self-diffusion coefficients, D_2 and D, reveals the following features of 'normal' diffusion of substitutional impurities:

– The diffusivities of impurities, D_2, lie in a relatively narrow band around self-diffusivities, D, of the solvent atoms. The following limits apply in the temperature range between the melting temperature T_m and about $2/3\, T_m$:

$$1/100 \leq D_2/D \leq 100 \,. \tag{19.1}$$

– The pre-exponential factors mostly lie in the interval

$$0.1 < \frac{D^0(solute)}{D^0(self)} < 10 \,. \tag{19.2}$$

– The activation enthalpies of impurity and self-diffusion, ΔH_2 and ΔH, are not much different:

$$0.75 < \frac{\Delta H_2}{\Delta H} < 1.25 \,. \tag{19.3}$$

Substitutional impurities in other fcc metals (Cu, Au, Ni) and in the hcp metals (Zn, Cd) behave similarly as in Ag (for references see the chapter of LE CLAIRE AND NEUMANN in the data collection [1]). Like self-diffusion of the solvent, diffusion of substitutional impurities occurs via the vacancy mechanism. For impurity diffusion in a very dilute alloy it is justified to assume that solute atoms are isolated, i.e. interaction with other solute atoms (formation of solute pairs, triplets, etc.) is negligible. The theory of vacancy-mediated diffusion of substitutional impurities takes into account three aspects:

1. The formation of solute-vacancy pairs: we remind the reader of the LOMER equation introduced in Chap. 5, which shows that the probability of a vacancy occupying a nearest-neighbour site of a substitutional impurity is given by

$$p = C_{1V}^{eq} \exp\left(\frac{G^B}{k_B T}\right) = \exp\left(\frac{S_{1V}^F - S^B}{k_B}\right) \exp\left(-\frac{H_{1V}^F - H^B}{k_B T}\right) \quad (19.4)$$

where $G^B = H^B - TS^B$ is the Gibbs free energy of solute-vacancy interaction, C_{1V}^{eq} the atomic fraction of vacancies in the pure matrix in thermal equilibrium. H_{1V}^F and S_{1V}^F denote the formation enthalpy and entropy of a monovacancy. For $G_B > 0$ the interaction is attractive, for $G_B < 0$ it is repulsive.

2. In contrast to the case of self-diffusion in the pure matrix, for impurity diffusion it is necessary to consider several atom-vacancy exchange rates. Five types of exchanges, between vacancy, impurity and host atoms have been introduced in LIDIARDS 'five-frequency model' (see Chap. 7).

3. The impurity correlation factor is then no longer a constant depending on the lattice geometry as in the case of self-diffusion. It depends on the various jump rates of the vacancy (see Chap. 7).

From the atomistic description developed in Chap. 8, we get for the impurity diffusion coefficient of vacancy-mediated diffusion in cubic Bravais lattices

$$D_2 = f_2 a^2 \omega_2 p = f_2 a^2 \omega_2 C_{1V}^{eq} \exp\left(\frac{G^B}{k_B T}\right), \quad (19.5)$$

where a is the lattice parameter, f_2 the impurity correlation factor, and ω_2 the vacancy-impurity exchange rate.

To be specific, we consider in the following fcc solvents. As discussed in Chap. 7, within the framework of the 'five-frequency model' the correlation factor can be written as [3]

$$f_2 = \frac{\omega_1 + (7/2)F_3\omega_3}{\omega_2 + \omega_1 + (7/2)F_3\omega_3}. \quad (19.6)$$

The various jumps rates in an fcc lattice are illustrated in Fig. 7.4. For convenience we remind the reader of their meaning: ω_1 is the rotation rate of the solute-vacancy complex, w_3 and w_4 denote rates of its dissociation or association. The escape probability F_3 is a function of the ratio w/w_4, where w denotes the vacancy jump rate in the pure matrix. It is also useful to remember that in detailed thermal equilibrium according to

$$\frac{w_4}{w_3} = \exp\left(\frac{G^B}{k_B T}\right) \qquad (19.7)$$

the dissociation and association rates are related to the Gibbs free energy of binding of the vacancy-impurity complex, $G^B = H^B - TS^B$. Equation (19.5) can then be recast to give

$$D_2 = f_2 a^2 \nu^0 \exp\left(\frac{S_{1V}^F - S^B + S_2^M}{k_B}\right) \exp\left(-\frac{H_{1V}^F - H^B + H_2^M}{k_B T}\right), \qquad (19.8)$$

where H^B and S^B denote the binding enthalpy and entropy of the vacancy-impurity complex and H_2^M and S_2^M the enthalpy and entropy of the vacancy-impurity exchange jump w_2. Thus, the activation enthalpy of impurity diffusion is given by

$$\Delta H_2 = H_{1V}^F - H^B + H_2^M + C, \qquad (19.9)$$

where

$$C = -k_B \frac{\partial \ln f_2}{\partial (1/T)}. \qquad (19.10)$$

This term describes the temperature dependence of the impurity correlation factor[1].

It is evident from Eqs. (19.5), (19.7), and Chap. 17 that the ratio of the diffusion coefficients of impurity and self-diffusion can be written as

$$\frac{D_2}{D} = \frac{f_2}{f} \frac{w_2}{w} \frac{w_4}{w_3}. \qquad (19.11)$$

This expression shows that the diffusion coefficient of a substitutional impurity differs from the self-diffusion coefficient of the pure solvent for three reasons, namely: (i) correlation effects, because $f_2 \neq f$, (ii) differences in the atom-vacancy exchange rates between impurity and solvent atoms, because $w_2 \neq w$, and (iii) interaction between impurity and vacancy, because $w_4/w_3 \neq 1$ or $G^B \neq 0$.

Since solute and solvent atoms are located on the same lattice and since the diffusion of both is mediated by vacancies, the rather small diffusivity dispersion (see Fig. 19.1) is not too surprising. It reflects the high efficiency of screening of point charges in some metals, which normally limits the vacancy-impurity interaction enthalpy H^B to values between 0.1 and 0.3 eV. Such values are small relative to the vacancy formation enthalpies (see Chap. 5). Using

---

[1] Sometimes in the literature C is defined with the opposite sign.

$\Delta H = H_{1V}^F + H_{1V}^M$, we get for the difference of the activation enthalpies between impurity and self-diffusion

$$\Delta Q \equiv \Delta H_2 - \Delta H = -H^B + (H_2^M - H_{1V}^M) + C. \qquad (19.12)$$

A useful theoretical approach is the so-called *electrostatic model*, which associates ΔQ with the valence difference between solute and solvent. Relatively positive impurities, generally those of higher nominal valence than the solvent, tend to attract vacancies and, hence, to diffuse more rapidly and with lower activation enthalpies than self-diffusion. In the electrostatic model of LAZARUS [4] refined by LE CLAIRE [5], it is assumed that the excess charge ΔZe (e = electron charge) is responsible for ΔQ. Vacancy and impurity are considered to behave as point charges $-Ze$ and $Z_2 e$, respectively. In the Thomas-Fermi approximation, the excess charge of the impurity gives rise to a perturbing potential

$$V(r) = \alpha \frac{\Delta Ze}{r} \exp(-qr). \qquad (19.13)$$

This equation describes a screened Coulomb potential with a screening radius $1/q$, which is independent of ΔZ and can be calculated from the Fermi energy of the host. α is a dimensionless screening parameter, which depends on ΔZ [5]. In the electrostatic model, the interaction enthalpy H^B is equal to the electrostatic energy $V(d)$ of the vacancy located at a nearest-neighbour distance, d, of the impurity. For the calculation of the differences in the migration enthalpies in the second and third term of Eq. (19.12), LE CLAIRE describes the saddle-point configuration by two 'half-vacancies' located on two neighbouring sites. Each half-vacancy carries a charge $Ze/2$ at a distance $\frac{11}{16}d$ [5].

In a number of solvents, the quantity ΔQ varies indeed monotonically with the valence difference $\Delta Z = Z_2 - Z$ (Z_2= valence of impurity, Z= valence of solvent). Impurities with $Z_2 > Z$ ($Z_2 < Z$) diffuse faster (slower) than self-diffusion, the pertaining activation enthalpies fulfill $\Delta Q < 0$ ($\Delta Q > 0$), and the impurity diffusion coefficients increase with ΔZ. This behaviour is observed for the following solvents: noble metals and group-IIB hexagonal metals Zn and Cd (for references see [1]). Calculations of ΔQ based on the electrostatic model yield good agreement with experiments for impurity diffusion in these metals.

For transition metal solutes in noble metals and for other solvents such as the alkali metals, the divalent magnesium, and the trivalent aluminium the values of ΔQ calculated on the basis of a screened Coulomb potential are at variance with the experimentel values [5, 6]. There are several reasons for the failure of the electrostatic model:

(i) The choice of a screened Coulomb potential may not be appropriate for certain solute-solvent combinations. A self-consistent potential has an oscillatory form with so-called Friedel oscillations. The vacancy-solute

interaction can then, for example, be attractive at the nearest neighbour positions but repulsive at the saddle-point position.
(ii) A model based only on the difference in valence between solute and solvent atoms disregards the effects of atomic size.
(iii) Finally, the electrostatic model can only predict ΔQ. It says nothing about pre-exponential factors.

Various attempts have been made in the literature to improve the theory of impurity diffusion. For details the reader may consult older textbooks [6, 8] and the review of NEUMANN AND TUIJN [2].

19.1.2 Impurity Diffusion in Al

Aluminum like the noble metals has also an fcc structure. In contrast to the monovalent noble metals, Al is trivalent. The diffusion of impurities in Al is remarkably different from that of impurities in noble metals. Figure 19.2 shows an Arrhenius diagram of impurity diffusion in Al together with Al self-diffusion. A striking feature of this figure is that most transition metals have extremely low diffusivities with respect to Al self-diffusion. In addition, they have high activation enthalpies, which are at variance with the simple form of the electrostatic model mentioned above. They also have very high pre-exponential factors, which violate the limits of Eq. (19.2). Non-transition elements have diffusivities similar or slightly higher than self-diffusion and show only small diffusivity dispersion. Their activation enthalpies are similar to those of Al self-diffusion, almost independent of their valence.

The pressure dependence for diffusion of several impurities in aluminium has been studied by RUMMEL ET AL. [10] and THÜRER ET AL. [11] (see also Chap. 8). The activation volumes of non-transition element diffusers (Zn, Ge) are close to one atomic volume (Ω) and not much different from the activation volume of self-diffusion reported by BEYELER AND ADDA [12]. However, the transition elements (Co, Mn) are diffusers with high activation volumes between 2.7 and 1.67 Ω. These findings have been attributed to differences in vacancy-impurity interaction between transition and non-transition elements [10].

The large ΔH_2 values of transition element solutes according to Eq. (19.9) indicate a strong repulsion between solute and vacancy ($H^B < 0$) and/or a large activation enthalpy H_2^M for solute-vacancy exchange jumps. Ab initio calculations, based on the local density functional theory, of the solute-vacancy interaction energy for 3d and 4sp and for 4d and 5sp solutes in aluminium have been performed by HOSHINO ET AL. [13]. They have demonstrated that for 3d and 4d impurities this interaction is indeed strongly repulsive, whereas for 4sp and 5sp impurities it is weakly attractive. Unfortunately, according to the author's knowledge *ab initio* calculations of H_2^M and of the activation volumes are not available.

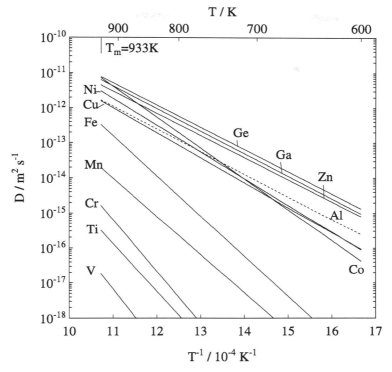

Fig. 19.2. Diffusion of several impurities in Al and self-diffusion of Al (*dashed line*) according to [1, 9]

19.2 Impurity Diffusion in 'Open' Metals – Dissociative Mechanism

The fast diffusion of gold in lead was already reported by ROBERTS-AUSTEN [14], one of the pioneers of solid-state diffusion (see Chap. 1). After the measurements of self- and impurity diffusion in lead by VON HEVESY and his coworkers [15, 16], the extreme rapidity of gold diffusion was appreciated. Since this pioneerig work the interest in the Au/Pb and analogous systems continues to the present day. Lead is still the most extensively studied metallic solvent where fast impurity diffusion plays an important role.

Figure 19.3 shows an Arrhenius diagram of impurities in Pb together with self-diffusion (for references see the chapter of LE CLAIRE AND NEUMANN in [1] and the review of NEUMANN AND TUIJN [2]). Some solutes (e.g., Tl, Sn) show 'normal' behaviour. However, noble metals, Ni-group elements, and Zn have diffusivities which are three or more orders of magnitude faster than self-diffusion. Fast diffusion of 3d transition metals is also observed in some other polyvalent metals (In, Sn, Sb, Ti, Zr, Hf), and for noble metal diffusion

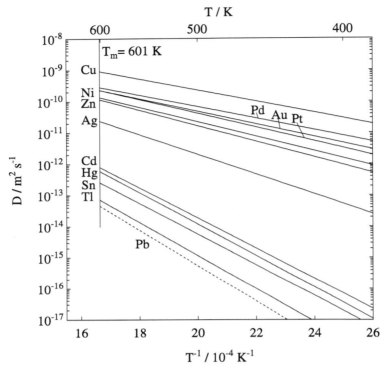

Fig. 19.3. Diffusion of impurities in Pb and self-diffusion of Pb (*dashed line*) according to [1, 2]

in Na. Figure 19.4 shows diffusion of several impurities in Na according to BARR AND ASSOCIATES [18–20]. Noble metal solutes are fast diffusers in the group-IVB metal Sn and in the group-IIIB metals In and Tl. Late transition elements in group-IVA metals (α-Ti, α-Zr, and α-Hf) are fast diffusers as well. Solvents that permit fast diffusion are sometimes denoted as 'open' metals [17]. 'Open' refers to the large ratio of atomic radii of solvent and solute. This solvent property appears to entail fast diffusion for solutes with relatively small radii.

Fast solute diffusion in open metals is thought to be exceptional, because the diffusivities lie far above the range expected for the vacancy mechanism. On the other hand, the still relatively large atom size ratio (e.g., $R_{Au}/R_{Pb} \approx$ 0.83) would seem to preclude large interstitial occupancies. Fast diffusion of solutes that are mainly dissolved on substitutional sites has been attributed to the dissociative mechanism (see, e.g., WARBURTON AND TURNBULL [21]). This mechanism operates for solutes which have a certain (small) interstitial component with a high mobility in the solvent metal (see Chap. 6). Such solutes are called hybrid solutes. As represented by equation

19.2 Impurity Diffusion in 'Open' Metals – Dissociative Mechanism

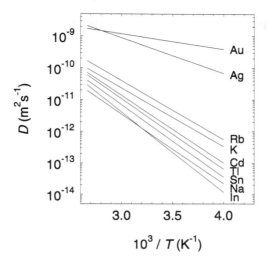

Fig. 19.4. Diffusion of impurities in Na and self-diffusion of Na according to BARR AND ASSOCIATES [18–20]

$$B_i + V \rightleftharpoons B_s \qquad (19.14)$$

B solutes are distributed over both interstitial (B_i) and substitutional (B_s) sites. Provided that local equilibrium in Eq. (19.14) is established, the concentrations fulfill the law of mass action

$$\frac{C_i C_V}{C_s} = K(T) = \frac{C_i^{eq} C_V^{eq}}{C_s^{eq}} \qquad (19.15)$$

where C_i, C_s, and C_V denote the molar fractions of interstitial solute, substitutional solute, and vacancies, respectively. $K(T)$ is a constant which depends on temperature. The superscript *eq* indicates thermal equilibrium. A metal with a normal content of dislocations has a sufficient abundance of vacancy sources or sinks to keep vacancies in equilibrium. The effective diffusivity of a solute is then given by

$$D_{eff} = \frac{D_i C_i^{eq}}{C_i^{eq} + C_s^{eq}} + \frac{D_V C_V^{eq}}{C_i^{eq} + C_s^{eq}}, \qquad (19.16)$$

where D_i denotes the diffusivity of the solute in its interstitial state and D_V the diffusivity of the vacancy-solute complex. Substitutional transport by vacancies is negligible if $C_i^{eq} D_i \gg D_V C_V^{eq}$ is fulfilled. Then, the effective diffusivity of solutes is given by [21, 22]

$$D_{eff} \approx \frac{D_i C_i^{eq}}{C_i^{eq} + C_s^{eq}} \qquad (19.17)$$

For solutes with dominating interstitial solubility, i.e. for $C_i^{eq} \gg C_s^{eq}$, the effective diffusivity approaches the diffusivity of the solute in its interstitial state:

$$D_{eff} \approx D_i. \tag{19.18}$$

This corresponds to the direct interstitial mechanism, which is relevant for the interstitial diffusers C, N and O (see Sect. 18.1). For solutes with dominating substitutional solubility, i.e. for $C_s^{eq} \gg C_i^{eq}$, the effective diffusivity

$$D_{eff} \approx \frac{D_i C_i^{eq}}{C_s^{eq}} \tag{19.19}$$

contains the factor

$$\frac{C_i^{eq}}{C_s^{eq}} = \exp\left(\frac{-G_i}{k_\mathrm{B} T}\right). \tag{19.20}$$

Here G_i denotes the Gibbs free energy of forming an interstitial solute from a substitutional one. The rather wide diffusivity dispersion of fast solute diffusers can be largely attributed to this factor.

The phenomenon of fast diffusion is also well-known for hybrid solutes in the semiconducting elements Si and Ge (see, e.g., [23, 24]) and is discussed in detail in Chap. 25. It has been attributed to interstitial-substitutional exchange mechanisms, which occur in two basic versions (see Chaps. 6 and 25). The dissociative mechanism was suggested in 1956 by FRANK AND TURNBULL for Cu in Ge [25]. The kick-out mechanism, which involves self-interstitials, was originally proposed by GÖSELE, FRANK AND SEEGER for Au diffusion in Si [26]. In the meantime, further fast diffusing foreign elements in Si (e.g., Pt, Zn) were identified as kick-out diffusers [22, 28].

From a chemical viewpoint the elemental semiconductors Si and Ge are group-IV elements such as the 'open' metals Pb and Sn. Actually, the concepts growing out of studies of fast diffusion in semiconductors (see part IV of this book) have influenced the interpretation of fast diffusion in metals.

19.3 Solute Diffusion and Solvent Diffusion in Alloys

Alloying one element with another can significantly affect diffusion. In a homogeneous binary A-B alloy a tracer self-diffusion coefficient for A atoms, D_A^*, and another one for B atoms, D_B^*, can be measured. In this section, we discuss the variation of solute and solvent diffusion with the solute concentration. The major effects of alloying on diffusion are to introduce vacancy-solute and solute-solute interactions and to change the vacancy jump rates.

Let us consider, for example, the influence of solute atoms on solvent diffusion: in a very dilute alloy most of the solvent atoms are not near a solute atom; the mobility of such solvent atoms agrees with that of atoms in the pure solvent. Depending on the interaction energy, a solute attracts or repels vacancies and the jump rates of the solvent atoms in the vicinity of the solute atom are altered. This influences the solvent diffusion coefficient. As the solute concentration increases, the number of affected solvent atoms increases. Depending on the energy landscape experienced by the vacancy in the vicinity of the solute, certain solutes accelerate solvent diffusion and others slow it down.

In dilute alloys containing small atomic fractions X_B of the solute B the *self-diffusion coefficient of the solvent* can be written as

$$D_A^*(X_B) = D_A^*(0) \exp(bX_B), \qquad (19.21)$$

which for $X_B < 1\%$ can be approximated by

$$D_A^*(X_B) = D_A^*(0)[1 + b_1 X_B + \ldots]. \qquad (19.22)$$

The term $b_1 X_B$ accounts for the change in the jump rates of those solvent atoms neighbouring solute atoms. Quadratic terms in X_B correspond to the effect of pairs of B atoms. b_1 is sometimes denoted as the *linear enhancement factor*.

An analogous expression holds for the *solute diffusion coefficient*:

$$D_B^*(X_B) = D_B^*(0)[1 + B_1 X_B + \ldots]. \qquad (19.23)$$

$D_B^*(0)$ is the solute diffusion coefficient for infinite dilution, i.e. the impurity diffusion coefficient discussed in the previous section. The term B_1 expresses the influence of solute-solute pairs.

The *linear enhancement factor of solvent diffusion*, b_1, has attracted particular interest. It is related to the various jump rates in the neighbourhood of the solute. Using the five-frequency model for fcc solvents, LIDIARD [30] and HOWARD AND MANNING [31] have deduced the following expression

$$\begin{aligned}b_1 &= -18 + \frac{\omega_4}{\omega}\left[4\frac{\chi_1}{f}\frac{\omega_1}{\omega_3} + 14\frac{\chi_2}{f}\right] \\ &= -18 + \left[4\frac{\chi_1}{f}\frac{\omega_1}{\omega} + 14\frac{\chi_2}{f}\frac{\omega_3}{\omega}\right]\exp\left(\frac{G_B}{k_b T}\right).\end{aligned} \qquad (19.24)$$

Here f is the tracer correlation factor of self-diffusion in the pure solvent (only ω-jumps). The quantities χ_1 and χ_2 are partial correlation factors tabulated in [31]. Equation (19.24) is based on the assumption that the jump rates are independent of the solute concentration. It shows that an increase of solvent diffusivity can arise from an increase of the vacancy concentration due to vacancy-solute interaction and/or from an increase in the solvent jump frequencies near the solute (e.g., for $\omega_1 > \omega$).

Using Eqs. (19.5), (19.6), and (19.24), b_1 can be expressed as a function of the ratio $D_B^*(0)/D_A^*(0)$:

$$b_1 = -18 + 4\frac{D_B^*(0)}{D_A^*(0)}\frac{f}{1-f_2}\frac{4\chi_1(\omega_1/\omega_3) + 14\chi_2}{f(4\omega_1/\omega + 14F_3)}. \qquad (19.25)$$

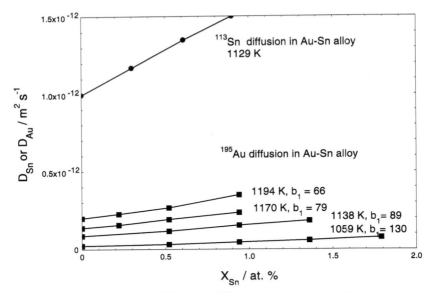

Fig. 19.5. Tracer diffusion of ^{113}Sn and ^{195}Au in dilute Au-Sn solid-solution alloys according to HERZIG AND HEUMANN [29]

This expression shows that faster diffusing solutes $(D_B^* > D_A^*)$ have the tendency to increase the solvent diffusivity. This effect is mainly due to an increase of the vacancy content. The last quotient in Eq. (19.25) has usually a value not much differnet from unity. Thus, with values of D_B^*/D_A^* around 10 and $f_2 \approx 0.5$, b_1 lies between 0 and 100. For slow diffusing solutes, b_1 tends to be negative but never less than -18.

We illustrate the influence of solute concentration on solvent and solute diffusion in a dilute fcc solid solution by an experimental example. The diffusion of the radiotracers ^{113}Sn and ^{195}Au in pure Au and in a dilute Au-Sn alloy is displayed for various temperatures and Sn concentrations in Fig. 19.5. In pure Au, the Sn solute is about one order of magnitude faster than Au self-diffusion. Au-diffusion in the alloy is enhanced by the presence of Sn additions [29]. The corresponding enhancement factors are listed in Fig. 19.5. As is the case in this figure, the experimental values of solvent and solute enhancement factors have often the same sign.

We refrain from discussing further details and refer the reader to the textbooks of PHILIBERT [6] and HEUMANN [8] and a more recent paper by BELOVA AND MURCH [32] and the references therein. .

References

1. A.D. Le Claire, G. Neumann, *Diffusion of Impurities in Solid Metallic Elements*, Chap. 3 in: *Diffusion in Solid Metals and Alloys*, H. Mehrer (Vol. Ed.),

References

1. Landolt-Börnstein, Numerical Data and Functional Relationships in Science and Technology, New Series Vol. III/26, Springer-Verlag, (1990)
2. G. Neumann, C. Tuijn, *Impurity Diffusion in Metals*, Scitec Publications Ltd., Uetikon-Zürich, Switzerland, 2002
3. J.R. Manning, *Diffusion Kinetics for Atoms in Crystals*, van Nostrand Comp., Princeton, 1968
4. D. Lazarus, Phys. Rev. **93**, 973 (1954)
5. A.D. Le Claire, Philos. Mag. **7**, 141 (1962)
6. J. Philibert, *Atom Movements – Diffusion and Mass Transport in Solids*, Les Editions de Physique, Les Ulis, 1991
7. R.M. Cotts, Ber. Bunsenges. Phys. Chem. **76**, 760 (1972)
8. Th. Heumann, *Diffusion in Metallen*, Springer-Verlag, Berlin, 1992
9. G. Rummel, Th. Zumkley, M. Eggersmann, K. Freitag, H. Mehrer, Z. Metallkd. **86**, 121 (1995)
10. G. Rummel, Th. Zumkley, M. Eggersmann, K. Freitag, H. Mehrer, Z. Metallkd. **86**, 131 (1995)
11. A. Thürer, G. Rummel, Th. Zumkley, K. Freitag, H. Mehrer, Phys. Stat. Sol. (a) **143**, 535 (1995)
12. M. Beyeler, Y. Adda, J. Phys. (Paris) **29**, 345 (1968)
13. T. Hoshino, R. Zeller, P.H. Dederichs, Phys. Rev. B **53**, 8971 (1996)
14. W.C. Roberts-Austen, Phil. Trans. Roy. Soc. **187**, 383 (1896)
15. J. Groh, G. von Hevesy, Ann. Physik **63**, 85–92 (1920)
16. J. Groh, G. von Hevesy, Ann. Physik **65**, 216 (1921)
17. G.M. Hood, Defect and Diffusion Forum **95–98**, 755 (1993)
18. L.W. Barr, J.N. Mundy, F.A. Smith, Philos. Mag. **13**, 1299 (1966)
19. L.W. Barr, J.N. Mundy, F.A. Smith, Philos. Mag. **20**, 390 (1969)
20. L.W. Barr, F.A. Smith, in: *DIMETA-62, Diffusion in Metals and Alloys*, F.J. Kedves, D.L. Beke (Eds.), Trans Tech Publications, Switzerland, 1983, p. 365
21. W.K. Warburton, D. Turnbull, in: *Diffusion in Solids – Recent Developments*, A.S. Nowick, J.J. Burton (Eds.), Academic Press, 1975, p. 171
22. W. Frank, U. Gösele, H. Mehrer, A. Seeger, *Diffusion in Silicon and Germanium*, in: *Diffusion in Crystalline Solids*, G.E. Murch, A.S. Nowick (Eds.), Academic Press, 1984, p. 63
23. A. Seeger, K.P. Chik, Phys. Stat. Sol. **29**, 455 (1968)
24. N.A. Stolwijk, H. Bracht, in: *Diffusion in Semiconductors and Non-Metallic Solids*, D.L. Beke (Vol.-Ed.), Landolt-Börnstein, Numerical Data and Functional Relationships in Science and Technology, New Series, Group III: Condensed Matter, Vol. 33, Springer-Verlag, 1998, p. 2–1
25. F.C. Frank, D. Turnbull, Phys. Rev. **104**, 617 (1956)
26. U. Gösele, W. Frank, A. Seeger, Appl. Phys. A **23**, 361 (1980)
27. N.A. Stolwijk, G. Bösker, J. Pöpping, Defect and Diffusion Forum **194–199**, 687 (2001)
28. H. Bracht, N.A. Stolwijk, H. Mehrer, Phys. Rev. B. **52**, 16542 (1995)
29. Chr. Herzig, Th. Heumann, Z. für Naturforschung **27a**, 1109 (1972)
30. A.B. Lidiard, Philos. Mag. **5**, 1171 (1960)
31. R.E. Howard, J.R. Manning, Phys. Rev. **154**, 5612 (1967)
32. I.V. Belova, G.E. Murch, Philos. Mag. **33**, 377 (2003)

20 Diffusion in Binary Intermetallics

20.1 General Remarks

Intermetallics are compounds of metals or of metals and semimetals. The crystal structures of intermertallics are different from those of their constituents. This definition includes intermetallic phases and ordered alloys. Intermetallics have a wide spectrum of properties ranging between metallic and non-metallic. Some intermetallics are interesting functional materials others have attracted attention as structural materials for high-temperature applications. For an introduction to the field see, e.g., the books of WESTBROOK AND FLEISCHER [1] and of SAUTHOFF [2].

Intermetallic-based materials have been used for a long time, based essentially on their chemical properties, on their high hardness and wear resistance. Examples are some dental alloys, amalgams, jewelry, and coatings. There are a large number of engineering alloys that find applications because of some special physical or chemical properties – for example, magnetic behaviour, superconductivity, or chemical stability in corroding atmospheres. The last few decades have seen tremendous world-wide efforts in intermetallics, which largely focused on aluminides with some smaller efforts on silicides. The prime goals are high-temperature or power-generation applications, such as aeroengines, high-temperature edges on aircraft wings and rocket fins, automobile engine valves, turbochargers and so on. Important properties in all these applications are the ability to withstand high temperature and aggressive oxidising or corrosive environments, as well as other properties such as low weight or inertia. This has triggered important materials developments and a flood of conferences. It is impossible to make reference to more than a few of the books and review papers that cover this effort.

An atomistic understanding of diffusion in intermetallics in terms of defect structure and diffusion mechanisms is more complex than for metallic elements. An inspection of a collection of phase diagrams [3] shows the following: intermetallics crystallise in a variety of structures with ordered atom distributions. Examples are the B2, $D0_3$, $L1_2$, $L1_0$, $D0_{19}$, B20, $C11_b$, and Laves-phase structures. Some intermetallics are ordered up to the melting temperature, others undergo order-disorder transitions because entropy favours a less ordered or even a random arrangement of atoms at high temperatures. We know intermetallic phases with wide phase fields and others which exist as

342   20 Diffusion in Binary Intermetallics

line compounds. Some intermetallics occur for certain stoichiometric compositions, others are observed for off-stoichiometric compositions only. Some phases compensate off-stoichiometry by vacancies, others by antisite atoms. Some phases display small, others large differences between the diffusivities of the components. Some non-cubic intermetallics are highly anisotropic in their diffusion behaviour.

Let us first illustrate structures of some of the more common intermetallics (see Figs. 20.1 and 20.2):

The **B2 (or CsCl) structure** is cubic with the approximate composition AB. The B2 structure can be derived from the bcc lattice, if the two primitive cubic sublattices are occupied by different kinds of atoms. Examples are FeAl, CoAl, NiAl, CoGa, PdIn, CuZn, AuZn, AuCd, and in an intermediate temperature range NiMn.

The **D0$_3$ (or Fe$_3$Si) structure** is cubic with the approximate composition A$_3$B. The D0$_3$ structure can be considered also as an ordered structure derived from the bcc lattice. In Fig. 20.1, A atoms occupy white and grey sites, B atoms occupy black sites. D0$_3$ order is observed, for example, in Fe$_3$Si

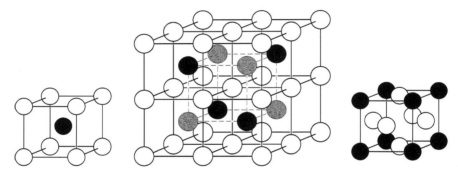

Fig. 20.1. Ideally ordered structures of some cubic intermetallics: B2 (*left*), D0$_3$ (*middle*), L1$_2$ (*right*)

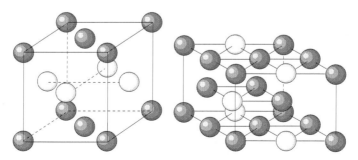

Fig. 20.2. Ideally ordered structures of titanium aluminides: L1$_0$ (*left*), D0$_{19}$ (*right*)

over a wide temperature range, in Fe_3Al below about 825 K, and in Cu_3Sn, Cu_3Sb, and Ni_3Sb at high temperatures.

The **$L1_2$ (or Cu_3Au) structure** is cubic with the approximate composition A_3B. The $L1_2$ structure is an ordered structure derived from the fcc lattice. In Fig. 20.1, A atoms occupy white sites, B atoms occupy black sites. Examples are the Ni-based compounds Ni_3Al, Ni_3Ga, and Ni_3Ge. The prototype Cu_3Au shows $L1_2$ order only at low temperatures; above 660 K it is disordered fcc.

The **$L1_0$ (or CuAu) structure** is tetragonal with the approximate composition AB. The $L1_0$ structure can be considered as an ordered fcc structure with A and B atoms occupying sequential (001)-planes. The structure has a slight tetragonal distortion in the $\langle 001 \rangle$-direction. An example of technologically interest for structural applications is β-TiAl. Other $L1_0$ systems such as NiPt, FePt, CoPt, FePd and NiMn are studied because they are candidtaes for high-density magnetic storage media due to their high magnetic anisotropy related to their tetragonal anisotropy.

The **$D0_{19}$ structure** is hexagonal with the approximate composition A_3B. It can be considered as an ordered structure derived from the hcp lattice. In Fig. 20.2, the majority atoms occupy white sites, B atoms occupy black sites. Ti_3Al crystallises in this structure.

The diffusion behaviour of intermetallics is of interest for the production of these materials and for their practical use, especially when high-temperature applications are intended. Whereas self- and solute diffusion in pure metals is thoroughly investigated and reasonably well understood, systematic diffusion studies for intermetallics are still relatively scarce, although considerable progress has been achieved [4–13].

Just as in metallic elements, self-diffusion is the most basic diffusion process in alloys and compounds as well. Studies of self-diffusion utilise such tiny amounts of tracer atoms (see Sect. 13.3) of the diffusing species. Then, in practice the chemical composition of the sample does *not* change due to diffusion. In a binary system, two tracer self-diffusion coefficients – one for A atoms and another one for B atoms – can be determined. Such complete studies have been performed only for a limited number of binary intermetallics. Table 20.1 compiles binary intermetallics with B2, $D0_3$, $L1_2$, $L1_0$, $D0_{19}$, C15, and $C11_b$ structures for which self-diffusion data are available. In some cases self-diffusion of both components has indeed been studied. For the aluminides of Ni, Fe, and Ti impurity diffusion of some solutes has been studied, since for Al no suitable tracer is available. The intention was to mimic self-diffusion. Similarly, for Fe_3Si a radioisotope of Ge was used to simulate Si diffusion, since Si has only a very short-lived isotope with a 2.6 hours half-life. The validity of this approach has been confirmed by diffusion studies with an enriched stable Si isotope and SIMS profiling. For the $C11_b$ structured $MoSi_2$ diffusion of the short-lived ^{31}Si isotope and of the chemically related ^{71}Ge as well as diffusion of ^{99}Mo has been investigated.

Table 20.1. Self-diffusion in several intermetallics. For the underlined elements data of tracer self-diffusion and of substitutional substitutes (in brackets) are available. The latter data provide 'good estimates' for self-diffusion of the Al and Si components. For references see, e.g., [7, 8, 11] and this chapter

Structure	Intermetallic
B2	CuZn, AuCd, AuZn, CoGa, PdIn, FeCo, NiAl(Ga), FeAl(Zn,In), AgMg, NiGa, NiMn
$L1_2$	Ni_3Al, Ni_3Ge, Ni_3Ga, Co_3Ti, Pt_3Mn, Cu_3Au (disordered)
$D0_3$	Fe_3Si(Ge), Cu_3Sn, Cu_3Sb, Ni_3Sb, Fe_3Al
$L1_0$	TiAl(Ga, Ge), FePt
$D0_{19}$	Ti_3Al(Ga,In,Ge)
C15	Co_2Nb (cubic Laves phase)
$C11_b$	$MoSi_2$ (Ge), both principal tetragonal directions

20.2 Influence of Order-Disorder Transitions

Order-disorder alloys are characterised by an ordered arrangement of atoms at low temperatures, which becomes progressively disordered as the temperature is raised until the long-range order disappears at a critical temperature. Ordering implies the arrangement of atoms on distinct sublattices of an intermetallic. Disordering occurs by an exchange of atoms, i.e. by the creation of antisite atoms. The proportion of antisite atoms rises rapidly as temperature approaches the critical temperature, where the distinction between the sublattices disappears. Ordering transitions are accompanied by anomalies in many physical properties including diffusion.

An order-disorder transition occurs, for example, between the β- and β'-brass phases of the Cu-Zn system. Below the order-disorder transition (at about 741 K) CuZn shows B2 order (β'-brass). At high temperatures the disordered A2 structure (β-brass) is formed. The pioneering but still valid work of KUPER ET AL. [17] on self-diffusion of ^{64}Cu and ^{65}Zn in CuZn is displayed in Fig. 20.3. The influence of the order-disorder transition on the diffusion behaviour of both components is visible as a change in slope of the Arrhenius plot. The activation enthalpies of both components obey the inequality

$$\Delta H_{B2} > \Delta H_{A2}. \tag{20.1}$$

Figure 20.3 also shows that occurrence of order impedes the diffusion of both components in a similar way.

The Fe-Co system undergoes an order-disorder transition from a disordered A2 high-temperature phase to an ordered B2 phase at lower temperatures. At the equiatomic composition the transformation temperature is 1003 K. Similar effects of the B2-A2 transition as in the CuZn system were observed for diffusion of Fe and Co in FeCo by FISHMAN ET AL.[18, 19]. The measurements of FISHMAN ET AL. and at diffusivities of about $10^{-19}\,\mathrm{m^2\,s^{-1}}$.

20.2 Influence of Order-Disorder Transitions

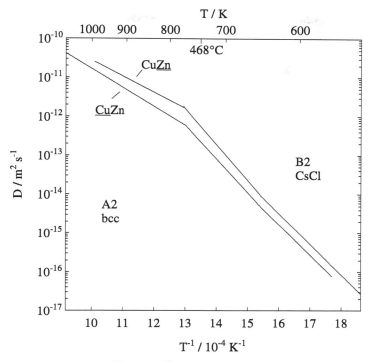

Fig. 20.3. Self-diffusion of ^{64}Cu and ^{65}Zn in CuZn according to KUPER ET AL. [17]

Diffusion studies have been extended by IIJIMA AND COWORKERS [20, 21] down to 2×10^{-21} m^2 s^{-1}. The ratio D_{Fe}/D_{Co} exeeds unity by not more than 20 % above about 940 K corresponding to diffusivities around 10^{-20} m^2 s^{-1}. This ratio becomes slightly smaller than unity in the B2 phase.

The diffusion of Fe in Fe$_3$Al has been studied over a wide temperature range by TÖKEI ET AL. [24] and by EGGERSMANN AND MEHRER [23]. Fe$_3$Al undergoes two ordering transitions at about 825 K from D0$_3$ order to B2 order and to the disordered A2 structure near 1000 K. At the critical temperatures the slope of the Arrhenius diagram changes slightly and the activation enthalpies obey the following sequence

$$\Delta H_{D0_3} > \Delta H_{B2} > \Delta H_{A2} . \tag{20.2}$$

The activation enthalpy is highest for the structure with the highest degree of order and lowest for the disordered structure. The effect of ordering transitions is, however, less pronounced in Fe$_3$Al than in CuZn.

20.3 B2 Intermetallics

Intermetallics with B2 structure form one of the largest group of intermetallics with a broad variation of chemical, physical, and mechanical properties. The classical B2 phase is CuZn (β-brass). It is a prototype of B2 phases consisting of noble metals and group-II metals. The B2 phase NiAl has attracted much interest for its potential in high-temperature applications. The B2 aluminides of Fe and Co are closely related to NiAl since Ni, Fe, and Co can substitute for each other in the B2 structure. The B2 structure is also adopted by compounds of group-VIII and group-IV transition metals.

Much of the diffusion work on intermetallics concerns B2 phases (see Table 20.1). Figure 20.3 reveals a fairly general feature of self-diffusion in B2 intermetallics. We recognise that Zn in β-brass diffuses only slightly faster than Cu and that the ratio D_{Zn}/D_{Cu} never exceeds 2.3 [17]. For equiatomic FeCo, the ratio D_{Fe}/D_{Co} is always close to unity [18–21]. This type of 'cou-

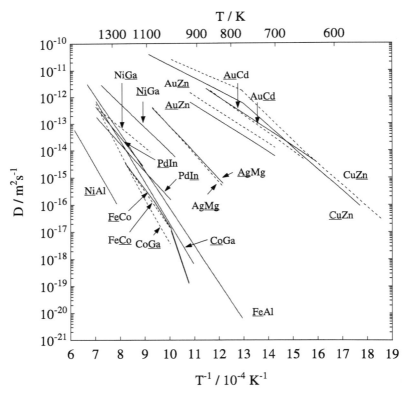

Fig. 20.4. Self-diffusion in B2 structure intermetallics from [7]. The diffusing species is *underlined*

pling' between the diffusivities of the components seems to be typical for B2 phases. It can be observed in Fig. 20.4 for practically all B2 compounds, for which both constituents have been investigated. In some cases (e.g., NiGa, CoGa) the bounds for the diffusivity ratio are somewhat wider than in the cases of CuZn and FeCo. However, in any case the difference between the diffusivities is less than one order of magnitude. This 'coupling' between the diffusivities of the components indicates that diffusion of both atomic species is mediated by the same defect(s).

20.3.1 Diffusion Mechanisms in B2 Phases

It is generally recognised that diffusion in B2 phases is mediated by vacancy-type defects. An intriguing question is, how do these defects move in a structure with long-range order? The B2 structure consists of two primitive cubic sublattices (see Fig. 20.1). In the completely ordered state of a stoichiometric B2 compound, A atoms occupy one sublattice and B atoms the other. This implies that each A atom is surrounded by 8 B atoms on nearest-neighbour sites and vice versa. If A and B atoms are distributed at random, a body-centered cubic (bcc) structure (A2 structure) is obtained. When atomic diffusion in a B2 compound would take place by random interchanges between vacancy-type defects and atoms of both sublattices, migrating defects would leave trails of antisite defects (A_B and B_A) behind and finally cause complete disordering of the compound. Vacancy-type defects should, therefore, move in a highly correlated way to preserve the equilibrium degree of long-range order. In order to maintain order, disordered regions must either be avoided or compensated during the diffusion process.

If the ordering energy is very high, sublattice diffusion of each component via second nearest-neighbour jumps is conceivable. Diffusion of the components in ionic crystals and in simple oxides occurs by sublattice diffusion (see Chap. 26 and [14]). However, sublattice diffusion cannot be the dominating mechanism in B2 intermetallics since separate diffusion on each sublattice diffusion hardly would entail similar diffusivities of the components.

Ingenious order-retaining mechanisms, for which diffusion of both components is coupled, have been proposed for B2 intermetallics:

- **Six-jump-cycle (6JC) mechanism:** This mechanism is an order-retaining vacancy mechanism, which was originally proposed by ELCOCK AND MCCOMBIE [25, 26] and advocated later by HUNTINGTON [27] and others [28, 31]. A vacancy trajectory of 6 consecutive nearest-neighbour jumps displaces atoms in such a way that after the cycle is completed the order is re-established. Several possibilities for six-jump-vacancy-cycles are illustrated in Fig. 20.5.
  Because of the participation of both species in the 6JC with a fixed ratio of the number of jumps the diffusivities of both components cannot be very different. In a highly ordered state, the ratio of the diffusivities of

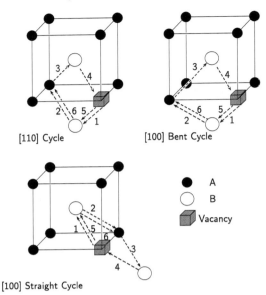

Fig. 20.5. Schematic illustration of six-jump vacancy cycles in the B2 structure. The *arrows* show vacancy jumps; the *numbers* indicate the jump sequence

both components, D_A/D_B, lies within the following fairly narrow limits:

$$\frac{1}{q} < \frac{D_A}{D_B} < q. \tag{20.3}$$

q was first estimated to be 2 [29, 5] and by including correlation effects was later slightly corrected to $q = 2.034$ [30]. The upper (lower) limit is attained when vacancies are preferentially formed on the B (A) sublattice

As the chemical composition deviates from the stoichiometric one and as disorder increases at high temperatures, antisite atoms appear. As shown by BELOVA AND MURCH [32], interaction of the six-jump-cycles with antisite atoms remarkable widens the limits of Eq. (20.3). Thus, in a B2 alloy with some disorder values of D_A/D_B beyond the limits of Eq. (20.3) cannot be considered as an indication that the 6JC mechanism does not operate.

- **Triple-defect mechanism:** In a B2 compound triple-defect disorder can occur according to the reaction

$$V_A + V_B \rightleftharpoons 2V_A + A_B. \tag{20.4}$$

V_A (V_B) denotes a vacancy on the A (B) sublattice and A_B an A atom on the B sublattice (see also Chap. 5). Triple-defect disorder does not change the composition. Instead of forming equal numbers of vacancies

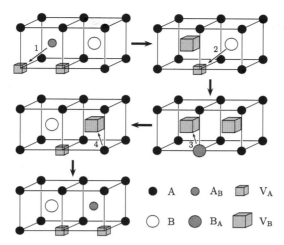

Fig. 20.6. Illustration of the triple-defect diffusion mechanism in the B2 structure. The *arrows* show vacancy jumps; the *numbers* indicate the jump sequence

on both sublattices, two vacancies on one sublattice and an antisite atom on the other sublattice can appear. Triple-defect formation according to Eq. (20.4) is favoured in intermetallics with high formation enthalpies of V_B vacancies.

Vacancies and antisite defects can associate to form bound triple defects (see also Chap. 5). A triple-defect mechanism involving bound triple defects was proposed by STOLWIJK ET AL. [33] for the B2 compound CoGa. The triple-defect mechanism in CoGa was attributed to two nearest-neighbour jumps of Co atoms and to a next-nearest neighbour jumps of Ga atoms. Detailed calculations for NiAl predict that an Al atom performs two nearest-neighbour jumps instead of one second-nearest neighbour jump [34]. Figure 20.6 shows the triple-defect mechanism with this

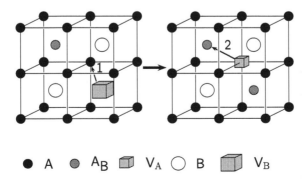

Fig. 20.7. Illustration of the antistructural-bridge (ASB) mechanism. The *arrows* show vacancy jumps; the *numbers* indicate the jump sequence

modification. The ratio of the diffusivities for this mechanism lies within the following limits [35]:

$$1/13.3 < D_A/D_B < 13.3 \, . \qquad (20.5)$$

The triple-defect mechanism is closely related to the vacancy-pair mechanism (see below). The configurations which appear after jumps 1 and 3 of Fig. 20.6 are nearest-neighbour vacancy pairs.

- **Antistructural-bridge (ASB) mechanism:** This mechanism was proposed by KAO AND CHANG [38] and is illustrated in Fig. 20.7. As a result of the two jumps indicated, the vacancy and the antisite atom exchange their position. For a B2 phase with some substitutional disorder, antisite defects can act as 'bridges' to establish low energy sequences for vacancy jumps.

  It is important to note that the ASB mechanism has a percolation threshold. Long-range diffusion via the ASB mechanism requires a sufficient concentration of antistructure atoms to reach the percolation threshold. A relatively high threshold was estimated from purely geometrical arguments [38]. Monte Carlo simulations yielded lower values for the percolation threshold of about 6 % [36, 37]. Such antistructure atom concentrations can indeed occur in B2 intermetallics with a wide phase field like NiAl (see below).

- **Vacancy-pair mechanism:** A bound pair of vacancies, i.e. a vacancy in one sublattice and a scond vacancy on a neighbour site of the other sublattice, can mediate diffusion of both components by successive correlated next-nearest-neighbour jumps. Whereas this mechanism has some rele-

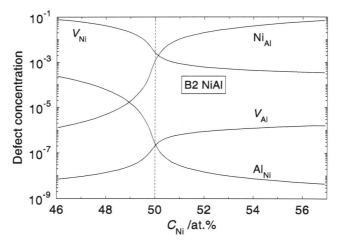

Fig. 20.8. Defect site fractions in B2 NiAl as a function of composition at 0.75 T_m from [39]

vance for ionic crystals such as alkali halides (see Chap.26), it is unlikely for B2 intermetallics.

It seems that in those B2 compounds, which are composed of a group VIIIB metal (Co, Fe, Ni, Pd, etc.) and a group IIIA metal (Al, Ga, In, etc.), the triple-defect mechanism is important. By contrast, B2 phases composed of a noble metal (Cu, Ag, Au) and a divalent metal (Mg, Zn, Cd) and FeCo are considered as candidates for the six-jump-cycle mechanism. Clearly, the antistructural-bridge (or antisite bridge) mechanism becomes more important at larger deviations from stoichiometry, because of its percolation threshold.

20.3.2 Example B2 NiAl

The phasefield of B2 NiAl is fairly wide. It extends from about 45 % Ni on the Al-rich side to about 65 % Ni on the Ni-rich side [3]. Theoretical calculations of defect concentrations performed for various intermetallics have been summarised by HERZIG AND DIVINSKI [39]. The concentrations of defects

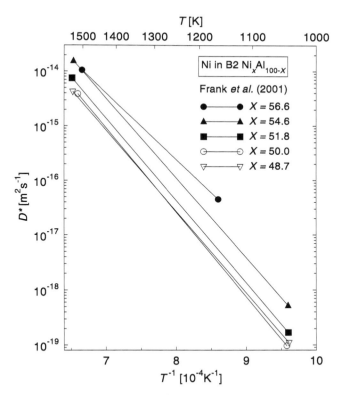

Fig. 20.9. Ni tracer diffusion in B2 NiAl at various compositions X according to FRANK ET AL. [48] and DIVINSKI AND HERZIG [49]

in NiAl are shown in Fig. 20.8. NiAl reveals a triple-defect type of disorder: structural Ni vacancies (V_{Ni}) are the dominating defects on the Ni-lean side, whereas Ni antisite atoms (Ni_{Al}) dominate on the Ni-rich side of the stoichiometric composition. Moreover, vacancies form mainly on the Ni sublattice whereas the concentration of vacancies on the Al sublattice (V_{Al}) is remarkably smaller even on the Al-lean side.

Ni diffusion in B2 NiAl alloys has been measured at various compositions on both Al- and Ni-rich sides and over wide temperature intervals by FRANK ET AL. [48] and reviewed in a paper on NiAl interdiffusion by DIVINSKI AND HERZIG [49][1]. These data are displayed in Fig. 20.9 for various compositions. The diffusivity increases notably on the Ni-rich side of the stoichiometric composition. It is practically independent of composition on the Al-rich side in spite of the considerable amount of structural Ni vacancies (see Fig. 20.8).

Theoretical studies of the atomic mechanism using embedded atom potentials showed that the triple-defect mechanism dominates self-diffusion

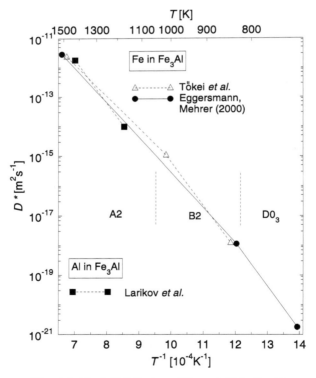

Fig. 20.10. Self-diffusion of Fe and Al in Fe$_3$Al

---

[1] Older measurements of Ni diffusion in NiAl [40] are very likely influenced by grain-boundary contributions [39] and are not considered here.

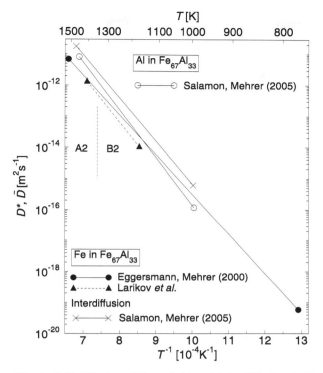

Fig. 20.11. Self-diffusion of Fe and Al and interdiffusion in Fe$_2$Al

on the Al-rich side and for stoichiometric NiAl. The widely abundant isolated Ni vacancies do not contribute significantly to Ni diffusion, because their motion via the six-jump-cycle mechanism is energetically unfavourable. With increasing Ni content, after reaching the percolation threshold, the antistructural-bridge mechanism on the Ni-rich side leads to an increase in the Ni diffusivity [48].

20.3.3 Example B2 Fe-Al

The phasefield of B2 order in the Fe-Al system is fairly extended [3]. B2 order exists between about 22 and 50 at.% Al. In contrast to NiAl, B2 order does hardly extend to compositions on the Al-rich side of stoichiometry. At higher temperatures an order-disorder transition to the disordered A2 structure occurs. The corresponding transition temperature increases with increasing Al content.

Some tracer data for ^{26}Al in aluminides are available from the work of LARIKOV ET AL. [22]. Tracer measurements of Fe self-diffusion were carried out by TÖKEI ET AL. [24] and by EGGERSMANN AND MEHRER [23]. Interdiffu-

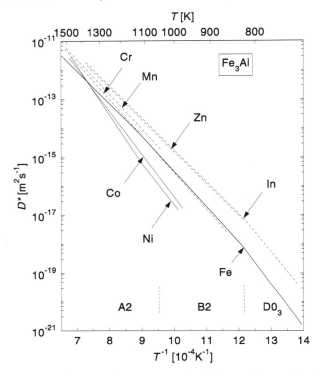

Fig. 20.12. Solute diffusion of Zn, In, Ni, Co, Mn, and Cr in Fe$_3$Al according to [23, 51]. Fe self-diffusion in Fe$_3$Al is also shown for comparison

sion in the whole B2 phasefield of Fe-Al alloys has been studied by SALAMON AND MEHRER [50]. These authors used the Darken-Manning equation (see Chap. 10), the Kirkendall shift, calculated thermodynamic factors, and Fe tracer data of the Fe-Al system and deduced Al tracer diffusivities for alloys with the approximate compositions Fe$_3$Al, Fe$_2$Al, and FeAl. Some of the results for diffusion in iron-aluminides are shown in Figs. 20.10 and 20.11.

Fe$_3$Al reveals A2 disordered, B2 ordered, and D0$_3$ ordered structures with decreasing temperature. As already indicated in Eq. (20.2), the increase in the degree of order results in an increase of the activation enthalpy, which can be seen in Fig. 20.10. Self-diffusion in Fe$_2$Al and FeAl has been investigated almost exclusively in the B2 phase region. For all three compositions the diffusivities of Fe and Al are not much different indicating a coupled diffusion of both components.

Solute diffusion in Fe-Al alloys has also been investigated. Typical results for ternary alloying elements in Fe$_3$Al are compiled in Fig. 20.12 and compared with Fe self-diffusion. Zn and In are incorporated on Al sites and diffuse slightly faster than self-diffusion of both components Fe and Al [23]. Ni and

20.4 L1$_2$ Intermetallics

Co substitute Fe atoms. They are slower diffusers and have higher activation enthalpies than self-diffusion. The diffusivities of Mn and Cr are both fairly similar to Fe self-diffusion [51].

20.4 L1$_2$ Intermetallics

In completely ordered L1$_2$ compounds, each A atom is surrounded by 8 A atoms and 4 B atoms on nearest-neighbour sites (see Fig. 20.1). In contrast to this situation, a B atom faces only A atoms on nearest-neighbour sites. This implies that the sublattice of the majority component A is interconnected by nearest-neighbour bonds, whereas this is not the case for the sublattice of the minority component B. Vacancy motion restricted to the majority sublattice can promote diffusion of A atoms as illustrated in Fig. 20.13. Diffusion of B atoms on its own sublattice requires jump lengths larger than the nearest-neighbour distance, which are energetically unfavourable. Another possibility is the formation of antisite defects and diffusion via vacancies of the majority sublattice.

Perhaps the best known L1$_2$ intermetallic is Ni$_3$Al. It has been used as a strengthening phase in Ni-base superalloys for a long time. The Ni$_3$Al phasefield in the Ni-Al system exists on both sides of the stoichiometric compositions, but in a faily narrow composition interval.

The concentrations of defects in Ni$_3$Al taken from the review [39] are shown in Fig. 20.14. Ni$_3$Al belongs to the antistructural-defect type of intermetallics, in which antisite atoms (Ni$_{Al}$ and Al$_{Ni}$) are preferentially formed to accommodate deviations from stoichiometry. Vacancies are mainly formed on the Ni sublattice. Their concentrations are similar to thermal vacancy concentrations in pure Ni at the same homologous temperature. Vacancy formation on the Al sublattice is energetically less favourable.

Ni diffusion has been studied using tracer techniques by BRONFIN ET AL. [41], HOSHINO ET AL. [42], SHI ET AL. [43], and FRANK ET AL. [44].

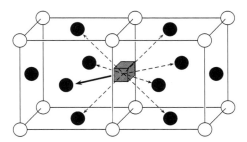

Fig. 20.13. Schematic illustration of the sublattice vacancy mechanism in the majority sublattice of an L1$_2$ structured intermetallic. *Full circles*: majority atoms; *open circles*: minority atoms

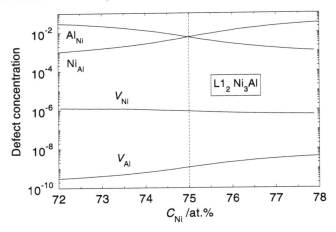

Fig. 20.14. Defect site fractions in L1$_2$ structured Ni$_3$Al as a function of composition at 0.75 T_m from [39]

Unfortunately, diffusion studies on Ni$_3$Al, as for other aluminides, suffer from the lack of a suitable radiotracer for Al. On the other hand, interdiffusion coefficients across the phase field of Ni$_3$Al were measured by IKEDA ET AL.[45] and WATANABE ET AL. [46]. Using the Darken-Manning equation and the Kirkendall shift, FUJIWARA AND HORITA [47] deduced Al tracer diffusivities. It was found that the Ni and Al diffusivities are not much different. Suitable substitutes for Al (e.g., Ge and Ga) have been studied (see Table 20.1 and [39]) and support this finding.

Diffusion in the L1$_2$ compounds Ni$_3$Ge and Ni$_3$Ga has also been studied. Fortunately, in these cases suitable radiotracers for both constituents are available. As can be seen from Fig. 20.16, diffusion of the majority component Ni in Ni$_3$Ge is indeed significantly faster than that of the minority component Ge. Experiments on Ni$_3$Ga revealed a trend similar to the case of Ni$_3$Ge, but the difference of the diffusivities is not so large [52]. For Ni$_3$Al only Ni self-diffusion is indicated. According to the above reasoning the ratio of the two tracer diffusivities, D_{Ni}/D_{Al}, in Ni$_3$Al is not much different from unity.

It is quite natural that diffusion of the majority component in L1$_2$ compounds occurs by a sublattice vacancy mechanism. The diffusion coefficient is expressed as

$$D_A = \frac{2}{3}a^2 f C_V^{eq} \omega, \quad (20.6)$$

where a is the lattice parameter, C_V^{eq} the concentration of vacancies in the majority sublattice, and ω the vacancy jump rate. The random walk properties and the tracer correlation factor for sublattice diffusion of the majority component in L1$_2$ compounds via the vacancy mechanism have been discussed by KOIWA ET AL. [53]. A value of $f = 0.6889$ has been reported for the tracer correlation factor.

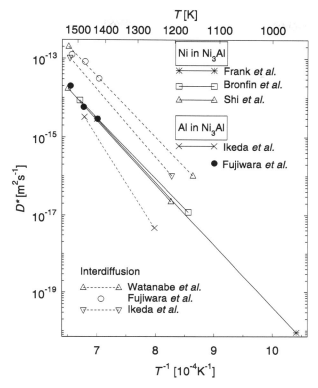

Fig. 20.15. Self-diffusion in L1$_2$ structured Ni$_3$Al according to [39]

The diffusion mechanism of the minority elements in L1$_2$ compounds is less obvious. As can be seen from Fig. 20.16 and from the discussion of diffusion in Ni$_3$Al, the tracer diffusivities of the minority elements in these compounds can vary from very different to similar of those of the majority elements. The diffusivity of Ge in Ni$_3$Ge is rather low, whereas the diffusivity of Ga in Ni$_3$Ga and very likely the diffusivity of Al in Ni$_3$Al are not much different from the respective majority components. Possible mechanisms are discussed in [8]. Minority elements most likely diffuse as antisite atoms in the majority sublattice.

20.5 D0$_3$ Intermetallics

A prominent example of a D0$_3$ intermetallic is Fe$_3$Si. Its phase field is located between the stoichiometric composition and Fe-rich compositions up to about 82 at.% Fe. Information about the diffusion of both constituents and of Ge diffusion is available. Fe$_3$Al also shows D0$_3$ order but only at fairly low

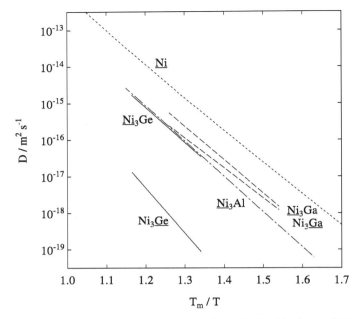

Fig. 20.16. Self-diffusion in the $L1_2$ intermetallics Ni_3Ge, Ni_3Ga, and Ni_3Al. The temperature scale is normalised to the corresounding melting temperatures. For comparison self-diffusion in Ni is also shown. For references see [7]

tempertaures. Fe_3Al is formed on cooling by ordering reactions in the solid state that transform the bcc ordered solid solution, which is stable above about 1000 K, into a B2 phase and then at about 800 K into the $D0_3$ structure. Diffusion studies in $D0_3$ ordered Fe_3Al are difficult due to the fairly low diffusivities (see above). Some high-temperature intermetallics (Cu_3Sn, Ni_3Sn, ...) crystallise in the $D0_3$ structure. Only for Cu_3Sn diffusion of both constituents has been investigated (see Table 20.1).

The majority sublattice (A sublattice) in $D0_3$ compounds, similar to the $L1_2$ structure, is interconnected by nearest-neighbour bonds, whereas this is not the case for the B sublattice (see Fig. 20.1). An A atom can diffuse within its own sublattice via nearest-neighbour jumps. If B atoms migrate within their own sublattice, their jump vector would be a third-nearest neighbour jump with respect to the bcc unit cell. An alternative for the diffusion of B atoms are nearest-neighbour jumps, which create B antisite defects. Then, B atoms can diffuse as 'imperities' in the majority sublattice. Both options require higher activation enthalpies for diffusion of B atoms.

Figure 20.17 shows an Arrhenius plot of Fe- and Ge tracer diffusion for three compositions of Fe_3Si according to GUDE AND MEHRER [54]. The data cover rather wide temperature ranges mostly within the $D0_3$ phasefield. One experiment with the short-lived isotope ^{31}Si confirmed that Ge and Si diffuse

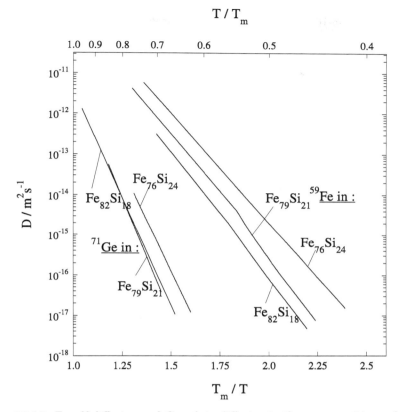

Fig. 20.17. Fe self-diffusion and Ge solute diffusion in three compositions of the D0$_3$ phase Fe$_3$Si according to GUDE AND MEHRER [54]. The temperatures are normalised to the corresponding liquidus temperatures. A slight influence of the paramagnetic-ferromagnetic transition can be seen for Fe diffusion in Fe$_{79}$Si$_{21}$ and Fe$_{82}$Si$_{18}$

at very similar rates [54]. Diffusion studies of Si and Ge after ion implantation using SIMS profiling also showed that Ge is a 'good' substitute to mimic Si diffusion [55]. The most salient features of Fig. 20.17 are: (i) The asymmetry between the fast Fe diffusion and the relatively slow Ge or Si diffusion is large. (ii) The Fe diffusivity increases with Si content and Fe diffuses fastest in the nearly stoichiometric compound.

Positron annihilation studies by SCHAEFER AND COWORKERS [56] have shown that the increase in Fe diffusivity is accompanied by a comparable increase of the content in thermal vacancies. In addition, Mössbauer experiments on stoichiometric Fe$_3$Si by SEPIOL AND VOGL [57] have demonstrated that the atomic jump vector of Fe atoms is in agreement with nearest-neighbour jumps in the Fe sublattice. These findings clearly support a sublattice vacancy mechanism for Fe diffusion. The diffusion mechanism of Ge and Si is not completely clear. Ge diffusion is less affected by the increase of vacancy concentration with increasing Si content than Fe diffusion.

For the high-temperature phase Cu_3Sn (see Table 20.1) diffusion data for both constituents are also available. As in the case of Fe_3Si the majority component is significantly faster than the minority component.

20.6 Uniaxial Intermetallics

For non-cubic materials a diffusion ansiotropy can be expected as dicussed in Chap. 2. Diffusion in uniaxial materials, such as intermetallics with a tetragonal, hexagonal or trigonal axis, is described by a diffusivity tensor with two principal components. In general, the diffusivity parallel to the axis, $D_\parallel$, and the diffusivity perpendicular to the crystal axis, $D_\perp$, are different. A comprehensive study of diffusion then requires measurements on single crystals in at least two independent directions. Experiments on polycrystalline samples conceal the anisotropy effects and reveal average values of the diffusivities only. Nevertheless, they can provide useful information, if single crystals are not available.

According to the author's knowledge, diffusion studies on single cystalline intermetallics are very scarce. A few studies concern $L1_0$ intermetallics such as γ-TiAl and FePt. Another well-studied example concerns the tetragonal $C11_b$ structured material molybdenum disilicide.

20.6.1 $L1_0$ Intermetallics

As already discussed at the beginning of this chapter, the $L1_0$ structure shown in Fig. 20.2 is tetragonal and has an approximate composition AB. Its atomic arrangement can be considered as an ordered fcc structure with A and B atoms occupying sequential (001)-planes. An example of technologically interest for structural applications is β-TiAl. Other $L1_0$ intermetallics such as NiPt, FePt, CoPt, FePd, and NiMn are candidates for high-density magnetic storage media due to their high magnetic anisotropy related to their tetragonal anisotropy.

γ**-TiAl:** This intermetallic has attracted much attention as a structural material for high-temperature applications because of its mechanical properties and its low density. The tracer diffusivities of ^{44}Ti [58, 59] and of several substitutional impurities [60–62] have been studied in polycrystalline γ-TiAl.

Tracer diffusion studies on single crystals of γ-TiAl have been reported by NAKAJIMA AND ASSOCIATES [63, 64]. The self-diffusivities of ^{44}Ti and diffusivities of the impurities Fe, Ni, and In have been studied. As In is isoelectonic to Al, its diffusivities are belived to be good estimates for the Al self-diffusivities of the material. Figure 20.18 shows the diffusivities of Ti and

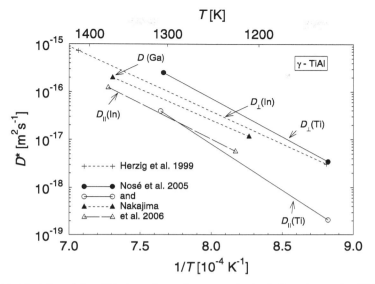

Fig. 20.18. Ti and In diffusion along the two principal directions of γ-TiAl according to NAKAJIMA AND ASSOCIATES [63, 64]. Diffusion of Ga is shown for polycrystalline γ-TiAl according to HERZIG ET AL. [60]

In. In both cases, the diffusion is faster perpendicular to the tetragonal axis than parallel to it. For Ti self-diffusion the anisotropy ratio $D_\perp/D_\parallel$ is almost one decade, whereas for In it is slightly smaller. As can be expected, diffusion along the layers of the L1$_0$ structure is faster than along the tetragonal axis. A detailed model has been suggested by NOSE ET AL. [64]. Diffusion of Ga, which like In is isoelectronic to Al has been studied by HERZIG ET AL. [60] in polycrystalline γ-TiAl. The Ga diffusivities are similar to those of In diffusion.

NiMn: Equiatomic NiMn thin films are of interest for use in magnetoresistive sensors. In these sensors, exchange coupling between an antiferromagnetic film, such as NiMn, is required to 'pin' the magnetisation within the ferromagnetic layers. Diffusion information is of importance, because interdiffusion is a primary long-term failure mechanism of magnetoresistive sensors.

The phase diagram of the Ni-Mn system [65] shows that equiatomic NiMn occurs in three different crystal structures as the temperature varies. At temperatures below about 1000 K the L1$_0$ structure forms. At intermediate temperatures between about 1000 and 1180 K the alloy shows B2 order and above 1180 K NiMn is a disordered fcc alloy.

Self-diffusion of Ni and Mn tracers has been investigated by PETELINE ET AL. [66] for polycrystalline samples. The data are shown in the Arrhenius diagram of Fig. 20.19. The diffusivities of Mn are higher by factors of 3 to 5 than those of Ni in both the fcc and the B2 structures. There is more than

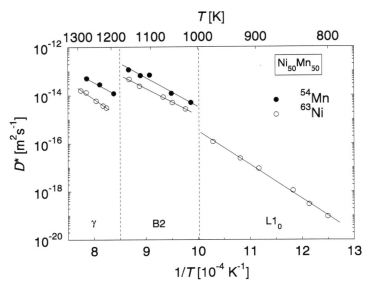

Fig. 20.19. Self-diffusion of ^{54}Mn and ^{63}Ni in polycrystalline, equiatomic NiMn according to PETELINE ET AL. [66]

one decade increase of the diffusivities of both components upon the phase transition from the high-temperature fcc to the B2 structure in the intermediate temperature region. This increase is in agreement with the general observation that diffusion in bcc related structures is faster than in close-packed fcc structures (see, for example, self-diffusion in fcc and bcc Fe in Fig. 17.9). There is also a decrease of the Ni diffusivity upon the phase transition from the B2 phase to the low-temperature $L1_0$ phase. As mentioned in the introduction to this chapter, the $L1_0$ phase can be considered as a slightly distorted, ordered structure of an fcc lattice. Since the experiments were performed on polycrystalline samples nothing can be said about the diffusion anisotropy in the $L1_0$ NiMn phase.

20.6.2 Molybdenum Disilicide ($C11_b$ structure)

Molybdenum disilicide ($MoSi_2$) is a highly stoichiometric compound with a very high melting temperature. It has found widespread use as a functional material in heating elements of high-temperature furnaces for temperatures up to 2000 K, because of its advantageous electrical properties and its excellent oxidation resistance. It is considered as a material with high potential for high-temperature structural applications as well [2, 67]. Molybdenum disilicide is an intermetallic with non-cubic crystal structure. It is a material which exhibits an extraordinary large diffusion asymmetry between Si and

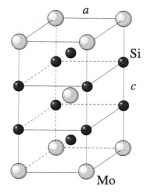

Fig. 20.20. Unit cell of tetragonal MoSi$_2$ with the lattice parameters a and c

Mo similar to that of the components of ceramic materials. In addition, MoSi$_2$ shows a considerable diffusion anisotropy.

The tetragonal unit cell of the C11$_b$ structure is shown in Fig. 20.20. As mentioned above, a tetragonal crystal has two principal components of the diffusion tensor. Two diffusivities, $D_\|$ and $D_\perp$, parallel and perpendicular to the tetragonal axis need to be considered for each component and for the diffusion of solute elements.

The structure of MoSi$_2$ can be considered as a sequential stacking of one Mo layer followed by two Si layers. Each Mo atom resides in a 'cage' of 10 Si atoms whereas each Si atom has five Si and five Mo neighbours. With respect to the diffusivity ratio of the two components, this structure is a good candidate for the Cu$_3$Au rule dicussed below. For MoSi$_2$ the rule suggests that Si self-diffusion and diffusion of solutes in the Si sublattice is much faster than Mo diffusion.

Diffusion of ^{31}Si, ^{99}Mo, and ^{71}Ge tracers has been measured on MoSi$_2$ monocrystals in both principal directions using radiotracer techniques by SALAMON AND COWORKERS [68, 69]. An Arrhenius diagram is displayed in Fig. 20.21. For both principal directions, Si diffusion is indeed many orders of magnitude faster than Mo diffusion:

$$D_\|(\text{Si}), D_\perp(\text{Si}) \gg D_\|(\text{Mo}), D_\perp(\text{Mo}). \tag{20.7}$$

This complies with the observation that the activation enthalpies of Si diffusion are remarkably smaller than those of Mo diffusion. This large diffusion asymmetry of the components suggests that Si and Mo diffuse independently. Figure 20.21 also shows that Ge and Si diffusion are not much different.

Diffusion in MoSi$_2$ reveals a fairly large anisotropy. For both components and for Ge the diffusivity perpendicular to the tetragonal axis is faster than parallel to it (Fig. 20.21):

$$D_\perp(\text{Si})/D_\|(\text{Si}) \gg 1 \quad \text{and} \quad D_\perp(\text{Mo})/D_\|(\text{Mo}) \gg 1. \tag{20.8}$$

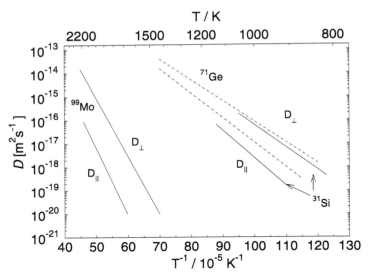

Fig. 20.21. Mo, Si, and Ge diffusion along the two principal directions of MoSi$_2$ according to SALAMON AND MEHRER [68]

The anisotropy ratios are about 100 for Mo diffusion and around several ten for Si diffusion:

Positron annihilation studies by SPRENGEL ET AL. [71] indicate that thermal vacancies are formed on the Si sublattice of MoSi$_2$. This observation and the large diffusion asymmetry suggest that Si diffusion occurs via thermal vacancies on the Si sublattice. The anisotropy of Si diffusion indicates that vacancies are considerably more mobile within the Si layers perpendicular than parallel to the tetragonal axis. Correlation effects of vacancy-mediated self-diffusion in the Si sublattice have been treated in [70]. The slightly faster diffusion of Ge as compared to Si suggests an attractive interaction between Ge and vacancies of the Si sublattice.

According to the author's knowledge, no information is available about vacancies in the Mo sublattice except that their formation definitely requires more energy that the formation of Si vacancies. It is possible that Mo diffusion occurs via thermal vacancies in the minority sublattice. Another possibility is that Mo atoms invade the majority sublattice and diffuse as antisite atoms. The high energy to form Mo antisite atoms would then be a substantial contribution to the activation enthalpy of Mo diffusion.

20.7 Laves Phases

A very numerous group of intermetallic phases are the Laves phases, which belong to the family of Frank-Kaspar phases with topologically close-packed

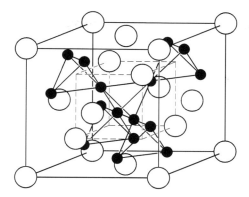

Fig. 20.22. C15 type cubic Laves phase Co_2Nb. *Full circles* represent majority atoms and *open circles* minority atoms

structures [2]. More than 900 binary and ternary Laves phases are known including more that 360 binary Laves phases. Binary Laves phases can be formed if the effective atomic radii of the two components exhibit a ratio of approximately $\sqrt{3/2} \approx 1.25$, which allows a high packing density [72]. Laves phases have an approximate composition AB_2 and crystallise in three intimately related structures: C15 (cubic, Mg_2Cu type), C14 (hexagonal, $MgCu_2$-type), and C36 (hexagonal, $MgNi_2$-type).

Information about diffusion in Laves phases is very scarce [4]. According to the author's knowledge, the only study of self-diffusion for both constituents concerns the cubic C15-type phase with the approximate composition Co_2Nb. The C15 structure of the cubic Laves phase Co_2Nb is shown in Fig. 20.22. In the completely ordered state, the unit cell would contain 8 A atoms and 16 B atoms. The structure consists of two interpenetrating sublattices where the larger A atoms (open circles) form a diamond-type sublattice. The smaller B atoms (full circles) form a network of tetrahedra being joined to each other at their vertices. Co_2Nb mainly exists on the Co-rich side of the stoichiometric composition. Co atoms are considerably smaller than Nb atoms. This suggests that the deviation from stoichiometry on the Co-rich side is compensated by Co-antisite atoms.

Figure 20.23 shows the self-diffusion data according to DENKINGER AND MEHRER [73]. The Co diffusivity has been measured for two compositions, namely $Nb_{31}Co_{69}$ and $Nb_{29}Co_{71}$. Both compositions lie inside the phasefield of the C15 phase of the Co-Nb system [3]. The Nb diffusivity is reported for $Nb_{31}Co_{69}$ only.

Some conclusions from Fig. 20.23 are the following: (i) diffusion of the majority component Co is significantly faster than that of the minority component Nb. This is in accordance with the Cu_3Au rule (see below). Similarities to diffusion in the $L1_2$ phases, which are also close-packed structures, suggest that self-diffusion in Co_2Nb also occurs via thermal vacancies. (ii) The rela-

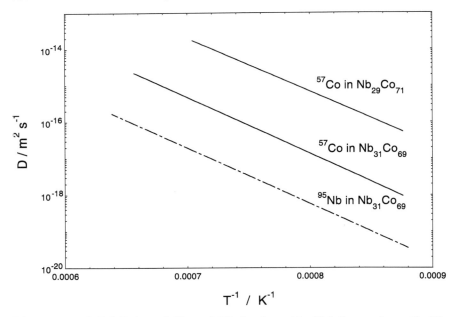

Fig. 20.23. Self-diffusion of Co and Nb in the cubic C15 Laves phase Co_2Nb according to DENKINGER AND MEHRER [73]

tively slow self-diffusion of the large Nb atoms is probably restricted to the Nb sublattice. (iii) The increase of the Co diffusivity with increasing Co concentration can be attributed to the occurrence of additional diffusion paths via anti-structural bridges due to additional Co antisite atoms [73].

20.8 The Cu_3Au Rule

An empirical rule suggested by D'HEURLE AND COWORKERS [74, 75], called the Cu_3Au rule[2], provides often a good guess for self-diffusion in non-equiatomic intermetallics. It states that in compounds of type A_mB_n, where the ratio m/n is 2 or greater, the majority element diffuses faster than the minority element:

$$D_A > D_B \quad \text{or} \quad D_A \gg D_B. \tag{20.9}$$

Here D_A (D_B) denote the tracer diffusion coefficients of the majority (minority) component.

---
[2] This name is a bit misleading. According to the author's knowledge, tracer diffusion experiments of both components in the ordered Cu_3Au phase are not available. Due to the fairly low ordering temperature of this alloy, the diffusivities are very low and thus difficult to measure.

For geometrical reasons discussed above, A_3B intermetallics with $L1_2$ or DO_3 structure are good candidates to test the validity 'Cu$_3$Au' rule. The $L1_2$ phases Ni$_3$Ge and Ni$_3$Ga clearly confirm this rule (see Fig. 20.16). However, Ni$_3$Al is at the border line. The asymmetry of diffusion between majority and minority component in DO_3 phases is in full accordance with the Cu$_3$Au rule (see Fig. 20.17). Diffusion in molybdenum disilicide and in the cubic Laves phase Co$_2$Nb also confirm the rule very well (see Figs. 20.21 and 20.23).

The Cu$_3$Au rule is explained by a diffusion mechanism where atoms of the majority component can diffuse by energetically favourable jumps in their own sublattice. Atoms of the minority component diffuse either via energetically unfavourable jumps to more distant sites on their own sublattice or as antisite atoms on the wrong sublattice.

Of course, the Cu$_3$Au rule should be applied to those intermetallics only in which diffusion occurs via vacancies. Phases such as carbides, nitrides, and metal hydrides, where one of the two elements is sufficiently small to occupy interstitial sites, must be considered separately.

References

1. J.H. Westbrook, R.L. Fleischer (Eds.), *Intermetallic Compounds*, Vol.1, *Principles*, Vol. 2 *Practise*, J. Wiley and Sons Ltd., 1995
2. G. Sauthoff, *Intermetallics*, VCH Weinheim, 1995
3. T.B. Massalski (Ed.), *Binary Alloy Phase Diagrams*, Am. Soc. for Metals, Metals Park, Ohio, 1986
4. H. Bakker, *Self-diffusion in Binary Alloys and Intermediate Phases*, Chap. 4 in: *Diffusion in Solid Metals and Alloys*, H. Mehrer (Vol. Ed.), Landolt-Börnstein, Numerical Data and Functional Relationships in Science and Technology, New Series, Group III: Crystal and Solid State Physics, Vol. 26, Springer-Verlag, 1990. p. 213
5. H. Bakker, *Tracer Diffusion in Concentrated Alloys*, in: *Diffusion in Crystalline Solids*, G.E. Murch, A.S. Nowick (Eds.), Academic Press, Orlando, 1984, p. 189
6. H. Wever, Defect and Diffusion Forum **83**, 55 (1992)
7. H. Mehrer, Materials Transactions, JIM, **37**, 1259 (1996)
8. M. Koiwa, H. Numakura, S. Ishioka, Defect and Diffusion Forum **143–147**, 209 (1997)
9. H. Mehrer, F. Wenwer, *Diffusion in Metals*, in: *Diffusion in Condensed Matter*, J. Kärger, P. Heitjans, R. Haberlandt (Eds.), Verlag Vieweg, Braunschweig, 1998, p.1
10. H. Nakajima, W. Sprengel, K. Nonaka, Intermetallics **4**, 517 (1996)
11. H. Mehrer, Chr. Herzig, in: *Advances in Science and Technology 29, Mass and Charge Transport in Inorganic Materials: Fundamentals to Devices*, P. Vicenzini, V. Buscaglia (Eds.), Techna Srl. 2000, p.187
12. Chr. Herzig, S.V. Divinski, St. Frank, T. Przeorski, Defect and Diffusion Forum **194–199**, 317 (2001)
13. H. Mehrer, *Diffusion: Introduction and Case Studies in Metals and Binary Alloys*, in: *Diffusion in Condensed Matter – Methods, Materials, Models*, P. Heitjans, J. Kärger (Eds.), Springer-Verlag, 2005, p.3

14. J. Philibert, *Atom Movements – Diffusion and Mass Transport in Solids*, Les Editions de Physique, Les Ulis, 1991
15. R.M. Cotts, Ber. Bunsenges. Phys. Chem. **76**, 760 (1972)
16. Th. Heumann, *Diffusion in Metallen*, Springer Verlag, Berlin, 1992
17. A.B. Kuper, D. Lazarus, J.R. Manning, C.T. Tomizuka, Phys. Rev. **104**, 1536 (1956)
18. S.G. Fishman, D. Gupta, D.S. Lieberman, Phys. Rev. B **2**, 1451 (1970)
19. S.G. Fishman, R.N. Jeffrey, Phys. Rev. **B 120**, 4424 (1971)
20. Y. Iijima, Ch.-G. Lee, Acta Metall. Mater. **43**, 1183 (1995)
21. H. Nitta, Y. Iijima, K. Tanaka, Y. Yamazaki, Ch.-G. Lee, T. Matsuzaki, T. Watanabe, Mater. Sci. Eng. **A 382**, 243, 250 (2004)
22. L.N. Larikov, V.V. Geichenko, V.M. Gal'chenko, *Diffusion Processes in Ordered Alloys*, translated from Russion, Amerind Publ. Co. Ltd, New Dehli, 1981
23. M. Eggersmann, H. Mehrer, Philos. Mag. A **80**, 1219 (2000)
24. Zs. Tökei, J. Bernardini, P. Gas, D.L. Beke, Acta Mater. **45**, 541 (1997)
25. E.W. Elcock, C.W. McCombie, Phys. Rev. **109**, 605 (1958)
26. E.W. Elcock, Proc. Phys. Soc. **73**, 250 (1959)
27. H.B. Huntington, N.C. Miller, V. Nerses, Acta Metall. **9**, 749 (1961)
28. P. Wynblatt, Acta Metall. **47**, 2437 (1967)
29. H.A. Domian, H.I. Aaronson, in: *Diffusion in Body-Centered Cubic Metals*, American Society for Metals, 1965, p. 209
30. M. Arita, M. Koiwa, S. Ishioka, Acta Metall. **37**, 1363 (1989)
31. R. Drautz, M. Fähnle, Acta Mater. **47**, 2437 (1999)
32. I.V. Belova, G.E. Murch, Philos. Mag. **A 82**, 269 (2002)
33. N.A. Stolwijk, M. van Gend, H. Bakker, Philos. Mag, **A 42**, 283 (1980)
34. S. Frank, S.V. Divinski, U. Södervall, Chr. Herzig, Acta Mater. **49**, 1399 (2001)
35. H. Bakker, N.A. Stolwijk, M.A. Hoetjes-Eijkel, Philos. Mag. **A 43**, 251 (1981)
36. S.V. Divinski, L.N. Larikov, J. Phys.: Condensed Matter **35**, 7377 (1997)
37. I.V. Belova, G.E. Murch, Intermetallics **6**, 115 (1998)
38. C.R. Kao, Y.A. Chang, Intermetallics **1**, 237 (1993)
39. Ch. Herzig, S. Divinski, *Diffusion in Intermetallic Compounds*, in: *Diffusion Processes in Advanced Technological Materials*, D. Gupta (Ed.), William Andrew, Inc., 2005
40. G.F. Hancock, B.R. McDonnel, Phys. Stat. Sol. **A 4**, 143 (1971)
41. M.B. Bronfin, G.S. Bulatov, I.A. Drugova, Fiz. Metal. Metalloved. **40**, 363 (1975)
42. K. Hoshino, S.J. Rothman, R.S. Averback, Acta Metall. **36**, 1271 (1988)
43. Y. Shi, G. Frohberg, H. Wever, Phys. Stat. Sol. **A 191**, 361 (1995)
44. S. Frank, U. Södervall, Chr. Herzig, Phys. Stat. Sol. **B 191**, 45 (1995)
45. T. Ikeda, A. Almazouzi, H. Numakura, M. Koiwa, W. Sprengel, H. Nakajima, Acta Mater. **46**, 5369 (1998)
46. M. Watanabe, Z. Horita, M. Nemoto, Defect and Diffusion Forum **143–147**, 345 (1997)
47. K. Fujiwara, Z. Horita, Acta Mater. **50**, 1571 (2002)
48. S. Frank, S.V. Divinski, U. Södervall, Chr. Herzig, Acta Mater. **49**, 1399 (2001)
49. S.V. Divinski, Chr. Herzig, Defect and Diffusion Forum **203–205**, 177 (2002)
50. M. Salamon, H. Mehrer, Z. Metallkd. **96**, 1 (2005)
51. S. Peteline, E.M. Tanguep Njiokep, S. Divinski, H. Mehrer, Defect and Diffusion Forum **216–217**, 175 (2003)

52. K. Nonaka, T. Arayashiki, H. Nakajima, A. Almazouzi, T. Ikeda, K. Tanaka, H. Numakura, M. Koiwa, Defect and Diffusion Forum **143–147**, 209 (1997)
53. M. Koiwa, S. Ishioka, Philos. Mag. A **48**, 1 (1983)
54. A. Gude, H. Mehrer, Philos. Mag. A **76**, 1 (1997)
55. M. Wellen, B. Fielitz, G. Borchardt, S. Weber, S. Scherrer, H. Mehrer, H. Baumann, B. Sepiol, Defect and Diffusion Forum **194–199**, 499 (2001)
56. E.A. Kümmerle, K. Badura, B. Sepiol, H. Mehrer, H.-E. Schaefer, Phys. Rev. B **52**, R6947 (1995)
57. B. Sepiol, G. Vogl, Phys. Rev. Lett. **71**, 731 (1995)
58. S. Kroll, N.A. Stolwijk, Chr. Herzig, H. Mehrer, Defect and Diffusion Forum **95–98**, 865 (1992)
59. Chr. Herzig, T. Przeorski, Y. Mishin, Intermetallics **7**, 389 (1999)
60. Chr. Herzig, M. Friesel, D. Derdau, S.V. Divinski, Intermetallics **7**, 1141 (1999)
61. Y. Iijima, C.G. Lee, S.E. Kim, High Temp. Mater. Process. **18**, 305 (1999)
62. Y. Mishin, Chr. Herzig, Acta Mater. **48**, 589 (2000)
63. T. Ikeda, H. Kadowski, H. Nakajima, Acta Mater. **49** 3475 (2001)
64. Y. Nose, N. Terashita, T. Ikeda, H. Nakajima, Acta Mater. **54**, 2511 (2006)
65. L. Ding, P.F. Ladwig, X. Yan, Y.A. Chang, Appl. Phys. Letters **80**, 1186 (2002)
66. S. Peteline, H. Mehrer, M.-L. Huang, Y.A. Chang, Defect and Diffusion Forum **237–240**, 352 (2005)
67. M. Yamaguchi, H. Inui, K. Ito, Acta Mater **48**, 307 (2000)
68. M. Salamon, A. Strohm, T. Voss, P. Laitinen, I. Rihimäki, S. Divinski, W. Frank, J. Räisänen, H. Mehrer, Philos. Mag. **84**, 737 (2004)
69. M. Salamon, H. Mehrer, Z. Metallkd. **96**, 833 (2005)
70. S. Divinski, M. Salamon, H. Mehrer, Philos. Mag. **84**, 757 (2005)
71. W. Sprengel, X.Y. Zhang, H. Inui, H.-E. Schaefer, Verhandl. DPG **5**, 339 (2003)
72. D.J. Thoma, J.H. Perepezko, J. Alloys and Compounds **224**, 330 (1995)
73. M. Denkinger, H. Mehrer, Philos. Mag. **A 80**, 1245 (2000)
74. F.M. d'Heurle, P. Gas, J. Philibert, Solid State Phenomena **41**, 93 (1995)
75. F.M. d'Heurle, P. Gas, C. Lavoie, J. Philibert, Z. Metallkd. **95**, 852 (2004)

21 Diffusion in Quasicrystalline Alloys

21.1 General Remarks on Quasicrystals

The quasicrystalline state is a third form of solid matter besides the crystalline and amorphous state. The atomic positions are ordered, but with five-, eight-,
ten-, or twelvefold rotational symmetries, which violate canonical rules of classical crystallography. These symmetries forbid a periodic structure and, instead, enforce quasiperiodicity. Quasicrystalline alloys can also be considered as complex intermetallics. Most quasicrystals are indeed composed of metallic components. Usually, quasicrystals have crystalline intermetallics as neighbour phases.

The discovery of the first quasicrystal, $Al_{86}Mn_{14}$, was announced by SHECHTMAN ET AL. in 1984 [1]. In the early days, only metastable quasicrystalline alloys, such as $Al_{86}Mn_{14}$ and other Al-TM alloys (TM = transition metal), were produced. Since these times a large number of quasicrystalline systems are known and progress in the understanding of their physical properties has been achieved [2–5]. Some of them crystallise in quasiperiodic lattices upon slow cooling and exist as stable phases. Stable quasicrystals are mostly formed in ternary or higher component systems (see Table 21.1). The first stable quasicrystals were discovered by TSAI AND COWORKERS are all Al-dominated. These were icosahedral Al-Cu-Fe [6], decagonal Al-Ni-Co [7], and icosahedral Al-Pd-Mn [8]. The first Al-free icosahedral phase was found in 1993 in the Zn-Mg-Y system [9]. Meanwhile, several Zn-Mg-RE quasicrystals (RE = rare earth elements Y, Ho, Er, Gd, Tb, or Dy) are known. They are prototypes of non-Al-based quasicrystalline phases.

After the growth of large single-quasicrystals was mastered, one could start to check structure and atomic dynamics with elastic and inelastic scattering methods. Transport measurements revealed fairly high electric resistivities in Al-based alloys. Mechanical experiments showed brittleness and hardness at low temperatures, but ductility with extreme plasticity at high temperatures. Other observations, for example low friction, high corrosion and oxidation resistance, low wetting of surfaces, and strength of composites with quasicrystals promised technological applications [11, 12]. Quasicrystals are a prominent example for systems whose properties are mainly determined by their structure.

Table 21.1. Examples of some stable quasicrystalline phases

Icosahedral quasicrystals	Decagonal quasicrystals	Dodecagonal quasicrystals
$Al_{70}Pd_{21}Re_9$	$Al_{64}Co_{16}Cu_{20}$	$Ta_{1.6}Te$
$Al_{70}Pd_{21}Mn_9$	$Al_{71.5}Co_{14}Ni_{14.5}$	
$Al_{62}Cu_{25}TM_{13}$ (TM = Fe, Ru, Os)	$Al_{69.8}Pd_{18.1}Mn_{12.3}$	
$Ni_{17}Ti_{41.5}Zr_{41.5}$	$Al_{71}Fe_5Ni_{24}$	
$Ga_{20}Mg_{43}Zn_{37}$	$Al_{40}Fe_{15}Ge_{20}Mn_{25}$	
$Al_{37}Cu_{10.8}Li_{32.2}$		
$Al_{43}Mg_{43}Pd_{14}$		
$Zn_{64.2}Mg_{26.4}Ho_{9.4}$		
$Zn_{60.7}Mg_{30.6}Y_{8.7}$		

Fig. 21.1. Icosahedral single-quasicrystal (Zn-Mg-Ho) with the habit of a dodecahedron. A dodecahedron has 12 pentagon-shaped faces and 20 vertices

Nowadays, quasicrystals are no longer laboratory curiosities. Quasicrystalline phases such as icosahedral Al-Pd-Mn, icosahedral Zn-Mg-RE, and decagonal Al-Ni-Co can be grown in the form of centimeter-size single-quasicrystals with high structural quality by Bridgman-, Czochralski-, and flux-growth techniques [15–17]. Figure 21.1 shows an icosahedral Zn-Mg-Y quasicrystal grown by the flux-growth technique. Its habit is a dodecahedron – one of the five regular (Platonian) polyhedra[1].

Quasicrystalline structures are fairly complex, because there is no periodicity. In addition, most quasicrystals are ternary, quaternary, or multicomponent alloys. Quasicrystals have rotational order and sharp diffraction peaks. Quasiperiodic space tiling procedures such as two-dimensional Penrose tiling and its three-dimensional analogues [18], projections or cuts from higher di-

---

[1] The five Platonian polyhedra are: tetrahedron, cube, octahedron, icosahedron, dodecahedron.

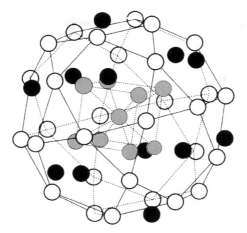

Fig. 21.2. The pseudo-Mackay cluster suggested for icosahedral Al-Pd-Mn [19]. It consists of a central cube (*filled grey circles*), an icosahedron (*filled black circles*), and an icosidodecahedron (*open circles*)

mensional periodic lattices, and cluster covering models have been developed to describe quasi-crystallography (see [2] and Chap. 2 in [5]).

In order to illustrate the structural complexity of quasicrystals we mention just one example: the first structure determinations of icosahedral Al-Pd-Mn by diffraction techniques has been published by BOUDARD ET AL. in 1992 [19]. The authors found evidence for the presence of so-called pseudo-Mackay clusters (Fig. 21.2). In their structure model, the most simple cluster consists of Al atoms on the central cube and on the 30 vertices of the icosidodecahedron, whereas Pd and Mn atoms are distributed on the 12 vertices of the icosahedron. However, other decorations have also been suggested in the literature. A further property of the Al-Pd-Mn quasicrystal structure is its self-similarity based on an inflated cluster hierarchy [20]. There seems to be no completely accepted structure model for icosahedral Al-Pd-Mn [10].

21.2 Diffusion Properties of Quasicrystals

Diffusion in solids – irrespective of whether they are crystalline, quasicrystalline, or amorphous – is an important topic of materials science. Thorough diffusion studies on single quasicrystals have been performed on the icosahedral phases Al-Pd-Mn and Zn-Mg-RE and on the decagonal phase Al-Ni-Co. Tracer techniques have mainly been applied and the diffusers were either labeled by their radioactivity or, when SIMS profiling was employed, by their atomic mass. As already stated, both icosahedral Al-Pd-Mn and decagonal Al-Ni-Co are Al dominated. Tracer diffusion of Al has not been studied. This is because natural Al is monoisotopic, which forbids SIMS studies, and a suitable radioisotope for tracer studies is missing[2]. Instead, diffusion of several solutes was investigated, especially in icosahedral Al-Pd-Mn. Single crystals should be used for basic studies to avoid complications due to diffusion along grain boundaries (see Chap. 32).

The diffusion coefficient in a quasicrystal, like in a crystal, is a symmetric second rank tensor. Icosahedral quasicrystals are highly symmetric, diffusion is isotropic, and the diffusion coefficient is a scalar quantity. Decagonal quasicrystals are uniaxial and the diffusion coefficient has two principal components, like in uniaxial crystals. One component corresponds to diffusion parallel to the decagonal axis and the other one to diffusion perpendicular to the axis. A thorough diffusion study on decagonal quasicrystals requires diffusivity measurements on monocrystalline samples with two different orientations.

Diffusion in quasicrystalline alloys has been reviewed by NAKAJIMA AND ZUMKLEY [13] and by MEHRER ET AL. [14]. In this chapter, we summarise the state-of-the-art in this area, compare diffusion in quasicrystals with diffusion in related crystalline metals, and discuss possible diffusion mechanisms.

21.2.1 Icosahedral Quasicrystals

Icosahedral Al-Pd-Mn: The information on the phase diagram for the ternary Al-Pd-Mn system is rather complete and has been reviewed by LÜCK [21]. Al-Pd-Mn was the first system for which stable icosahedral as well as decagonal quasicrystals were detected [22]. The formation of the icosahedral phase occurs in a ternary peritectic reaction from the melt. Conventional procedures permit the growth of single crystals [15]. The formation of the decagonal phase is sluggish, and the determination of the phase diagram in the decagonal region is difficult.

Icosahedral Al-Pd-Mn contains about 70 at. % Al (see Table 21.1). The contents of transition elements Pd and Mn are about 21 and 9 at. %, respectively. Icosahedral Al-Pd-Mn is not a stoichiometric compound. It possesses a relatively narrow phase field, which widens with increasing temperature, but its width never exceeds a few percent [21].

Al-Pd-Mn is the quasicrystalline material for which the largest body of diffusion data is available. These data are displayed in the Arrhenius diagram of Fig. 21.3. Self-diffusion of the Mn and Pd components as well as solute diffusion of the transition elements Co, Cr, Fe, Ni, and Au and of the non-transition elements Zn, Ga, In, and Ge has been investigated. For detailed references, the reader may consult the already mentioned reviews [13, 14]. Inspection of Fig. 21.3 shows that the diffusivities can be grouped into two major categories:

---

[2] The radiosisotope ^{26}Al has a half-life of 7×10^5 years. Its specific activity is very low, its production requires an accelerator and is very expensive.

21.2 Diffusion Properties of Quasicrystals   375

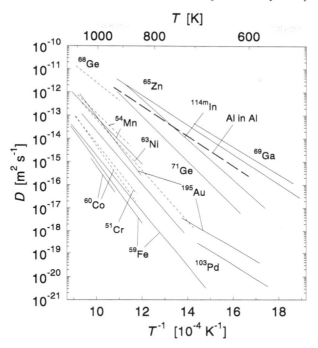

Fig. 21.3. Tracer diffusion in single-crystals of icosahedral Al-Pd-Mn according to MEHRER ET AL. [14]. Self-diffusion in Al is indicated as a *long-dashed line*

- Non-transition elements are relatively fast diffusers. Their diffusivities are comparable, within one order of magnitude, to the self-diffusivity of metallic aluminium (long-dashed line). Ga and Zn were studied to mimic Al self-diffusion in the quasicrystal. Ga is isoelectronic to Al. Zn diffusion is believed to be close to Al self-diffusion, since in metallic Al it is only about a factor of 2 faster than Al [23].
- Transition elements are slow, in some cases extremely slow diffusers. For example, diffusion of Fe at 700 K is about 7 orders of magnitude slower than Ga and Zn diffusion. The diffusion enthalpies of Fe, Co, and Cr are high, almost twice as large as for Ga and Zn. The pre-exponential factors of the transition elements are two to three orders of magnitude larger than those of Zn and Ga [14].

Most of the diffusivities in Fig. 21.3 follow linear Arrhenius behaviour in the whole temperature range investigated. For Pd and Au diffusion, studied after implantation of the radioisotopes, a kink was reported in the Arrhenius diagram around 773 K by FRANK AND COWORKERS [25, 26]. The Arrhenius parameters in the high-temperature regime are similar to those of the other transition elements, whereas in the low-temperature regime very low values of the activation parameters have been reported.

The pressure dependence of diffusion of the non-transition element Zn and of the transition element Mn has been studied [27]. The following activation volumes were deduced:

$$\Delta V(Zn) = 0.74\,\Omega \quad \text{at} \quad 776\,\text{K}, \tag{21.1}$$

$$\Delta V(Mn) = 0.67\,\Omega \quad \text{at} \quad 1023\,\text{K}. \tag{21.2}$$

Ω denotes the mean atomic volume of icosahedral Al-Pd-Mn, which in the case of Al-based Al-Pd-Mn is not much different from the atomic volume of Al. It is remarkable that there is no significant difference between the activation volumes of the transition and the non-transition element indicating that the diffusion mechanism of both elements is basically the same. The magnitudes and signs of these activation volumes are similar to those observed for metals, which are attributed to vacancy-mediated diffusion (see Chap. 8).

Diffusion in crystalline materials is mediated by point defects (Chaps. 6, 17, 19, 20, 26). In metallic elements and alloys, vacancy-type defects are responsible for the diffusion of matrix atoms and of substitutional solutes. In quasicrystals, as in crystalline alloys, vacancies are present in thermal equilibrium as demonstrated in positron annihilation studies by SCHAEFER AND COWORKERS [28, 29]. Phasons are additional point defects which are specific to quasicrystalline materials [30]. It has been suggested theoretically by KALUGIN AND KATZ [31] that, in addition to vacancies, phason flips might contribute to some extend to self- and solute diffusion in quasicrystals.

Remembering that Al is the major component in icosahedral Al-Pd-Mn, one is prompted to compare its diffusion data to those of aluminium. Figure 21.4 shows an Arrhenius diagram of Al self-diffusion together with the diffusivities of those transition and non-transition elements which were also studied in icosahedral Al-Pd-Mn. A comparison of Fig. 21.3 and 21.4 supports the following conclusions:

1. Diffusion of Zn and Ga in icosahedral Al-Pd-Mn and in Al obey very similar Arrhenius laws. This suggests that Zn and Ga diffusion in icosahedral Al-Pd-Mn is vacancy-mediated as well and restricted to the Al-subnetwork of the quasicrystalline structure (Fig. 21.2).
2. The diffusivities of Zn and Ga are close to self-diffusion of Al. The activation enthalpies and pre-exponential factors of Ga ($113\,\text{kJ}\,\text{mol}^{-1}$, $1.2 \times 10^{-5}\,\text{m}^2\,\text{s}^{-1}$) and Zn ($121\,\text{kJ}\,\text{mol}^{-1}$, $2.7 \times 10^{-5}\,\text{m}^2\,\text{s}^{-1}$) in icosahedral Al-Pd-Mn [14] and those of Al self-diffusion in metallic Al ($123.5\,\text{kJ}\,\text{mol}^{-1}$, $1.37 \times 10^{-5}\,\text{m}^2\,\text{s}^{-1}$ [24]) are very similar. This suggests that Zn and Ga diffusion provide indeed good estimates for Al self-diffusion in icosahedral Al-Pd-Mn. Al self-diffusion could not be investigated due to the lack of a suitable Al tracer (see above).
3. Diffusion of solutes in icosahedral Al-Pd-Mn and in Al are both characterised by a 'wide spectrum' of diffusivities, ranging from very slow diffusing transition elements to relatively fast diffusing non-transition elements. Self-diffusion and diffusion of substitutional solutes in Al are both

21.2 Diffusion Properties of Quasicrystals   377

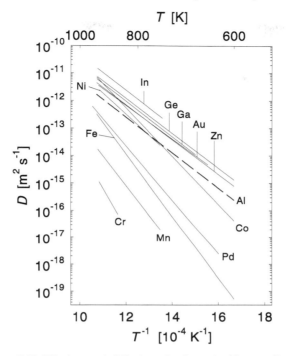

Fig. 21.4. Self-diffusion and diffusion of solutes in Al according to [14]

mediated by vacancies. The striking similarities between the spectra of solute diffusion in both materials, apart from minor differences in detail, are a strong argument in favour of a vacancy-type mechanism in icosahedral Al-Pd-Mn as well. As discussed in Chap. 19, the slow diffusion of transition elements in Al can be attributed to a repulsive interaction between vacancy and solute.

In the low-diffusivity regime, the diffusivities of Pd and Au in icosahedral Al-Pd-Mn are distinctly higher than expected from an extrapolation of the Arrhenius-laws corresponding to the high-diffusivity regime. The low preexponential factors found for Pd and Au diffusion in this regime are orders of magnitude too small to be reconcilable with diffusion mechanisms operating in crystalline solids. A tentative explanation attributes this low-diffusivity regime to phason-flip assisted diffusion [14, 25, 26]. This explanation is intimately related to the quasicrystalline structure. Hence, it cannot work for substitutional solutes, like the normal diffusers Zn or Ga, residing almost exclusively in the Al subnetwork. However, it does work for intermediate diffusers such as Au and Pd by promoting their interchange between the subnetworks.

Icosahedral Zn-Mg-RE: Zn-Mg-RE alloys (RE = rare earth element) are the prototype of non-Al-based quasicrystalline alloys. In these phases, Zn is the base element with a content of about 60 at. % (Table 21.1). These quasicrystals are considered to belong to the Frank-Kaspar type phases, which are characterised by an electron to atom ratio of about 2.1. Their structureal units are Bergman clusters [32], which are formed by dense stacking of tetrahedra, relating them to the Frank-Kaspar phases [33]. Three quasicrystalline phases have been found: a face-centered icosahedral (fci) phase, a simple cubic icosahedral (si) phase, and a phase with decagonal structure. Crystalline structures which are related to the quasicrystalline phases are found as well (for references see, e.g., [16]).

Tracer diffusion experiments of ^{65}Zn have been performed on icosahedral $Zn_{64.2}Mg_{26.4}Ho_{9.4}$ and $Zn_{60.7}Mg_{30.6}Y_{8.7}$ quasicrystals [34] grown by the top-seeded solution-growth technique [16]. Diffusion data are displayed in Fig. 21.5 together with tracer diffusion of ^{65}Zn in a related crystalline Zn-Mg-Y phase with hexagonal structure. It is not surprising that Zn diffusion

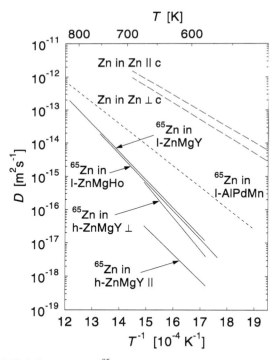

Fig. 21.5. Self-diffusion of ^{65}Zn in icosahedral $Zn_{64.2}Mg_{26.4}Ho_{9.4}$ and $Zn_{60.7}Mg_{30.6}Y_{8.7}$ quasicrystals and in a related hexagonal phase (h-ZnMgY) according to GALLER ET AL. [34]. *Dashed lines*: self-diffusion in Zn parallel and perpendicular to its hexagonal axis; *dotted line*: Zn diffusion in icosahedral Al-Pd-Mn

in the closely related Zn-Mg-Ho and Zn-Mg-Y quasicrystals proceeds at almost identical rates. Its diffusion is, however, slower than self-diffusion in metallic Zn. This difference can be attributed only partly to the different melting temperatures of Zn (693 K) and Zn-Mg-Ho (863 K). In a temperature scale normalised to the respective melting temperatures, Zn self-diffusion in hexagonal Zn is still about one order of magnitude faster. Zn diffusion in icosahedral Al-Pd-Mn (dotted line) proceeds about one order of magnitude faster than Zn diffusion in the Zn-based quasicrystals. A comparison between Zn diffusion in the Zn-based quasicrystals and the ternary crystalline compound with similar composition reveals similar diffusivities of Zn diffusion in particular for diffusion perpendicular to the hexagonal axis. These facts and the magnitude of the pre-exponential factors for Zn diffusion in both quasicrystals (Zn-Mg-Ho: 3.9×10^{-3} m^2 s^{-1}; Zn-Mg-Y: 3.3×10^{-3} m^2 s^{-1}) are hints that diffusion in these materials is vacancy-mediated as well.

21.2.2 Decagonal Quasicrystals

Decagonal phases belong to the category of two-dimensional quasicrystals. They exhibit quasiperiodic order in layers perpendicular to the decagonal axis, whereas they are periodic parallel to the decagonal axis [35]. The ternary systems Al-CoAl-NiAl and Al-CoAl-CuAl are characterised by the formation of decagonal phases; no icosahedral phases are observed [21]. The field of the decagonal phase is elongated with respect to a widely varying Ni to Co ratio. In contrast, only a slight variation of the Al content is possible without leaving the phase field with quasicrystalline order.

Diffusion of several tracers has been studied on oriented single crystals parallel and perpendicular to the decagonal axis and also in polycrystals. Diffusion data for decagonal Al-Ni-Co are summarised in Fig. 21.6. Self-diffusion of both minority components (solid lines) has been studied on single crystals of $Al_{72.6}Ni_{10.5}Co_{16.9}$ using ^{57}Co and ^{63}Ni as radiotracers by KHOUKAZ ET AL. [37, 38]. Diffusion of ^{63}Ni has been studied by the same authors studied on polycrystals of the composition $Al_{70.2}Ni_{15.1}Co_{14.7}$. Data obtained by NAKAJIMA AND COWORKERS for diffusion of ^{60}Co in decagonal $Al_{72.2}Ni_{11.8}Co_{16}$ are shown as dashed lines [13, 36]. Employing SIMS profiling, diffusion of Ga has been studied to mimic Al diffusion by GALLER ET AL. [34].

Co diffusion data are available for two slightly different alloys. Co diffusion depends only weakly on composition. Diffusion of Co and Ni in both principal directions obey Arrrhenius laws. It is remarkable that no significant deviation from Arrhenius behaviour has been detected down to the lowest temperatures. In a temperature scale normalised with the melting temperatures, the diffusivities of Co and Ni in decagonal Al-Ni-Co and the (vacancy-mediated) self-diffusivities of metallic Co and Ni [39] are fairly close to each other. Diffusion of Ga is several orders of magnitude faster than diffusion of the transition elements Co and Ni. As in icosahedral Al-Pd-Mn, the non-transition

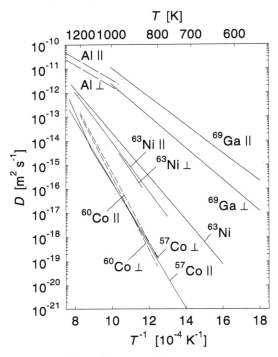

Fig. 21.6. Self-diffusion of ^{65}Ni, ^{60}Co, and ^{57}Co in decagonal Al-Ni-Co quasicrystals from [14, 34, 36]. Monte Carlo simulations of Al diffusion are also shown [40]

element diffuses significantly faster than the transition elements. Diffusivities deduced from molecular dynamic simulations for Al self-diffusion in Al-Ni-Co by GÄHLER AND HOCKER [40] also shown in Fig. 21.6. The magnitude of the Al diffusivities and the sign of the diffusion anisotropy are similar to those of Ga diffusion, supporting the view that Ga is indeed suitable to mimic Al self-diffusion.

The combination of crystalline and quasicrystalline order makes decagonal Al-Ni-Co particularly interesting from the viewpoint of diffusion mechanisms. Quasicrystalline order exists only in layers perpendicular to the decagonal axis, whereas periodic (crystalline) order prevails parallel to the decagonal axis. If a diffusion mechanism would dominate, which is specific to quasiperiodic order, one should expect that diffusion within the layers perpendicular to the decagonal axis is faster than parallel to it. Phasons are specific to quasiperiodic order and phason-mediated diffusion should cause a diffusion anisotropy with slower diffusion in the direction of the decagonal axis. Figure 21.6 shows that for Co and Ni diffusion the anisotropy is very small. For ^{60}Co diffusion in $Al_{72.2}Ni_{11.8}Co_{16}$ the anisotropy is even opposite to the expectation for phason-dominated diffusion in the whole temperature range [13]. This is even more so for Ga diffusion, where diffusion parallel to decagonal axis is clearly faster than perpendicular to the axis.

The magnitude of the diffusion anisotropy Co and Ni in decagonal Al-Ni-Co is similar to the anisotropies reported for uniaxial metals (see Chap. 17). For self-diffusion in the hexagonal metals Be, Mg, Zn, and Cd and for tetragonal In and Sn, the diffusivities parallel and perpendicular to the axis, differ by not more than a factor of 2 (see Chap. 17 and [39]). Whereas in Zn and Cd, diffusion parallel to the hexagonal axis is slightly faster, the opposite is true in Be and Mg. The similarities between the small diffusion anisotropies of uniaxial metals and the decagonal quasicrystals provides additional evidence for vacancy-mediated diffusion in decagonal Al-Ni-Co.

References

1. D. Shechtman, I.A. Blech, J.W. Cahn, Phys. Rev. Lett. **53**, 1951 (1984)
2. C. Janot, *Quasicrystals – a Primer*, Clarendon Press, Oxford, 1994
3. Z. Stadnik (Ed.), *Physical Properties of Quasicrystals*, Springer-Verlag, 1999
4. J.-B. Suck, M. Schreiber, P. Häussler (Eds.), *Quasicrystals – an Introduction to Structure, Physical Properties, and Applications*, Springer-Series in Materials Science, Vol. 55, 2002
5. H.-R. Trebin (Ed.), *Quasicrystals – Structure and Physical Properties*, Wiley-VCH, 2003
6. A.P. Tsai, A. Inoue, T. Masumoto, Jap. J. Appl. Phys. **36**, L1505 (1987)
7. A.P. Tsai, A. Inoue, T. Masumoto, Acta Met. **37**, 1443 (1989)
8. A.P. Tsai, A. Inoue, Y. Yokoyama, T. Masumoto, Mater. Trans. Japan Inst. Met. **31**, 98 (1990)
9. Z.P. Luo, S. Zhang, Y. Tang, D. Zhao, Scripta Metal. et Mat. **28**, 1513 (1993)
10. H.-U. Nissen, C. Beeli, *Electron Microscopy and Surface Investigations of Quasicrystals*, Chap. 6 in [4]
11. J.-M. Dubois, *Bulk and Surface Properties of Quasicrystalline Materials and their Potential Applications*, Chap. 28 in [4]
12. J.-M. Dubois, *Useful Quasicrystals*, World Scientific Publ. Comp., Singapore, 2005
13. H. Nakajima, Th. Zumkley, Defect and Diffusion Forum **194–199**, 789 (2001)
14. H. Mehrer, R. Galler, W. Frank, R. Blüher, A. Strohm, *Diffusion in Quasicrystals*, Chap. 4.1 in [5]
15. M. Feuerbacher, C. Thomas, K. Urban, *Single Quasicrystal Growth*, Chap. 1.1 in [5]
16. R. Sterzel, E. Uhrig, E. Dahlmann, A. Langsdorf, W. Assmus, *Preparation of Zn-Mg-RE Quasicrystals and Related Compounds (RE = Y, Ho, Er, Dy)*, Chap. 1.3 in [5]
17. P. Gille, R.-U. Barz, L.M. Zhang, *Growth of Decagonal Al-Ni-Co and Al-Co-Cu Quasicrystals by the Czochralski Method*, Chap. 1.5 in [5]
18. D. Levine, P.J. Steinhardt, Phys. Rev. Lett. **53**, 2477 (1984)
19. M. Boudard, M. de Boissieu, H. Vincent, G. Heger, C. Beeli, H.-U. Nissen, R. Ibberson, M. Audier, J.-M. Dubois, C. Janot, J. Phys: Cond. Matter **4**, 10149 (1992)

20. C. Janot, Phys. Rev. B **73**, 181 (1996)
21. R. Lück, *Production of Quasicrystalline Alloys and Phase Diagrams*, p. 222 in [4], 2002
22. C. Beeli, H.-U. Nissen, J. Robadey, Philos. Mag. Lett. **63**, 87 (1991)
23. A.D. Le Claire, G. Neumann, *Diffusion of Impurities in Metallic Elements*, Chap. 3 in: *Diffusion in Solid Metals and Alloys*, H. Mehrer (Vol. Ed.), Landolt Börnstein, New Series, Group III: Crystal and Solid State Physics, Vol. 26, Springer-Verlag, 1990
24. R. Messer, S. Dais, D. Wolf, in: Proc. 18th Ampere Congress, P.S. Allen, E.R. Andrew, C.A. Bates (Eds.), Nottingham, England, 1974
25. R. Blüher, P. Scharwächter, W. Frank, H. Kronmüller, Phys. Rev. Lett. **80**, 1014 (1998)
26. W. Frank, R. Blüher, I. Schmich, in: *Mass and Charge Transport in Inorganic Materials – Fundamentals to Devices*, P. Vincencini, V. Buscaglia (Eds.), Techna, Faenza, 2000, p. 205
27. H. Mehrer, R. Galler, Mat. Res. Soc. Symp. Proc. **553**, 67 (1999)
28. F. Baier, H.-E. Schaefer, Phys. Rev. B **66**, 064208 (2002)
29. W. Sprengel, F. Baier, K. Sato, X.Y. Zhang, K. Reimann, R. Würschum, R. Sterzel, W. Assmus, F. Frey, H.-E. Schaefer, *Vacancies, Atomic Processes, and Structural Transformations in Quasicrystals*, Chap. 4.5 in [5]
30. H.-R. Trebin, *Phasons, Dislocations, and Cracks*, Chap. 12 in [4]
31. P.A. Kalugin, A. Katz, Europhys. Lett **21**, 921 (1993)
32. G. Bergman, J.L.T. Waugh, L. Pauling, Acta Cryst. **10**, 254 (1957)
33. F.C. Frank, J.S. Kaspar, Acta Cryst. **11**, 184 (1958); **12**, 483 (1959)
34. R. Galler, E. Uhrig, W. Assmus, S. Flege, H. Mehrer, Defect and Diffusion Forum **237–240**, 358 (2005)
35. W. Steurer, T. Haibach, p. 51 in [3]
36. Th. Zumkley, J.Q. Guo, A.-P. Tsai, H. Nakajima, Mat. Sci. A **294–296**, 702 (2000)
37. C. Khoukaz, R. Galler, H. Mehrer, P.C. Canfield, I.R. Fisher, M. Feuerbacher, Mater. Sci. Eng. A **294–296**, 697 (2000)
38. C. Khoukaz, R. Galler, M. Feuerbacher, H. Mehrer, Defect and Diffusion Forum **194–199**, 867 (2001)
39. H. Mehrer, N. Stolica, N.A. Stolwijk, *Self-Diffusion of Solid Metallic Elements*, Chap. 2 in: *Diffusion in Solid Metals and Alloys*, H. Mehrer (Vol. Ed.), Landolt Börnstein, New Series, Group III: Crystal and Solid State Physics, Vol. 26, Springer-Verlag, 1990
40. F. Gähler, S. Hocker, J. Non-Cryst. Solids **334–335**, 308 (2004)

Part IV

Diffusion in Semiconductors

22 General Remarks on Semiconductors

The present chapter and the two subsequent ones deal with diffusion in the elemental semiconductors Si and Ge. Semiconducting materials play a major rôle in high-tech equipment used in industry and in daily life. Silicon (Si) is the most important semiconductor for the fabrication of microelectronic devices such as memory and processor chips for computers and solar cells for energy production in photovoltaic devices. Germanium (Ge) constitutes the base material for γ-radiation detectors. Gallium arsenide (GaAs) and other compounds of group III and group V elements of the periodic table are mainly used in opto-electronic and high-frequency devices such as solid-state lasers in compact-disc players and receivers in cellular phones. SiC is a semiconductor with large band gap and has potential for applications in devices that must operate at high temperatures, high frequencies, and under irradiation.

We remind the reader that both Si and Ge crystallise in the diamond structure. Many III-V compounds like GaAs occur in the zinc blende structure, which is closely related to the diamond structure. In both structures, the Bravais lattice is face-centered cubic and the structure can be created by the translation of two atoms at the positions (0,0,0) and $\frac{a}{4}(1,1,1)$, where a is the cubic lattice parameter. In the diamond structure, both positions are occupied by the same type of atoms; in the zinc blende structure, the basis is formed by two different types of atoms (Fig. 22.1). In both cases the

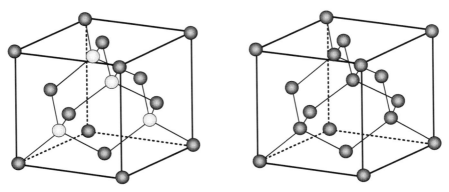

Fig. 22.1. Diamond structure of Si and Ge (*right*) and zinc blende structure (*left*)

Table 22.1. Crystal structure, lattice parameters, and band gaps of Si, Ge, and of some III-V compound semiconductors according to SHAW [1]

Crystal	Structure	Lattice parameter in nm	Band gap at 300 K in eV, type of band gap
Silicon (Si)	Diamond	0.543095	1.120, indirect
Germanium (Ge)	Diamond	0.564613	0.663, indirect
Gallium nitride (GaN)	Wurtzite	a: 0.3111, c: 0.4978	3.7, direct
Gallium phosphide (GaP)	Zinc blende	0.54512	2.261, indirect
Gallium arsenide (GaAs)	Zinc blende	0.56532	1.435, direct
Gallium antimonide (GaSb)	Zinc blende	0.60959	0.72, direct
Indium phosphide (InP)	Zinc blende	0.58687	1.351, direct
Indium arsenide (InAs)	Zinc blende	0.60584	0.35, direct
Indium antimonide (InSb)	Zinc blende	0.64794	0.280, direct

coordination number is 4, each atom is surrounded by a tetrahedron of four neighbouring atoms, and the bonding is covalent. The crystal structure, the lattice constant, the ambient temperature band gap energy, and the type of band gap are listed for the elemental group IV semiconductors and for some III-V compound semiconductors in Table 22.1. We note that both Si and Ge are semiconductors with indirect band gaps, whereas most compound semiconductors have direct band gaps. We also note that the packing density of atoms in semiconductor structures is considerably lower than in close-packed metals. Silicon-carbide (SiC) crystallises in various polytypes and has a very high melting temperature of 2545 °C. Depending on the polytype the band gap lies between 2.39 and 3.26 eV.

22.1 'Semiconductor Age' and Diffusion

Periods of mankind are named after materials: stone age, bronze age, and iron age. In the 1970s the number of publications about semiconductors outnumbered for the first time those about steels and some people started to denote the present period as the 'semiconductor age'. This development was initiated in 1945, when a research group at the Bell Telephone Laboratories in the United States was established to focus on a better understanding of semiconductors. Vacuum tube technology had fully matured at that time but it had become also clear that the short life and the high power consumption of tubes would limit further progress in telephony and other electronic endeavours. The transistor effect was discovered in 1947 by WILLIAM B. SHOCKLEY, JOHN BARDEEN AND WALTER H. BRATTAIN, three members of the Bell Labs group, who received the Nobel prize in physics in 1956 '... *for their studies of semiconductors and the invention of the transistor effect.'* This invention triggered one of the most remarkable odysseys in the history of mankind, from the point-contact transistor via grown-junction transistors, alloy transistors, mesa transistors to planar transistor and numerous other microelectronic devices. This journey continues until the present day. Semiconductors produced a revolution in mankind at least as profound as the introduction of steam engines and steel. Nowadays, semiconductor electronics pervades our life and has an impact on everything that we do at work or at home.

There was some basis for an understanding of the physics of semiconductor materials already in the 1940s. The concept of band gaps existed. Two types of conduction, already named n-type and p-type, had been identified and attributed to the presence of certain impurities (nowadays called dopants) in very small concentrations. What were called p-n junctions had been found within ingots formed by melting and refreezing the purest silicon then commercially available. In pure semiconductor crystals all valence electrons are used to form bonds of a completely occupied valence band, which is separated by the band gap from the conduction band. The latter is empty at low temperatures. Thermal excitation across the band gap can create electrons in the conduction band and 'missing electrons in the valence band', which are called holes. Electrons and holes give rise to the intrinsic carrier density and the associated conductivity of a semiconductor. The carrier density increases according to a Boltzmann factor containing the gap energy.

Although the first transistor – and about ten years later the first integrated circuits – were made not from Si but from its sister element Ge, it was soon understood that Si would be a better transistor material than Ge for most applications. This mainly resulted from the higher band gap of Si (see Table 22.1). In Ge at room temperature, the thermal generation of minority carriers led to substantial reverse currents in p-n junctions. The reverse current in Si was orders of magnitude smaller and made the material a superior rectifier.

However, at the beginning of the odyssey there was much uncertainty, much still unknown. The highest purity of semiconductors available was orders of magnitude above the purity that was eventually needed. Semiconductor materials were polycrystals and frequently used in powder form. The chemist JAN CZOCHRALSKI (1885–1953), who was born in West Prussia, Germany, (now part of Poland), developed the Czochralski (CZ) process. This process has been used since the 1950s to grow semiconductor single crystals. The most serious problem with Si was that critical chemical and metallurgical processes all took place at higher temperatures than with Ge. These problems were solved and oriented single crystals of silicon, germanium and of III-V compounds could be grown. Nowadays, high-purity and high-perfection CZ silicon single crystals 30 cm in diameter and 2 m long, weighing 150 kg are commercial standard and are used in the production of silicon wafers. The production of larger crystals is on the way.

Foreign atoms incorporated in the crystal lattice are vital for the electronic properties of semiconductors. The group III elements boron (B), aluminium (Al), gallium (Ga), indium (In) in Si and Ge and Zn in GaAs act as acceptors since they can pick an electron from the valence band by creating a hole. The donor elements phosphorus (P), arsenic (As), antimony (Sb), bismuth (Bi) in Si and Ge can provide electrons to the conduction band. The conductivity (resistivity) of a semiconductor can be changed by the process of doping by several orders of magnitude, about eight orders in the case of silicon (Fig. 22.2). The in-diffusion of doping elements into the semiconductor surface is an important process during device fabrication, whereas background doping is performed by adding the doping element to the semiconductor melt during the crystal growth process.

A further important observation was made at Bell Laboratories in 1955. A thin layer of silicon dioxide grown by thermal oxidation on the surface of silicon prior to diffusion could mask the diffusion of certain donor and acceptor elements into Si. It was also discovered that diffusion would occur unimpeded through windows etched into the oxide layer. Somewhat later it was shown that certain photoresists deposited at the oxide surface prevent etching of the oxide. Hence, optical exposure of the resist by projection or through contact masks could be used to create precise window patterns in

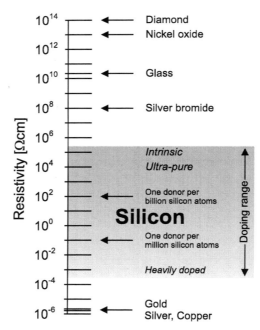

Fig. 22.2. Resistivity of various materials and silicon of various doping levels

the oxide and in turn provide precise control of the areas in which diffusion would occur. This combined process of photolithography and oxide masking (photoresist process) has since been developed to a precision that can be controlled to sub-micrometer scales. This development complements the precision of the depth control of junctions diffused into the Si surface, providing the means to control the fabrication of devices in three dimensions to the precision of a fraction of a micron. These advances also ended the rôle of Ge as a major player. No material was found that would provide diffusion masking for Ge.

Transistor densities on integrated circuits manufactured on Si wafers have increased exponentially over decades, by a succession of lithography techniques, which are used in chip making to project patterns on wafers. GORDON MOORE, a co-founder of the Intel Company, says '... *the number of elements on a microchip and the switching speed double every eighteen month*' – a rule that has worked for over 40 years.

The greatest impact of the III-V compound semiconductors has not been in the areas dominated by Si. Its applications depend one unique properties of these compounds. One class of applications is based on the transfer of high-mobility electrons from the conduction-band minimum to the low-mobility indirect minimum such as that found in GaAs and other direct band gap semiconductors. One rapidly evolving technology based on electroluminescence is optoelectronics, where Si and Ge still cannot compete with direct band gap III-V compounds. Electroluminescence is the emission of light associated with the application of a small voltage. In 1962, the observation of high quantum efficiencies in the infrared for GaAs and in the visible for GaP, and the achievement of laser action in GaAs resulted in considerable research and development activity on III-V compounds. It has been demonstrated later that high-brightness light emitting diodes (LED's) based on III-V GaN compound devices can be used for interior and exterior applications in automobiles and in traffic lights.

Silicon-carbide is a material with high potential for some niche applications due to its high melting temperature, its large band gap and its high thermal conductivity. These applications include high-temperatures up to $1000\,^\circ$C, high voltages, and applications under conditions of particle irradiation. Recent improvements of the quality and size of SiC wafers are promising for SiC microelectronic devices.

We end this section by drawing the reader's attention to literature about the significance of semiconductors [2–5].

22.2 Specific Features of Semiconductor Diffusion

The advent of microelectronic and optoelectronic devices has strongly stimulated the interest in diffusion processes in semiconductors, because solid-state diffusion is used as a fundamental process in their manufacture. Elements from different parts of the periodic table have been studied with respect to their diffusion behaviour for various reasons:

- Shallow dopants such as B, P in Si, As, Ga in Ge and Zn in GaAs determine the conductivity type of the material.
- The light elements O and C are major contaminants and difficult to eliminate during chemical processing and crystal growth. In Czochralski-grown Si single crystals, oxygen plays a crucial rôle for impurity gettering.
- Some impurities introduce deep levels in the forbidden gap of the semiconductors. Such impurities like Au in Si are useful for effective electron-hole recombination in fast switching devices. Another example is Cr, which is used for electronic compensation of hard to avoid donors and acceptors in GaAs.
- Fast diffusing metallic impurities are feared for their detrimental effects on device performance. Well-known examples are Fe and Ni in Si and Cu in Ge. Without taking severe precautions, these metals are unintentionally introduced during high-temperature processes.
- Other metallic elements are deliberately deposited on semiconductor surfaces to produce ohmic or rectifying contacts.
- Last but not least, several elements are of interest from a fundamental point of view:
    - Diffusion of self-atoms is the most basic diffusion process. Therefore, its understanding is important.
    - The diffusion of *hybrid solutes* such as Au, Pt, Zn, Ir in Si and Cd, Zn in GaAs can provide information on native point defects. Hybrid solutes dissolve on substitutional and interstitial sites of the host crystal. Their diffusivity in the interstitial state is much higher than in the substitutional state.

To be able to predict and control diffusion steps is important for the design and development of semiconductor devices. If diffusion exhibited only simple behaviour following from Fick's laws for constant diffusivity, the prediction of the diffusion behaviour would then be a straightforward task, limited only by the accuracy of the diffusion data. That diffusion in semiconductors can be considerably more complicated became apparent in the early 1960s and continues to the present day. The scientific and technological problems that have emerged are associated with the fact that in semiconductors particular factors influence diffusion. Such factors are:

1. The covalent bonding entails relatively high formation enthalpies of intrinsic point defects such as vacancies and self-interstitials. As a consequence, the concentrations of point defects in thermal equilibrium are orders of magnitude lower than in metals (see Chap. 5).
2. The open structure of the diamond and zinc blende lattice favours interstitial incorporation of self- and foreign atoms. For this reason, interstitial

22.2 Specific Features of Semiconductor Diffusion

diffusion and interstitial-substitutional exchange diffusion of hybrid solutes via the dissociative or kick-out mechanisms, (see Chaps. 6 and 25) are more prominent in semiconductors than in metals.

3. Beside vacancies, self-interstitials may act as diffusion vehicles. It is generally accepted that shallow dopants in silicon and host atoms exchange with both types of intrinsic point defects. This so-called dual mechanism gives rise to a variety of complex diffusion phenomena.
4. The covalent bonding between the host atoms limits the solubility of most foreign atoms to ppm ranges. This also implies that small amounts of solute substance can act as inexhaustible diffusion sources.
5. Commercially available semiconductors are extremely pure. Impurity levels are commonly below about 0.1 ppm. Therefore, stringent precautions must be taken in order to reduce impurity contamination during sample preparation and diffusion treatments.
6. Some semiconductor materials can be grown as single crystals with very high perfection. Dislocations densities vary from values as low as 10^3 to 10^4 cm^2 (GaAs) to virtually zero (Si). Germanium crystals with dislocation densities in the range 10^3 to 10^5 cm^{-2} are quite common. Grain boundaries and dislocations, which usually help attain the equilibrium concentration of point defects, are often absent. Then, only free surfaces can act as sinks or sources for the annihilation or creation of point defects.
7. Electronic effects play an important rôle in semiconductors:
    a) The electric-field effect accounts for the diffusion drift, which is due to the electric field built-up by inhomogeneous distributions of dopants.
    b) The Fermi-level effect describes changes in the equilibrium concentration of intrinsic point defects or in the solute solubilities, caused by shifts of the Fermi level. These changes are due to the fact that point defects and solute atoms can occur in various charge states (neutral and ionised). Shifts of the Fermi level originate either by background doping or can be induced by the diffusing species itself.
    c) Self-doping during the diffusion of a solute, which acts as dopant, can lead to non-Fickian diffusion profiles. Due to the larger band gap the Fermi-level effect is more important in GaAs than in Si and Ge. For the elemental semiconductors it is detectable in diffusion mainly at high doping concentrations.
    d) Ion-pairing effects arise from the Coulomb interaction between charged solute atoms and intrinsic point defects.
8. Surface reactions proceeding under certain ambient conditions may strongly influence solute diffusion. Oxidation-enhanced or -retarded diffusion is a well-known example. The crucial rôle of self-interstitials injected into silicon by the oxidising SiO_2/Si interface is a widely accepted example.
9. Point defects such as vacancies and self-interstitial, if created or annihilated for example in dissociative or kick-out reactions with hybrid solutes,

can occur in under- or oversaturation (see Chap. 25). This is because grain boundaries and dislocations, which usually help attain the equilibrium concentration of point defects, are often absent in perfect semiconductor crystals.

A comprehensive collection of diffusion data in Si and Ge, Si-Ge, and SiC alloys has been assembled by STOLWIJK AND BRACHT [6]. Diffusion data for silicides are summarised by GAS AND D'HEURLE [7] and those for compound semiconductors by DUTT AND SHARMA [8]. Solubility data for solutes in Si have been collected by BRACHT AND STOLWIJK and for solutes in Ge by STOLWIJK AND BRACHT and can be found in [10]. Chemical diffusion data in inhomogeneous semiconductor compounds are summarised by BRUFF AND MURCH [9]. The subjects of diffusion and point defects in semiconductors have been treated in a textbook [1] and many reviews [11–20].

References

1. D. Shaw (Ed.), *Atomic Diffusion in Semiconductors*, Plenum Press, 1973
2. I.M. Ross, *The Foundation of Semiconductor Age*, Physics Today, 34–47 (1997)
3. H. Queisser, *The Conquest of the Microchip*, Havard Univerity Press, 1998
4. M. Eckert, H. Schubert, *Crystals, Electrons, Transistors*, AIP, 1997
5. M. Riordan, L. Hoddeson, *Crystal Fire*, W.W. Norton and Co., 1997
6. N.A. Stolwijk, H. Bracht, *Diffusion in Silicon, Germanium and their Alloys*, in: *Diffusion in Semiconductors and Non-Metallic Solids*, D.L. Beke (Vol. Ed.), Landolt-Börnstein, New Series, Group III, Vol. 33, Subvolume A, Springer-Verlag, Berlin, 1998
7. P. Gas, F.M. d'Heurle, *Diffusion in Silicides*, in: *Diffusion in Semiconductors and Non-Metallic Solids*, D.L. Beke (Vol. Ed.), Landolt-Börnstein, New Series, Group III, Vol. 33, Subvolume A, Springer-Verlag, Berlin, 1998
8. M.B. Dutt, B.L. Sharma, *Diffusion in Compound Semiconductors*, in: *Diffusion in Semiconductors and Non-Metallic Solids*, D.L. Beke (Vol. Ed.), Landolt-Börnstein, New Series, Group III, Vol. 33, Subvolume A, Springer-Verlag, Berlin, 1998
9. C.M. Bruff, G.E. Murch, *Chemical Diffusion in Semiconductors*, in: *Diffusion in Semiconductors and Non-Metallic Solids*, D.L. Beke (Vol. Ed.), Landolt-Börnstein, New Series, Group III, Vol. 33, Subvolume A, Springer-Verlag, Berlin, 1998
10. M. Schulz (Vol. Ed.), *Impurities and Defects in Group IV Elements and III-V Compounds*, Landolt-Börnstein, New Series, Group III, Vol. 22b, Springer-Verlag, Berlin, 1989
11. A. Seeger, K.P. Chik, *Diffusion Mechanisms and Point Defects in Silicon and Germanium*, Phys. Stat. Sol. **29**, 455–542 (1968)
12. D. Shaw, *Self- and Impurity Diffusion in Ge and Si*, Phys. Stat. Sol. (b) **72**, 11–39 (1975)
13. A.F. Willoughby, *Atomic Diffusion in Semiconductors*, Rep. Progr. Phys. **41**, 1665–1705 (1978)

14. W. Frank, U. Gösele, H. Mehrer, A. Seeger, *Diffusion in Silicon and Germanium*, in: *Diffusion in Crystalline Solids*, G.E. Murch, A.S. Nowick (Vol. Eds.), Academic Press Inc., 1984
15. P.M. Fahey, P.B. Griffin, J.D. Plummer, *Point Defects and Dopant Diffusion in Silicon*, Rev. Mod. Phys. **61**, 289–384 (1989)
16. N.A. Stolwijk, *Atomic Transport in Semiconductors: Diffusion Mechanisms and Chemical Trends*, Defect and Diffusion Forum **95–98**, 895–916 (1993)
17. S.M. Hu, Mater. Sci. Eng. **R13**, 105 (1994)
18. N.A. Stolwijk, G. Bösker, J. Pöpping, *Hybrid Impurity and Self-Diffusion in GaAs and Related Compounds: Recent Progress*, Defect and Diffusion Forum **194–199**, 687–702 (2000)
19. H. Bracht, *Diffusion and Intrinsic Point-Defect Properties in Silicon*, MRS Bulletin, 22–27. June 2000
20. T.Y. Tan, U. Gösele, *Diffusion in Semiconductors*, in: *Diffusion in Condensed Matter – Methods, Materials, Models*, P. Heitjans, J. Kärger (Eds.), Springer-Verlag, 2005, p. 165

23 Self-diffusion in Elemental Semiconductors

Compared with metals, self-diffusion in semiconductors is a very slow process. For the elemental semiconductors this is illustrated in Fig. 23.1, in which the self-diffusivities of Si and Ge are compared with those of typical metals. At the melting temperature T_m, self-diffusion in Si and Ge is at least four orders of magnitude slower than in metals. This difference increases at lower temperatures due to the relatively large activation enthalpies of the semiconducting materials. Generally speaking, the origin of these differences and of others to be discussed in the present chapter and in the two subsequent ones lies in the difference between metallic and covalent bonding.

The basic question is the same as in the case of metals: What are the mechanisms of self-diffusion? Does it occur by a vacancy or an interstitialcy mechanism (see Chap. 6)? The usual way to answer this question is to determine the mobility and the equilibrium concentration of point defects sep-

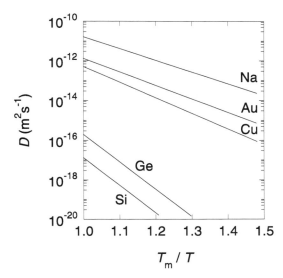

Fig. 23.1. Self-diffusion of Si and Ge and of some metals (Cu, Au, Na) in a homologous temperature scale

arately as functions of temperature and to check whether they are able to account for the measured values of self-diffusion. In the case of metals, direct techniques such as differential dilatometry, positron annihilation, or quenching from high temperatures (see Chap. 5) have thrown considerable light on point defects and diffusion. Up to now, it has not been possible to deduce reliable information on the concentration of equilibrium point defects in semiconductors by any of the direct techniques.

Very likely, these 'conventional' techniques fail for semiconductors, because the equilibrium concentrations of intrinsic point defects are extremely small, presumably less or much less than one ppm near the melting temperature. This implies that point defect concentrations are below the present detection limits of the above mentioned techniques. This is in accordance with the low self-diffusivities, which can be attributed largely to low equilibrium concentrations of intrinsic defects.

For Si and Ge, the above questions had to be answered by employing less direct approaches. In this context, the study and thorough analysis of foreign atom diffusion turned out to be a useful tool. For example, this tool permitted to separate the contributions of vacancies and self-interstitials to the self-diffusion coefficient of silicon (see below).

23.1 Intrinsic Point Defects and Diffusion

Unlike in metals, in semiconductors not only vacancies (V) but also self-interstitials (I) are relevant for atomic transport. The thermal concentration of self-interstitials, C_I^{eq}, and their diffusivity, D_I, are described by

$$C_I^{eq} = \exp\left(\frac{S_I^F}{k_B}\right) \exp\left(-\frac{H_I^F}{k_B T}\right) \quad \text{and}$$

$$D_I = g_I a^2 \nu_I^0 \exp\left(\frac{S_I^M}{k_B}\right) \exp\left(-\frac{H_I^M}{k_B T}\right), \quad (23.1)$$

where H_I^F and H_I^M are the formation and migration enthalpies, S_I^F and S_I^M denote the respective entropies of self-interstitial formation and migration, a is the lattice parameter, $g_I = 1/4$ is the geometry factor for interstitialcy diffusion, and ν_I^0 an attempt frequency of the order of the Debye frequency. The thermal concentration of vacancies, C_V^{eq}, and their diffusivity, D_V, are described by

$$C_V^{eq} = \exp\left(\frac{S_V^F}{k_B}\right) \exp\left(-\frac{H_V^F}{k_B T}\right) \quad \text{and}$$

$$D_V = g_V a^2 \nu_V^0 \exp\left(\frac{S_V^M}{k_B}\right) \exp\left(-\frac{H_V^M}{k_B T}\right), \quad (23.2)$$

where H_V^F and H_V^M are the formation and migration enthalpies, S_V^F and S_V^M denote the respective entropies. $g_V = 1/8$ is the geometry factor for vacancy diffusion in the diamond structure and ν_V^0 the vacancy attempt frequency.

Vacancies and self-interstitials in semiconductors may carry electronic charge due to the energy levels of the defects in the band gap (see Chap. 5). At room temperature, the relative abundancies among the differently charged point defects can be readily changed by n- or p-doping of the base material. By contrast, at the elevated temperatures usually encountered in diffusion experiments, doping levels as high as 10^{25} m^{-3} are necessary to render Ge or Si electronically extrinsic (see, e.g., the textbook of SZE [1]). In GaAs lower dopant concentrations of 10^{24} m^{-3} are already enough for extrinsic behaviour at 1200 K. This is a consequence of the higher band gap energy of GaAs and of the associated lower intrinsic carrier concentrations compared to Ge and Si. In what follows, we concentrate on elemental semiconductors. Then, it is often sufficient to consider intrinsic conditions.

Despite their weak sensitivity in elemental semiconductors to doping, C_I^{eq} and C_V^{eq} as well as $C_I^{eq} D_I$ and $D_V^{eq} D_V$ contain contributions of several charge states [2, 3]:

$$C_I^{eq} = C_{I^0}^{eq} + C_{I^+}^{eq} + C_{I^{2+}}^{eq} + \ldots \tag{23.3}$$

$$C_V^{eq} = C_{V^+}^{eq} + C_{V^0}^{eq} + C_{V^-}^{eq} \ldots, \tag{23.4}$$

and

$$C_I^{eq} D_I = C_{I^0}^{eq} D_{I^0} + C_{I^+}^{eq} D_{I^+} + C_{I^{2+}}^{eq} D_{I^{2+}} + \ldots \tag{23.5}$$

$$C_V^{eq} D_V = C_{V^+}^{eq} D_{V^+} + C_{V^0}^{eq} D_{V^0} + C_{V^-}^{eq} D_{V^-} + \ldots . \tag{23.6}$$

$I^0, I^+, I^{2+}, V^+, V^0, V^-, \ldots$ denote charge states of self-interstitials and vacancies. Unfortunately, it is still not unambiguously known which configurations dominate at various temperatures in Si [3]. On the other hand, there is more information on vacancy charge states in Ge [5] (see below). However, for Ge and Si self-diffusion in intrinsic material it often suffices to utilise the overall transport product of self-interstitials, $C_I^{eq} D_I$, relative to that of vacancies, $C_V^{eq} D_V$.

The transport products are intimately connected with the tracer diffusivities of host atoms [4–6]. Taking into account contributions of the vacancy and of the interstitialcy mechanism to self-diffusion, the tracer diffusivity is given by

$$D^* = f_I C_I^{eq} D_I + f_V C_V^{eq} D_V + \underbrace{D_{ex}}_{\approx 0} . \tag{23.7}$$

$f_I = 0.7273$ is the correlation factor for the interstitialcy mechanism and $f_V = 0.5$ for the vacancy mechanism in the diamond lattice (see Table 7.2). D_{ex} stands for a possible contribution from a direct exchange of crystal atoms. However, first-principles calculations of self-diffusion coefficients have shown that D_{ex} plays no rôle [26] and no experimental evidence has been found for direct exchange. In the following we omit this term in Eq. (23.7).

The self-interstitial can be the major diffusion vehicle under certain conditions, e.g., in Si at high temperatures or in GaAs under heavy p-doping. This does not necessarily mean that the self-interstitial concentration in thermal equilibrium, C_I^{eq}, is higher than the equilibrium concentration of vacancies, C_V^{eq}. What counts from the standpoint of diffusion is the magnitude of the transport product $C_I^{eq} D_I$ relative to $D_V^{eq} D_V$. A low defect concentration may be overcompensated by a high defect mobility. This is the case for Si, where $C_I^{eq} \ll C_V^{eq}$ but nevertheless $C_I^{eq} D_I > D_V^{eq} D_V$ holds at high temperatures. By contrast, vacancies are the major diffusion vehicles in Ge (see below).

23.2 Germanium

Self-diffusion studies in germanium have been performed by radioactive tracers (^{71}Ge or ^{68}Ge). LETAW ET AL. [7] and VALENTA AND RAMASASHTRY [8] used radiotracer techniques in combination with either mechanical or chemical sectioning techniques. In the latter case, the effect of heavy doping was also studied. WIDMER AND GUNTER-MOHR [9] employed neutron activated natural Ge, in which ^{71}Ge is the most abundant radioisotope, and either the Steigmann or the Gruzin technique (see Chap. 13) in combination with grinder sectioning. CAMPBELL [10] studied the simultaneous diffusion of the radioisotopes ^{71}Ge and ^{77}Ge at 900 and 925 °C and deduced the isotope effect therefrom. VOGEL ET AL. [11] and WERNER ET AL. [12] used ^{71}Ge and employed a sputtering technique for serial sectioning (see Chap. 13). In [12] the influence of n- and p-doping and the effect of hydrostatic pressure on self-diffusion was also investigated. FUCHS ET AL. [13] used stable ^{70}Ge/^{74}Ge isotope heterostructures grown by molecular-beam epitaxy and SIMS profiling after interdiffusion.

The overall agreement between self-diffusion data of various authors is good (Fig 23.2). The temperature dependences are well described by Arrhenius laws and the activation parameters obtained from the measurements of different groups (D^0: 7.8×10^{-4} to 44 ×10^{-4} m^2 s^{-1}; ΔH: 2.97 eV to 3.14 eV) are not much different. The pre-exponential factors of Ge are larger than the D^0 values for typical metals (see Chap. 17). The corresponding self-diffusion entropy is about 10 k_B for Ge, whereas for close-packed metals values between 1 and 3 k_B are common.

It is commonly agreed (see, e.g., [4–6]) that self-diffusion in Ge occurs by a vacancy mechanism. This interpretation is among other observations (see also Chaps. 24 and 25) supported by the following ones:

– Measurements of the dependence of the diffusion coefficient on the isotope mass (isotope effect) have been performed by CAMPBELL using the isotope

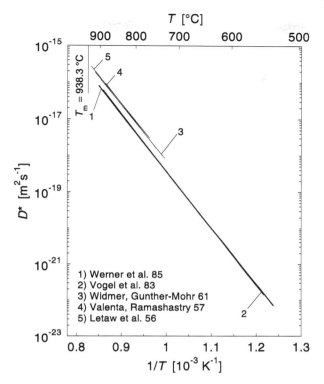

Fig. 23.2. Self-diffusion coefficients of Ge measured by tracer methods [7–9, 11, 12]

pair ^{71}Ge/^{77}Ge [10]. The following values of the isotope effect parameter E (see Chaps. 24 and 25) have been reported:

$$E = f\Delta K = 0.25 \text{ to } 0.3 \tag{23.8}$$

The correlation factor f for a vacancy mechanism in the diamond structure is 0.5. Hence, the interpretation in terms of a vacancy mechanism requires values of the kinetic energy factor $\Delta K = 0.5$ to 0.6. Such values are smaller than the values of $\Delta K \approx 0.9$, which are typical for monovacancies in close-packed metals. ΔK decreases with the number of atoms with which the diffusing atom shares its kinetic energy in the saddle point. It has been concluded that in close-packed metals the vacancy is rather localised with only small relaxation of neighbouring atoms whereas in germanium it is more relaxed [5]. This is in accordance with the relatively large diffusion entropy mentioned above.
– The doping dependence of Ge self-diffusion has been attributed to the vacancy mechanism with contributions of neutral and charged vacancies.

Diffusion-mediating defects in semiconductors occur in various charge states (see above and Chap. 5). The equilibrium concentration of charged defects depends on the position of the Fermi level. A vacancy in germanium has an acceptor level in the lower half of the band gap [4]. If self-diffusion proceeds via neutral vacancies (V^0) and singly charged vacancies (V^-), the total tracer self-diffusion coefficient D^* is given by

$$D^* = D^*_{V^0} + D^*_{V^-}, \tag{23.9}$$

where $D^*_{V^0}$ denotes the contribution of neutral vacancies and $D^*_{V^-}$ that of charged vacancies. These contributions are given by

$$D^*_{V^0} = f_V g_V a^2 \omega_{V^0} C^{eq}_{V^0} \quad \text{and}$$
$$D^*_{V^-} = f_V g_V a^2 \omega_{V^-} C^{eq}_{V^-}, \tag{23.10}$$

where f_V is the correlation factor, g_V a geometrical factor, and a the lattice parameter. For the vacancy mechanism in the diamond lattice, we have $f_V = 0.5$ and $g_V = 1/8$. $C^{eq}_{V^0}$ and $C^{eq}_{V^-}$ denote the concentrations of neutral and charged vacancies in thermal equilibrium, ω_{V^0} and ω_{V^-} the exchange jump rates between tracer atom and neutral and charged vacancies, respectively.

For non-degenerate semiconductors it is readily shown (see Chap. 5) that

$$\frac{C^{eq}_{V^-}(n)}{C^{eq}_{V^-}(n_i)} = \frac{n}{n_i}, \tag{23.11}$$

where n and n_i are the free electron densities in the doped and intrinsic material, respectively. Combining Eqs. (23.9), (23.10), and (23.11) yields for the ratios of self-diffusion under extrinsic and intrinsic conditions

$$\frac{D^*(n)}{D^*(n_i)} = \frac{D^*_{V^0}(n_i)}{D^*(n_i)} + \frac{n}{n_i} \frac{D^*_{V^-}(n_i)}{D^*(n_i)}. \tag{23.12}$$

This equation predicts a linear dependence of the tracer diffusivity on the ratio n/n_i. It is based on the assumption that the vacancy occurs in a neutral and a single-acceptor state. The occurrence of additional charge states would lead to additional terms in Eq. (23.12) with factors $(n/n_i)^r$ for an acceptor level and to $(n_i/n)^r$ for a donor level, where r is the multiplicity of the ionised state.

Figure 23.3 shows the ratio of Ge self-diffusion in doped germanium to that in intrinsic germanium, $D^*(n)/D^*(n_i)$, *versus* the ratio of the free electron densities, n/n_i, according to [12]. The diffusivity increases with n-doping and decreases with p-doping. In addition, Fig. 23.3 shows that vacancies with a single-acceptor state are responsible for Ge self-diffusion and that additional levels are not essential for the description of the doping dependence of self-diffusion.

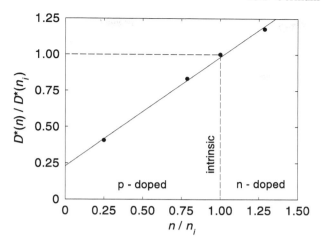

Fig. 23.3. Doping dependence of Ge self-diffusion according to WERNER ET AL. [12]

- Further information about the nature and properties of point defects involved in the diffusion process has been deduced from the effect of pressure on diffusion (see Chap. 8). Measurements of Ge self-diffusion under pressure are reported in [12]. The activation volumes are comparatively small and vary from 0.24 to 0.41 atomic volumes (Ω) as the temperature increases from 876 to 1086 K (Fig. 23.4). Values of the activation volume for self-diffusion of gold are shown for comparison [14]. These larger values are typical for vacancy-mediated diffusion in close-packed metals. The lower values for Ge support the concept that self-diffusion in Ge occurs via vacancies, which are more relaxed or 'spread-out' than in close-packed materials. The positive sign very likely excludes self-interstitials as the defects responsible for Ge self-diffusion. For an interstitialcy mechanism the activation volume should be negative (see Chap. 8).

The experimental activation volume increases with temperature. This increase can be attributed to the fact that neutral and negatively charged vacancies with differnet activation volumes contribute to self-diffusion. The activation volume of the neutral vacancy contribution ($\Delta V_{V^0} = 0.56\,\Omega$) is larger than that of the negatively charged vacancy ($\Delta V_{V^-} = 0.28\,\Omega$) [12]. The experimental activation volume is an average value weighted with the relative contributions of the two types of vacancies to the total self-diffusivity;

$$\Delta V = \Delta V_{V^0} \frac{D^*_{V^0}}{D^*} + \Delta V_{V^-} \frac{D^*_{V^0-}}{D^*} \,. \tag{23.13}$$

Since the contribution of charged defects becomes more significant with decreasing temperature, the effective activation volume decreases.

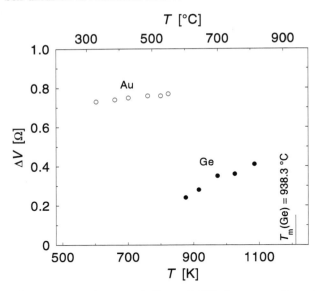

Fig. 23.4. Activation volumes of Ge self-diffusion according to WERNER ET AL. [12]. For comparison activation volumes of Au self-diffusion are also shown [14]

23.3 Silicon

Natural silicon has three stable isotopes with the following abundances: 92.2 % ^{28}Si, 4.7 % ^{29}Si, and 3.1 % ^{30}Si. Self-diffusion studies on silicon are aggravated by the fact that the only accessible radioisotope, ^{31}Si, has a half-life of 2.6 hours. This limits the time for a diffusion experiment. Nevertheless, in view of the great importance of silicon as base material for microelectronic devices several attempts have been made to apply tracer methods using either the radioisotope ^{31}Si, the stable enriched isotope ^{30}Si, or isotopically controlled heterostructures. GHOSHTAGORE [15] evaporated enriched ^{30}Si layers onto Si wafers, performed neutron activation of the sample after diffusion annealing and subsequent chemical sectioning for profile analysis. PEART [16] evaporated the radiotracer ^{31}Si and utilised mechanical sectioning. FAIRFIELD AND MASTERS [17] studied ^{31}Si diffusion in intrinsic and doped Si and sectioned the samples by chemical etching. MAYER ET AL. [18] and HETTICH ET AL. [19] sputter-deposited thin layers of neutron activated Si (containing ^{31}Si) and used sputter sectioning for profile analysis. HIRVONEN AND ANTTILA [20] and DEMOND ET AL. [21] implanted a layer of ^{30}Si and studied its diffusional broadening by a proton beam utilising the resonant nuclear reaction ^{30}Si$(p,\gamma)^{31}$Si. KALINOWSKI AND SEGUIN [22] evaporated layers of ^{30}Si and studied in-diffusion profiles by SIMS. The study by BRACHT ET AL. [23] employed stable silicon isotope heterostructures with highly enriched ^{28}Si

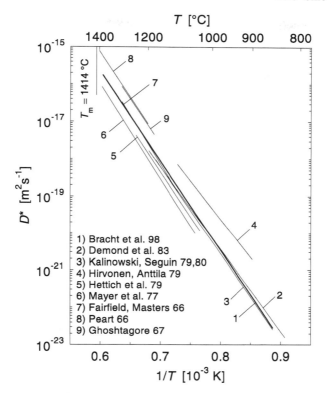

Fig. 23.5. Self-diffusion coefficients of Si measured by tracer methods [15–23]

layers. The heterostructures were grown by chemical vapour deposition on natural Si wafers. Diffusion profiles of ^{29}Si and ^{30}Si isotopes were determined by SIMS.

The data of the various authors are shown in Fig. 23.5. The results are less consistent than those on Ge self-diffusion (see Fig. 23.2) and far less consistent than those on self-diffusion in metallic elements (see, e.g., [24]). This is surprising since in most of the above mentioned studies virtually dislocation-free Si single crystals of high purity were used. The reason for these discrepancies are not completely clear. Oxygen is one of the main impurities in both Czochralski-grown and float-zone single crystals. It is well-known that oxidation at the surface or oxide formation or dissolution causes deviations of intrinsic point defects from their thermal equilibrium concentrations. Implantation damage may be an additional reason for discrepancies in those experiments where ^{30}Si was implanted prior to diffusion. All earlier experiments either suffer from the short half-life of ^{31}Si or from the natural abundance of stable ^{30}Si, when enriched ^{30}Si was used as tracer.

Drawbacks of the earlier experiments have been avoided in the already mentioned study with isotope heterostructures by BRACHT ET AL. [23]. Neither a short-lived radioisotope nor the background of the natural ^{30}Si limited the experiment. An influence of the surface can very likely be also excluded, since diffusion is observed near an interface of the isotope heterostructure. Even at the lowest temperature no doping effects could be detected. Thus, it is believed that these data reflect intrinsic behaviour of Si self-diffusion. Values of the self-diffusion coefficient cover about seven orders of magnitude and the widest temperature range of all studies. Accordingly, the self-diffusivity of intrinsic silicon is described by [23]

$$D^* = 0.53 \exp\left(-\frac{4.75 \text{ eV}}{k_B T}\right) \text{ m}^2 \text{ s}^{-1}. \tag{23.14}$$

Measurements of the tracer self-diffusion coefficient alone are not definitive in establishing the self-interstitial and vacancy contributions to self-diffusion. Additional information is needed and can be obtained, for example, from the analysis of self- and foreign-atom diffusion experiments involving non-equilibrium concentrations of intrinsic point defects. Below, we summarise the main conclusions that have been drawn from such experiments concerning Si self-diffusion [25].

The native point defect which predominantly mediates Si self-diffusion is the self-interstitial. This conclusion is consistent with experimental data for the transport product $C_I^{eq} D_I$ deduced from diffusion of hybrid foreign elements (see Chap. 25). The self-interstitial contribution, $C_I^{eq} D_I$, is well-known from diffusion studies of hybrid foreign atom diffusers (Au: [27], Zn: [28]). Therefore, one can extract the vacancy term, $C_V^{eq} D_V$ in Eq. (23.7), from the total self-diffusivity. From an analysis of Zn diffusion in Si BRACHT ET AL. [28] obtain:

$$C_I^{eq} D_I = 0.298 \exp\left(-\frac{4.95 \text{ eV}}{k_B T}\right) \text{ m}^2 \text{ s}^{-1}. \tag{23.15}$$

Using this expression, a fit of Eq. (23.7) to the most reliable self-diffusion data [23] yields the vacancy transport product as:

$$C_V^{eq} D_V = 0.92 \times 10^{-4} \exp\left(-\frac{4.14 \text{ eV}}{k_B T}\right) \text{ m}^2 \text{ s}^{-1}. \tag{23.16}$$

The temperature dependences of the transport products, $C_I^{eq} D_I$ and $C_V^{eq} D_V$, are shown in Fig. 23.6 and compared with the self-diffusion data [23]. The agreement between D^* and $f_I C_I^{eq} D_I + f_V C_V^{eq} D_V$ (taking into account the correlation factors) implies that self-diffusion is mediated by both self-interstitials and vacancies. $C_I^{eq} D_I$ equals $C_V^{eq} D_V$ at about 890 °C and not at temperatures between 1000 and 1100 °C as had been assumed earlier [4, 5]. At temperatures above 890 °C the self-interstitial contribution dominates, whereas at lower temperatures vacancy-mediated diffusion dominates.

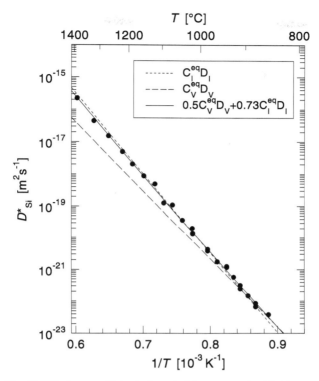

Fig. 23.6. Si self-diffusion coefficients (*symbols*) compared with the self-interstitial and vacancy transport products, $C_I^{eq} D_I$ and $C_V^{eq} D_V$, according to BRACHT ET AL. [23]

Slightly different activation parameters for the transport products are reported in [6] based on earlier studies.

The enthalpy and entropy of self-interstitial mediated diffusion obtained from Eq. (23.15) are:

$$H_I^{SD} = H_I^F + H_I^M = 4.95\,\text{eV} \quad \text{and} \quad S_I^{SD} = S_I^F + S_I^M \approx 13.2\,k_\text{B}\,. \quad (23.17)$$

Those for vacancy-mediated diffusion are

$$H_V^{SD} = H_V^F + H_V^M = 4.14\,\text{eV} \quad \text{and} \quad S_V^{SD} = S_V^F + S_V^M \approx 5.5\,k_\text{B}\,. \quad (23.18)$$

Theoretical calculations of intrinsic point defect properties are in good agreement with these values. TANG ET AL. [29] obtained from their tight-binding molecular dynamic studies $H_I^{SD} = (5.18 \pm 0.2)\,\text{eV}$, $H_V^{SD} = (4.07 \pm 0.2)\,\text{eV}$, and $S_I^{SD} = 14.3\,k_\text{B}$. The first principles calculations of BLÖCHL ET AL. [26] yield $S_V^F = (5 \pm 2)\,k_\text{B}$ leaving for S_V^M values between 1 and 2 k_B.

In conclusion, we can state that experiments and theory suggest that Si self-diffusion is mediated by simultaneous contributions of self-interstitials and vacancies. Self-interstitials dominate at high temperatures whereas vacancies take over at lower temperatures.

The doping dependence of Si self-diffusion [5] allows the conclusion that neutral as well as positively and negatively charged defects are involved in self-diffusion. However, the data are not accurate enough to determine the inidvidual terms in Eqs. (23.5) and (23.6). Since the total tracer diffusivity as well as the transport products consist of several terms, Eqs. (23.15) and (23.16) can only be approximations holding for a limited temperature range.

References

1. S.M. Sze, *Physics of Semiconductor Devices*, Wiley, New York, 1967
2. S.T. Pantelides, Defect and Diffusion Forum **75**, 149 (1991)
3. U. Gösele, T.Y. Tan, Defect and Diffusion Forum **83**, 189 (1992)
4. A. Seeger, K.P. Chik, *Diffusion Mechanisms and Point Defects in Silicon and Germanium*, Phys. Stat. Sol. **29**, 455–542 (1968)
5. W. Frank, U. Gösele, H. Mehrer, A. Seeger, *Diffusion in Silicon and Germanium*, in: *Diffusion in Crystalline Solids*, G.E. Murch, A.S. Nowick (Eds.), Academic Press, 1984
6. T.Y. Tan, U. Gösele, *Diffusion in Semiconductors*, in: *Diffusion in Condensed Matter – Methods, Materials, Models*, P. Heitjans, J. Kärger (Eds.), Springer-Verlag, 2005
7. H. Letaw Jr., W.M. Portnoy, L. Slifkin, Phys. Rev. **102**, 636 (1956)
8. M.W. Valenta, C. Ramsastry, Phys. Rev. **106**, 73 (1957)
9. H. Widmer, G.R. Gunter-Mohr, Helv. Phys. Acta. **34**, 635 (1961)
10. R. Campbell, Phys. Rev. B **12**, 2318 (1975)
11. G. Vogel, G. Hettich, H. Mehrer, J. Phys. C **16**, 6197 (1983)
12. M. Werner, H. Mehrer, H.D. Hochheimer, Phys. Rev. B **32**, 3930 (1985)
13. H.D. Fuchs, W. Walukiewicz, E.E. Haller, W. Dondl, R. Schorer, G. Abstreiter, A.I. Rudnev, A.V. Tikhomirov, V.I. Oshogin, Phys. Rev. B **51**, 16817 (1995)
14. M. Werner, H. Mehrer, in: *DIMETA-82, Diffusion in Metals and Alloys*, F.J. Kedves, D.L. Beke (Eds.), Trans Tech Publications, Diffusion and Defect Monograph Series No. 7, 392 (1983)
15. R.N. Ghostagore, Phys. Rev. Lett. **16**, 890 (1966)
16. R.F. Peart, Phys. Status Solidi **15**, K119 (1966)
17. J.M. Fairfield, B.J. Masters, J. Appl. Phys. **38**, 3148 (1967)
18. H.J. Mayer, H. Mehrer, K. Maier, Inst. Phys. Conf. Series **31**, 186 (1977)
19. G. Hettich, H. Mehrer, K. Maier, Inst. Phys. Conf. Series **46**, 500 (1979)
20. J. Hirvonen, A. Anttila, Appl. Phys. Lett. **35**, 703 (1979)
21. F.J. Demond, S. Kalbitzer, H. Mannsperger, H. Damjantschitsch, Phys. Lett. A **93**, 503 (1983)
22. L. Kalinowski, R. Seguin, Appl. Phys. Lett. **35**, 211 (1979); Erratum: Appl. Phys. Lett. **36**, 171 (1980)
23. H. Bracht, E.E. Haller, R. Clark-Phelps, Phys. Rev. Lett. **81**, 393 (1998)
24. H. Mehrer, N. Stolica, N.A. Stolwijk, *Self-diffusion in Solid Metallic Elements*, Chap. 2 in: *Diffusion in Solid Metals and Alloys*, H. Mehrer (Vol.Ed.), Landolt-Börnstein, New Series, Group III: Crystal and Solid State Physics, Vol. 26, Springer-Verlag, 1990, p.32

25. H. Bracht, MRS Bulletin, June 2000, 22
26. P.E. Blöchl, E. Smargiassi, R. Car, D.B. Laks, W. Andreoni, S.T. Pantelides, Phys. Rev. Lett **70**, 2435 (1993)
27. N.A. Stolwijk, B. Schuster, J. Hölzl, H. Mehrer, W. Frank, Physica (Amsterdam) **116 B&C**, 335 (1983)
28. H. Bracht, N.A. Stolwijk, H. Mehrer, Phys. Rev. B **52**, 16542 (1995)
29. M. Tang, L. Colombo, T. Diaz de la Rubbia, Phys. Rev B **55**, 14279 (1997)

24 Foreign-Atom Diffusion in Silicon and Germanium

Diffusion of foreign atoms in Si and Ge is very important from a technological point of view. Due to its complexity it provides a challenge from a scientific point of view as well. Group-III elements B, Al, Ga and group-V elements P, As, Sb are a special class of foreign elements known as *dopants*. Dopants are easily ionised and act as donors or acceptors. Their solubility is fairly high compared to most other foreign elements except group-IV elements. Diffusion of dopants plays a vital rôle in diffusion-doping to create p-n junctions of microelectronic devices. Diffusion also controls the incorporation of 'unwanted' foreign atoms, e.g., of the metal atoms Fe, Ni, and Cu during thermal annealing treatments of device fabrication. A detailed understanding of the diffusion behaviour of unwanted foreign atoms is of technological significance to keep the contamination of electronic devices during processing to a harmless state. Oxygen diffusion and growth of SiO_2 precipitates play a crucial rôle in gettering processes of unwanted foreign elements. Other foreign atoms like Au in Si are used for tuning the minority-carrier lifetime.

Dopants are incorporated in substitutional sites of the host lattice, some foreign elements dissolve in interstitial sites only, others are *hybrid* foreign elements, which are dissolved on substitutional and on interstitial sites. The very high mobility of the interstitial fraction can dominate diffusion of hybrid elements. In addition, the interstitial-substitutional exchange reactions – either the dissociative or the kick-out reaction – can lead to non-Fickian diffusion profiles of hybrid diffusers.

The diffusion behaviour of foreign elements is largely determined by the type of solution, i.e. whether they are located at substitutional or interstitial sites, or a mixture of both. In what follows, we consider first solubilities of foreign elements and their site preference. Then, we give a brief review of the diffusion of foreign elements in Ge and Si and classify them according to their site occupancy and diffusivity. The theoretical framework of the relatively complex diffusion patterns of hybrid solutes involving interstitial-substitutional exchange reactions is postponed to Chapt. 25.

24.1 Solubility and Site Occupancy

The solubility of a foreign element is the maximum concentration which can be incorporated in the host solid without forming a new phase. Solid solubilities are temperature-dependent as represented by the solvus or solidus lines of the phase diagram. The solubility limit is defined with respect to a second phase. For most foreign elements in solid Ge or Si at high temperatures equilibrium is achieved with the liquid phase. At lower temperatures the reference phase usually is the solid foreign element or a compound of the foreign element. When the foreign atom is volatile, the saturated host crystal is in equilibrium with a vapour. In this case, the solubility depends not only on the temperature but also on the vapour pressure.

For solid-solid equilibria and for solid-vapour equilibria the temperature dependence of the solubility limit is usually described by an Arrhenius relation containing a solution enthalpy. If a solid-liquid equilibrium is involved, the behaviour is more complex. The right-hand side of Fig. 24.1 shows the normal variation of a solubility giving rise to a maximum at the eutectic temperature. A frequently encountered case is the so-called *retrograde solubility*, illustrated on the left-hand side of Fig. 24.1. This phenomenon implies that the maximum solubility is achieved at a temperature which lies below the melting temperature of the host crystal but above the eutectic temperature. Below this maximum temperature, the solubility can often but not always be approximated by an Arrhenius relation.

Solubility data of foreign elements have been collected for Si by SCHULZ [6], for Ge by STOLWIJK [7] and have been updated for Ge by STOLWIJK AND BRACHT [8] and for Si by BRACHT AND STOLWIJK [9]. Depending on the foreign element, the solubility can vary over orders of magnitude: B, P, As in Si and Al, Ga, Sn in Ge can be incorporated to several percent into the host crystals. Other dopant elements, such as Ga, Sb, Li in Si and As and Sb in Ge

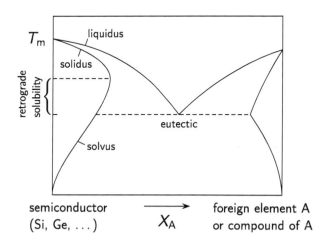

Fig. 24.1. Schematic phase diagram of a semiconductor and a foreign element (or a compound of a foreign element) illustrating the phenomenon of retrograde solubility

have maximum solubilities in the range of 10^{-3} atomic fractions. Noble metal impurities, iron group impurities, nickel group impurities, cobalt group impurities, and Zn have very low solubilities in the range of 10^{-6} atomic fractions.

Generally speaking, every foreign atom A may occupy both substitutional A_s and interstitial A_i sites in the lattice. The equilibrium solubility of either configuration, C_s^{eq} and C_i^{eq}, is determined by the pertinent enthalpy and entropy of solution, which refers to some external phase of the foreign atom. Dealing with equilibrium solubilities one should be aware of the the following features [10]:

- For most foreign atoms one atomic configuration dominates in the well accessible temperature range of diffusion studies between about $2/3\, T_m$ and $0.9\, T_m$:
  - Dopants are dissolved in substitutional sites and diffuse with the aid of intrinsic point defects.
  - Foreign atoms with interstitial site preference ($C_i^{eq} \gg C_s^{eq}$) diffuse via a direct interstitial mechanism. Examples are group-I and some group-VIII elements.
  - Foreign atoms with substitutional site preference ($C_i^{eq} \ll C_s^{eq}$), but some minor interstitial fraction are interstitial-substitutional exchange diffusers (hybrid solutes). Examples are some noble metals and further elements mainly from neighbouring groups of the noble metals.
- The equilibrium solubilities on substitutional and interstitial sites, C_s^{eq} and C_i^{eq}, depend sensitively on the solute, the solvent, and on temperature. C_s^{eq} and C_i^{eq} can have a dissimilar variation with temperature. This leads to appreciable changes in the interstitial to substitutional concentration ratio C_i^{eq}/C_s^{eq} with temperature. There are some foreign elements for which the dissimilar variation and the decreasing solubility near the melting temperature leads to intersections of the solubilities of C_s^{eq} and C_i^{eq}. This is illustrated for the best studied example Cu in Ge in Fig. 24.2.
- In general, foreign atoms in semiconductors carry electronic charges. Therefore, the position of the Fermi level affects their solubility. A concomitant phenomenon is that the ratio C_i^{eq}/C_s^{eq} may change with background doping, because of the different electronic structure of A_i and A_s. Dramatic effects have been observed for 3d transition elements in Si [11, 12]. For example, in intrinsic Si the predominant species of cobalt, Co_i, is a donor. By contrast, in heavily phosphorous doped Si acceptor-like Co_s becomes more abundant.
- In what follows, we confine ourselves to intrinsic conditions. This implies that the intrinsic carrier concentration at the diffusion temperature, $n_i(T)$, exceeds the maximum concentration of electrically active foreign atoms. This is usually fulfilled for host crystals having doping levels not higher than 10^{18} cm^{-3}.[1]

---
[1] In shallow dopant diffusion experiments on, e.g., GaAs the relatively small $n_i(T)$ in conjunction with the high solubility of dopants causes a shift of the Fermi level

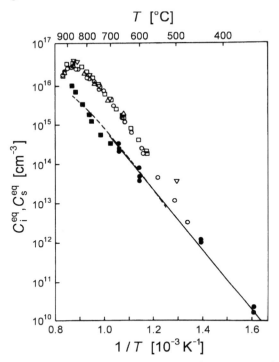

Fig. 24.2. Equilibrium solubilities of substitutional (*open symbols*) and interstitial (*filled symbols*) Cu in Ge according to [10]

- As already mentioned, when the diffuser is supplied from a vapour phase the solubilities depend on the vapour pressure. However, according to the mass action law the ratio C_i^{eq}/C_s^{eq} will not change as long as both A_i and A_s are dissolved as isolated atoms. Pairs and larger clusters of foreign atoms are only formed upon slow cooling. Despite the fixed value of C_i^{eq}/C_s^{eq} at any temperature, the absolute magnitude of the A_i and A_s concentration may influence the diffusion behaviour. This is the case for diffusion of hybrid elements, which diffuse via the kick-out and/or via the dissociative mechanism treated in detail in Chap. 25.

24.2 Diffusivities and Diffusion Modes

Figures 24.3 and 24.4 provide overviews of the diffusivities of technologically important foreign elements in Ge and Si, respectively. In both figures self-

away from the mid of the band gap at the diffusion temperature. In an extreme case like diffusion of the acceptor type dopant Zn in GaAs ($C_s^{eq} \gg n_i(T)$), the Zn_s concentration equals the hole density. This is called self-doping.

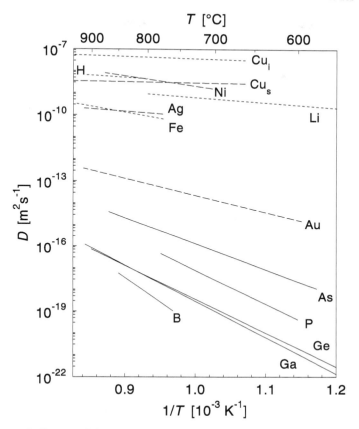

Fig. 24.3. Diffusion of foreign atoms in Ge compared with Ge self-diffusion according to [3]: *Solid lines* represent diffusion of elements that are incorporated on substitutional sites and diffuse via the vacancy mechanism. *Long-dashed lines* represent diffusion of hybrid elements, which are mainly dissolved on substitutional sites; their diffusion proceeds by the dissociative mechanism via a minor fraction in an interstitial configuration (Au, Ag, Ni, Cu). The *short-dashed lines* represent diffusion of elements that diffuse via a direct interstitial mechanism (H, Li [16, 17]). The *short-dashed line on top* shows the diffusivity deduced for interstitial Cu

diffusion and diffusion of substitutional atoms (mainly dopants and for Si also C) are represented by solid lines; diffusivities of foreign elements with interstitial site preference and direct interstitial diffusion are represented by short-dashed lines; hybrid diffusing elements are represented by long-dashed lines. Outstanding features of both figures are the grouping of the diffusivities of dopant elements in a range around or moderately higher than self-diffusion and the fact that the diffusivities of interstitial and hybrid foreign atoms are several orders of magnitude higher. A comprehensive and critical collection

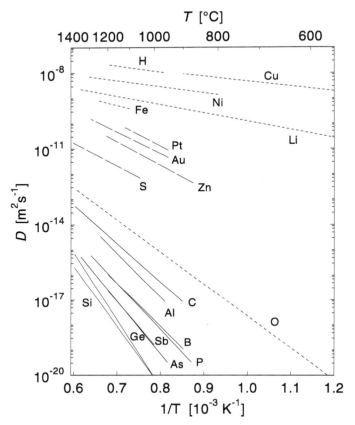

Fig. 24.4. Diffusion of foreign atoms in Si compared with Si self-diffusion according to [4]: *Solid lines* represent diffusion of elements that are incorporated on substitutional sites and diffuse via the vacancy or interstitialcy mechanism (B, Al, Ge, P, As, Sb, As, C). *Long-dashed lines* represent diffusion of hybrid elements, which are mainly dissolved on substitutional sites: their diffusion proceeds via a minor fraction in an interstitial configuration (Au, Pt, S, Zn). The *short-dashed lines* indicate the elements (H [14], Li [16–18], Cu, Fe, Ni, O) that diffuse via the direct interstitial mechanism

of diffusion data for Si and Ge and Ge-Si alloys has been performed by STOLWIJK AND BRACHT [13].

24.2.1 Interstitial Diffusion

Foreign atoms with interstitial site preference can hop from one interstitial site to another. In other words, they diffuse by the direct interstitial mechanism (see Chap. 6). For interstitial diffusion in semiconductors the following features are relevant:

- The diffusion coefficients are very high, commonly in the range between 10^{-7} to 10^{-10} m² s⁻¹ near the melting temperature. The decrease of the diffusivity with decreasing temperature is moderate due to relatively small activation enthalpies between 0.5 and 1.5 eV. Typical examples are H or Li in Ge and H, Li, Cu, Ni, and Fe in Si.
- Usually, interstitial foreign atoms show donor character. This relates to the tendency not to interact with the host atoms by adopting the smallest possible closed-electron-shell configuration.
- Since intrinsic point defects are not involved in interstitial diffusion there is no connection with self-diffusion of the semiconductor host.

Hydrogen: Hydrogen plays a significant rôle in Si technology. It is capable of passivating electrically active defects. The passivation of dislocations and grain boundaries is important for polycrystalline Si used in solar cells. Both donors and acceptors can be passivated. Hydrogen diffuses as an interstitial, either in a neutral form or as a proton. Between room temperature and 650 °C the diffusivity of hydrogen is much slower than values extrapolated from the high-temperature data shown in Fig. 24.4. This low diffusivity has been attributed to the formation of hydrogen molecules [19].

Oxygen: Oxygen is a technologically very important foreign element in Si. Although interstitial oxygen atoms are neutral, they play a crucial rôle in obtaining high device yields. In Czochralski (CZ) silicon, oxygen is incorporated during the growth process from the quartz crucible in concentrations around 10^{24} m⁻³. This concentration can be several orders of magnitude higher than the concentration of electrically active dopants introduced during the fabrication of microelectronic circuits. Since in most Si crystals the grown-in oxygen concentration exceeds the oxygen solubility at lower temperatures, SiO₂ precipitates form by post-growth diffusion processes. These precipitates act as sinks for unintentionally introduced fast diffusing metallic contaminants that must be excluded from the active regions. This process is called *intrinsic gettering*. Controlling the location and size of SiO₂ precipitates and their precursors requires a detailed knowledge of the solubility and diffusivity of oxygen.

The solubility limit of oxygen, C_i^{eq}, has been determined from infrared absorption measurements [21] to be

$$C_i^{eq} = 3 \times 10^{-2} \exp\left(-\frac{1.03 \text{ eV}}{k_\text{B}T}\right)$$
$$= 1.53 \times 10^{27} \exp\left(-\frac{1.03 \text{ eV}}{k_\text{B}T}\right) \text{ m}^{-3}. \quad (24.1)$$

Oxygen dissolves interstitially in a bond-centered configuration shown in Fig. 24.5 (see [22–24]). The oxygen diffusivity has been determined over wide

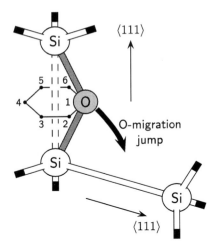

Fig. 24.5. Bond-centered configuration of the oxygen interstitial in the Si lattice according to FRANK ET AL. [2]

temperature ranges ($350 \leq T \leq 1300^0$C) from internal friction measurements, from in-diffused ^{18}O profiles determined by SIMS, and from stress-induced dichroism (see [13, 25] for references). It is given by

$$D_i = 1.3 \times 10^{-5} \exp\left(-\frac{2.53\,\text{eV}}{k_\text{B}T}\right) \text{ m}^2\text{s}^{-1}. \tag{24.2}$$

Slightly different values for the activation enthalpy (2.44 eV) have also been reported in the literature [34]. The activation enthalpy for oxygen diffusion is high compared to hydrogen and to metallic interstitial diffusers. An oxygen atom in Si occupies the bond-centered interstitial position and forms bonds with two Si atoms (Fig. 24.5). The Si-O-Si bridge around the preferred ⟨111⟩ direction is kinked. Above room temperature the rotation of this kinked bond occurs rapidly. However, the elementary step of the migration of oxygen interstitial involves a change into another ⟨111⟩ direction. Hence long-range diffusion requires the breaking and making of Si-O bonds. At temperatures around 450 °C, interstitial oxygen forms electrically active agglomerates called 'thermal donors' [20].

24.2.2 Dopant Diffusion

Group-III and group-V dopants (and also group-IV elements) are incorporated on regular lattice sites. Their diffusivities are not dramatically different from self-diffusion of the host crystals. This is a strong hint that diffusion of dopants is mediated by the same intrinsic defects as self-diffusion. These are vacancies in the case of Ge and self-interstitials plus vacancies in the case of Si. The interaction of dopants (donors or acceptors) with intrinsic defects enters the expression for the dopant diffusivity.

Germanium: Germanium has lost its leading rôle for electronic devices several decades ago (see Chap. 22). It is nowadays mainly used in detectors for γ spectroscopy and in Si-Ge alloys. Therefore, most of the diffusion work on Ge has been performed decades ago. Furthermore, dopant and self-diffusion in Ge are consistently explained in terms of the vacancy mechanism. No self-interstitial contribution is involved, which makes the interpretation of diffusion processes in Si more complex.

Inspection of Fig. 24.3 shows that diffusion of group-III elements is slower (B) or similar (Ga) to self-diffusion of Ge. By contrast, the diffusivities of the group-V elements (P, As) are one to two orders of magnitude larger than the self-diffusion coefficients. The elements Al, In and Sb, not shown in the figure, reveal the same trends [13].

This behaviour is consistent with the picture that self-diffusion in Ge occurs via a vacancy mechanism with contributions mainly from neutral and negatively charged vacancies (see Chap. 23). The features of dopant diffusion and its doping dependence have been taken as evidence that dopants in Ge diffuse by a vacancy mechanism as well [1–4, 34]. An ionised donor atom carries a positive charge. Coulomb attraction enhances the probability to find a negatively charged vacancy in the neighbourhood of a donor atom. Ionised acceptor atoms carry a negative charge and the Coulomb interaction with negatively charged vacancies is repulsive. Acceptor diffusion is mediated mainly by neutral vacancies. Perhaps the small dopant B is an exception. As a result of its slow diffusion, boron is a fairly stable dopant.

Silicon: Dopant diffusion in Si has been reviewed in theory and experiment by FAHEY ET AL. [26] and a collection of diffusion data has been performed by STOLWIJK AND BRACHT [13]. As discussed in Chap. 23, self-diffusion in Si is dominated by the vacancy mechanism at low temperatures and by the interstitialcy mechanism at high temperatures. Contrary to the case of Ge, the diffusion of the common dopants (B, P, As, Sb) in Si is always faster than self-diffusion (see Fig. 24.4), irrespective of whether the atom has a smaller (B, P) or larger size (As, Sb) than Si. This is an indication that diffusion of dopants is mediated by vacancies (V) *and* self-interstitials (I). The diffusion of dopants can be represented by the point defect reactions

$$A_s + V \Longleftrightarrow AV, \quad (V \Rightarrow V^+, V^0, V^-, \ldots) \tag{24.3}$$
$$A_s + I \Longleftrightarrow AI \quad (I \Rightarrow I^+, I^0, I^-, \ldots). \tag{24.4}$$

Intrinsic defects in various charge states ($V^+, V^0, V^-; I^+, I^0, I^-$) approach substitutional foreign atoms A and form defect-dopant pairs according to their specific and charge-dependent interaction. Correspondingly, the dopant diffusivity can be composed of several contributions

$$D_A = D_A(AV) + D_A(AI) \qquad (24.5)$$
$$= D_A(AV^+) + D_A(AV^0) + D_A(AV^-)$$
$$+ D_A(AI^+) + D_A(AI^0) + D_A(AI^-) + \ldots, \qquad (24.6)$$

where $D_A(AV)$ and $D_A(AI)$ denote the total contributions of vacancies and self-interstitials to the dopant diffusivity D_A. Both terms are composed of contributions from dopant-defect pairs with self-interstitials and vacancies in their various charge states. Each single term in Eq. (24.6) is a product, which contains a geometrical factor, the lattice parameter squared, the equilibrium concentration of the defect, a Boltzmann factor with the nearest-neighbour interaction energy of the dopant-defect pair, and a correlation factor. Correlation factors for vacancy-mediated diffusion in the diamond lattice treated in Chap. 7. The correlation factor can be expressed in terms of the jump rates of the defect in the neighbourhood of the dopant. For the interstitialcy mechanism of dopant elements detailed calculations of the correlation effects of AI pairs are not known to the author.

Coulombic interaction between positively charged donors or negatively charged acceptors and charged defect configurations provides a contribution to defect-dopant interaction. Considering lattice distortions small dopants attract self-interstitials, whereas bigger dopants are more attractive for vacancies than for self-interstitials.

Experimental determinations of the activation enthalpies of dopants indicate that in Si the activation enthalpy of self-diffusion can be about 1 eV greater than the activation enthalpy of dopant diffusion [27, 26, 13]. Since dopant diffusion proceeds via the same mechanisms as self-diffusion, a key question is what activation enthalpies can be expected for the vacancy and for the interstitialcy mechanism:

– The energetics of the vacancy mechanism of dopant diffusion for the diamond structure has been pointed out by Hu [27, 28]. An enthalpy diagram of the vacancy diffusion process is displayed in Fig. 24.6. In this figure, H_V^M is the migration enthalpy of an unperturbed vacancy, H_{AV}^M the activation enthalpy for the vacancy-dopant exchange jump, H_{AV}^B the total binding enthalpy of the vacancy-dopant pair, and H_{AV}^{B3} the binding enthalpy when vacancy and dopant atom are separated to the third coordination shell. For long-range migration an AV pair must partially dissociate to at least third-nearest neighbour sites and return to the foreign atom via a different path to complete a diffusion step, otherwise the correlation factor would be zero (see Chap. 7).

An analysis along the lines discussed for solute diffusion in metals (see Chap. 19) yields for the activation enthalpy ΔH_{AV} of dopant diffusion

$$\Delta H_{AV} = H_V^F - H_{AV}^B + H_{AV}^M + C_{AV}, \qquad (24.7)$$

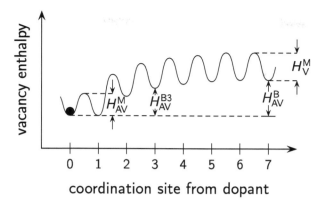

Fig. 24.6. Vacancy enthalpy as a function of the coordination site away from a substitutional dopant (*full circle*)

where the term

$$C_{AV} = -k_B \frac{\partial \ln f_{AV}}{\partial (1/T)} \tag{24.8}$$

is caused by the temperature dependence of the correlation factor. Taking into account the activation enthalpy of vacancy-mediated self-diffusion, $\Delta H_V = H_V^F + H_V^M$, we get

$$\Delta H_V - \Delta H_{AV} = H_V^M - H_{AV}^M + H_{AV}^B - C. \tag{24.9}$$

According to [28] $H_{AV}^B - C \approx H_{AV}^{B3}$. For attractive interaction (Coulombic and/or elastic), the binding enthalpy H_{AV}^{B3} is positive (see Fig. 24.6). Then, the activation enthalpy of self-diffusion is indeed larger than that of dopant diffusion by the same mechanism.
- Less formal analysis has been directed at the atomistic picture of dopant diffusion via the interstitialcy mechanism. However, similar arguments as for vacancy-mediated diffusion can be used.
- Phosphorous in-diffusion profiles show a tail, first observed by YOSHJIDA ET AL. [29], in which the P diffusivity is much higher (about a factor of 100) than expected from isoconcentration studies. In n-p-n transistors, in which high-concentration P diffusion is used for emitter diffusion, the diffusion of the base dopant B below the diffused region is similarly enhanced. This is denoted as the 'emitter-push effect'. The diffusion of B, P, and Ga in buried layers many microns away from the P diffused region is also enhanced. In contrast, diffusion of Sb is retarded under the same conditions. These phenomena are due to a supersaturation of self-interstitials, associated with an undersaturation of vacancies, induced by high-concentration in-diffusion of phosphorous.

The appearance of so-called kink-and-tail profiles can be attributed to the changeover from the vacancy mechanism at high P concentrations near the surface. Self-interstitials dominate P diffusion in the deeper profile regions [30, 34]. Numerical simulations of B, P, and As in-diffusion profiles on the basis of a unified model, taking into account the vacancy and the interstitialcy mechanism and various charge states of the defects [31, 34], are able to clarify the origin of the 'anomalous' diffusion profiles of P.

24.2.3 Diffusion of Hybrid Foreign Elements

Several foreign elements such as Au, Pt, and Zn in Si and Cu in Ge diffuse via interstitial-substitutional exchange mechanisms. Such foreign elements are denoted as hybrid diffusers. Au and Pt are used in Si power devices to improve the frequency behaviour. This is due to the fact that these elements can reduce the life-times of minority carriers in Si, because they introduce energy levels close to the middle of the band gap. In contrast, Au and Pt are undesirable contaminants in microelectronic devices and have to be avoided or removed. For these reasons, diffusion of Au and Pt in Si has been investigated intensively. Zn is technologically not important in Si. However, scientifically it served as a fine example to explore interstitial-substitutional diffusion.

The high diffusivities of hybrid elements in Ge and Si (see Figs. 24.3 and 24.4) have two major causes:

(i) A non-negligible fraction of these foreign atoms resides in interstitial sites of the host crystal (see, e.g., [15, 10]).
(ii) The migration enthalpies controlling jumps of these atoms from interstice to interstice are small, due to the weak coupling to the host lattice.

Diffusion of hybrid foreign atoms is described on the basis of interstitial-substitutional exchange mechanisms, either the dissociative or the kick-out mechanism or both. A detailed mathematical description of hybrid diffusion is given in Chap. 25. For the moment it may suffice to mention two limiting cases, which emerge from the mathematical description:

– **Defect-controlled Diffusion of Hybrid Solutes**: For appropriate experimental conditions the full set of equations for interstitial-substitutional exchange diffusion (see Chap. 25) can be reduced to a single Fick-like diffusion equation with an effective diffusion coefficient

$$D_{eff}^{I+V} = \frac{C_I^{eq} D_I}{C_s^{eq}} \left(\frac{C_i^{eq}}{C_s(x)} \right)^2 + \frac{C_V^{eq} D_V}{C_s^{eq}}. \qquad (24.10)$$

This effective defect-controlled diffusivity is composed of a C_s-dependent self-interstitial contribution and a C_s-independent vacancy contribution. Each contribution is associated with the transport products and of self-interstitials, $C_I^{eq} D_I$, and of vacancies, $C_V^{eq} D_V$, respectively. D_{eff}^{I+V} depends on the actual concentration C_s. Therefore, diffusion profiles develop that cannot be described by normal Fickian behaviour with concentration-independent diffusivity.

The defect-controlled mode for diffusion of hybrid solutes is expected for crystals with low densities of dislocations and other extended defects, which act as internal sources and sinks for intrinsic defects. Then, the concentrations of point defects during the diffusion process deviate from their equilibrium concentrations and thus determine the incorporation of foreign atoms. As already mentioned, the transport product of self-interstitials, $C_I^{eq} D_I$, in Fig. 23.6 has been derived from diffusion studies of the hybrid solute Zn in Si.

- **Interstitial-controlled Diffusion of Hybrid Solutes**: Thermal equilibrium of self-interstitials and vacancies during the diffusion process can be maintained, if enough sources and sinks for point defects are available. For example, dislocations with sufficient densities can provide sources and sinks. Under such conditions the diffusion of hybrid solutes is controlled by the following effective diffusivity

$$D_{eff}^i = \frac{C_i^{eq} D_i}{C_i^{eq} + C_s^{eq}} \approx \frac{C_i^{eq} D_i}{C_s^{eq}}. \tag{24.11}$$

We note that D_{eff} is independent of the actual concentration and normal Fickian behaviour can be expected in the diffusion profiles. The effective diffusivity is controlled by the transport product, $C_i^{eq} D_i$, which reflects the in-flow of foreign interstitials.

24.3 Self- and Foreign Atom Diffusion – a Summary

A rather sophisticated picture of the diffusion in Ge and Si has evolved in the literature. In Chaps. 23 and 24 we have sketched the present state-of-the-art. Important steps in understanding self- and foreign atom diffusion in Ge and Si were the discoveries of the dissociative and the kick-out mechanisms (see also Chap. 25). These concepts not only clarified the relative complex diffusion behaviour of hybrid elements but also yielded quantitative informations on the rôle of vacancies and self-interstitials for self-diffusion. The reader may appreciate the following summary:

1. Self-diffusion in Ge occurs by the vacancy mechanism with contributions from neutral and negatively charged vacancies.
2. The interstitialcy and the vacancy mechanism contribute to self-diffusion in Si. Interstitialcy-mediated diffusion dominates at high temperatures and vacancy-mediated diffusion at lower temperatures. The cross-over temperature is near 890 °C.
A subdivision of self-diffusion into the contributions of self-interstitials, $C_I^{eq} D_I$, and of vacancies, $C_V^{eq} D_V$, has been established for Si (Fig. 23.6). A reliable subdivison of separate equilibrium concentrations (C_I^{eq}, C_V^{eq}) and diffusivities of vacancies and self-interstitials (D_I, D_V) is on the horizon. There seems to be agreement in the literature that vacancies are

more abundant than self-interstitials ($C_V^{eq} \gg C_I^{eq}$) at least near the melting temperature. On the other hand, the self-interstitial is more mobile than the vacancy ($D_V \ll D_I$) [32–34].
3. Foreign elements that mainly diffuse via the direct interstitial mechanism are the following ones: H, Li, and Fe in Ge and H, Li, Fe, Cu, and Ni in Si.
4. The relatively slow interstitial diffusion of O in Si is explained by the bond-centered position of interstitial O, whose motion requires the breaking and making of silicon-oxygen bonds.
5. Dopant elements in Ge diffuse via a vacancy mechanism. Donor diffusion has a significant contribution from negatively charged vacancies, since Coulomb attraction enhances the probability to form donor-vacancy pairs. Perhaps, the small dopant B is an exception.
6. Dopant elements in Si diffuse via a combination of the vacancy and the interstitialcy mechanism. The fractional interstitialcy component of group-V elements decreases with increasing atomic radius. The relatively large Sb atoms diffuse almost exclusively via vacancies. The group-III elements B, Al, and Ga diffuse predominantly via the interstitialcy mechanism. The 'anomalous' in-diffusion profiles of P in Si have been attributed to a changeover from vacancy-mediated to self-interstitial mediated diffusion.
7. Hybrid foreign elements are incorporated predominantly on substitutional sites. A minor fraction occupies interstitial sites. Interstitial-controlled and defect-controlled diffusion modes of hybrid solutes can be studied under appropriate experimental conditions. The interstitial-controlled effective diffusivity is very high; the defect-controlled effective diffusivity is lower but still high as compared to self-diffusion of the host crystal. For the kick-out mechanism the self-interstitial-controlled effective diffusivity is strongly dependent on the substitutional concentration of the hybrid element.
   a) Cu, Ag, and Au are examples of hybrid solutes in Ge. Their diffusion is attributed to the dissociative mechanism.
   b) Au, Pt, Zn, and S are hybrid solutes in Si. Their diffusion proceeds mainly via the kick-out mechanism.

References

1. A. Seeger, K.P. Chik, Phys. Stat. Sol. **29**, 455–542 (1968)
2. W. Frank, U. Gösele, H. Mehrer, A. Seeger, *Diffusion in Silicon and Germanium*, in: *Diffusion in Crystalline Solids*, G.E. Murch, A.S. Nowick (Eds.), Academic Press, 1984
3. H. Bracht, Materials Science in Semiconductor Processing **7**, 113 (2004)
4. H. Bracht, MRS Bulletin 2000, p.22
5. N.A. Stolwijk, Phys. Rev. B **42**, 5793 (1990)

6. M. Schulz, *Silicon*, in: *Impurities and Defects in Group IV Elements and III-V Compounds*, Landolt-Börnstein, New Series, Group III: Crystal and Solid State Physics, Vol. 22: Semiconductors, Subvolume B, M. Schulz (Vol.Ed.), Springer-Verlag, 1989, p. 207
7. N.A. Stolwijk, *Germanium*, in: *Impurities and Defects in Group IV Elements and III-V Compounds*, Landolt-Börnstein, New Series, Group III: Crystal and Solid State Physics, Vol. 22: Semiconductors, Subvolume B, M. Schulz (Vol.Ed.), Springer-Verlag, 1989, p. 439
8. N.A. Stolwijk, H. Bracht, *Germanium*, in: *Impurities and Defects in Group IV Elements, IV-IV and III-V Compounds*, Landolt-Börnstein, New Series, Group III: Condensed Matter, Vol. 41: Semiconductors, Subvolume A2, Part α: Group IV Elements, M. Schulz (Vol.Ed.), Springer-Verlag, 2002, p. 382
9. H. Bracht, N.A. Stolwijk, *Silicon*, in: *Impurities and Defects in Group IV Elements, IV-IV and III-V Compounds*, Landolt-Börnstein, New Series, Group III: Condensed Matter, Vol. 41: Semiconductors, Subvolume A2, Part α: Group IV Elements, M. Schulz (Vol.Ed.), Springer-Verlag, 2002, p. 77
10. N.A. Stolwijk, Defect and Diffusion Forum **95–98**, 895 (1993)
11. D. Gilles, W. Schröter, W. Bergholz, Phys. Rev. B **41**, 5770 (1990)
12. W. Schröter, M. Seibt, D. Gilles, in: Mater. Science and Technology, Vol.4, *Electronic Structure and Properties of Semiconductors*, W. Schröter (Ed.), VCH, Weinheim 1991
13. N.A. Stolwijk, H. Bracht, *Diffusion in Silicon, Germanium and their Alloys*, in: Landolt-Börnstein, New Series, Group III: Condensed Matter, Vol. 33: *Diffusion in Semiconductors and Non-Metallic Solids*, Subvolume A: *Diffusion in Semiconductors*, D.L. Beke (Vol.Ed.), Springer-Verlag, 1998, p. 2–1
14. A. van Wieringen, N. Warmholtz, Physica **22**, 849 (1956)
15. E.R. Weber, Appl. Phys. A **30**, 1 (1983)
16. C.S. Fuller, J.A. Ditzenberger, Phys. Rev. **91**, 193 (1953)
17. C.S. Fuller, J.C. Severiens, Phys. Rev. **96**, 225 (1954)
18. E.M. Pell, Phys. Rev. **119**, 1014 (1960)
19. S. Pearton, J.W. Corbett, T.S. Shi, Appl. Phys. **A 43**, 153 (1967)
20. W. Kaiser, H.L. Fritsch, H. Reiss, Phys. Rev. **112**, 1546 (1958)
21. R.A. Craven, in: *Semiconductor Si 1981*, H.R. Huff, R.J. Kriegler, Y. Takeichi (Eds.), The Electrochem. Soc., Pennington, New Jersey, 1981, p.254
22. W. Kaiser, Phys. Rev. B **105**, 1751 (1957)
23. C. Haas, J. Phys. Chem. Sol. **15**, 108 (1960)
24. J.W. Corbett, R.S. McDonald, G.D. Watkins, J. Phys. Chem. Sol. **25**, 873 (1964)
25. R.C. Newman, Defect and Diffusion Forum **143–147**, 9903 (1997)
26. P.M. Fahey, P.B. Griffin, J.D. Plummer, Rev. Modern Physics **621**, 289 (1989)
27. S.M. Hu, in: *Diffusion in Semiconductors*, D. Shaw (Ed.), Plenum Press, London and New York, 1973, p. 217
28. S.M. Hu, Phys. Stat. Sol. (b) **60**, 595 (1973)
29. M. Yoshida, E. Arai, H. Nakamura, Y. Terunuma, J. Appl. Phys. **45**, 1489 (1974)
30. U. Gösele, in: *Microelectronic Materials and Processes*, R.A. Levy (Ed.), Kluwer Academic, Dordrecht 1989, p. 588
31. M. Uematsu, J. Appl. Phys. **62**, 2228 (1997)
32. S.M. Hu, Materials Science and Engineering **R13**, 105 (1994)

33. H. Bracht, N.A. Stolwijk, H. Mehrer, Phys. Rev. B **52**, 16542 (1995)
34. T.Y. Tan, U. Gösele, *Diffusion in Semiconductors*, in: *Diffusion in Condensed Matter – Methods, Materials, Models*, P. Heitjans, J. Kärger (Eds.), Springer-Verlag, 2005

25 Interstitial-Substitutional Diffusion

Hybrid foreign atoms (A) are mainly dissolved as substitutional atoms (A_s) but a minor fraction is located on interstitial sites (A_i) of the host crystal. An isolated substitutional atom A_s surrounded by the atoms of the host crystal in a perfect neighbourhood cannot move to neighbouring lattice sites except via exchange with intrinsic point defects. The equilibrium concentrations of intrinsic point defects are very low in Si and Ge (see Chaps. 23 and 24). Hence, pure substitutional diffusion entails a low diffusivity comparable to that of self-diffusion. To break out from its position a foreign atom may also escape into an interstitial site and move as an interstitials A_i. The interstitial transport can be fast enough to be responsible for diffusion of A atoms, taking into account that A_s changes over to A_i and vice versa.

In terms of quasi-chemical reactions there are two possibilities for substitutional-interstitial exchange of hybrid solutes:

$$A_i + V \rightleftharpoons A_s \tag{25.1}$$

and

$$A_i \rightleftharpoons A_s + I. \tag{25.2}$$

The first equation describes the *dissociative reaction* involving vacancies (V) and the second one describes the *kick-out reaction* involving self-interstitials (I). Figure 6.8 in Chap. 6 provide schematic illustrations of both mechanisms. The diffusivity of hybrid solutes in the interstitial configuration, D_i, is much higher than in the substitutional configuration, D_s. Under such conditions, the incorporation of A atoms occurs by fast diffusion of A_i and subsequent change-over to A_s. The corresponding diffusion modes are the dissociative mechanism, suggested initially for Cu diffusion in Ge by FRANK AND TURNBULL [1], and the kick-out mechanism proposed first to explain Au diffusion in Si by GÖSELE ET AL. [2]. In the meantime, positive identifications of interstitial-substitutional exchange diffusion have been made in several other cases (see Sects. 25.2, 25.3, and Chap. 24).

25.1 Combined Dissociative and Kick-out Diffusion

This section is based on the assumption that both mechanisms operate simultaneously. In two subsequent sections, we consider the two mechanisms separately together with experimental examples. Electric charge effects are ignored, which implies that the species involved are uncharged and the concentrations do not markedly exceed the intrinsic carrier density $n_i(T)$ at the diffusion temperature T.[1] The diffusion and the reactions of the four species A_i, A_s, V, and I are described by four coupled differential equations, each one containing a Fickian term and reaction terms between the species themselves and between sources and sinks for the intrinsic defects. The full set of partial differential equations for modelling interstitial-substitutional exchange diffusion in crystals is given by:

$$\frac{\partial C_i}{\partial t} = D_i \frac{\partial^2 C_i}{\partial x^2} + k_{-1}C_s - k_{+1}C_i C_V + k_{-2}C_s C_I - k_{+2}C_i \quad (25.3)$$

$$\frac{\partial C_s}{\partial t} = D_s \underbrace{\frac{\partial^2 C_s}{\partial x^2}}_{D_s \approx 0} + k_{+1}C_i C_V - k_{-1}C_s + k_{+2}C_i - k_{-2}C_s C_I \quad (25.4)$$

$$\frac{\partial C_V}{\partial t} = D_V \frac{\partial^2 C_V}{\partial x^2} + k_{-1}C_s - k_{+1}C_i C_V + K_V \left(1 - \frac{C_V}{C_V^{eq}}\right) \quad (25.5)$$

$$\frac{\partial C_I}{\partial t} = D_I \frac{\partial^2 C_I}{\partial x^2} - k_{-2}C_s C_I + k_{+2}C_i + K_I \left(\frac{C_I}{C_I^{eq}} - 1\right) \quad (25.6)$$

These equations describe the diffusion of hybrid foreign atoms and their reactions according to Eqs. (25.1) and (25.2) based on individual diffusion coefficients D_i, D_s, D_V, and D_I, which are independent of the actual concentration. C_X ($X = i, s, V, I$) represents the concentration of the particular species (measured in site fractions[2]) as function of position x and diffusion time t. The quantities K_V and K_I describe the strengths of sources or sinks for vacancies or self-interstitials, respectively. If dislocations act as sinks or sources, which is the case in most applications, their strengths can be expressed as

$$K_V = \gamma_V \rho D_V C_V^{eq} \quad \text{or} \quad K_I = \gamma_I \rho D_I C_I^{eq}, \quad (25.7)$$

where ρ is the dislocation density. The constants γ_V and γ_I characterising the dislocation arrangements are of the order of unity [3]. $k_{+1}(k_{+2})$ and $k_{-1}(k_{-2})$ denote the forward and backward reaction rates of the dissociative (kick-out) reaction, respectively. These reaction rates are related via laws of mass action to the equilibrium concentrations C_X^{eq}:

---

[1] This assumption is appropriate for Ge and Si but not for GaAs due to its higher gap energy.
[2] If number densities are used for C_X instead of site fractions, a factor C^0 must be included whenever C_X and C_X^{eq} stand alone in Eqs. (25.3) to (25.6) and in the laws of mass action. C^0 is the number density of host lattice site: $C^0 = 5 \times 10^{28} \mathrm{m}^{-3}$ for Si.

25.1 Combined Dissociative and Kick-out Diffusion

$$\frac{k_{+1}}{k_{-1}} = \frac{C_s^{eq}}{C_i^{eq} C_V^{eq}}, \qquad (25.8)$$

$$\frac{k_{+2}}{k_{-2}} = \frac{C_s^{eq} C_I^{eq}}{C_i^{eq}}. \qquad (25.9)$$

These ratios are constants, which depend on temperature.

Simplifications of Eqs. (25.3) to (25.6), which are usually justified under conditions of diffusion experiments [4, 5], concern local equilibrium between the various species. For local equilibrium between the partners of Eqs. (25.1) or (25.2), the laws of mass action can be written as

$$\frac{C_s}{C_i C_V} = \frac{C_s^{eq}}{C_i^{eq} C_V^{eq}} \qquad (25.10)$$

and

$$\frac{C_i}{C_s C_I} = \frac{C_i^{eq}}{C_s^{eq} C_I^{eq}}. \qquad (25.11)$$

In addition, as indicated in Eq. (25.4), the pure substitutional diffusivity, D_s, can often be neglected for hybrid diffusers.

Even with these simplifications, the coupled nonlinear system of partial diffusion-reaction Eqs. (25.3) to (25.6) cannot be solved analytically in full generality. Particular solutions $C_X(x,t)$ are determined by the pertinent initial and boundary conditions. Often, numerical methods of computational physics must be used to derive solutions. In the following, we consider first the two limiting cases of practical importance already given in Chap. 24. Then, we report an example for an intermediate case, which requires numerical treatment.

25.1.1 Diffusion Limited by the Flow of Intrinsic Defects

Hybrid diffusion controlled by the slow flow products (or transport products) of self-interstitials and/or vacancies emerges, if the relationship

$$C_I^{eq} D_I + C_V^{eq} D_V \ll C_i^{eq} D_i \qquad (25.12)$$

is fulfilled. This condition can be realised in material that is completely or virtually free of inner sinks and sources of point defects. Dislocation-free Si wafers are commercially available[3]. Such wafers are the base materials for the fabrication of microelectronic devices. Ge and GaAs materials are available at least with very low values of the dislocation density. If dislocations (and other extended defects) are absent, the sink and source terms in Eqs. (25.5) and (25.6) can be omitted, i.e. $K_V = K_I = 0$. Then, the interstitial-substitutional

---

[3] This situation is very different from metals. Carefully grown metal single crystals still have dislocation densities of about 10^{10} m^{-2}.

exchange reactions and the pertinent establishment of the A_i equilibrium are the fastest processes and Eq. (25.3) can be replaced by

$$C_i \approx C_i^{eq}. \tag{25.13}$$

This approximation may be approximately true either for in-diffusion in a finite region beneath the surface of thick crystals or throughout the entire width of a thin wafer, provided that the diffusion time is not too short.

Taking into account the exchange reactions via local equilibria, Eqs. (25.10) and (25.11), equations (25.4), (25.5), and (25.6) can be simplified to

$$\frac{\partial C_s}{\partial t} = \frac{\partial}{\partial x}\left(D_{eff}^{I+V} \frac{\partial C_s}{\partial x}\right). \tag{25.14}$$

This is a diffusion equation with an effective diffusion coefficient given by[4]

$$D_{eff}^{I+V} = \underbrace{\frac{C_I^{eq} D_I}{C_s^{eq}} \left(\frac{C_s^{eq}}{C_s}\right)^2}_{D_{eff}^{I}} + \underbrace{\frac{C_V^{eq} D_V}{C_s^{eq}}}_{D_{eff}^{V}}. \tag{25.15}$$

We recognise in Eq. (25.15) the diffusivity for defect-controlled diffusion of hybrid solutes given in Eq. (24.10). D_{eff}^{I+V} contains a self-interstitial component, D_{eff}^{I} (kick-out mechanism), and a vacancy component, D_{eff}^{V} (dissociative mechanism). For in-diffusion in dislocation-free material, the kick-out term leads to a supersaturation of self-interstitials and the dissociative term to an undersaturation of vacancies. Let us point out some further interesting features of the combined diffusion described by Eq. (25.14):

– The defect-limited diffusion mode of hybrid solutes is closely related to self-diffusion of the host (see Chap. 23). The tracer self-diffusion coefficient

$$D^* = f_I C_I^{eq} D_I + f_V C_V^{eq} D_V \tag{25.16}$$

equals the sum of the flow products of self-interstitials and vacancies modified by their correlation factors f_I and f_V, respectively. This equation provides a valuable key to deduce these contributions from diffusion experiments of hybrid solutes (see Chap. 23).

---

[4] The assumptions made to derive Eq. (25.15) are: $\sqrt{C_V^{eq} C_I^{eq}} \ll C_s$ and $C_V^{eq} \ll C_s^{eq}$. These are not severe restrictions, because the equilibrium concentrations of point defects in semiconductors are very small compared to the substitutional solubility of hybrid diffusers. As explicitly shown in the case of kick-out diffusion by GÖSELE ET AL. [2], the use of the local equilibrium condition Eq. (25.11) also imposes the restriction $C_s \geq \sqrt{C_I^{eq} C_s^{eq}}$. Therefore, Eq. (25.15) can be used with virtually no further loss of generality.

25.1 Combined Dissociative and Kick-out Diffusion

– If both mass action laws for the dissociative reaction, Eq. (25.10), and the kick-out reaction, Eq. (25.11), are fulfilled

$$C_I C_V = C_V^{eq} C_I^{eq} \tag{25.17}$$

holds automatically. This means that the equilibrium between vacancies and self-interstitials can be established via the reactions of Eqs. (25.1) and (25.2) even if the direct reaction of Frenkel pair annihilation and creation

$$V + I \rightleftharpoons 0 \tag{25.18}$$

is hampered, for example, by high activation barriers for spontaneous Frenkel-pair recombination or formation.
– The self-interstitial component D^I_{eff} of Eq. (25.15) depends strongly on the actual local concentration $C_s(x,t)$, whereas the vacancy component D^V_{eff} is independent of concentration. The concentration dependence of D^I_{eff} provides an important key for the identification of kick-out diffusion from experimental determined concentration profiles (see below).
– Under practical diffusion conditions, usually one of the two terms on the right-hand side of Eq. (25.15) predominates. Such positive identification has been made, for example, for Cu in Ge and Co in Nb (dissociative diffusion) and for Au, Pt, and Zn in Si (kick-out diffusion). Some key results that led to these conclusions are presented in Sects. 25.2 and 25.3.

25.1.2 Diffusion Limited by the Flow of Interstitial Solutes

Hybrid diffusion controlled by the low flow products of foreign interstitials emerges if the inequality

$$C_i^{eq} D_i \ll C_I^{eq} D_I + C_V^{eq} D_V \tag{25.19}$$

is fulfilled. In this case, Eqs. (25.3) and (25.4) together with the local equilibrium conditions yield for the substitutional concentration a normal diffusion equation with an effective diffusion coefficient

$$D^i_{eff} = \frac{C_i^{eq} D_i}{C_i^{eq} + C_s^{eq}} \approx \frac{C_i^{eq} D_i}{C_s^{eq}}. \tag{25.20}$$

This effective diffusivity is independent of the actual concentration, C_s. The approximation made on the right-hand side, $C_i^{eq} \ll C_s^{eq}$, is readily fulfilled for most hybrid diffusers. We recognise that Eq. (25.20) is the interstitial-controlled diffusivity of hybrid solutes given in Eq. (24.11). This diffusion mode can be expected under the following conditions:

1. The interstitial flow product of hybrid atoms is smaller than the combined point defect flow products. This implies that

$$C_V \approx C_V^{eq} \quad \text{and} \quad C_I \approx C_I^{eq}. \tag{25.21}$$

This condition is virtually maintained during in-diffusion by the large flow product of intrinsic defects as compared to that of foreign interstitials. This holds even for dislocation-free substrates.

2. Equation (25.20) can also hold for crystals with high dislocation densities, regardless of the validity of Eq. (25.19). This is because the presence of sinks (sources) for self-interstitials (vacancies) shortcircuits the flows of intrinsic point defects, whereas the foreign interstitials have to cover a long distance from the surface of the crystal. Under such conditions, the equilibrium concentration of intrinsic point defects can be established almost instantaneously everywhere. Then point defect equilibrium is virtually maintained during in-diffusion, although the fast penetration of A_i with the subsequent changeover to A_s tends to create a supersaturation (undersaturation) of self-interstitials (vacancies).

Let us consider as a simple example, diffusion of a hybrid solute limited by its interstitial flow product[5]. During in-diffusion of foreign atoms into a thick sample from an inexhaustible source at the surface $x = 0$, the surface concentration $C_s(0)$ is kept at C_s^{eq}. Then, an erfc-type diffusion profile typical of a concentration-independent diffusion coefficient is expected (see Chap. 3 and [6, 7]):

$$C_s = C_s^{eq}\text{erfc}\left(\frac{x}{2\sqrt{D_{eff}^i t}}\right). \qquad (25.22)$$

We emphasise that the same result is obtained for the kick-out and for the dissociative mechanism. Hence, in experiments involving high sink and source densities or a priori slow interstitial transport both mechanism cannot be distinguished. The same holds true for in-diffusion into a thin wafer.

25.1.3 Numerical Analysis of an Intermediate Case

So far, the basic concepts of interstitial-substitutional diffusion have been elucidated with the aid of limiting cases. In practice, however, one is often confronted with more complex 'intermediate' cases. Then, numerical simulations can provide a more complete picture. Figure 25.1 shows the result of computer simulations within the kick-out model [4] using the soft ware package ZOMBIE [8]. The numbers chosen for the numerical simulation are representative for in-diffusion of Zn in Si at 1380 °C for 280 s. The in-diffusion profiles of Fig. 25.1 (a) represent distributions of substitutional atoms, $C_s(x)$, in dislocation-free material for various magnitudes of the interstitial flow product, $C_i^{eq}D_i$, relative to a fixed value of the self-interstitial flow product,

---

[5] In case of a highly dislocated sample, we further assume that dislocations act as sources or sinks of intrinsic defects only. Trapping of foreign atoms is not considered.

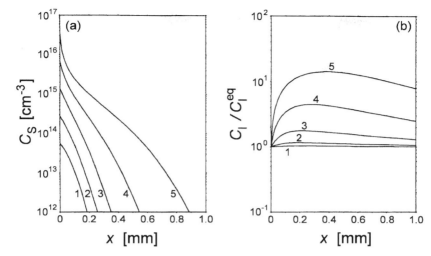

Fig. 25.1a,b Computer simulation of in-diffusion into a thick sample via the kick-out mechanism according to STOLWIJK [4]. The *left part* (**a**) shows the concentration of substitutional foreign atoms. The *right part* (**b**) shows the associated self-interstitial supersaturations. The values of the ratio $\alpha_j \equiv C_I^{eq} D_I / C_i^{eq} D_i$ are: $\alpha_1 = 1/25, \alpha_2 = 1/5, \alpha_3 = 1, \alpha_4 = 5, \alpha_5 = 25$, with the indices j listed in the diagrams

$C_I^{eq} D_I$. Figure 25.1 (b) reveals the corresponding supersaturation of self-interstitials induced by the kick-out reaction (25.2), which mainly proceeds from right to left in the case of in-diffusion. The lowest profile in Fig. 25.1 (a) and (b) is representative of diffusion limited by the flow product of foreign interstitials, $C_i^{eq} D_i$, according to Eq. (25.19). This profile is of the erfc-type and the self-interstitial concentration is practically at its equilibrium value. The upper profiles in Figs. 25.1 (a) and (b) visualise solutions of Eq. (25.14), with $D_{eff}^{I+V} \equiv D_{eff}^{I}$ generating a convex near-surface shape for $C_s(x)$ due to the factor $(C_s^{eq}/C_s)^2$ and a significant self-interstital supersaturation. At greater depths, however, the $C_s(x)$ profile changes its shape to concave, which relates to the violation of $C_i \approx C_i^{eq}$ adopted in the derivation of Eq. (25.14). Figures 25.1 (a) and (b) also exhibit intermediate cases, where $C_i^{eq} D_i$ and $C_I^{eq} D_I$ are not too much different.

25.2 Kick-out Mechanism

25.2.1 Basic Equations and two Solutions

Kick-out diffusion involves the interchange of the diffusing hybrid atoms between substitutional and interstitial sites with the aid of self-interstitials. If we follow GÖSELE ET AL.[2] and assume local thermal equilibrium, this type of diffusion simplifies to the following set of equations:

$$\frac{\partial C_I}{\partial t} = D_I \frac{\partial^2 C_I}{\partial x^2} + \frac{\partial C_s}{\partial t} + K_I \left(\frac{C_I}{C_I^{eq}} - 1\right) \qquad (25.23)$$

$$\frac{\partial C_i}{\partial t} = D_i \frac{\partial^2 C_i}{\partial x^2} - \frac{\partial C_s}{\partial t} \qquad (25.24)$$

$$\frac{C_i}{C_s C_I} = \frac{C_i^{eq}}{C_s^{eq} C_I^{eq}}. \qquad (25.25)$$

As discussed in the previous section, the *a priori* interstitial-limited diffusion as well as diffusion in highly dislocated material imply that self-interstitial equilibrium is virtually maintained, i.e. that $C_I \approx C_I^{eq}$ holds. Then, Eqs. (25.24) and (25.25) yield for the substitutional concentration a normal diffusion equation with the diffusion coefficient given by Eq. (25.20). The lowest $C_s(x)$ profile in Fig. 25.1 provides an example of such a case.

The self-interstitial controlled limit emerges for $C_I^{eq} D_I \ll C_i^{eq} D_i$. For negligible sink and source density, the K_I-term in Eq. (25.23) can be omitted. Then, the kick-out reaction and the establishment of the A_i equilibrium are the fastest processes. Using the local mass-action law Eq. (25.25) and replacing Eq. (25.24) by $C_i \approx C_i^{eq}$, we get from Eqs. (25.23) and (25.25) the following non-linear partial differential equation:

$$\frac{\partial C_s}{\partial t} = \frac{\partial}{\partial x} \left[\underbrace{\left(\frac{C_s^{eq} C_I^{eq} D_I}{C_s^2}\right)}_{D_{eff}^I} \frac{\partial C_s}{\partial x} \right]. \qquad (25.26)$$

The effective diffusion coefficient D_{eff}^I is the self-interstitial limited diffusivity of Eq. (25.15). In the usual case of in-diffusion, the kick-out reaction leads to a supersaturation of self-interstitials. Then, the incorporation of hybrid foreign atoms on substitutional sites is limited by the outflow of self-interstitials, $C_I^{eq} D_I$. The concentration dependence of D_{eff}^I gives rise to distinct features of the kick-out diffusion. For example, its $1/C_s^2$-dependence speeds up diffusion at low concentrations and leads to convex-shaped diffusion profiles. In what follows, we mention two particular solutions of Eq. (25.26), which play a rôle in experiments discussed in the next section.

1. **Diffusion into a Dislocation-free Thick Specimen**: In-diffusion of hybrid atoms from the surface $x = 0$ of a thick specimen is described by Eq. (25.26) with the boundary condition $C_s(x = 0, t) = C_s^{eq}$ and the initial condition $C_s(x, t = 0) = C_s^0$. For a dislocation-free crystal SEEGER

AND COWORKERS [9, 10] derived an analytical solution in parametric form. The approximate solution for $C_s^0 \ll C_s^{eq}$ is

$$C_s = \frac{C_s^{eq}}{1 + |a_0|\sqrt{C_s^{eq} x^2 / D_I C_I^{eq} t}}, \qquad (25.27)$$

where

$$a_0 \exp a_0^2 = -\frac{C_s^{eq}}{2\sqrt{\pi} C_s^0}. \qquad (25.28)$$

In-diffusion profiles of kick-out diffusion into a thick, dislocation-free crystal differ radically from the erfc-type diffusion expected for a concentration-independent diffusivity. This finding played a decisive rôle in revealing the diffusion mechanism of Au, Pt, and Zn in Si (see next section).

2. **Diffusion into a Dislocation-Free Wafer**: Next, we consider diffusion into a wafer of thickness d and take advantage of the wafer symmetry by choosing the wafer center at $x = 0$ and the wafer surfaces at $x = \pm d/2$. For a dislocation-free wafer a solution of Eq. (25.26) has been derived for the boundary condition $C_s(x = \pm d/2, t) = \infty$. Such a solution is useful for not too long diffusion times and in the vicinity of the wafer center. GÖSELE ET AL [2] have shown that for $C_s^0 = 0$ solutions of the form

$$C_s(x,t) = X(x)\sqrt{D_I t} \qquad (25.29)$$

fulfill Eq. (25.26), if $X(x)$ satisfies the ordinary differential equation

$$2 C_I^{eq} C_s^{eq} \frac{d^2 X^{-1}}{dx^2} + X = 0. \qquad (25.30)$$

The solution of Eq. (25.30), with the above mentioned boundary condition, may be written in implicit form

$$\pm \frac{\sqrt{\pi C_I^{eq} C_s^{eq}}}{X(0)} \mathrm{erf}\sqrt{\ln[X(x)/X(0)]} = x \qquad (25.31)$$

with

$$X(0) = 2\sqrt{\pi C_I^{eq} C_s^{eq}}/d. \qquad (25.32)$$

Hence, the A_s concentration in the center of the wafer is given by

$$C_s^m(t) = X(0)\sqrt{D_I t} = \frac{2}{d}\sqrt{\pi C_s^{eq} C_I^{eq} D_I t}. \qquad (25.33)$$

For $C_s^0 \neq 0$ but otherwise equal assumptions, the expression

$$C_s^m(t) = \frac{C_s^0}{\sqrt{C_s^{eq} C_I^{eq}} \mathrm{erf}\left[C_s^0 d / 4\sqrt{C_s^{eq} C_I^{eq} D_I t}\right]} \qquad (25.34)$$

is valid [10], which reduces to Eq. (25.33) for $C_s^0 \to 0$, because for small values of z the error function may be replaced by $\mathrm{erf}(z) \approx 2z/\sqrt{\pi}$.

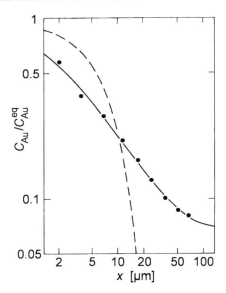

Fig. 25.2. In-diffusion of Au into a thick dislocation-free Si specimen according to STOLWIJK ET AL. [13]. The *solid line* represents a fit of the kick-out mechanism. The *dashed line* is an attempt to fit a complementary error function predicted for a concentration-independent diffusivity

25.2.2 Examples of Kick-Out Diffusion

Au and Pt Diffusion in Silicon: Diffusion of Au in Si has been studied by several authors. For a list of references see [10]. For example, WILCOX ET AL. [11, 12] report that the diffusion profiles in thick, dislocation-free samples differ considerably from the erfc-type shape. Accurate measurements by STOLWIJK ET AL. [13], using neutron-activation analysis and grinder sectioning for depth-profiling, confirmed the non-erfc nature of Au profiles in Si. A comparison between an experimental Au-profile and the predictions of the kick-out mechanism, Eq. (25.27), and an erfc-profile is shown in Fig. 25.2. The kick-out model permits a successful fit.

STOLWIJK ET AL. [13] also investigated the diffusion of Au into thin, dislocation-free Si wafers and found U-shaped diffusion profiles displayed in Fig. 25.3. Similar observations were reported by HILL ET AL. [15], who used the spreading-resistance technique for depth profiling (see Chap. 16). U-shaped profiles are in qualitative agreement with both the kick-out and the dissociative mechanism. A distinction between the two mechanisms is possible via a quantitative analysis of the profile shape. Figure 25.4 shows a comparison of an experimental profile from Fig. 25.3 with the relationship

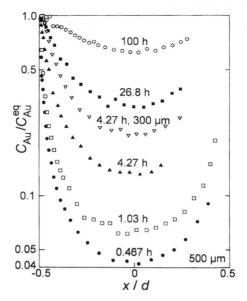

Fig. 25.3. In-diffusion of Au into dislocation-free Si wafers for various annealing times at 1273 K according to [13]. The Au concentration C_{Au} is given in units of the solubility C_{Au}^{eq}; the penetration depth x as a fraction of the wafer thickness d. For one penetration profile of the two 4.27 h anneals $d \approx 300$ μm, otherwise $d \approx 500$ μm

$$\mathrm{erf}\left(\ln\sqrt{\frac{C_s}{C_s^m}}\right) = \frac{2|x|}{d} \qquad (25.35)$$

predicted by Eqs. (25.31) and (25.32). Very good agreement is obvious.

Diffusion and solubility of Pt in dislocation-free Si has also been studied using both neutron activation analysis plus mechanical sectioning as well as the spreading-resistance technique by HAUBER ET AL. [16]. The results suggest that diffusion of Pt in Si is also dominated by the kick-out mechanism.

Zn Diffusion in Si: Diffusion of Zn in Si provides a well-studied example of interstitial-substitutional exchange diffusion dominated by the kick-out mechanism. Both cases – in-diffusion limited by the defect flow product or by the flow product of interstitial Zn – have been observed in the work of BRACHT ET AL. [17]. Figure 25.5 shows concentration profiles of Zn_s measured on dislocation-free and on highly dislocated Si. Both wafers were simultaneously exposed to Zn vapour from an elemental Zn source in a closed quartz ampoule. This source yields at both wafer surfaces a concentration of $C_{Zn_s}^{eq} \approx 2.5 \times 10^{16}$ cm^{-3}, which corresponds to the solubility limit of Zn in Si in equilibrium with the vapour phase from a pure Zn source. Also shown is a Zn_s profile in dislocation-free Si obtained at the same temperature but using a 0.1 molar solution of Zn in HCl as diffusion source, providing a much

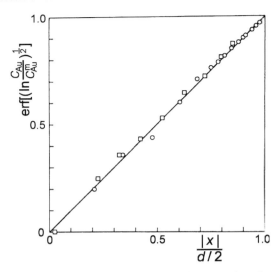

Fig. 25.4. Comparison of Au diffusion into a dislocation-free Si wafer (1.03 h at 1273 K) with the prediction of the kick-out mechanism (*solid line*) [13]. *Circles*: $x < 0$; *squares* $x > 0$

lower Zn vapour pressure. Then, the boundary concentration observed at both surfaces of the wafer is $C_{Zn_s}^{eq} \approx 4.5 \times 10^{14}$ cm^{-3}, which is a factor of 55 smaller than the solubility limit reached in equilibrium with an elemental Zn source. Figure 25.5 demonstrates that Zn diffusion in Si depends sensitively on the defect structure as well as on the prevailing ambient conditions. The diffusion profile in dislocation-free Si (*crosses*) is convex, whereas the profile in the highly dislocated wafer, apart from the central region, is concave (*squares*). Also the profile obtained for lower Zn concentrations is convex (*circles*). The kick-out mechanism yields a consistent description of all three profiles in Fig. 25.5, as illustrated by the solid lines.

In addition, BRACHT ET AL. [17] studied the time evolution of Zn incorporation in dislocation-free and in highly dislocated Si for various temperatures applying a specially designed short-time diffusion method. Typical examples are displayed in Fig. 25.6. The solid lines in the *top* part show the result of successful fitting of all experimental profiles based on the kick-out mechanism. The solid lines in the *bottom* part represent complementary error functions.

1. *Dislocation-free Si and high boundary concentration*: In dislocation-free material no sinks and sources for self-interstitials exist. High boundary concentrations of Zn maintain

$$C_i^{eq} D_i \gg C_I^{eq} D_I \,, \qquad (25.36)$$

representing self-interstitial limited flow. The supply of Zn$_i$ from the surface occurs more rapidly than the decay of self-interstitial supersatura-

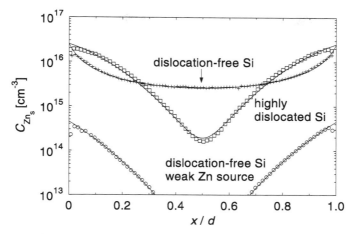

Fig. 25.5. In-diffusion of substitutional Zn measured on dislocation-free (*crosses*) and highly dislocated (*squares*) Si samples; Zn was diffused simultaneously at 1115 °C for 2880 s into both samples, using metallic Zn as vapour source. The lower profile (*circles*) represents in-diffusion into a dislocation-free wafer also at 1115 °C using a Zn solution in HCl as diffusion source. After BRACHT ET AL. [17]

tion by out-diffusion to the surface. For long diffusion times the condition $C_i \approx C_i^{eq}$ is already fulfilled, whereas the self-interstitial distribution still deviates from equilibrium. In this regime, the diffusion equation Eq. (25.26) holds. Then, the Zn_s concentration in the center of the wafer increases according to Eq. (25.33). In this long-time diffusion regime the Zn_s profiles in a semilogarithmic representation are distinctly U-shaped like the profile (*crosses*) in Fig. 25.5 and the Au-profiles in Fig. 25.3. These features reflect that Zn_s and Au_s incorporation is controlled by the out-diffusion of self-interstitials.

2. *Dislocation-free Si and low boundary concentration*: During Zn-diffusion the Dirichlet surface condition, $C_i(0,t) = C_i^{eq}$, is maintained by a constant Zn vapour pressure in the diffusion ampoule. This vapour pressure also determines the Zn_s equilibrium concentration C_s^{eq}, which is proportional to C_i^{eq}, whereas the properties D_i, C_I^{eq} and D_I are not influenced by the ambient conditions. Therefore, at sufficiently low boundary concentration the relationship

$$C_i^{eq} D_i \ll C_I^{eq} D_I \qquad (25.37)$$

is fulfilled. As a consequence, the kick-out reaction is not able to establish self-interstitial supersaturation and $C_I \approx C_I^{eq}$ holds during the diffusion process. Then, the incorporation of Zn is limited by the flow product of Zn interstitials with a diffusivity given by Eq. (25.20). Under these conditions diffusion profiles are described by complementary error functions (25.22), like the lower profile in Fig. 25.5.

3. *Highly dislocated Si*: In Si with a high dislocation density a supersaturation of self-interstitials can decay by annihilation at these defects. The profiles in the *bottom* part of Fig. 25.6 were obtained on plastically deformed Si with a dislocation density of about 10^{12} m^{-2}. This corresponds

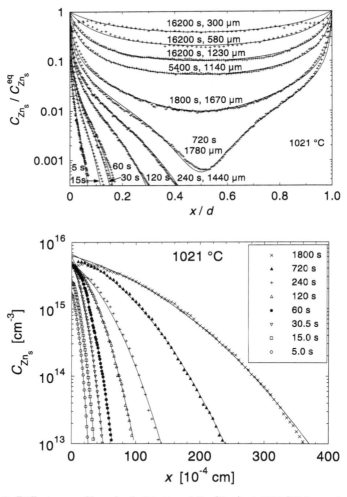

Fig. 25.6. Diffusion profiles of substitutional Zn (Zn_s) at 1021 °C for various diffusion times according to BRACHT ET AL. [17]. *Top*: Dislocation-free Si wafers; *solid lines* show calculated profiles based on the kick-out model using one set of parameters C_s^{eq} and $C_I^{eq} D_I$. *Bottom*: Highly dislocated Si; *solid lines* show fitting with complementary error functions

to a mean distance between dislocations of $x_d \approx 1\ \mu$m. Then, at penetration distances x for which

$$x^2 > x_d^2 \frac{C_i^{eq} D_i}{C_I^{eq} D_I} \tag{25.38}$$

holds, the annihilation of self-interstitials occurs more rapidly than the supply of Zn interstitials. Virtually no self-interstitial supersaturation is created due to the kick-out reaction. For Zn diffusion at the temperatures of the experiments $C_i^{eq} D_i / C_I^{eq} D_I \geq 100$ [17]. This yields $x > 5\ \mu$m as the penetration depth where self-interstitials mainly annihilate at internal sinks. In the regime $x < 5\ \mu$m, the surface acts as the major sink. As a consequence, the incorporation of Zn is limited by the flow product of Zn interstitials. The corresponding diffusivity Eq. (25.20) is concentration-independent and, indeed, the profiles are well described by complementary error functions.

25.3 Dissociative Mechanism

25.3.1 Basic Equations

Dissociative diffusion involves the interchange of hybrid solutes between interstitial and substitutional sites with the aid of vacancies. Assuming local thermal equilibrium for the dissociative reaction, Eqs. (25.3) to (25.6) simplify to the following set of equations:

$$\frac{\partial C_V}{\partial t} = D_V \frac{\partial^2 C_V}{\partial x^2} - \frac{\partial C_s}{\partial t} + K_V \left(1 - \frac{C_V}{C_V^{eq}}\right) \tag{25.39}$$

$$\frac{\partial C_i}{\partial t} = D_i \frac{\partial^2 C_i}{\partial x^2} - \frac{\partial C_s}{\partial t} \tag{25.40}$$

$$\frac{C_s}{C_V C_i} = \frac{C_s^{eq}}{C_V^{eq} C_i^{eq}}. \tag{25.41}$$

For material free of internal sinks and sources, we have $K_V = 0$. Then, the dissociative reaction and the establishment of the A_i equilibrium are the fastest processes. Using Eq. (25.41) and replacing Eq. (25.40) by $C_i = C_i^{eq}$, we arrive at a diffusion equation with the effective diffusion coefficient

$$D_{eff}^V = \frac{C_V^{eq} D_V}{C_V^{eq} + C_s^{eq}} \approx \frac{C_V^{eq}}{C_s^{eq}} D_V, \tag{25.42}$$

which is independent of the actual concentration C_s. The approximation $C_V^{eq} \ll C_s^{eq}$ is applicable since the solubility of hybrid atoms is usually much larger than the equilibrium vacancy concentration. Equation (25.42) is the diffusivity limited by the vacancy-flow introduced in Eq. (25.15). During indiffusion the dissociative reaction, Eq. (25.1), creates an undersaturation of vacancies and the incorporation of the hybrid atoms on substitutional sites is limited by the influx of vacancies form the surface of the crystal. This is why the product, $C_V^{eq} D_V$, determines the effective diffusion coefficient of foreign atoms.

High sink and source densities imply that vacancy equilibrium can be established almost instantaneously everywhere, i.e. $C_V \approx C_V^{eq}$. In this case, Eqs. (25.40) and (25.41) yield a diffusion equation with the effective diffusion coefficient of Eq. (25.20). As mentioned above, the same result is obtained for the kick-out mechanism if the corresponding assumption $C_I = C_I^{eq}$ is made. We recall that in experiments involving high dislocation densities, the kick-out and the dissociative mechanism cannot be distinguished on the basis of the shape of the diffusion profiles.

25.3.2 Examples of Dissociative Diffusion

Cu, Ag, and Au in Germanium: In the early days of semiconductor technology (see Chap. 22), Cu became known as an unwanted foreign atom, which – even in small concentrations – may give rise to detrimental effects. Extensive investigations of solubility and diffusion of Cu in Ge had been performed already in the 1950s by FULLER ET AL. [22, 23] and by WOODBERRY AND TYLER [24]. In Chap. 23 we have seen that the vehicles of self-diffusion in Ge are vacancies. Thus, it is not a surprise that the dissociative mechanism, which requires vacancies, has been proposed by FRANK AND TURNBULL [1] to explain the very fast diffusion of Cu in Ge. Later on, the falsification of the dissociative mechanism as the main process for Au diffusion in Si raised doubts whether the conclusion that the dissociative mechanism operates in Ge remains tenable, since the early experimental data were of limited accuracy.

In order to test this question, the solubility and diffusion of Cu, Ag, and Au in Ge has been studied on high-purity dislocation-free and dislocated Ge single crystals by STOLWIJK ET AL. [25] and BRACHT ET AL. [26]. The experiments showed that the penetration rate of Cu in highly dislocated Ge is much faster than in virtually dislocation-free material (see Fig. 25.7), whereas the diffusion of Ag and Au is not influenced by the presence of dislocations. The effective diffusivities of Cu, Ag, and Au and their substitutional solubilities, C_s^{eq}, have been measured and used to determine either the vacancy flow product, $C_V^{eq} D_V$, or the flow products of hybrid interstitials, $C_i^{eq} D_i$. The flow product of vacancies is obtained from Eq. (25.42) as

$$C_V^{eq} D_V = D_{eff}^V C_s^{eq}, \qquad (25.43)$$

provided that the incorporation of the hybrid solute is indeed limited by the vacancy flow. The flow product of interstitial solutes is obtained from Eq. (25.20) as

$$C_i^{eq} D_i = D_{eff}^i C_s^{eq}, \qquad (25.44)$$

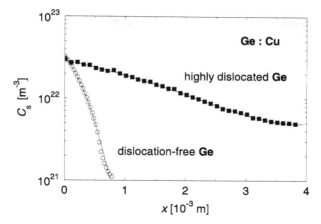

Fig. 25.7. Comparison of Cu penetration profiles in almost dislocation-free Ge (1126 K, 900 s) and in a crystal with high dislocation density (1124 K, 780 s) according to [17]

if the inflow of interstitial solutes is slower the vacancy flow. We recall that for vacancy-mediated self-diffusion the flow of vacancies times the vacancy correlation factor, f_V, equals the tracer self-diffusion coefficient:

$$D^* = f_V C_V^{eq} D_V . \tag{25.45}$$

For a vacancy mechanism in the diamond structure we have $f_V = 0.5$ (see Chap. 7).

Figure 25.8 shows a comparison of the 'triple products', effective diffusivities × correlation factor × substitutional solubility, for Cu, Ag, and Au diffusion in Ge. The figure reveals that this product for Cu in virtually dislocation-free Ge coincides with the tracer self-diffusion coefficient of Ge measured by tracer techniques (see Chap. 23). This agreement confirms both the dissociative model for Cu diffusion and the interpretation of Ge self-diffusion in terms of a vacancy mechanism. The product for Cu diffusion in highly dislocated material yield values that are characteristic of $C_i^{eq} D_i$. It is seen from Fig. 25.8 that the interstitial flow product of Cu in Ge is 1 to 3 orders of magnitude higher than the flow product of Ge vacancies. This observation is in accordance with the requirement that vacancy-flow limited diffusion of Cu in Ge occurs for

$$C_i^{eq} D_i(Cu) \gg C_V^{eq} D_V(Ge) . \tag{25.46}$$

The flow products of foreign interstitials, $C_i^{eq} D_i$, for all three noble metals solutes are given by [26]:

$$C_i^{eq}D_i(Cu) = 6.1 \times 10^{-8} \exp\left(-\frac{1.64 \text{ eV}}{k_B T}\right) \text{ m}^2\text{s}^{-1}, \quad (25.47)$$

$$C_i^{eq}D_i(Ag) = 4.3 \times 10^{-8} \exp\left(-\frac{2.30 \text{ eV}}{k_B T}\right) \text{ m}^2\text{s}^{-1}. \quad (25.48)$$

$$C_i^{eq}D_i(Au) = 1.3 \times 10^{-5} \exp\left(-\frac{2.98 \text{ eV}}{k_B T}\right) \text{ m}^2\text{s}^{-1}. \quad (25.49)$$

These equations are represented by solid-line fits in Fig. 25.8. In contrast to Cu, the corresponding products for Ag and Au are the same, irrespective whether Ge crystals with low or high dislocation densities are used. Furthermore, the flow products of Ag and Au interstitials lie one to two orders of magnitude below the tracer self-diffusivity of Ge. This shows that the vacancy flow in Ge proceeds *a priori* faster than the flow of Ag and Au interstitials. As a consequence, the latter process controls the Ag and Au penetration rate, i.e.

$$C_i^{eq}D_i(Ag), C_i^{eq}D_i(Au) \ll C_V^{eq}D_V(Ge). \quad (25.50)$$

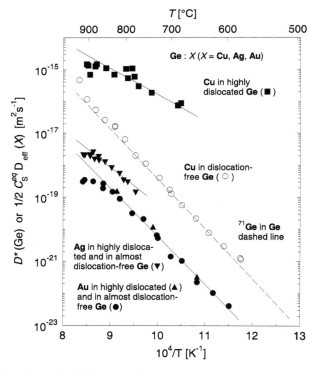

Fig. 25.8. Products of solubility × effective diffusivity × correlation factor for Cu, Ag, and Au diffusion in Ge with various dislocation densities compared to the Ge tracer diffusivity according to [17]

25.3 Dissociative Mechanism

Dissociative Diffusion of Hybrid Solutes in Metals: The phenomenon of fast diffusion of solute atoms is not restricted to semiconductor hosts. The fast diffusion of Au in Pb was recognised as early as 1896 by ROBERTS-AUSTEN [27], one of the pioneers of solid-state diffusion (see Chap. 1). After VON HEVESY AND COWORKERS [30, 31] had studied self-diffusion of Pb, it became evident that diffusion of Au in Pb is a very rapid process.

We recall Fig. 19.3, where an Arrhenius diagram of foreign atom diffusion in Pb together with self-diffusion is presented. Some solutes in lead (Tl, Sn, ...) show 'normal' behaviour. Others such as noble metals, Ni-group elements, and Zn have diffusivities which are three or more orders of magnitude faster than self-diffusion. We have denoted lead and some further polyvalent metallic hosts as 'open' metals [28]. The term 'open' refers to the large ratio between atomic and ionic radius of the solvent. This property leads to fast solute diffusion for solutes with relatively small atomic radii. As discussed in Chap. 19, fast diffusion of some 3d transition metals is observed in further polyvalent metals such as In, Sn, Sb, Ti, Zr, and Hf. Noble metal solutes are fast diffusers in the group-IVB metal tin and in the group-IIIB metals In and Tl. In addition, fast diffusion is reported for the late transition elements Fe, Co, Ni in the group-IVA metals α-Ti, α-Zr, and α-Hf.

Vacancies are the defects that mediate self-diffusion and diffusion of substitutional solutes in metals (see Chaps. 17 and 19). The fast diffusion of solutes in metals has been attributed by WARBURTON AND TURNBULL to the dissociative mechanism [29]. In contrast to semiconductor hosts, metals are not available as dislocation-free material. Metallic crystals usually have fairly high grown-in dislocation densities. These provide enough sources and sinks for vacancies to maintain vacancy equilibrium, i.e. $C_V \approx C_V^{eq}$. Therefore, usually only the diffusion mode limited by the interstitial flow product of hybrid solutes can be observed. If substitutional transport by vacancies is negligible, the effective diffusivity of solutes is given by Eq. (25.20).

Co in niobium: The system Co in Nb is an example for a metallic system for which the dissociative mechanism had been proposed in 1976. The only evidence for this suggestion in the work of PELLEG was the rapid diffusion of Co, when compared to Nb self-diffusion [32].

Diffusion of Co in Nb has been re-examined by WENWER AND COWORKERS [33, 34]. These authors had in mind that a clear cut identification of dissociative diffusion is the observation of the diffusion mode limited by the vacancy flow. Their study had to cope with the fact that Nb crystals are not available in dislocation-free form. Their crystals contained about 10^6 dislocations per cm^2. Fortunately, in the near-surface region the rôle of dislocations becomes insignificant, because the two-dimensional surface represents the dominating source and sink for vacancies. The width of a virtually dislocation-free zone is comparable to the mean distance between dislocations. For a dislocation density of about 10^{10}m^{-2} this corresponds to about 10 µm. With this in mind, careful radiotracer experiments of ^{60}Co diffusion

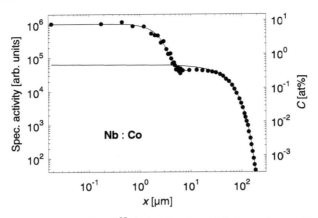

Fig. 25.9. Penetration profile of ^{60}Co in Nb after 10.3 days of annealing at 1422 K in double-logarithmic representation according to [34]

in Nb single crystals were conducted in such a way that both the surface-near region and greater penetration depths could be resolved. Figure 25.9 shows a typical penetration profile measured with ^{60}Co as radiotracer [34]. Diffusion has produced a Co distribution with two distinct stages. The solid line fitted to the deep profile part is a complementary error function. The near-surface portion of the profile can be fitted by a complementary error function as well but a little better by a Gaussian profile.

Within the dissociative model the two-stage Co distributions have a straightforward interpretation:

Deep profile stage: At deeper penetrations, the distance to adjacent dislocations (vacancy sources) is shorter than the distance to the surface (foreign-atom source). Beyond a certain depth, this difference is large enough to overcompensate the smallness of the vacancy flow product as compared to the Co interstitial flow product ($C_V^{eq} D_V(Nb) \ll C_i^{eq} D_i(Co)$). Thus, at great depths the situation corresponds to one of high dislocation density and the Co diffusivity is given by Eq. (25.20). Diffusion coefficients deduced from fits of complementary error functions to the deep stages of the experimental Co profiles agree well with those of previous authors (see, e.g., Fig. 5 in [33]). This shows that in the earlier work dissociative diffusion limited by the flow of Co interstitials had been measured.

Shallow profile stage: At shallow penetrations, vacancies are more efficiently supplied from the surface than from dislocations. On the other hand, transferring a vacancy to a shallow depth x takes more time than transferring a Co interstitial to the same position. This is so, because the vacancy flow is slower than the flow of Co interstitials, whereas both species must travel the same distance. Under such circumstances, the incorporation of Co is limited by the vacancy in-flow according to Eq. (25.42). To check this interpretation the authors [33, 34] deduced the product of vacancy-flow times correlation factor ($f_V = 0.727$ for a bcc lattice, see Chap. 7). A comparison between the so-obtained values with tracer self-diffusion coefficients of Nb from literature showed good agreement (see, e.g., Fig. 4 in [33]). This supports the interpretation that Co diffusion in Nb occurs via the dissociative mechanism.

Conclusions: The salient points from the discussion of the above examples can be summarised as follows:

1. Interstitial-substitutional diffusion of hybrid foreign atoms via the dissociative mechanism is established for Cu, Ag, and Au in Ge. The interstitial flow product of Cu in Ge is larger than the flow product of Ge vacancies. This feature leads to pronounced differences of the Cu diffusion behaviour in Ge, with low, intermediate and high dislocation densities. The interstitial flow product of Ag and Au atoms in Ge is smaller than the Ge vacancy flow product. Hence penetration and shape of the diffusion profiles do not depend on the dislocation density. The diffusivity of these atoms is determined by the small fractions and high mobilities in interstitial sites
2. Interstitial-substitutional diffusion also occurs for some fast diffusing solutes in 'open' metals. This is the case for the polyvalent solvents Pb, Sn, In, Sb. Ti, Zr, and Hf. Usually, fast diffusion in metals is limited by the flow product of the interstitial fraction of hybrid atoms. This is because in metals a sufficiently high density of dislocations usually warrants quasi-instantaneous supply of vacancies during diffusion experiments.
3. Both limiting modes of the dissociative mechanism have been observed for Cu in Ge and for Co in Nb. Foreign atom diffusion limited by the vacancy flow product has been established for Co in the near-surface region of moderately dislocated Nb crystals and for Cu diffusion in virtually dislocation-free Ge crystals.

References

1. F.C. Frank, D. Turnbull, Phys. Rev. **104**, 617 (1956)
2. U. Gösele, W. Frank, A. Seeger, Appl. Phys. A **23**, 361 (1980)
3. W. Meyberg, W. Frank, A. Seeger, H.A. Peretti, M.A. Mondino, Crystal Lattice Defects and Amorphous Matter **10**, 1 (1983)
4. N.A. Stolwijk, Phys. Rev. B **42**, 5793 (1990)
5. N.A. Stolwijk, Defect and Diffusion Forum **95–98**, 895 (1993)
6. H.S. Carslaw, J. C. Jaeger, *Conduction of Heat in Solids*, Oxford University Press, Oxford, 1959
7. J. Crank, *The Mathematics of Diffusion*, 2nd edition, Oxford University Press, Oxford, 1975
8. W. Jüngling, P. Pichler, S. Selberherr, E. Guerrero, H.W. Pötzl, IEEE Trans. Electron. Devices ED **32**, 156 (1985)

9. A. Seeger, Phys. Stat. Sol. (a) **61**, 521 (1980)
10. W. Frank, U. Gösele, H. Mehrer, A. Seeger, *Diffusion in Silicon and Germanium*, in: *Diffusion in Crystalline Solids*, G.E. Murch, A.S. Nowick (Eds.), Academic Press, 1984
11. W.R. Wilcox, T.J. LaChapelle, J. Appl, Phys. **35**, 240 (1964)
12. W.R. Wilcox, T.J. LaChapelle, D.H. Forbes, J. Electrochem Soc. **111**, 1377 (1964)
13. N.A. Stolwijk, B. Schuster, J. Hölzl, H. Mehrer, W. Frank, Physica **116 B**, 35 (1983)
14. N.A. Stolwijk, J. Hölzl, W. Frank, E.R. Weber, H. Mehrer, Appl. Phys. A **39**, 37 (1986)
15. M. Hill, M. Lietz, T. Sittig, J. Electrochem Soc. **129**, 1579 (1982)
16. J. Hauber, W. Frank, N.A. Stolwijk, Mater. Science Forum **38–41**, 707 (1989)
17. H. Bracht, N.A. Stolwijk, H. Mehrer, Phys. Rev. B **52**, 16542 (1995)
18. U. Gösele, W. Frank, in: *Defects in Semiconductors*, J. Narajan, T.Y. Tan (Eds.), North-Holland, New York, 1981, p. 53
19. U. Gösele, F. Morehead, F. Föll, W. Frank, H. Strunk, in: *Semiconductor Silicon 1981*, H.R. Huff, R.J. Hriegler, Y. Takeishi (Eds.), The Electrochem. Soc., Pennington, 1981, p. 766
20. U. Gösele, F. Morehead, W. Frank, A. Seeger, Appl. Phys. Lett. **38**, 157 (1981)
21. A. Seeger, W. Frank, Appl. Phys. **27**, 171 (1982)
22. C.S. Fuller, J.D. Struthers, Phys. Rev. **87**, 526 (1952)
23. C.S. Fuller, J.D. Struthers, J.A. Ditzenberger, K.B. Wolfstirn, Phys. Rev. **93**, 1182 (1954)
24. H.H. Woodberry, W.W. Tyler, Phys. Rev. **105**, 84 (1957)
25. N.A. Stolwijk, W. Frank, J. Hölzl, S.J. Pearton, E.E. Haller, J. Appl. Phys. **57**, 5211 (1985)
26. H. Bracht, N.A. Stolwijk, H. Mehrer, Phys. Rev. B **43**, 14465 (1991)
27. W.C. Roberts-Austen, Phil. Trans. Roy. Soc. **187**, 383 (1896)
28. G.M. Hood, Defect and Diffusion Forum **95–98**, 755 (1993)
29. W.K. Warburton, D. Turnbull, in: *Diffusion in Solids – Recent Developments*, A.S. Nowick, J.J. Burton (Eds.), Academic Press, 1975, p. 171
30. J. Groh, G. von Hevesy, Ann. Physik **63**, 85 (1920)
31. J. Groh, G. von Hevesy, Ann. Physik **65**, 216 (1921)
32. J. Pelleg, Philos. Mag. **33**, 165 (1976)
33. F. Wenwer, N.A. Stolwijk, H. Mehrer, Z. Metallkd. **80**, 205 (1989)
34. N.A. Stolwijk, F. Wenwer, H. Bracht, H. Mehrer, Proc. of the NATO ASI Series on: *Diffusion in Materials*, A.S. Laskar, J.L. Bocquet, G. Brebec, C. Monty (Eds.), Kluwer Academic Publishers, 1990, p. 297

Part V

Diffusion and Conduction in Ionic Materials

26 Ionic Crystals

26.1 General Remarks

In this chapter, we consider ionic crystals. The reader may recall that the forces between atoms in ionic crystals are largely classical (Coulombic) and that a well-developed theory of ionic crystals was established before the advent of quantum mechanics. Only the repulsive interaction between ions had to wait for a non-classical explanation. From a crystallographic point of view simple ionic crystals are characterised by two sublattices: a sublattice of cations and a sublattice of anions. As a consequence of Coulombic interactions between the ions the structure is completely ordered in the sense that a cation cannot enter the anion sublattice and vice versa. If diffusion occurred this way, ions would then occupy sites surrounded by ions of the same charge. The increase in electrostatic energy of this configuration over the normal situation is so great that this will not happen. Diffusion of a given ion is restricted to its own sublattice.

Ionic crystals are solids where diffusion and ionic conduction have been studied extensively to probe diffusion-mediating defects and their migration, defect-defect interactions, and defect-impurity interactions. The general theory of diffusion and most of the physical phenomena discussed in the metal chapters of this book apply equally well to ionic crystals as long as the two sublattices can be treated separately. For example, the cation and anion sublattices of the NaCl structure are both fcc. Hence correlation effects of diffusion for fcc metals and NaCl sublattices are alike. However, pairs of anion and cation vacancies can also contribute to the diffusion flux and thus couple the sublattices.

Because of their special electrical properties, several new effects are found in ionic crystals that are not found in metals. A topic specific of ionic materials is the relation between the ionic conductivity and tracer diffusion. This topic has been studied extensively and a relatively complete picture has emerged. Since ionic crystals are compounds, there are constraints on the concentration of defects. At constant composition a compound can only accommodate an increase in the concentration of cation vacancies (say, on heating) if there is a corresponding increase in anion vacancies or cation interstitials or both. Coulomb forces, play an important rôle and require charge neutrality throughout the bulk of the crystal apart from regions near sources and sinks of defects. In alkali and silver halides, which are considered in this chapter, the neutrality constraint is the same as that imposed by the structure. In any case, the formation of thermally generated point defects is more complex than in metallic elements.

Figure 26.1 shows the conductivity of various alkali and silver halides. Grouped together on the left-hand or high-temperature side we find alkali halide crystals. They are insulators at room temperature but have significant ionic conductivity within a range of a few hundred degrees Kelvin below their melting temperatures. Furthermore, the conductivity of the alkali halides is strongly temperature dependent; it changes by about 3 % per Kelvin and the corresponding activation enthalpies are about 2 eV. To the right in Fig. 26.1, we move to lower temperatures. Cesium and ammonium chloride have ionic conductivities that are similar in magnitude to most other alkali halides but are less strongly temperature-dependent, with activation enthalpies around 1.2 eV. With conductivities higher by about three orders of magnitude, we

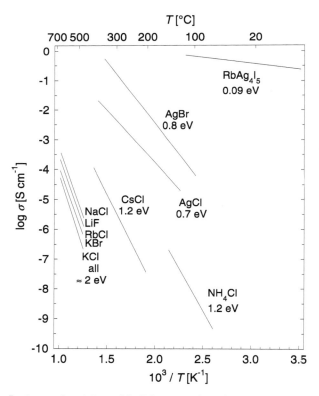

Fig. 26.1. Ionic conductivity of halide crystals. The corresponding activation enthalpies are listed. For comparison the conductivity of the fast ion conductor $RbAg_4I_5$ is also shown

find the silver halides AgCl and AgBr. In the far upper right-hand corner of the figure, near room temperature, we recognise the solid RbAg$_4$I$_5$ with a conductivity some four or five orders of magnitude higher than AgCl. The conductivity of this compound shows a small temperature variation with an activation enthalpy of only 0.09 eV. The phenomenon of fast ion conduction, of which RbAg$_4$I$_5$ is an example, is discussed in Chap. 27.

Our understanding of diffusion and ionic conduction in ionic crystals rests on voluminous work. Foundations were laid down in the early work of FRENKEL [1], SCHOTTKY AND WAGNER [2–4], and MOTT AND LITTLETON [5]. In the meantime, chapters in textbooks [6–10] and many reviews [11–16] summarise various aspects of the field. A comprehensive collection of diffusion data for alkali and alkaline earth halides has been given by BENIÈRE [17]. In the present chapter, we focus on those aspects that are required to understand diffusion and ionic conduction in alkali and silver halides.

26.2 Point Defects in Ionic Crystals

The ions in simple ionic crystals are mobile due to the presence of point defects. Figure 26.2 shows a two-dimensional representation of an ionic crystal perturbed by several types of point defects. The concentration of defects in a pure crystal is small. Even close to the melting temperature, the defect site fraction remains less than 1 %. Impurities with a valence different from

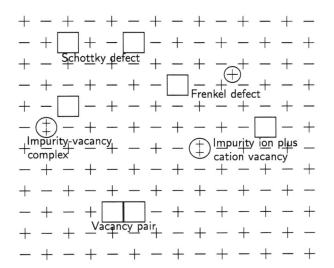

Fig. 26.2. Examples of point defects in ionic crystals: Schottky defects, Frenkel defects, divalent cation impurity and cation vacancy, complex of divalent cation and cation vacancy, vacancy pair

the host lattice, either added intentionally as doping elements or present inevitably in crystals of limited purity, enhance the concentration of point defects for charge neutrality reasons. In any case, diffusion and ionic conduction can be attributed to the presence and the mobility of point defects. In what follows, we consider explicitly ionic crystals with the formula CA (C=cation, A=anion).

26.2.1 Intrinsic Defects

As discussed in Chap. 5, Schottky disorder consisting of cation vacancies (V_C) and anion vacancies (V_A) prevails in alkali halides (Fig. 26.2). In thermal equilibrium the product of the site fractions of cation vacancies, C_{V_C}, and of anion vacancies, C_{V_A}, is a constant that depends on temperature via Eq. (5.38), which we repeat for convenience:

$$C_{V_C} C_{V_A} = \exp\left(-\frac{G_{SP}}{k_B T}\right) = K_{SP}(T). \tag{26.1}$$

G_{SP} denotes the Gibbs free energy for the formation of a *Schottky pair* (cation vacancy plus anion vacancy). The Gibbs energy can be split up via $G_{SP} = H_{SP} - T S_{SP}$ into enthalpy and entropy of pair formation. $K_{SP}(T)$ is called the *Schottky product*.

Frenkel pairs of the cation sublattice (Frenkel disorder), i.e. cation interstitials, I_C, and cation vacancies, V_C, are the predominant defects in AgCl and AgBr (Table 26.1). Here we refer the reader to Eq. (5.34), which can be written as:

$$C_{V_C} C_{I_C} = \exp\left(-\frac{G_{FP}}{k_B T}\right) \equiv K_{FP}(T). \tag{26.2}$$

G_{FP} denotes the Gibbs free energy for the formation of Frenkel pairs. Again, according to $G_{FP} = H_{FP} - T S_{FP}$, a formation enthalpy and a formation entropy of Frenkel pairs can be introduced. $K_{FP}(T)$ is the (cationic) Frenkel product. Frenkel pairs in the anion sublattice form in fluorite-related halides (PbF_2, $SrCl_2$, ...) and can be treated in an analogous way. This type of defects is sometimes called anti-Frenkel defects.

We emphasise that in the derivation of Eqs. (26.1) and (26.2) it is not necessary to assume $C_{V_C} = C_{V_A}$ and $C_{V_C} = C_{I_C}$, respectively. Equations (26.1) and (26.2) can be considered as mass action laws. The Schottky and Frenkel products describe the defect concentrations at thermal equilibrium. The products are valid irrespective of the defect's rôle as a majority or minority defect. As with any ion product, if the concentration of one type of defect is perturbed (e.g., by doping), the concentration of the other defect must adjust to keep the product constant. Unlike atoms in chemical reactions, vacancies and self-interstitials are not conserved. In thermal equilibrium their concentration is a function of temperature. Further features of interest are the following:

1. In undoped crystals with Schottky disorder, charge neutrality requires equal site fractions of cation and anion vacancies. Then

$$C_{V_C}(0) = C_{V_A}(0) = \exp\left(\frac{S_{SP}}{2k_B}\right)\exp\left(-\frac{H_{SP}}{2k_BT}\right) = \sqrt{K_{SP}(T)}. \quad (26.3)$$

2. In undoped crystals with Frenkel disorder, charge neutrality requires equal numbers of vacancies and interstitials. For Frenkel disorder in the cation sublattice we have

$$C_{V_C}(0) = C_{I_C}(0) = \exp\left(\frac{S_{FP}}{2k_B}\right)\exp\left(-\frac{H_{FP}}{2k_BT}\right) = \sqrt{K_{FP}(T)}. \quad (26.4)$$

In Eqs. (26.3) and (26.4) the zero in the arguments of the site fractions refers to intrinsic conditions. Frenkel disorder in the anion sublattice can be treated in an analogous way.

3. If the formation enthalpies of Schottky and Frenkel pairs are similar, cation vacancies, anion vacancies, and cation interstitials will be present. The relative concentrations of the various defects must then be determined by satisfying Eq. (26.1) and Eq. (26.2) simultaneously, and in addition the condition of charge neutrality which requires that

$$C_{V_C} = C_{V_A} + C_{I_C}, \quad (26.5)$$

provided all ions have the same charge.

4. Heterovalent doping of ionic crystals means that some of the matrix ions are replaced by foreign ions of different valence. For example, if Ca^{2+} ions replace a few Na^+ ions in NaCl, an equal number of cation vacancies must be added to preserve charge neutrality. Thus, the equilibrium concentration of defects is determined by the arrangement which maintains charge neutrality and at the same time satisfies Eqs. (26.1) or (26.2) (see below).

Vacancy-pairs: Relative to the perfect lattice cation and anion vacancies bear opposite charges and attract each other by Coulomb forces. When two

Table 26.1. Dominant type of disorder in ionic crystals

Material	Structure	Dominant disorder
Alkali halides (NaCl, KCl, ...)	NaCl	Schottky
Alkaline earth oxides	NaCl	Schottky
AgCl, AgBr	NaCl	Cation Frenkel
CsCl, TlCl, TlBr	CsCl	Schottky
BeO, MgO, ...	Wurtzite	Schottky
Alkaline earth fluorites, CeO_2, ThO_2	CaF_2	Anion Frenkel

vacancies form a nearest-neighbour pair, V_P, in addition to Coulomb energy, binding energy is also gained through lattice relaxation. It is then convenient to consider the pair as a distinct defect, which is formed by the reaction of a negatively charged cation vacancy, V_C^-, and a positively charged anion vacancy, V_A^+, according to[1]

$$V_C^- + V_A^+ \rightleftharpoons V_P . \qquad (26.6)$$

The vacancy-pair (Fig. 26.2) is a neutral defect and the concentration of pairs is given by

$$C_P = C_{V_C} C_{V_A} Z_P \exp\left(\frac{\Delta G_P}{k_B T}\right) = Z_P \exp\left(-\frac{G_{SP} - \Delta G_P}{k_B T}\right) , \qquad (26.7)$$

where Z_P is the number of distinct orientations of the pair ($Z_P = 6$ in the NaCl structure). ΔG_P is the Gibbs free energy released when two isolated vacancies form a pair. Since vacancy-pairs are neutral entities, their motion does not contribute to ionic conduction. However, pairs contribute to diffusion.

26.2.2 Extrinsic Defects

Doping Effects: In ionic crystals, additional defects are created by introducing aliovalent impurities (dopants), which differ in charge from the host ions. For example, if a divalent cation impurity i is substituted in the cation sublattice of a NaCl crystal, a cation vacancy must be present to compensate for the excess positive charge of the impurity. Depending on the temperature, the additional cation vacancy will be either isolated or, at low temperatures, associated with the impurity (Fig. 26.2).

The effect of aliovalent impurities on conductivity and diffusion of ionic solids has been the subject of many studies. It provides a powerful tool for investigating the types and relative mobilities of defects. The power of the technique comes from the fact that there are definite relations between the doping concentration and the defect concentrations. In practice, doping often means doping with divalent cations because it is easier from an experimental point of view than doping with aliovalent anions. This is a consequence of the higher solubility of divalent cations as compared to divalent anions in halide crystals.

Let us consider explicitly an alkali halide (e.g., NaCl) doped with a divalent alkaline earth halide (e.g., $CaCl_2$ or $SrCl_2$). In this case, the divalent cation is accompanied by one extra cation vacancy V_C (Fig. 26.2). The relation between the atomic fractions of impurities, C_i, and those of defects is then

---

[1] In the physico-chemical literature the so-called Kröger-Vink notation is used. In the present case this would correspond to: $V_C^- \equiv V_C'$ and $V_A^+ \equiv V_A'$.

$$C_i + C_{V_A} = C_{V_C}. \qquad (26.8)$$

As a consequence, the defect concentration is varied by the impurity concentration. In an undoped NaCl-type crystal, according to Eq. (26.3) the concentrations of cation vacancies, $C_{V_C}(0)$, and anion vacancies, $C_{V_A}(0)$, must be equal. If divalent impurities are added, the concentrations of cation and anion vacancies are different, but the Schottky product, Eq. (26.1), is still fulfilled. Using Eq. (26.8) to substitute for C_{V_A} yields

$$C_{V_C}(C_{V_C} - C_i) = \exp\left(-\frac{G_{SP}}{k_B T}\right) = C_{V_C}^2(0) = C_{V_A}^2(0). \qquad (26.9)$$

Equation (26.9) assumes as a useful approximation that the impurities and the defects are all distributed at random. It can be rewritten to give a quadratic equation in C_{V_C}. The physically acceptable root is

$$C_{V_C} = \frac{C_i}{2}\left[1 + \left(1 + \frac{4C_{V_C}^2(0)}{C_i^2}\right)^{1/2}\right]. \qquad (26.10)$$

This equation simplifies in two limiting cases:

1. *Intrinsic region:* In this region we have $C_{V_C}(0) \gg C_i$. Then,

$$C_{V_C} \approx C_{V_C}(0). \qquad (26.11)$$

   This case applies for undoped material or, since $C_{V_C}(0)$ increases exponentially with temperature, it may also apply to doped materials at elevated temperatures.

2. *Extrinsic region:* In this region we have $C_{V_C}(0) \ll C_i$. Then,

$$C_{V_C} \approx C_i. \qquad (26.12)$$

   Hence the total concentration of cation vacancies is fixed by the dopant concentration. Since $C_{V_C}(0)$ decreases exponentially with decreasing temperature, and since no material is absolutely free from multivalent impurities, there is always a temperature below which $C_{V_C}(0) \ll C_i$ is fulfilled.

Impurity-vacancy Complexes: At very low temperatures impurity atoms and cation vacancies form impurity-vacancy complexes. A divalent impurity attracts vacancies bearing the opposite effective charge. When these two defects are in their most stable configuration, they are considered as a separate defect, an associated impurity-vacancy pair (iV_C). Unless the impurity is small compared to the size of the ion for which it substitutes, the most stable configuration is a nearest-neighbour configuration. If it is small, the next-nearest-neighbour configuration is more stable [18, 19]. The association and dissociation reaction can be written as

$$i^+ + V_C^- \rightleftharpoons iV_C. \qquad (26.13)$$

The divalent impurity ion (i^+) provides an excess of one unit of positive charge relative to the perfect crystal, while the cation vacancy (V_C^-) – a missing cation – introduces one unit of negative charge. Neutral impurity-vacancy complexes form by association. In the extrinsic region, thermal vacancies can be neglected and the total vacancy concentration equals the analytical doping concentration C_i. The concentration of free vacancies is

$$C_{V_C^-} = C_i - C_{iV_C} = C_i(1-p), \qquad (26.14)$$

where C_{iV_C} is the concentration of complexes and

$$p \equiv \frac{C_{iV_C}}{C_i} \qquad (26.15)$$

is an abbreviation for the fraction of impurity ions associated with a vacancy. p is called the *degree of association*. The site fraction of unpaired impurities, i^+, is $C_i - C_{iV_C}$. Hence the law of mass action for Eq. (26.13) yields

$$\frac{C_{iV_C}}{(C_i - C_{iV_C})(C_i - C_{iV_C})} = Z_{iV_C} \exp\left(\frac{\Delta G_{iV_C}}{k_B T}\right) \qquad (26.16)$$

or

$$\frac{p}{(1-p)^2} = C_i Z_{iV_C} \exp\left(\frac{\Delta G_{iV_C}}{k_B T}\right). \qquad (26.17)$$

Z_{iV_C} denotes the number of distinct orientations of the iV_C complexes ($Z_{iV_C} = 12$ for nearest-neighbour complexes; $Z_{iV_C} = 6$ for next-nearest neighbour complexes in the NaCl structure). ΔG_{iV_C} is the Gibbs energy of association.

26.3 Methods for the Study of Defect and Transport Properties

Defect properties in ionic crystals are usually infered from studies of the ionic conductivity, tracer diffusion of cations and anions, isotope effect, heterodiffusion, electromigration, dielectric and anelastic relaxation. In Table 26.2 we have listed some of the experiments that have been or can be made on ionic crystals to provide information about defect properties. The experimental procedures for the measurement of tracer diffusion, isotope effect, and dc conductivity are discussed in Chaps. 9, 13, and 16, respectively, and need not be repeated here.

Tracer techniques can also be applied to study diffusion in an electric field (electromigration). In an experiment suggested by CHEMLA [20] one measures, at the same time, the diffusion coefficient from the spreading-out of the tracer distribution (Gaussian solution for a thin tracer layer) as well as the mobility from the drift of the Gaussian due in an electric field. The

26.3 Methods for the Study of Defect and Transport Properties

Table 26.2. Experimental techniques for the determination of transport properties in ionic crystals. For definition of the symbols see Sect. 26.4

Type of Measurement	Properties obtained
Conductivity of pure crystal	$\sigma = \sigma_C + \sigma_A$; D_σ
Conductivity of crystal doped with heterovalent cations	Mobility of cation vacancies; $H_{V_C}^M$
Conductivity of crystal doped with heterovalent anions	Mobility of anion vacancies; $H_{V_A}^M$
Cation tracer diffusion in pure crystal	$D_C^*(total) = D_C^* + D_{CP}^*$
Anion tracer diffusion in pure crystal	$D_A^*(total) = D_A^* + D_{AP}^*$
Cation tracer isotope effect	Information on D_{CP}^*
Anion tracer isotope effect	Information on D_{AP}^*
Cation tracer diffusion in electric field	Information on D_C^*
Anion tracer diffusion in electric field	Information on D_A^*
Impurity diffusion in cation sublattice	Information on defect complexes
Impurity diffusion in anion sublattice	Information on defect complexes
Dielectric relaxation	Reorientation of defect complexes
Anelastic relaxation	Reorientation of defect complexes

free vacancies drift along the field because of their electric charge, thereby giving an independent measure of their mobility, whereas vacancy pairs are unaffected by the field since they are neutral. Unfortunately, the method is technically very difficult. In order to maintain an electric field in an ionic crystal, a steady dc current must pass through the electrodes. This often causes deteriorations at the crystal-electrode contact interface. Nevertheless, electromigration measurements have been made for a few ionic crystals and have furthered our knowledge about these materials.

Complexes of aliovalent impurities and vacancies possess electric and mechanical dipole moments. Such entities profoundly affect many of the properties of an ionic crystal, including electrical as well as mechanical ones. Methods of dielectric and anelastic relaxation (see Chap. 14) may be used to observe the reorientation of the complex in an externally applied electric or stress field, respectively. These measurements give direct information about the kinetics of the reorientation process. Combining this information with information on the kinetics of migration from diffusion measurements provides a rather complete picture of the various ionic motions processes, which take place in the presence of aliovalent impurities. A review of this topic has been given by NOWICK [13].

26.4 Alkali Halides

26.4.1 Defect Motion, Tracer Self-diffusion, and Ionic Conduction

The physical interpretation of numerous experimental studies about ionic conductivity, tracer self-diffusion of cations and anions, isotope effect, electromigration, impurity diffusion, and relaxation performed on alkali halides requires a model of the perfect crystal perturbed mainly by the four kinds of point defects shown in Fig. 26.2: Schottky defects, heterovalent impurity ions, impurity-vacancy complexes, and vacancy-pairs. Most frequently, defects are observed not as a consequence of their static properties, but because of their mobility.

Tracer Self-diffusion: Diffusion mainly occurs by exchange jumps of ions with vacancies: alkali ions exchange with cation vacancies, halide ions exchange with anion vacancies. Neglecting for the moment contributions of vacancy-pairs, the tracer self-diffusion coefficients of cations and of anions, D_C^* and D_A^*, are given by

$$D_C^* = f_V a^2 C_{V_C} \omega_{V_C}, \qquad (26.18)$$

$$D_A^* = f_V a^2 C_{V_A} \omega_{V_A}. \qquad (26.19)$$

f_V is the correlation factor for the vacacy mechanism (see Chap. 7) and a the lattice parameter[2]. The jump rate of point defects has been discussed in Chap. 5. Derivations based either on absolute rate theory or on many-body theory of equilibrium statistics have resulted in an expression which can be written for the jump rate of cation vacancies, ω_{V_C}, as

$$\omega_{V_C} = \nu_C^0 \exp\left(-\frac{G_{V_C}^M}{k_B T}\right) = \nu_C^0 \exp\left(\frac{S_{V_C}^M}{k_B}\right) \exp\left(-\frac{H_{V_C}^M}{k_B T}\right) \qquad (26.20)$$

and for the jump rate of anion vacancies, ω_{V_A}, as

$$\omega_{V_A} = \nu_A^0 \exp\left(-\frac{G_{V_A}^M}{k_B T}\right) = \nu_A^0 \exp\left(\frac{S_{V_A}^M}{k_B}\right) \exp\left(-\frac{H_{V_A}^M}{k_B T}\right). \qquad (26.21)$$

ν_C^0 and ν_A^0 are the attempt frequencies for cation and anion jumps, $G_{V_C}^M$ and $G_{V_A}^M$ Gibbs free energies, $S_{V_C}^M, S_{V_A}^M$ and $H_{V_C}^M, H_{V_A}^M$ the corresponding entropies and enthalpies of vacancy migration (superscript M), respectively.

The vacancy fractions in Eqs. (26.18) and (26.19) depend on the purity of the material and on its temperature:

In the **intrinsic region**, where $C_{V_C} = C_{V_C}(0) = C_{V_A}(0) = C_{V_A}$, Eqs. (26.18) and (26.19) can be rewritten as

---

[2] Sometimes in the literature, the cation-anion distance is denoted by a. Then a factor of 4 must be included in Eqs. (26.18) and (26.19).

$$D_C^* = \underbrace{f_V a^2 \nu_C^0 \exp\left(\frac{S_{SP}/2 + S_{V_C}^M}{k_B}\right)}_{D_C^0} \exp\left(-\frac{H_{SP}/2 + H_{V_C}^M}{k_B T}\right), \quad (26.22)$$

$$D_A^* = \underbrace{f_V a^2 \nu_A^0 \exp\left(\frac{S_{SP}/2 + S_{V_A}^M}{k_B}\right)}_{D_A^0} \exp\left(-\frac{H_{SP}/2 + H_{V_A}^M}{k_B T}\right). \quad (26.23)$$

In this region, the variation of D_C^* and D_A^* with temperature stems from the fact that both the vacancy site fractions and the jump rates are Arrhenius activated. D_C^0 and D_A^0 are pre-exponential factors. The tracer diffusivities and the conductivity are independent of the purity of the specimen so that they are 'intrinsic' properties of the material. Apart from vacancy-pair corrections, the relevant activation enthalpies of self-diffusion of cations and anions, ΔH_C and ΔH_A, are:

$$\Delta H_C = \frac{H_{SP}}{2} + H_{V_C}^M, \quad (26.24)$$

$$\Delta H_A = \frac{H_{SP}}{2} + H_{V_A}^M. \quad (26.25)$$

In alkali halide crystals, cation vacancies are more mobile than anion vacancies, i.e. $\omega_{V_C} > \omega_{V_A}$. This is mainly a consequence of different sizes of the ions. Therefore, although both types of vacancies are present in equal numbers, the cation diffusivity in the intrinsic region is faster than the anion diffusivity:

$$D_C^* > D_A^*. \quad (26.26)$$

The difference is, however, not very large since it is a consequence of different defect mobilities only.

In the **extrinsic region** with divalent cation dopants, the concentration of cation vacancies is fixed, i.e. $C_{V_C} \approx C_i$. The equation for D_C^* then becomes

$$D_C^* = f_V a^2 C_i \omega_{V_C} = \underbrace{f_V a^2 C_i \nu_C^0 \exp\left(\frac{S_{V_C}^M}{k_B}\right)}_{D_C^{0'}} \exp\left(-\frac{H_{V_C}^M}{k_B T}\right). \quad (26.27)$$

In Eq. (26.27) we have assumed that the cation vacancies are not associated with impurities. This is justified for not too low temperatures. The pre-exponential factor $D_C^{0'}$ in Eq. (26.27) is much smaller than D_C^0 in Eq. (26.22) and it is proportional to the dopant concentration C_i. The anion vacancy concentration, C_{V_A}, is suppressed, since the Schottky product Eq. (26.1) demands that C_{V_A} is inversely proportional to the concentration of heterovalent cation impurities. For doping with divalent anion impurities, an equation analogous to Eq. (26.27) holds for anion diffusion.

The equations developed above cannot completely account for the experimental findings. The discrepancy is due to the existence of neutral **vacancy-pairs**. These contribute to self-diffusion but not to the ionic conductivity. The total tracer diffusivities, $D_C^*(total)$ and $D_A^*(total)$, are sums of the diffusivities mediated by isolated vacancies and by vacancy-pairs, D_{CP}^* and D_{AP}^*:

$$D_C^*(total) = D_C^* + D_{CP}^*, \qquad (26.28)$$
$$D_A^*(total) = D_A^* + D_{AP}^*. \qquad (26.29)$$

Interactions between cation and anion vacancies become appreciable at high vacancy concentrations. In thermal equilibrium the population of cation and anion vacancies increases with temperature according to Eq. (26.7). Thus, vacancy-pair formation in equilibrium is relevant in particular at high temperatures. Concomitantly, the pair contributions to tracer self-diffusion, D_{CP}^* and D_{AP}^*, increase with increasing temperature. They can be written as [11]:

$$D_{CP}^* = 2a^2 f_{CP}\omega_{CP} C_P, \qquad (26.30)$$
$$D_{AP}^* = 2a^2 f_{AP}\omega_{AP} C_P. \qquad (26.31)$$

These expressions contain rates for exchange jumps with cation or anion vacancies of the pair, ω_{CP} and ω_{AP}, and the relevant correlation factors for cation or anion tracer diffusion, f_{CP} and f_{AP}. Note that the correlation factors are different for cations and anions and different from those of isolated vacancies. As the lifetime of the vacancy pair is long compared to the time between atomic jumps, the pair motion involves both cation and anion jumps. The correlation factors are thus functions of the ratio ω_{CP}/ω_{AP}. More explicitly, the pair contributions can be written as

$$D_{CP}^* = 2a^2 f_{CP}\nu_{CP}^0 Z_P \exp\left(-\frac{G_{SP} - \Delta G_P + G_{CP}^M}{k_B T}\right), \qquad (26.32)$$

$$D_{AP}^* = 2a^2 f_{AP}\nu_{AP}^0 Z_P \exp\left(-\frac{G_{SP} - \Delta G_P + G_{AP}^M}{k_B T}\right). \qquad (26.33)$$

ν_{CP}^0 and ν_{AP}^0 denote attempt frequencies, whilst G_{CP}^M and G_{AP}^M are Gibbs energies of migration for cation and anion jumps of the pair, respectively.

Ionic Conductivity: When a solid is placed in an electrical circuit which maintains a voltage across it, a force is exerted on charged particles. In metals and semiconductors essentially all of the current is carried by the electrons. In ionic solids at elevated temperatures, electricity is conducted through the solid by the motion of ions. Cation and anion defects move so as to let current flow in the external circuit. To derive an equation relating the conductivity and the diffusion coefficient, it is necessary to account for the force exerted on the ions by the electric field. This reasoning led us to the Nernst-Einstein relation in Chap. 11. The total ionic dc conductivity, σ_{dc}, of an alkali halide

crystal is the sum of the contributions from cation and anion vacancies and can be written as:
$$\sigma_{dc} = \sigma_C + \sigma_A. \tag{26.34}$$
σ_C and σ_A denote the partial conductivities of the cation and the anion sublattice, respectively.

For ionic conductors, conductivity and diffusivity are related since the same ionic species are involved in charge and mass transport. As discussed in Chap. 11, the so-called *charge (or conductivity) diffusion coefficient* of ions, D_σ, is introduced. We then ascribe partial conductivities, σ_C and σ_A, to partial charge diffusivities, $D_{\sigma C}$ and $D_{\sigma A}$, via

$$D_{\sigma C} \equiv \frac{k_B T \sigma_C}{N_C q^2} \quad \text{and} \quad D_{\sigma A} \equiv \frac{k_B T \sigma_A}{N_A q^2}, \tag{26.35}$$

where N_C and N_A are the number densities of cations and anions, and q the charge of the ions. For alkali halides, we have $N_C = N_A = N_{ion}$ and we obtain a total charge diffusion coefficient via

$$D_\sigma \equiv D_{\sigma C} + D_{\sigma A} = \frac{k_B T \sigma_{dc}}{N_{ion} q^2}. \tag{26.36}$$

D_σ has the dimensions of a diffusion coefficient. However, it does not strictly correspond to any diffusion coefficient that can be measured by way of Fick's laws (see the remarks in Chap. 11).

A schematic Arrhenius plot of transport phenomena in an alkali halide crystal is shown in Fig. 26.3. In the extrinsic region, the number of vacancies is dominated by the impurity level and approaches the doping concentration. Then, the temperature dependence of σ_{dc} arises mainly from the migration enthalpy of the corresponding vacancies. These are cation (anion) vacancies in material doped with heterovalent cations (anions). In the intrinsic region, the tracer diffusivities of cations and anions are described by Eqs. (26.28) and (26.29), and the total charge diffusion by Eq. (26.36). Experimental data indicate that the behaviour in Fig. 26.3 is still somewhat idealised. For example, at low temperatures the conductivity deviates downward from the extrinsic line (see Fig.26.4). This is a consequence of the formation of impurity-vacancy complexes.

Haven Ratio: The Haven ratios for cations and anions, H_{RC} and H_{RA}, relate tracer and charge diffusivities of both sublattices via:

$$H_{RC} \equiv \frac{D_C^*(total)}{D_{\sigma C}} \quad \text{and} \quad H_{RA} \equiv \frac{D_A^*(total)}{D_{\sigma A}}. \tag{26.37}$$

Let us for the moment suppose that tracer diffusion and ionic conduction occur by isolated cation and anion vacancies. Then both Haven ratios are equal to the correlation factor of vacancy diffusion. For an fcc lattice we have $f_V = 0.781$. We then would arrive at

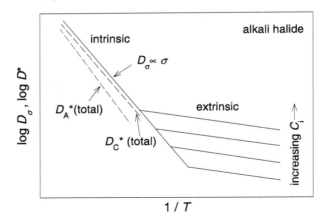

Fig. 26.3. Schematic diagram of charge diffusivity, D_σ, and tracer diffusivities of anions and cations, D_A^* and D_C^*, in alkali halides. *Parallel lines* in the extrinsic region correspond to different doping contents C_i

$$D_C^* + D_A^* = f_V(D_{\sigma C} + D_{\sigma A}) \equiv f_V D_\sigma = \frac{f_V k_B T \sigma_{dc}}{C_{ion} q^2}. \tag{26.38}$$

This equation is based on the assumption that only isolated cation and anion vacancies mediate diffusion. If vacancy-pairs act as additional vehicles of diffusion, deviations from Eq. (26.38) occur. Vacancy-pairs are neutral and do not contribute to the conductivity, but make contributions to tracer diffusion of both ionic species.

26.4.2 Example NaCl

The amount of experimental data available on transport properties of alkali halides and especially of NaCl is very large. Several reviews supply catalogues of experimental values for transport properties of alkali halide crystals [12–16, 21]. Rather than attempt to update these reviews, we illustrate some of the main features by presenting typical results for NaCl.

Conductivity and Self-diffusion: Figure 26.4 shows the conductivity of a NaCl single crystal doped with divalent Sr^{2+} ions according to BENIÈRE ET AL. [22]. An intrinsic and an extrinsic region of ion conduction can clearly be distinguished. The dashed lines represent extrapolations of the intrinsic and extrinsic parts and show where deviations occur. The slope of the intrinsic part (above about 650 °C) corresponds to an activation enthalpy, which equals half of the formation enthalpy of Schottky pairs plus the migration enthalpies of cation vacancies (see Eq. 26.24). The slope of the linear extrinsic part reflects the migration enthalpy of cation vacancies. At the low temperature end (below about 400 °C), a downward deviation of the data

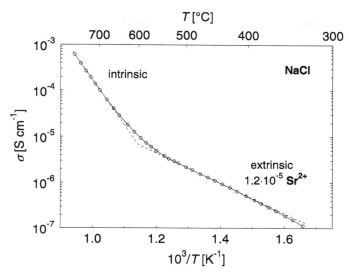

Fig. 26.4. Conductivity of a NaCl single crystal doped with a site fraction of 1.2×10^{-5} Sr^{2+} ions according to BENIÈRE ET AL. [22]

from the dashed line indicates that complexes between Sr^{2+} ions and cation vacancies are formed. In this low temperature region, cation vacancies must first acquire the additional energy needed to dissociate the vacancy-impurity pair. As a result, the activation enthalpy is again larger than the migration enthalpy of the vacancy.

Figure 26.5 illustrates the experimental Arrhenius behaviour for the total tracer diffusivities of Na [28], of Cl [27], and of the charge diffusion coefficient, D_σ, obtained from conductivity measurements by NELSON AND FRIAUF [28]. All data refer to the intrinsic range of NaCl and have been assembled by LASKAR [15]. Figure 26.5 shows that

$$D^*_{Na}(total) + D^*_{Cl}(total) > f_V D_\sigma . \tag{26.39}$$

The sum of the tracer diffusivities of cation and anion exceeds the product of charge diffusivity and correlation factor ($f_V = 0.781$ for an fcc lattice). We recall that the ionic conductivity is due to the motion of cation and anion vacancies. Neutral vacancy-pairs cannot contribute to the conductivity. The inequality, Eq. (26.39), reveals some vacancy-pair contribution to tracer diffusion. The diffusion data can be fully interpreted in terms of vacancy-mediated diffusion of cations and anions plus some contribution of vacancy-pairs. In NaCl vacancy-pairs contribute about 40 % to Na tracer diffusion near the melting point. Concomitantly, there is also a contribution of vacancy-pairs to anion tracer diffusion. Due to the correlation effects it is smaller than that of cation tracer diffusion.

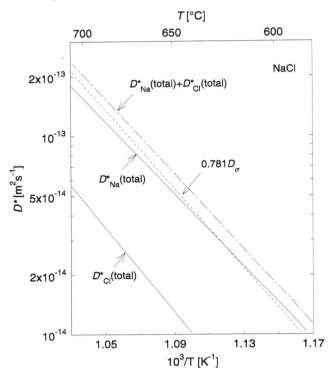

Fig. 26.5. Self-diffusion of ^{22}Na and ^{36}Cl in intrinsic NaCl. Also indicated is the product of charge diffusion coefficient D_σ and correlation factor $f_V = 0.781$. From LASKAR [15]

Isotope Effect and Electromigration: The isotope effect (see Chap. 9) of the radioisotope pair ^{22}Na/^{24}Na has been measured on single crystals of NaCl between 589 and 796 °C by ROTHMAN ET AL. [29] and by BARR AND LE CLAIRE [30]. The isotope effect parameter E decreases with increasing temperature from about 0.7 to 0.49, indicating two mechanisms of diffusion. For diffusion via isolated cation vacancies a value $E_C = f_V \Delta K_V$ with $f_V = 0.781$ is expected. ΔK_V is the kinetic energy factor for cation vacancy jumps. The vacancy-pair mechanism has a smaller correlation factor and hence its isotope effect parameter, E_{CP}, is smaller than E_C. For two mechanisms operating simultaneously, the measured isotope effect parameter, E, is a weighted average of the isotope effect parameters of the two mechanisms (see Chap. 9):

$$E = E_C \frac{D_C^*}{D_C^* + D_{CP}^*} + E_{CP} \frac{D_{CP}^*}{D_C^* + D_{CP}^*}. \qquad (26.40)$$

Since the pair contribution has the higher activation enthalpy relative to the contribution of single cation vacancies, E decreases with increasing temperature. According to ROTHMAN ET AL. [29] the pair contribution reaches 30–45 % near the melting point.

NELSON AND FRIAUF [28] studied tracer diffusion of Na in an external electric field ('Chemla' experiment). They placed a thin layer of ^{22}Na between two flat NaCl samples, which were thick compared to the diffusion length. A dc electric field was then applied perpendicular to the interface. The diffusion profile shifted in the field by an amount proportional to the mobility. By comparison with the ionic conductivity, a contribution of vacancy-pairs to tracer diffusion of about 40 % has been determined, which is in good agreement with the isotope effect measurements and the analysis of tracer and conductivity data.

Impurity Diffusion: The impurity diffusion coefficient in an fcc sublattice, D_2, can be written as (see Chap. 19)

$$D_2 = a^2 \omega_2 f_2 C_{V_C} \exp\left(\frac{\Delta G_{iV_C}}{k_B T}\right), \qquad (26.41)$$

where ω_2 is the jump rate for impurity-vacancy exchanges and ΔG_{iV_C} the Gibbs energy of impurity-vacancy association. f_2 denotes the impurity correlation factor. In the frame-work of the 'five-frequency model' introduced by LIDIARD, impurity diffusion in an fcc lattice is discussed in Chap. 7. The expression obtained for the correlation factor [35, 36] can be written as (see Eq. 7.37):

$$f_2 = \frac{2\omega_1 + 7\omega_3 F_3(\omega_4/\omega_{V_c})}{2\omega_2 + 2\omega_1 + 7\omega_3 F_3(\omega_4/\omega_{V_C})}. \qquad (26.42)$$

The jump rates ω_1, ω_2, ω_3, and ω_4 are illustrated in Fig. 7.4. In the present notation, the jump rate in the unperturbed host corresponds to the cation vacancy jump rate, i.e. $\omega \equiv \omega_{V_C}$. For tight binding between vacancy and impurity the dissociation jump rate of the complex, ω_3, is negligible and $f_2 \approx \omega_1/(\omega_1 + \omega_2)$.

It is useful to make a distinction between homo- and heterovalent impurities. For heterovalent impurities the vacancy-impurity association has a strong Coulombic contribution to the binding enthalpy, which is missing for homovalent impurities. In the latter case, the interaction is due to size and polarisability effects only.

Homovalent impurities: Tracer diffusion of homovalent alkali and halogen ions in alkali halides has been studied by several groups. For a review we refer the reader to [14, 31]. The Arrhenius diagram of homovalent impurities is similar to the one of self-diffusion and has also an intrinsic and an extrinsic part. Figure 26.6 shows diffusion of homovalent impurities in both sublattices of NaCl in the intrinsic range according to [32–34]. The diffusivities of homovalent impurities in the Na (Cl) sublattice lie in a relatively narrow band around Na (Cl) self-diffusion. The difference between impurity and self-diffusivities never exceeds a factor of five. In addition, the activation

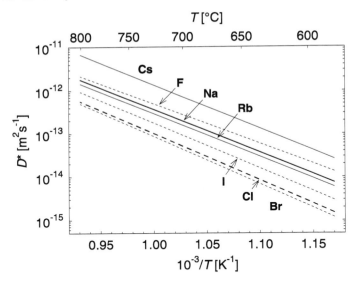

Fig. 26.6. Diffusion of the homovalent impurities Cs, Rb F, I, and Br in NaCl according to [32–34]. Self-diffusion of Na and Cl is also indicated for comparison

enthalpies of homovalent impurities are not much different from those of the self-diffusivities of Na or Cl, respectively. This is similar to normal impurity diffusion in fcc metals (see Chap. 19).

Heterovalent impurities: The association of heterovalent cations with vacancies can lead to a strong variation of the diffusion coefficients of heterovalent impurities with impurity concentration. This effect has been treated by HOWARD AND LIDIARD [40]. The theoretical variation of the impurity diffusion coefficient of divalent cations is shown in Fig. 26.7 as a function of impurity content. The strong variation of the diffusion coefficient at low concentrations results from the fact that the impurity can diffuse only when a vacancy is on a nearest-neighbour site. One finds

$$D_2 = \frac{8}{3}a^2\omega_2 f_2 \frac{p}{1+p}, \qquad (26.43)$$

where p is the degree of association given by Eq. (26.15). When p is unity, the diffusivity reaches a saturation value

$$D_2(sat) = \frac{4}{3}a^2\omega_2 f_2 \qquad (26.44)$$

and follows an Arrhenius law. Then, the activation enthalpy is given by the migration enthalpy of the impurity plus some correction term due to the temperature dependence of the correlation factor. For $p < 1$ the penetration profiles do not follow standard solutions of Fick's second law, because D_2 varies with concentration.

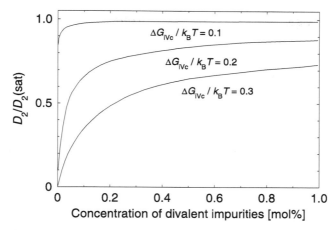

Fig. 26.7. Concentration dependence of the diffusion coefficient of divalent cations, D_2, relative to its saturation value, $D_2(sat)$, according to [39]. The curves refer to three values of the association enthalpy, ΔG_{iV_C}, normalised with $k_\mathrm{B}T$

26.4.3 Common Features of Alkali Halides

Let us summarise some characteristic features of diffusion and ionic conduction, which are common to alkali halides such as NaCl, KCl, RbCl, KBr, KI, ... :

1. The predominant intrinsic defects in alkali halides are Schottky defects (cation and anion vacancies). The formation enthalpy of Schottky defects is between about 2.0 and 2.3 eV.
2. Doping with heterovalent cations (anions) is compensated for charge neutrality reasons by the formation of cation (anion) vacancies.
3. The cation vacancy is slightly more mobile than the anion vacancy. Thus, the cation vacancy contribution to ionic conduction is larger than that of anion vacancies.
4. The ionic conductivity of an alkali halide crystal is a complicated function of temperature and impurity concentration. It has an intrinsic and an extrinsic region (Figs. 26.3 and 26.4). In the intrinsic region, equal numbers of thermal cation and anion vacancies contribute to the conductivity. In alkali halides doped with heterovalent cations (anions), the extrinsic conductivity is dominated by the motion of cation (anion) vacancies, which are formed for charge neutrality reasons. Complexes between heterovalent impurities and vacancies become significant in the extrinsic region at low temperatures.
5. Tracer self-diffusion of cations (anions) is mediated by cation (anion) vacancies and some additional contribution of neutral vacancy-pairs, which varies from one alkali halide to another. Cationic self-diffusion is slightly faster than anionic self-diffusion.

6. Near the melting temperature, the vacancy-pair contribution to the tracer diffusivity is approximately half of the contribution of isolated vacancies. The vacancy-pair contribution decreases with decreasing temperature.
7. Diffusion of homovalent impurities is not much different from self-diffusion in the same sublattice. The diffusion of heterovalent impurities is strongly concentration dependent.
8. The binding enthalpy between divalent cation impurities and cation vacancies is approximately 0.5 eV.

26.5 Silver Halides AgCl and AgBr

In this section, we consider the silver halides AgCl and AgBr. AgI is discussed in Chap. 27. It has a different structure and is a famous fast ion conductor for Ag ions. AgCl and AgBr have been extensively investigated because of their technological importance in the photographic process. Although their structure is the same as that of NaCl, their transport properties are very different. This can be seen in Fig. 26.1. The ionic conductivity of AgCl and AgBr is orders of magnitude higher than that of alkali halides. Let us first summarise a few salient points of defect and transport properties of AgCl and AgBr:

1. The predominant defects in AgCl and AgBr are cation Frenkel defects, i.e. Ag vacancies and interstitital Ag^+. The interstitital Ag^+ ions are more mobile than the Ag vacancies.
2. The ionic conductivity is mainly due to the motion of Ag^+ interstitials. Ag vacancies make a minor contribution.
3. Self-diffusion of Ag is due to Ag^+ interstitials and, to a lesser extent, due to Ag vacancies.
4. Tracer diffusion experiments of the anionic constituents, Cl and Br, indicate for both AgCl and AgBr that the anion diffusivity is three to four orders of magnitude lower than that of Ag^+ ions[3]. The anion diffusivity is mediated by anion vacancies. Their number is small, and they are part of Schottky pairs.
5. In the intrinsic region, the dominant diffusion mechanism of Ag is the interstitialcy mechanism. Several types of jumps, colinear and non-colinear ones contribute to Ag diffusion (see Fig. 26.8). Direct insterstitial jumps are negligible.
6. The migration enthalpy for Ag^+ interstitials for colinear and non-colinear jumps is very low (0.04 to 0.1 eV). This is probably the result of a quadrupolar deformation of the Ag^+ ion along the $\langle 110 \rangle$ direction.

---

[3] The fast ion conductor AgI is an extreme case, where the ratio between cation and anion diffusivity reaches about six orders of magnitude.

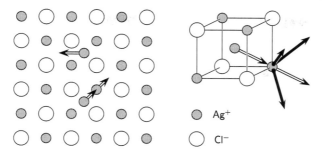

Fig. 26.8. *Left*: migration of interstitial Ag$^+$ ions via the direct interstitial and by the interstitialcy mechanism. *Right*: pathways for Ag$^+$ movements by the colinear (*double arrows*) and the non-colinear (*solid arrows*) interstitialcy mechanism

7. The diffusivity of cations and anions exhibit intrinsic and extrinsic regions. At low temperatures the diffusion of Ag takes place via extrinsic vacancies (see the discussion of doping effects below).

26.5.1 Self-diffusion and Ionic Conduction

Figure 26.9 shows a combined Arrhenius diagram of tracer diffusion and ionic conduction for AgCl and AgBr. We note that tracer diffusion of the anions is several orders of magnitude slower than that of the Ag cations. A silver tracer may diffuse through the crystal either by the vacancy mechanism or by the interstitialcy mechanism. The tracer self-diffusion coefficient is then given by

$$D^*_{Ag} = f_V a^2 \omega_{V_{Ag}} C_{V_{Ag}} + \sum_{I=colinear, non-colinear} f_I g_I a^2 \omega_{I_{Ag}} C_{I_{Ag}}, \quad (26.45)$$

where f_V and f_I are the correlation factors for vacancy and interstitialcy mechanisms. $\omega_{V_{Ag}}$ and $\omega_{I_{Ag}}$ denote the jump rates, $C_{V_{Ag}}$ and $C_{I_{Ag}}$ are the site fractions of Ag vacancies and interstitials, and g_I are geometrical factors.

Several possibilities for the migration of interstitial Ag$^+$ ions are illustrated in Fig. 26.8. In the direct interstitial mechanism, an interstitial Ag$^+$ jumps to an adjacent empty interstitial site. The indirect or interstitialcy mechanism involves a 'knock-on process', which requires the collective motion of (at least) two Ag ions. An interstitial Ag$^+$ ion causes one of its four Ag neighbours to move off its regular site into an adjacent interstitial site and then occupies the vacated lattice site itself (see also Chap. 6). As indicated in Fig. 26.8, one distinguishes between colinear and non-colinear interstitialcy jumps. The knocked-on Ag ion is displaced in a forward direction to the centre of any of the four neighbouring cells. If the second Ag ion moves in the same direction as the first Ag ion, the jump is called colinear. If it

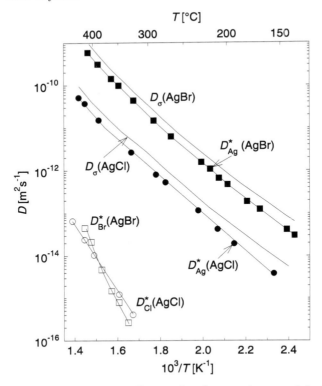

Fig. 26.9. Tracer self-diffusion coefficients for the constituents of AgCl [41] and AgBr [24, 42]. D_σ was calculated from the ionic conductivity via the Nernst-Einstein relation [37]

moves to any one of the three other forward positions, the jump is called non-colinear. Self-diffusion of Ag^+ is found to be due to vacancy jumps and due to colinear and non-colinear interstitialcy jumps. As already mentioned, direct interstitial jumps are negligible.

In such a case, the Haven ratio,

$$H_R = \frac{D^*_{Ag}}{k_B T \sigma_{dc}/(N_{ion} q^2)}, \qquad (26.46)$$

is a weighted average of several processes. Vacancy and interstitialcy contributions, the latter involving colinear and non-colinear interstitialcy jumps, contribute to diffusion and conductivity. H_R contains the correlation factors of the various contributions. A further contribution comes from the fact that the displacement of a tracer atom is smaller than that of the electric charge since (at least) two ions are displaced simultaneously in an interstitialcy process. FRIAUF [37] determined the Haven ratio from a comparison of Ag tracer

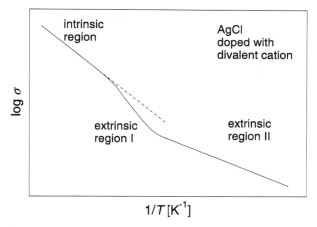

Fig. 26.10. Schematic illustration of the effect of divalent cationic impurity doping on the temperature dependence of the conductivity of AgCl

diffusion and ionic conductivity (see Fig. 26.9). The experimental Haven ratio increases with increasing temperature. These findings have been attributed to a mixture of colinear and non-colinear interstititalcy jumps with some minor contribution of Ag vacancies[4].

26.5.2 Doping Effects

The effect of aliovalent cation impurities in the extrinsic conductivity region of AgCl and AgBr is different from that observed for alkali halides: doping with divalent cations (say Cd^{2+}), like in alkali halides, increases the number of cation vacancies. However, as the Frenkel product, Eq. (26.2), of cation vacancies and Ag^+ interstitials is constant, the Ag^+ concentration must decrease with increasing Cd^{2+} concentration. Doping with divalent cations therefore reduces the concentration of the very mobile Ag^+ interstitials.

The resulting Arrhenius diagram of the ionic conductivity is illustrated schematically in Fig. 26.10. In the intrinsic region, the conductivity is independent of the dopant concentration. An extrinsic region at lower temperatures like in alkali halides is observed, but it is displaced downwards to lower conductivities. The degree of displacement increases with increasing doping concentration (extrinsic region I) until a minimum is reached (Fig. 26.11). At the conductivity minimum the contribution due to the more numerous but less mobile Ag vacancies equals that of the more mobile but less numerous Ag^+ interstitials. At still higher doping levels (extrinsic region II),

---
[4] The most frequent hopping process in AgBr are Ag ions, that hop from ordinary lattice sites into neighbouring interstitial sites and immediately back again. However, such hops neither contribute to tracer diffusion nor to the dc conductivity.

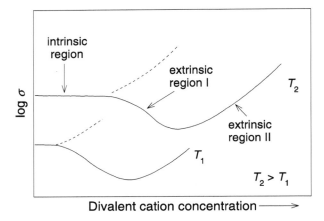

Fig. 26.11. Schematic illustration of the effect of divalent cationic impurity doping on the isothermal conductivity of AgCl. The *dashed lines* represent the effect of anionic impurity doping

the Ag vacancy conduction dominates and the conductivity increases with further doping. In the extrinsic region I, conduction via Ag^+ interstitials predominates and the conductivity decreases as the concentration of interstititals decreases with increasing doping concentration. In the extrinsic region II, Ag vacancy conduction takes over and the conductivity increases as the vacancy concentration increases with increasing doping concentration. The conductivity plot of Fig. 26.10 is still somewhat idealised. At high temperatures, interactions between defects become important and give an upward departure from the intrinsic slope. At low temperatures, like in the alkali halides, complexes between aliovalent impurities and vacancies form, causing downward curvature in the extrinsic region.

Some defect enthalpies deduced from measurements on AgCl and AgBr are listed in Table 26.3.

Table 26.3. Enthalpies of point defects in AgCl and AgBr according to BARR AND LIDIARD [11]

	AgCl	AgBr
Formation enthalpy of Frenkel pair	1.24 to 1.44 eV	1.06 eV
Migration enthalpy of cation vacancy	0.27 to 0.34 eV	0.34 eV
Migration enthalpy of Ag^+ interstitial	0.05 to 0.16 eV	0.15 eV
		colinear: 0.078 eV
		non-colinear: 0.225 eV

References

1. J.I. Frenkel, Z. Physik **35**, 652 (1926)
2. C. Wagner, W. Schottky, Z. Phys. Chem. B **11**, 163 (1931)
3. W. Schottky, Z. Phys. Chem. Abt B **29**, 335 (1935)
4. C. Wagner, Z. Phys. Chem. Abt B **38**, 325 (1938)
5. N.F. Mott, M.J. Littleton, Trans. Faraday Soc. **34**, 485 (1938)
6. Y. Adda, J. Philibert, *La Diffusion dans les Solides*, Presses Universitaires de France, Paris, 1966
7. P. Shewmon, *Diffusion in Solids*, 2nd edition, The Minerals, Metals and Materials Society, Warrendale, Pennsylvania, US, 1989
8. J. Philibert, *Atom Movements – Diffusion and Mass Transport in Solids*, Les Editions de Physique, Les Ulis, 1991
9. A.R. Allnatt, A.B. Lidiard, *Atomic Transport in Solids*, Cambridge University Press, 1993
10. A.R. West, *Basic Solid State Chemistry*, 2nd edition, John Wiley and Sons, 1999
11. L.W. Barr, A.B. Lidiard, *Defects in Ionic Crystals*, in: *Physical Chemistry – An Advanced Treatise*, Academic Press, New York, Vol. X, 1970
12. R.G. Fuller, *Ionic Conductivity (Including Self-Diffusion)*, in: *Point Defects in Solids*, J. M. Crawford Jr., L.M. Slifkin (Eds.), Plenum Press, 1972, p. 103
13. A.S. Nowick, *Defect Mobilities in Ionic Crystals Containing Aliovalent Ions*, in: *Point Defects in Solids*, J.M. Crawford Jr., L.M. Slifkin (Eds.), Plenum Press, 1972, p. 151
14. W.J. Fredericks, *Diffusion in Alkali Halides*, in: *Diffusion in Solids – Recent Developments*, A. S. Nowick, J.J. Burton (Eds.), Academic Press, 1975, p.381
15. A.L. Laskar, Proc. of the NATO ASI Series on *Diffusion in Materials*, A.S. Laskar, J.L. Bocquet, G. Brebec, C. Monty (Eds.), Kluwer Academic Publishers, 1990, p. 459
16. A.L. Laskar, *Diffusion in Ionic Solids: Unsolved Problems*, in: *Diffusion in Solids – Unsolved Problems*, G.E. Murch (Ed.), Trans Tech Publications, Zürich, Switzerland, 1992, p. 207
17. F. Benière, *Diffusion in Alkali and Alkaline Earth Halides*, in: *Diffusion in Semiconductors and Non-Metallic Solids*, D.L. Beke (Vol. Ed.), Landolt-Börnstein, New Series, Group III: Condensed Mattter, Vol. 33, Subvolume B1, Springer-Verlag, 1999
18. G.D. Watkins, Phys. Rev. **113**, 79 (1959)
19. G.D. Watkins, Phys. Rev. **113**, 91 (1959)
20. M. Chemla, PhD Thesis, Paris, 1954
21. S.M. Klotsman, I.P. Polikarpova, G.N. Tatarinova, A.N. Timofeev, Phys. Rev. B **38**, 7765 (1988)
22. F. Benière, M. Benière, M. Chemla, J. Phys. Chem. Sol. **31**, 1205 (1970)
23. M.D. Weber, R. Friauf, J. Phys. Chem. Sol. **30**, 407 (1969)
24. R.J. Friauf, Phys. Rev. **105**, 843 (1957)
25. H. Jain, O. Kanert, in: *Defects in Insulating Materials*, O. Kanert, J.-M. Spaeth (Eds.), World Scientific Publishing Comp., Ltd., Singapore, 1993
26. R.J. Friauf, J. Appl. Phys., Suppl. to Vol. **33**, 494 (1962)
27. N. Laurance, Phys. Rev. **120**, 57 (1960)
28. V.C. Nelson, R.J. Friauf, J. Phys. Chem. Sol. **31**, 825 (1970)

29. S.J. Rothman, N.L. Peterson, A.L. Laskar, L.C. Robinson, J. Phys. Chem. Sol. **33**, 1061 (1972)
30. L.W. Barr, A.D. Le Claire, Proc. Brit. Ceram. Soc. **1**, 109 (1964)
31. F. Benière, S.K. Sen, Philos. Mag. A **64**, 1167 (1991)
32. M. Benière, F. Benière, M. Chemla, J. Chim. Phys. **66**, 898 (1970); **67**, 1312 (1970)
33. M. Benière, F. Benière, C.R.A. Catlow, A.K. Shukla, C.N.R. Rao, J. Phys. Chem. Sol. **38**, 521 (1977)
34. S. Bandyopadhyay, S.K. Deb, Phys. Stat. Sol. (a) **129**, K27 (1985)
35. A.D. Le Claire, A.B. Lidiard, Philos. Mag. **1**. 518 (1956)
36. J.R. Manning, *Diffusion Kinetics for Atoms in Crystals*, Van Norstrand Comp., 1968
37. R.J. Friauf, J. of Appl. Phys. **33**, 494 (1962)
38. A.B. Lidiard, Philos. Mag. **46**, 1218 (1955)
39. R.E. Howard, A.B. Lidiard, J. Phys. Soc. Jap. (Suppl. II), **18**, 197 (1963)
40. R.E. Howard, A.B. Lidiard, Rep. Progr. Phys. **27**, 161–240 (1964)
41. W.D. Compton, R.J. Maurer, J. Phys. Chem. Solids **1**, 191 (1956)
42. D. Tannhauser, J. Phys. Chem. Solids **5**, 224 (1958)

27 Fast Ion Conductors

Fast ion conductors, sometimes referred to as superionic conductors or solid electrolytes, are solids with exceptionally high ionic conductivities over a reasonable temperature range, in some cases approaching the magnitude found in molten salts and aqueous solutions of strong electrolytes. Typical values of the conductivity of a fast ion conductor are in the range 10^{-3} to $10\,\mathrm{S\,cm^{-1}}$, which can be compared to the value achieved by normal ionic solids only at temperatures close to the melting point. For comparison, we mention that Cu at room temperature has an (electronic) conductivity of $6 \times 10^6\,\mathrm{S\,cm^{-1}}$. Alkali halide crystals near the melting point have $10^{-4}\,\mathrm{S\,cm^{-1}}$ ionic conductivity and a 0.1 n aqueous solution of NaCl has $10^{-2}\,\mathrm{S\,cm^{-1}}$.

Historically, fast ion conduction is an old phenomenon: high solid-state conductivity was first noted by the brilliant and perceptive English scientist MICHAEL FARADAY (1791–1867). He reported in 1833 that Ag_2S conducted electricity in the solid state and observed that hot PbF_2 also conducted electricity [1]. Around 1900, the German Nobel laureate in chemistry of 1920 WALTHER NERNST (1864–1941) observed that mixed oxides of ZrO_2 and Y_2O_3 glowed white-hot, when a current was passed through them at high temperatures. He attributed this to oxygen ion conduction and used it in a lamp known as the 'Nernst glower'. The unusual properties of AgI were studied for the first time in 1914 by TUBANDT AND LORENZ [2]. During their studies of electrical properties of silver halides, they observed that AgI, above $147\,^\circ\mathrm{C}$, has ionic conductivities comparable to the best conducting liquid electrolytes. The 1960s witnessed an increased interest in the field. A number of new fast ion conductors were reported, notably the Ag^+ ion conductors $RbAg_4I_5$ and related compounds [3, 4], the Na^+ ion conductor sodium β-alumina [5, 6], and the oxygen conductors based on stabilised zirconia. The oil crisis of the 1970s and the need for energy conservation further increased the number of studies on fast ion conduction with a major focus on battery and sensor technology, and on fuel cells.

Most ionic crystals, like NaCl or MgO, have low ionic conductivities. Although the ions undergo thermal vibrations, they can only escape from their lattice sites with the help of point defects, which are present in very small concentrations only (see Chaps. 5 and 26). Good solid electrolytes are an exception. Depending on the material, one component, either cationic or an-

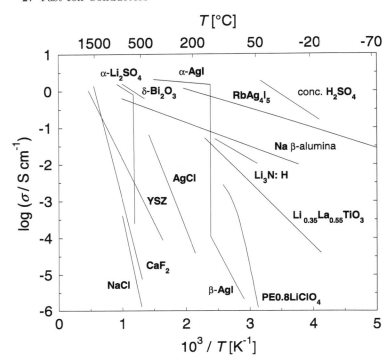

Fig. 27.1. Electrical dc conductivity of several fast ion conductors. Some ordinary solid electrolytes and concentrated H_2SO_4 are shown for comparison

ionic, is essentially free to move through the structure. Nowadays, many materials are known to exhibit fast ion behaviour. As a general rule, fast ion conductors have an 'open' crystal structure, which allows the rapid motion of ions. Thermal agitation is not needed to create point defects in the ion-conducting sublattice because large numbers of empty sites are available due to the structure. Some conductivity values of normal solid electrolytes, an optimised liquid electrolyte and established fast ion conductors are displayed in Fig. 27.1 to emphasise the magnitude of the effect. The diagram also illustrates the type of ions which exhibit fast ion conduction: ions with small formal charges and small Pauling radii (Li^+: 0.06 nm, Na^+: 0.095 nm, K^+: 0.133 nm, Ag^+: 0.126 nm, F^-: 0.136 nm, O^{2-}: 0.14 nm). High-conductivity materials have greatly expanded the range over which ionic transport phenomena have been observed. Besides solid-state batteries and solid oxide fuel cells (SOFC), applications include chemical sensors and electrochromic displays.

The literature on fast ion conduction is very extensive. The newcomer to this field can find information in textbooks on solid-state chemistry [8, 9], several books on fast ion conduction [10–17], a variety of reviews [18–29], and several conference proceedings [30–37]. Furthermore, proceedings of the biannual series on 'Solid State Ionics' are published in the journal with the same title [38]. A comprehensive collection of diffusion and conductivity data has been assembled by CHADWICK [7].

There are some general features of materials that are considered as fast ion conductors. First, the mobile ion is small and has a low formal charge (see above). Second, the crystalline conductors have an open lattice structure, such as the AgI or fluorite structure. Others reveal layered or channeled structures. However, fast ion conduction is not limited to crystals and there is a growing interest in ion-conducting glasses (see Chap. 30) and polymers. Since the range of materials is now extremely wide, various ways of classifying fast ion-conductors can be found in the literature. For example, fast ion conductors can be classified by the nature of the mobile ions, or in terms of the lattice structure, or in terms of device applications [25, 27]. From a purely academic viewpoint an approach based on the mechanism of conduction is appropriate [24]. In what follows we use a hybrid classification.

27.1 Fast Silver-Ion Conductors

This class of materials comprises compounds with a relatively fixed anion lattice and a three-dimensionally disordered cation lattice. Fast ion behaviour is often observed after a first-order transition between a normally conducting phase at low temperatures and a superionic phase at high temperatures. We mention first the simpler compounds of this type, where we can distinguish three different anion structures, namely bcc, fcc, and hcp structures. Then, we discuss the more complicated ones like $RbAg_4I_5$.

27.1.1 AgI and related Simple Anion Structures

A famous example is AgI: at low temperatures it crystallises in the cubic γ-phase and in the wurtzite structure (β-AgI) with normal ionic conductivity; above 147 °C the fast ion conducting α-AgI phase is formed. This phase transition increases the conductivity by several orders of magnitude as seen in Fig. 27.1. The sudden increase in conductivity is accompanied by a surprising increase in mass density. The I^- ions of α-AgI form a bcc lattice, while no definite sites can be assigned to the Ag^+-ions (Fig. 27.2). According to STROCK [41], a total number of 42 sites is in principle available for the two Ag^+-ions within the cubic cell, namely 6 octahedral, 12 tetrahedral, and 24 trigonal bipyramidal positions.

Neutron-diffraction studies on single crystals have been performed by CAVA, REIDINGER, AND WUENSCH [42] and silver density contours have been deduced from the scattering intensity. Figure 27.3 shows an example for the distribution of the Ag scattering density in a (001) plane of the crystal. Neutron scattering indicates that the equilibrium positions of silver are

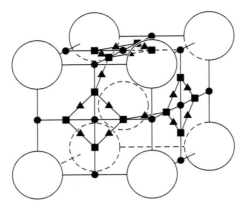

Fig. 27.2. Crystal structure of α-AgI. *Large circles*: I^- ions; *filled small circles*: octahedral sites; *filled squares*: tetrahedral sites; *filled triangles*: trigonal sites. Octahedral, tetrahedral, and trigonal sites can be used by Ag^+ ions

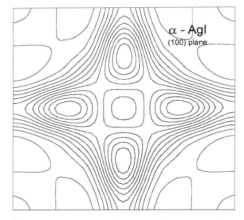

Fig. 27.3. Probabiliy distribution of Ag in α-AgI at 300 °C according to CAVA, REIDINGER, AND WUENSCH [42]

tetrahedral sites. The data also indicate that the Ag^+ ions are preferentially found in oblong ellipsoidal regions centered at the tetrahedral sites and extending in the directions of the neighbouring octahedral sites. This suggests that the motion of the Ag^+ ions is not completely liquid-like and that the $\langle 100 \rangle$ directions can be regarded as channel-like diffusion paths [29]. In other words, diffusion of Ag ions occurs mainly by jumps between neighbouring tetrahedral sites.

Besides α-AgI, the phases α-CuBr, α-Ag$_2$S, α-Ag$_2$Se, and α-Ag$_3$SI have bcc anion structures. The number of cations per bcc unit cell is two for α-AgI and α-CuBr, three for α-Ag$_3$SI, and four for α-Ag$_2$S and α-Ag$_2$Se. In

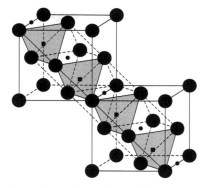

Fig. 27.4. Cation pathway in an fcc anion sublattice according to FUNKE [19]. *Filled squares*: tetrahedral sites; *small filled circles*: octahedral sites

contradistinction to α-AgI and α-CuBr, which have bcc anion lattices, the anion lattice of α-CuI is fcc. This is understandable from the ratios of the cationic and anionic radii of these compounds, as the cation sites provided by an fcc lattice are smaller in size when compared to those of a bcc lattice. The same systematic variation of the anion structure is observed in the case of the Ag and Cu chalcogenides; while α-Ag$_2$S and α-Ag$_2$Se still exhibit the bcc structure, α-Ag$_2$Te, α-Cu$_2$S and α-Cu$_2$Se have fcc arrangements.

Possible cation diffusion paths in fcc and hcp anion lattices have been discussed in [43]. In an fcc unit cell, there are 8 tetrahedral and 4 octahedral interstitial sites. The cation diffusion paths consist of alternating octahedral and tetrahedral sites. Cations jump from tetrahedron to octahedron to tetrahedron etc.. An almost linear pathway is illustrated in Fig. 27.4. Each anion tetrahedron shares four faces with four octahedra and each octahedron with eight tetrahedra. This structure provides a large variety of pathways through the anion lattice.

27.1.2 RbAg$_4$I$_5$ and related Compounds

There have been several attempts to obtain better Ag ion conduction. One was to stabilise α-AgI at lower temperatures. Another was to find new highly conducting phases by substitution. The most successful seems to be the partial replacement of Ag by Rb in α-RbAg$_4$I$_5$. This material has still today one of the highest ionic conductivities at room temperature (0.25 Scm^{-1}) of any known crystalline substance (see Fig. 27.1). Its electronic conductivity is negligibly small (about 10^{-9} S cm^{-1}). Some related compounds with similar properties are MAg$_4$I$_5$, with M = K, Cs, and NH$_4$.

The crystal structure of α-RbAg$_4$I$_5$ and its isomorphs is different from that of α-AgI and rather complex. The arrangement of the 20 iodine ions in the unit cell is similar to that of Mn atoms in the β-Mn structure and provides 56 tetrahedral voids for the 16 Ag^+ ions, while the 4 Rb^+ ions are immobilised at distorted octahedral environments of I^- ions [29]. Again there are many more available sites than Ag^+ ions to fill them. $RbAg_4I_5$ undergoes phase transitions at 209 K and at 122 K. The one at 209 K is a second order phase transition with a discontinuity in the temperature derivative of the conductivity, $d\sigma/dT$, while the conductivity is continuous. The transition at 122 K is first order, which entails a sudden change in conductivity of several orders of magnitude.

A disordered α-AgI-type structure can also be stabilised at low temperatures by a variety of cations, notably large alkalis, NH_4^+, and certain organic cations. Some examples, all of which have room temperature conductivities in the range of 0.02 Scm^{-1} to 0.2 Scm^{-1}, are $(NH_4)Ag_4I_5$, $[(CH_3)_4]_2Ag_{13}I_{15}$ and $PyAg_5I_6$, where Py^+ is the pyridinium ion $(C_5H_5NH)^+$. A range of anions may partially substitute for iodine to form, e.g., Ag_3SI, $Ag_7I_4PO_4$ and $Ag_6I_4WO_4$.

27.2 PbF$_2$ and other Halide Ion Conductors

Fast fluor-ion conduction in PbF_2, which has the fluorite structure (prototype CaF_2), was observed already by MICHAEL FARADAY. Several halides and oxides with the fluorite structure are very good anion conductors. Other alkaline earth fluorides, e.g., $SrCl_2$, and β-PbF_2 adopt this structure. They may be classified as fast ion condcutors at high temperatures, where they have high halogen ion conductivity. One of the best examples is PbF_2 with $\sigma \approx 5\,Scm^{-1}$ at about 500 °C. Above this temperature, the conductivity increases slowly and there is little, if any, change in conductivity on melting at 822 °C.

The fluorite structure consists of simple cubes of anions, half of them occupied by cations at the cube centers (Fig. 27.5). The sites available for interstitial F^- ions are at the centers of the set of unoccupied cubes. In creating an interstitial F^- ion, one corner F^- ion must leave its corner site and move into the body of the cube. Defect complexes probably form, but the details of the sites occupied are not fully known.

At low to moderate temperatures, fluorite-structured halides are like normal ionic solids; they contain low concentrations of anion Frenkel pairs. Only the anions are mobile. Most fluorites and anti-fluorites exhibit a broad specific heat anomaly which passes through a maximum temperature, T_c, a few hundred degrees below the melting temperature. In the same temperature regime as the thermal anomaly, the ionic conductivity increases rapidly to the extent that above T_c it reaches about 1 Scm^{-1}. The high temperature activation enthalpy is about 0.2 eV. This behaviour is attributed to a transition, which involves disordering of the anion sublattice, a transition which is called the *Faraday transition*.

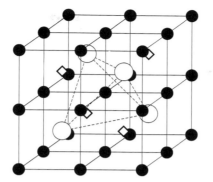

Fig. 27.5. Fluorite structure (prototype CaF_2): *Filled circles* represent anions and *open circles* cations. *Diamonds* represent sites for anion interstitials

27.3 Stabilised Zirconia and related Oxide Ion Conductors

The high-temperature cubic polymorph of zirconia (ZrO_2) has the fluorite structure as well. At room temperature, pure ZrO_2 is monoclinic. However, the fluorite structure can be stabilised by additions of Y_2O_3 or CaO. Such stabilised zirconias (e.g., yttrium stabilised zirconia = YSZ) are good O^{2-} ion conductors at high temperatures. This is because the formation of a solid solution between ZrO_2 and Y_2O_3 (or CaO) introduces vacant sites in the oxygen sublattice in order to preserve charge neutrality. For example, lime-stabilised zirconia (CSZ) has the formula $Ca_xZr_{(1-x)}O_{(2-x)}$ with $0.1 \leq x \leq 0.2$. One O^{2-} ion vacancy is created for each Ca^{2+} ion that is introduced.

Typical conductivities in stabilised zirconia (e.g., 85 mol% ZrO_2, 15% CaO) are about 5×10^{-2} S cm^{-1} at 1000 °C with activation enthalpies around 1.3 eV. At lower temperatures, stabilised zirconias have conductivities that are many orders of magnitude smaller than those of good Ag^+ and Na^+ ion conductors. The usefulness of zirconias stems from the fact that they are refractory materials, which can be used to very high temperatures and have good oxygen-ion conduction. CeO_2, HfO_2, and ThO_2 may also be doped heterovalently and are then good O^{2-} ion conductors as well.

Increasing the point defect concentration increases the ionic conductivity. A compound in which this occurs naturally is bismuth oxide, Bi_2O_3. This material has a solid-state phase transformation to a fluorite-structured δ-phase. In this structure, 25% of the anion sites are vacant. It is hardly surprising that due to the structural vacancies this compound has a very high O^{2-} conductivity [44]. The highest oxygen-ion conductivities are found in Bi_2O_3-based materials. However, most of these are readily susceptible to reduction, thus becoming mixed electron-ion conductors. Therefore, they cannot be used as solid electrolytes in reducing atmospheres or at low oxygen partial pressure.

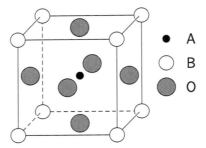

Fig. 27.6. Perovskite structure

There have been attempts to stabilise the high-temperature phase at lower temperatures by doping, e.g., with zirconia and vanadium oxide.

27.4 Perovskite Oxide Ion Conductors

Perovskites have the general formula ABO_3. The perovskite structure is illustrated in Fig. 27.6. The structure prototype is $CaTiO_3$ and has a primitive cubic unit cell. It contains one Ca^{2+} ion per unit cell, e.g., at the cube edges, one Ti^{4+} ion in the cube center, and O^{2-} ions at the face centers.

Perovskite type oxides based on $LaGaO_3$ are of considerable interest because of their high oxygen-ion conductivity. As for other materials, doping is a convenient strategy to increase the ionic conductivity of perovskite-type oxides. Lanthanum gallates doped with Sr on La sites and with Mg on Ga sites, $La_{(1-x)}Sr_xGa_{(1-y)}Mg_yO_{[3-(x+y)/2]}$ (LSGM), reach higher oxygen-ion conductivities than YSZ [47]. After optimising the single-phase composition of LSGM an oxide-ion conductivity of 0.15 Scm^{-1} at 800 °C is stable over time at any oxygen partial pressures between 10^{-23} and 1 atm [48]. This conductivity is comparable to that of YSZ at 1000 °C. Therefore, LSGM appears to be a more promising electrolyte than YSZ for solid oxide fuel cells operating below 800 °C. Cation diffusion in perovskites is known to be very slow. Nevertheless, one long term degradation effect may be due to a demixing of the electrolyte because of different cation diffusivities [49].

27.5 Sodium β-Alumina and related Materials

A family of phases with the general formula $M_2OnX_2O_3$, where n is in the range of 5 to 11, is denoted as β-alumina. M is a monovalent cation (alkali$^+$, Cu$^-$, Ag$^+$, Ga$^+$, In$^+$, Tl$^+$, NH_4^+, H_2O^+) and X is a trivalent cation (Al^{3+}, Ga^{3+}, or Fe^{3+}). The most important member of this family is sodium β-alumina with $M = Na^+$ and $X = Al^{3+}$, which has been long known as a byproduct of the glass-making industry. Interest in the β-aluminas began in the

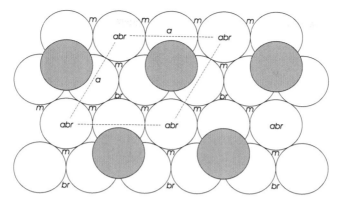

Fig. 27.7. Sites for Na$^+$ ions in the conduction plane of β-alumina. m: mid-oxygen position, br: Beevers-Ross site, abr: anti-Beevers-Ross site. *Open circles*: O^{2-}, *grey circles*: O^{2-} spacer ions

1960s with the pioneering work at the Ford Motor Company when YAO AND KUMMER detected that the Na$^+$ ions are very mobile at room temperature and above [5].

The high conductivity of monovalent ions in β-alumina is a consequence of its unusual crystal structure. It is built of close-packed layers of oxygen ions, stacked in three dimensions. Every fifth layer has three-quarters of its oxygens missing. The Na$^+$ ions reside in these oxygen-deficient layers and are easily mobile, because their radius is smaller than that of the O^{2-} ions. β-aluminas exist in two structural modifications, called β and β'', which differ in the stacking sequence of the layers. The β'' form occurs with Na-rich crystals where $n \approx 5-7$, whereas the β-form occurs for $n \approx 8-11$. Both structures are closely related to that of spinel (MgAl$_2$O$_4$) and may be regarded as being built of 'spinel blocks'. The blocks are four oxide layers thick and their oxygen layers are in cubic stacking sequence, separated by the oxygen-deficient layers of the conduction planes.

The atomic structure within the conduction plane has been the subject of much crystallographic work. The present understanding is as follows: sites available for Na$^+$ ions in the conduction plane of the β-modification are shown in Fig. 27.7. The conduction plane consists of close-packed layers of O^{2-} ions separated by pairs of O^{2-} ions. The 'spacer' O^{2-} ions (grey) are located in the conduction plane. Only one quarter of the available O^{2-} sites in the conduction plane are occupied, i.e. for every grey O^{2-} ion there are three empty sites. Na$^+$ ions can occupy three different sites: the 'mid-oxygen' positions (m), the 'Beevers-Ross' sites[1] (br), and the 'anti-Beevers-Ross' sites (abr). It appears that Na$^+$ ions spend most of their time in m and br sites,

---

[1] These sites were favoured in the original structure determination of BEEVERS AND ROSS.

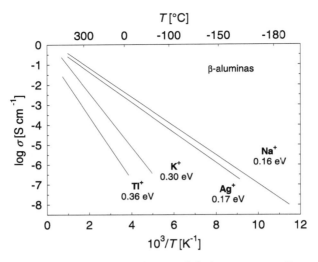

Fig. 27.8. Conductivities of some single crystal β-aluminas according to WEST [45]

but in order to undergo long-range migration they must pass through the *abr* sites, which are much smaller than the *m* and *br* sites.

The β-aluminas are two-dimensional conductors. Alkali ions can move easily within the conduction planes but cannot penetrate the dense spinel blocks. Most other monovalent ions also prefer the *br* and *m* sites in β-alumina, with the exception of Ag^+ and Tl^+ which prefer the *abr* sites. This is understandable, because Ag^+ and Tl^+ prefer covalent binding and sites of low oxygen coordination. The conductivities of various β-alumina single crystals (Fig. 27.8) parallel to the conduction plane fit Arrhenius equations over wide ranges of temperature. The conductivity is highest and the activation enthalpy lowest for Na^+ and Ag^+ β-alumina. With increasing cation size (K^+, Tl^+) the conductivity becomes lower, since the larger cations cannot move as easily in the conduction planes.

There are other layered materials in which the conductivity is two-dimensional. On the whole they have not been as thoroughly studied as the β-aluminas. An example of a three dimensional conductor is the Na^+-conductor $Na_3Zr_2PSi_2O_{12}$ [46], which is now referred to as NASICON (Na superionic conductor). Like β-alumina it is a ceramic material, but at 300 °C its conductivity is higher than that of β-alumina.

27.6 Lithium Ion Conductors

Materials that have high Li^+-ion conductivity are used as electrolytes in lithium batteries. The enormous, world-wide interest in such devices arises because cells containing Li anodes generally have a higher *emf* than corresponding cells containing, e.g., Na anodes. Thus, commercial lithium batteries currently have 4 V single cells with an anode containing Li metal and an intercalation cathode based on $LiCoO_2$ or on the spinel $LiCoMnO_4$. Solid-state lithium batteries have important applications in a variety of consumer and medical products. The batteries consist of cathodes that are crystalline or nanocrystalline oxide-based lithium intercalation compounds. At present, most Li cells still work with liquid, non-aqueous electrolytes such as $LiPF_6$ disolved in an organic solvent (see, however, Sect. 27.7). Sometimes the electrolyte is a glassy lithium phosphorous oxynitride ('Lipon') [50].

Conductivity data of some solid Li^+ ion conductors are shown in Fig. 27.1. Li_2SO_4 undergoes a phase transition at 572 °C and has a high conductivity around 1 Scm^{-1} in its high-temperature phase. Above that temperature, many substituted sulphates have been studied in attempts to reduce the temperature of the phase transition and thus preserve the fast-conducting α-polymorph even at lower temperatures. It seems that the α-polymorph cannot be stabilised at room temperature.

An example of a binary compound that exhibits two-dimensional ionic conductivity is lithium nitride (Li_3N). Its anisotropy is the result of the crystal structure [52]. It has a layered structure with sheets of 'Li_2N' alternating with layers of Li. Conductivity appears to occur primarily in the 'Li_2N' sheets by a Li vacancy mechanism. The conductivity of impure, H-containing Li_3N is higher than that of pure Li_3N. Hydrogen is tightly bound to N, forming NH units and leaving Li sites vacant in $Li_{3-x}NH_x$.

A family of Li-containing perovskites has high Li^+ ion conductivities around 10^{-3} Scm^{-1} at room temperature. These perovskites are based on $Li_{0.5}La_{0.5}TiO_3$, which does not exist in stoichiometric form but only as Li-deficient compound. It is formed by substitution of La^{3+} for $3Li^+$ to form $Li_{0.5-3x}La_{0.5+x}TiO_3$.

27.7 Polymer Electrolytes

Since their discovery in 1973 by WRIGHT AND COWORKERS [51], polymer electrolytes have attracted much attention because of their promising applications as ion-conducting materials. Polymer electrolytes are mixtures of polymers and salts, which are ionic conductors at moderate temperatures. The technological interest in polymer electrolytes stems from the work of ARMAND AND COWORKERS, who studied polyethylene oxide (PEO) and polypropylene oxide (PPO) salt complexes and highlighted the potential of these materials for battery applications [53]. The electrolyte is the heart of any battery. It must allow the passage of the ions, while blocking electron conduction between the active components of the battery. Indeed, Li-ion batteries, nowadays commonly used in laptop computers and in cellular phones, are based on polymer electrolytes containing a suitable Li salt [54].

Polymer electrolytes contrast sharply with the fast ion conducting materials based on ceramics, glasses, or inorganic crystals discussed above. Polymer electrolytes transport charge well only above their glass transition temperature. The conductivity of polymer electrolytes is of the order of 10^{-4} to $10^{-3}\,\mathrm{S\,cm^{-1}}$ and thus two to three orders of magnitude lower than the best fast ion conductors (see Fig. 27.1). This disadvantage is countered by their ease of processing as very thin films of only a few microns thickness. In addition, they have the advantage of being flexible. The flexible nature of these materials allows a space-efficient battery design of variable dimensions. The polymer electrolyte flexibility has the important advantage that volume changes in the cell can be accommodated during cycling without degradation of the interfacial contacts, which is often observed for crystalline or vitreous solid electrolytes [55].

Polymer electrolytes may be categorised into several classes according to electrolyte composition and morphology [56]. In what follows, we focus on PEO–salt systems, which belong to the most thoroughly investigated polymer electrolytes [55, 57, 58]. The state-of-the-art knowledge is restricted to a few established features [58]:

1. High ionic conductivity is observed in the amorphous phase of the polymer electrolyte. This relates to the fact that pure PEO (partially) crystallises at temperatures below about 65 °C. Similar crystallisation properties are also found in PEO–salt systems with not too high salt concentrations ($\approx$ one salt molecule per 30 O-atoms).
2. Long-range ionic motion is coupled to local motions of the polymer chain segments. This coupling is most prominent for the cations since these ions are usually coordinated by four to five ether oxygens. In fact, the cation-oxygen interaction is responsible for the main enthalpy contribution to the solvation of the salt in the polymer matrix. The cation translational motion is illustrated in Fig. 27.9. This schematic conveys the notion that cation motion proceeds through the 'making and breaking of bonds' between the cation and oxygen atoms of one or two locally mobile polymer chains.
3. Anions move faster than cations. The higher mobility of anions can be understood from their higher degree of freedom: they are not directly bound to the polymer chains (Fig. 27.9).

Despite numerous studies related to ionic conductivity, the understanding of the diffusion mechanisms in these electrolytes is still unsatisfactory. A major reason for this unsatisfactory situation is that conductivity measurements only yield the net effect of all mobile species. Only few publications in this field report the use of ion-specific techniques, by which the diffusion properties of cations and anions can be determined individually. One such technique is the pulsed-field nuclear magnetic resonance (see, e.g., [59]). Another powerful ion-specific method is radiotracer diffusion, which has been employed only on few polymer-salt systems [60–63]. Both techniques have pro-

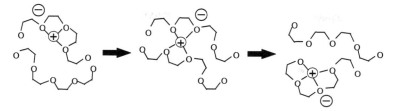

Fig. 27.9. Schematic illustration of ion solvation and migration in amorphous polymer electrolytes according to [62]

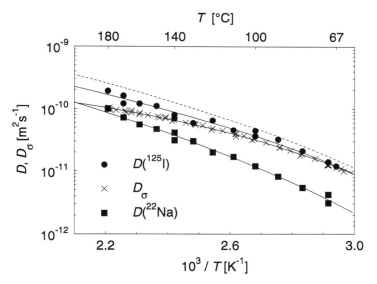

Fig. 27.10. Tracer diffusion coefficients of ^{22}Na and ^{125}I in an amorphous PEO–NaI polymer electrolyte compared to the charge diffusivity, D_σ, according to STOLWIJK AND OBEIDI [62, 63]. The *dashed line* is shown for comparison: it represents the sum $D(^{22}\text{Na}) + D(^{125}\text{I})$

vided unambiguous evidence that the anion is moving at least as fast as the cation.

As a typical example, we present results of STOLWIJK AND OBEIDI on a polymer-salt system consisting of PEO and NaI [62, 63]. These authors performed measurements of ^{22}Na and ^{125}I tracer diffusion and of the overall ionic conductivity. They also deduced the charge diffusivity, D_σ, from the dc conductivity via the Nernst-Einstein relation. Figure 27.10 compares the tracer diffusivities of both ions, $D(^{22}\text{Na})$ and $D(^{125}\text{I})$, with the charge diffusivity. The latter exhibits a downward curvature, characteristic of Vogel-Fulcher-Tammann behaviour frequently observed in the (supercooled) liquid state. The charge diffusivity falls below the sum of the tracer diffusivities. To explain this discrepancy the authors propose a model, which considers contributions from isolated cations and anions and neutral cation-anion pairs. The latter contribute to tracer transport of both ions but not to charge transport. The authors conclude that diffusivities increases in the order cation, anion, and ion pair. This sequence reflects the decreasing degree of coupling to the polymer matrix. This model reveals some analogy to the diffusion via cation and anion vacancies and neutral vacancy pairs in alkali halide crystals mentioned in Chap. 26.

References

1. M. Faraday, *Experimental Researches in Electricity*, R. and J.E. Taylor, London, 1939; see also *Faraday's Diaries 1820–1862*, volume II, H.G. Bell and sons, London, 1932; diary entries for 21st February 1833 and for 19th February 1833
2. C. Tubandt, E. Lorenz, Z. Phys. Chem. **87**, 513, 543 (1914)
3. J.N. Bradley, P.D. Green, Trans. Faraday Soc. **62**, 2069 (1966); **63**, 424 (1967)
4. B.B. Owens, G.R. Argue, Science **157**, 308 (1967)
5. Y.F.Y. Yao, J.T. Kummer, J. Inorg. Nucl. Chem. **29**, 2453 (1967)
6. J.T. Kummer, Progr. Solid State Chem. **7**, 141 (1972)
7. A.V. Chadwick, *Diffusion in Fast Ion Conducting Solids*, in: *Diffusion in Semiconductors and Non-Metallic Solids*, D.L. Beke (Vol. Ed.), Landolt-Börnstein, New Series, Group III: Condensed Mattter, Vol. 33, Subvolume B1, Springer-Verlag, 1999
8. A.R. West, *Solid-State Chemistry and its Applications*, John Wiley and Sons, 1984
9. A.R. West, *Basic Solid-State Chemistry*, 2nd edition, John Wiley and Sons, 1999
10. S. Geller (Ed.), *Solid Electrolytes*, Springer-Verlag, Berlin, 1977
11. P. Hagenmüller, W. van Gool (Eds.), *Solid Electrolytes*, Academic, New York, 1978
12. M.B. Salamon (Ed.), *Physics of Superionic Conductors*, Springer-Verlag, Berlin, 1979
13. T. Takahashi, A. Kozawa (Eds.), *Applications of Solid Electrolytes*, JEC Press, Cleveland, 1981
14. S. Chandra, *Superionic Solids*, Academic Press, New York, 1981
15. T.A. Wheat, A. Ahmad, A.K. Kuriakose (Eds.), *Progress in Solid Electrolytes*, Energy, Mines and Resources, Ottawa, 1983
16. A. Laskar, S. Chandra (Eds.), *Superionic Solids and Solid Electrolytes*, Academic, New York, 1990
17. A.M. Stoneham (Ed.), *Ionic Solids at High Temperatures*, World Scientific, Singapore, 1990
18. M. O'Keefe, B.G. Hyde, Philos. Mag. **33**, 219 (1976)
19. K. Funke, Progr. Solid St. Chem. **11**, 345 (1976)
20. W. Hayes, Contemp. Phys. **19**, 469 (1978)
21. J.B. Boyce, B.A. Habermann, Phys. Reports **51**, 519 (1979)
22. H. Schulz, Ann. Rev. Mat. Sci. **12**, 351 (1982)

23. B.C.H. Steele, in: *Mass Transport in Solids*, F. Benière, C.R.A. Catlow (Eds.), Plenum Press, New York, 1983, p. 537
24. C.R.A. Catlow, J. Chem. Soc. Faraday Trans. **86**, 1167 (1990)
25. A.V. Chadwick, in: *Diffusion in Materials*, A.L. Laskar, J.L. Bocquet, G. Brebec, C. Monty (Eds.), Kluwer Academic Publisher, Dordrecht, 1990, p. 489
26. S. Hoshino, Solid State Ionics **48**, 179 (1991)
27. A.V. Chadwick, *Diffusion in Fast Ion Conductors: Unsolved Problems*, in: *Diffusion in Solids: Unsolved Problems*, G.E. Murch (Ed.), Defect and Diffusion Forum **83**, 235 (1992)
28. A.V. Chadwick, *Diffusion Mechanisms in Fast Ion Conductors*, in: *Diffusion in Materials*, M. Koiwa, K. Hirano, H. Nakajima, T. Okada (Eds.), Defect and Diffusion Forum **95–98**, 1015 (1993)
29. K. Funke, Progr. Solid St. Chem. **22**, 111 (1993)
30. W. van Gool (Ed.), *Fast Ion Transport in Solids*, North-Holland, Amsterdam, 1973
31. G.D. Mahan, W.L. Roth (Eds.), *Fast Ion Transport in Solids*, Plenum Press, New York, 1976
32. P. Vashishta, J.N. Mundy, G.K. Shenoy (Eds.), *Fast Ion Transport in Solids – Electrodes and Electrolytes*, Elsevier North-Holland, 1979
33. J.W. Perram (Ed.), *The Physics of Superionic Conductors and Electrode Materials*, Plenum Press, New York, 1983
34. B. Bates, G.C. Farrington (Eds.), Proc. Int. Conf. on *Fast Ion Transport in Solids*, North-Holland, Amsterdam, 1981
35. M. Kleitz, B. Sapoval, D. Ravaine (Eds.), *Solid State Ionics-83*, North-Holland, New York, 1983
36. R.A. Huggins (Ed.), *Solid State Ionics-85*, North-Holland, New York, 1985
37. H. Schulz, W. Weppner (Eds.), *Solid State Ionics-87*, North-Holland, New York, 1987
38. The journal *Solid State Ionics* is published by Elsevier, North-Holland
39. R.G. Fuller, *Ionic Conductivity (Including Self-Diffusion)*, in: *Point Defects in Solids*, J. M. Crawford Jr. and L.M. Slifkin (Eds.), Plenum Press, 1972, p. 103
40. A.V. Chadwick, Defect and Diffusion Forum **95–98**, 1015 (1993)
41. L.W. Strock, Z. Phys. Chem B **25**, 411 (1934); and B **31**, 132 (1936)
42. R.J. Cava, F. Reidinger, B.J. Wuensch, Solid State Commun. **24** 411 (1977)
43. L.V. Azaroff, J. Appl. Phys. **32**, 1658, 1663 (1961)
44. H.A. Harwig, A.G. Gerards, J. Sol. St. Chem. **26**, 265 (1978)
45. A.R. West, *Solid-State Chemistry and its Applications*, John Wiley and Sons Ltd. 1984; and *Basic Solid-State Chemistry*, John Wiley and Sons, Ltd., 2nd edition, 1999
46. J.B. Goodenough, H.Y.P. Hong, J.A. Kafalas, Mater. Res. Bull. **11**, 203 (1976)
47. T. Takahashi, H. Iwahara, Y. Nagai, J. Appl. Electrochemistry **2**, 97 (1972)
48. J.B. Goodenough, K. Huang, in: *Mass and Charge Transport in Inorganic Materials – Fundamentals to Devices*, P. Vincenzini, V. Buscaglia (Eds.), Techna, Faenza, 2000, p. 3
49. O. Schulz, M. Martin, in: *Mass and Charge Transport in Inorganic Materials – Fundamentals to Devices*, P. Vincenzini, V.. Buscaglia (Eds.), Techna, Faenza, 2000, p. 83
50. J.B. Bates, N.J. Dudney, B. Neudecker, A. Ueda, C.D. Evans, Solid State Ionics **135**, 33 (2000)

51. D.E. Fenton, J.M. Parker, P.V. Wright, Polymer **14**, 589 (1973)
52. A. Rabenau, H. Schulz, J. Less Common Metals **50**, 155 (1976)
53. M.B. Armand, J. M. Chabagno, M.J. Duclot, in: *Fast Ion Transport in Solids – Electrodes and Electrolytes*, P. Vashishta, J.N. Mundy, G.K. Sheroy (Eds.), North-Holland, 1979, p. 131
54. P.V. Wright, MRS Bulletin **27**, 597 (2002)
55. F.M. Gray, *Solid Polymer Electrolytes – Fundamentals and Technolgical Applications*, VCH Publishers Inc., 1991
56. B. Scrosati, C.A. Vincent, MRS Bulletin **28**, 28 (2000)
57. MRS Bulletin, March (2000)
58. J.R. MacCallum, C.A. Vincent (eds.), *Polymer Electrolyte Reviews*, Elsevier Applied Science, London, 1987
59. N. Boden, S.A. Leng, I.M. Ward, Solid State Ionics **45**, 261 (1991)
60. A.V. Chadwick, J.H. Strange, M.R. Worboys, Solid State Ionics **9–10**, 1155 (1983)
61. D. Fauteux, M.D. Lupien, C.D. Robiotille, J. Electrochem. Soc. **143**, 2761 (1987)
62. N.A. Stolwijk, S. Obeidi, Defect and Diffusion Forum **237–240**, 1004 (2005)
63. N.A. Stolwijk, S. Obeidi, Phys. Rev. Lett. **93**, 125901 (2004)

Part VI

Diffusion in Glasses

28 The Glassy State

28.1 What is a Glass?

To the layman, glass is a transparent and lustrous solid that breaks easily. Yet a number of glass types, in particular metallic glasses, are opaque and not at all brittle. Also high-strength transparent glasses can be made by various techniques, some being used in bulletproof security glazing. At one time, glass was defined as 'an inorganic product of fusion which has been cooled to a rigid condition without crystallising'. Even this definition is too restrictive, as many organic glasses are known and melting is not the only means of glass making. The sol-gel process avoids the normally high temperatures employed for glass melting; so does formation of glassy metals via solid-state diffusion reaction of thin layered structures formed by evaporation or ball-milling. Chemical vapour deposition is another process that avoids fusion of constituent materials.

The outward appearance of glasses is essentially solid-like. Density, electrical, mechanical, and thermal properties of glasses are similar to those of corresponding crystalline substances. However, unlike crystals, glasses do not have a sharp, well-defined melting point. In the absence of applied forces and internal stresses, glasses are isotropic. Their isotropic physical properties make glasses resemble liquids. It follows that the atomic arrangement in glass re-

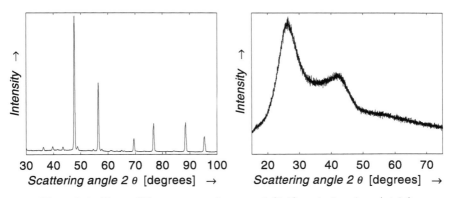

Fig. 28.1. X-ray diffractogram of a crystal (*left*) and of a glass (*right*)

sembles that of the disorder of corresponding liquids. We are left with the classical definition of GUSTAV TAMMANN (1861–1938) of a glass as *'a strongly undercooled liquid, i.e. a solid with liquid-like structure'* [1], or 'a non-crystalline solid' or simply 'an amorphous solid'. The amorphous characteristics are evidenced by X-ray diffraction (Fig. 28.1). This definition excludes substances that are micro- or nanocrystalline and display more or less sharp Bragg peaks in
X-ray analysis. For more detailed information we refer the reader to reviews by LUBORSKY [2], DAVIES [3], and CAHN [4] on metallic glasses, and textbooks by VOGEL [5], SHELBY [6], DOREMUS [7], and VARSHNEYA [8] on inorganic glasses.

28.2 Volume-Temperature Diagram

To get a clearer picture of the fundamentals of glass, it is useful to consider the volume-temperature relationship (V-T diagram) with respect to a liquid, a glass, and a crystal. Since enthalpy H and volume V behave in a similar fashion, the choice of the ordinate is somewhat arbitrary and we may also consider an H-T diagram. Such a diagram is displayed in Fig. 28.2. For its discussion let us envision a small volume of melt at a temperature well above the melting temperature T_m of the material. With decreasing temperature the atomic structure of the melt will gradually change and will be characteristic of the temperature at which the melt is held. The thermodynamic melting temperature is defined by the condition that the Gibbs free energies of the melt and the crystal are equal. Cooling down below T_m will convert the melt to the crystalline state, provided that the kinetics permits nucleation and growth of the crystalline phase. If this occurs the volume (enthalpy) will decrease abruptly in a first order phase transition to the value typical of the crystal[1]. Continued cooling of the crystal will result in a further decrease of volume (enthalpy) due to the thermal contraction (specific heat) of the crystal.

If the melt can be cooled below T_m without crystallisation, a supercooled melt is obtained, which is metastable with respect to crystallisation. Crystallisation is avoided only since cooling does not leave enough time for the formation of crystalline nuclei. Upon further cooling the structure of the liquid continues to remain in a metastable configurational equilibrium as the temperature decreases, but no abrupt change in volume (enthalpy) due to a phase transformation occurs. With increasing undercooling to temperatures between T_m and the fictive temperature T_f (defined below), the viscosity increases by about 15 orders of magnitude (see also Fig. 30.1, Chap. 30). The increase in viscosity becomes so large that the atoms can no longer rearrange

---

[1] For most materials the crystalline state has indeed a higher density. For example, water and Si are exceptions having higher densities in the liquid state.

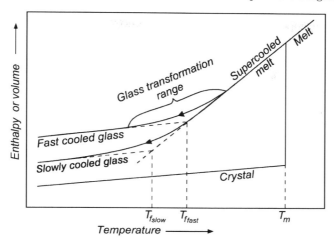

Fig. 28.2. Volume (or enthalpy) *versus* temperature diagram of a glass-forming liquid

to the equilibrium structure of the undercooled melt during the time allowed by the experiment. The structure of the undercooled melt begins to deviate from that which would develop if sufficient time were allowed to reach configurational equilibrium. This situation usually occurs for viscosities around 10^{12} Pa s (10^{13} Poise). Correspondingly, the volume (enthalpy) begins to deviate from the metastable equilibrium line, following a line of gradually decreasing slope (see Fig. 28.2), until it becomes dominated by the thermal expansion (specific heat) of the 'isoconfigurational melt'. When the viscosity is so high that the structure of the material becomes fixed, the 'mobile melt' has then become a rigid *glass*. A glass can be defined as an undercooled melt congealed to a rigid, isoconfigurational state. The temperature region lying between the limits where the volume (enthalpy) is either that of an equilibrated liquid or that of a glass is denoted as the *glass transformation region*.

Since the glass transformation region is controlled by kinetic factors, a slow cooling rate will allow the volume (enthalpy) to follow the (metastable) equilibrium line to lower temperatures. Such a glass will have a lower volume (enthalpy) than that obtained at a faster cooling rate. Its atomic arrangement will be that characteristic of the undercooled melt at lower temperatures. Although the glass transformation occurs over a more or less wide temperature range, it is convenient to define a temperature which allows the difference in thermal history between the two glasses to be expressed. The extrapolations of the undercooled melt and glass lines intersect at a temperature which in the literature on glasses is denoted as *fictive temperature* T_f. A glass produced at a slower cooling rate has a lower T_f, as indicated in Fig. 28.2.

It is convenient to introduce the concept of the *glass-transition temperature* (or *glass-transformation temperature*). The glass-transition temperature

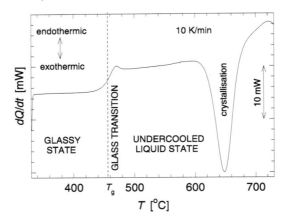

Fig. 28.3. Differential Scanning Calorimetry (DSC) thermogram of a $0.2(0.8Na_2O\ 0.2\ Rb_2O)\ 0.8B_2O_3$ glass measured at a heating rate of $10\,K/min$ from [9]. The glassy and undercooled liquid state are indicated. The strong exothermic signal (near $650\,°C$) corresponds to the crystallisation of the undercooled melt

T_g is used as an indication of the onset of the transformation of a glass to the undercooled melt during heating. This temperature is frequently determined by changes in thermal analysis curves (caloric glass-transition temperature) or in thermal expansion curves. T_g values obtained from these two methods are similar although not identical. In addition, T_g values are a function of the heating rate used during the analysis. An example for a measurement of T_g by differential scanning calorimetry (DSC) is shown in Fig. 28.3. A change in specific heat is observed around $460\,°C$. The inflexion point of the thermogram is usually taken as the caloric glass transition temperature T_g. Although T_g and the fictive temperature are not identical, the differences are usually small, i.e. not more than a few K. Therefore, T_g is a useful indicator for the approximate temperature, where the undercooled melt converts to a glass.

28.3 Temperature-Time-Transformation Diagram

Many of the liquids that would crystallise during normal cooling can be brought to the vitreous state by more rapid cooling. We should recognise that the question 'What materials can form glass?' is only academic: the correct question should be, 'at what rate should a given liquid be cooled to bring it into the glassy state?' It is now well established that most liquids, including water and metallic melts, can be vitrified provided that the rate of cooling is fast enough to avoid crystallisation. Crystallisation requires first the formation of a sufficient number of crystalline nuclei and in addition a measurable growth rate of the nuclei. To avoid crystallisation one needs

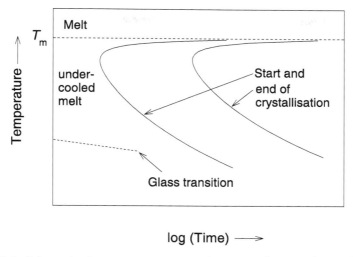

Fig. 28.4. Schematic time-temperature-transformation diagram (TTT diagram) for the crystallisation of an undercooled melt

to avoid either one or both of the two. Formation of a glass from the melt involves cooling in such a manner as to prevent crystallisation.

To put glass formation in the right perspective, we consider the time-temperature-transformation diagram (TTT diagram) for crystallisation of an undercooled melt (Fig. 28.4). At the thermodynamic melting temperature, T_m, the time necessary for crystallisation tends to infinity since the driving force to form crystalline nuclei vanishes. With increasing undercooling below T_m the nucleation rate increases, since the driving force for nucleation becomes larger. At large undercoolings the crystallisation rate is very sluggish, because diffusion and/or the growth rate of crystalline nuclei is very slow. As a consequence, for intermediate undercooling the crystallisation rate has a maximum in the TTT diagram, which looks like an ugly 'nose'. All combinations of heat treatment times and temperatures to the left of this curve yield samples in the undercooled liquid or glassy state, while all combinations of time and temperature to the right of this curve yield a partial or totally crystallised material. To form a glass, cooling must be so fast that the crystallisation nose can be avoided. Below the glass-transition region the undercooled melt becomes a glass.

The position of the 'nose' in the TTT diagram for crystallisation of an undercooled melt is a measure for the glass-forming ability of a material. For example, for conventional metallic glasses the time scale is in the 0.1 to 1 milliseconds range at the 'nose' of the nucleation curve. Rapid cooling with rates of about 10^6 Ks^{-1} is required to form such glasses. They can be manufactured for example by melt-spinning in the form of thin ribbons or sheets of several ten micron thickness only. The so-called *bulk metallic glasses* (see below) exhibit TTT-diagrams with a crystallisation 'nose' in the range between 1–100 seconds or more. This implies that these alloys have an exceptional glass-forming ability and that their undercooled melts are relatively stable. Their glass forming ability and processability is comparable to that of many silcate glasses.

28.4 Glass Families

Glasses are formed in a great variety of systems including manmade and natural glasses. In the following, we list major glass families (see also [5, 6, 8]):

1. **Vitreous silica (SiO_2)** is the most refractory glass in commercial use. In addition, it has a high chemical resistance to corrosion, a very low electrical conductivity, a near-zero coefficient of thermal expansion (about $5.5 \times 10^{-7} K^{-1}$), and good transparency for ultraviolet light. Because of the high cost of manufacture, the use of vitreous silica is mostly limited to astronomical mirrors, crucibles for melting high-purity silicon in Czochralsky crystal pulling devices, special laboratory ware, and high efficiency lamp envelopes. Optical fibers and insulating films in microelectronic devices are further important applications.
Bulk vitreous silica is obtained by melting high-purity quartz crystals at temperatures above 2000 °C. In a second technique, silicon tetrachloride is sprayed into an oxy-hydrogen flame or a water-vapour-free oxygen plasma. Silica vapours deposit on a substrate and are consolidated subsequently at about 1800 °C.
2. **Soda-Lime Silicate Glasses** contain large amounts of sodium and calcium oxide and can be produced by melting silica with sodium and calcium carbonates or nitrates. Soda-lime glasses are perhaps the least expensive and the most widely used of all the glasses made commercially. Most of the glass windows, beverage containers and envelopes of incandescent and fluorescent lamps are made from soda-lime glass. It is interesting to note that the telescope of GALILEO consisted of lenses that were made of one single glass, crown glass, a Na_2O-CaO-SiO_2 glass. The coloured borders obtained with theses lenses gave rise to the later demand for colour-free images, that is to make the lenses achromatic.
Soda-lime glasses have good chemical durability, high electrical resistivity, and good light transmission in the visible region. Because of its relatively high coefficient of thermal expansion (about 10^{-5} K^{-1}), soda-lime glasses are prone to thermal shock failures, which prevents its use in a number of applications.
Large-scale continuous melting of inexpensive batch materials such as soda ash (Na_2CO_3), limestone ($CaCO_3$), and sand at 1400 to 1500 °C makes it possible to form products at high speeds inexpensively. An example is the float-glass technique, which is used to produce window glass.

3. **Borosilicate Glasses** are obtained when SiO_2 and B_2O_3 and alkali carbonates such as Na_2CO_3 are used as batch materials. This family of glasses is of interest for their low thermal expansion coefficient (2 to 3×10^{-6} K^{-1}) and a high resistance to chemical attack. Household cooking untensils, laboratory glass ware, and automobile head lamps are examples of their usage. Borosilicate glasses can be produced in a manner similar to soda-lime glasses, but require slightly higher temperatures. The high cost of B_2O_3 makes them less competitive for common products. Most of the commercial borosilicate glasses, while transparent, are actually phase separated with a very fine scale morphology.
4. **Lead Silicate Glasses** contain PbO and SiO_2 as the major components with small amounts of soda and potash (K_2CO_3). These glasses are utilised for their high degree of brilliance. Their large 'working range' is useful to make art objects without frequent re-heating. PbO additions increase the fluidity (inverse viscosity) of glass and its wetability to oxide ceramics. The high concentration of lead oxide found in many glasses suggests that PbO does not act as a normal modifier oxide in the structure. High lead borosilicate glasses without additions of alkalis are used for electrical feedthrough components and in microelectronics.
5. **Aluminosilicate Glasses** contain, apart from SiO_2 and Al_2O_3, also varying amounts of alkali oxide and a moderate amount of alkaline earth oxides such as MgO and CaO. The low-alkali-containing aluminosilicates have high elastic moduli and a high resistance to chemical corrosion. They are used as load-bearing fiber component in fiber-reinforced plastics. The electrical resistance of alkali-free alkaline earth aluminosilicates is comparable to vitreous silica. These glasses are intermediate between soda-lime glasses and vitreous silica for refractoriness and thermal expansion (about 5×10^{-6} K^{-1}). A major commercial use of this family is in lamps involving the tungsten-halogen cycle, e.g., in automobile halogen headlamps.
6. **Non-Silica-Based Glasses** are oxide glasses lacking silica as a principal component. They do not have much commercial use. B_2O_3- and P_2O_5-based glasses are hygroscopic. However, their study is important towards enhancing our understanding of glass structure and properties. Non-silica oxide glasses with some commercial interest are (i) boro-aluminates with electrical resistivities exceeding that of silica and (ii) alkaline earth aluminates as a high temperature sealants and infra-red transmitting glasses.
7. **Amorphous Semiconductors** can be formed of Si, Ge, P, As, and in the family of tetrahedral glasses. The latter are compounds such as $CdGe_xAs_2$ where $x = 0$ to 1.2, and $Si_{1-x}H_x$, where $x = 0.1$ to 0.2. They retain their semiconducting behaviour in the glassy state. As discovered in the 1960s, some of these glasses display switchng between high- and low-conductivity states while remaining semiconducting [10], a property which at that time promised commercial use for computer memories.

Amorphous semiconductors are also photovoltaic materials, which makes them candidates at least for niche-market applications of solar-cell technology.

8. **Metallic glasses** are also referred to as *amorphous metals* or *glassy metals or alloys* (Chap. 29). They are composed either of two or more metals or of metals and metalloids. Metallic glasses are metals in the sense that their electrical, optical, and magnetic properties are typical of metals.

    Conventional metallic glasses can be made by rapid cooling from the molten state as reported for the first time by DUWEZ AND COWORKERS in 1960 [11]. These glasses are usually made in the form of thin ribbons using high-speed quenching techniques like melt-spinning [12]. Common examples are $Pd_{80}Si_{20}$, $Ni_{80}P_{20}$, and $Fe_{40}Ni_{40}P_{14}B_6$. The latter glassy metal is sold under the trademark 'Metglas'. As they are readily produced as thin ribbons, the primary commercial use of amorphous Fe-based ferromagnets is in electromagnetic devices such as relays and transformer-core laminations and in flexible magnetic shielding. These materials have very low hysteresis losses and at the same time about three times higher electrical resistivity than their crystalline counterparts such as Fe-Si alloys. As a result, their use in transformer core laminations can lead to as much as 30 % power savings.

9. **Bulk metallic glasses** are multicomponent alloys such as Zr-Ti-Cu-Ni-Be [13]. Bulk metallic glasses can be processed by common methods available in a foundry. One of the most stable metallic glasses, $Pd_{43}Cu_{27}Ni_{10}P_{20}$, has a critical cooling rate as low as 5×10^{-3} Ks^{-1}. Its glass-forming ability is almost comparable to that of silicate glasses. Bulk metallic glasses can be either of allmetallic or of metal-metalloid type. The commercial applications of bulk metallic glasses benefit from their excellent elastic properties, for example, in heads of golf clubs. Another advantage is their good thermo-plastic formability at moderate temperatures in the supercooled liquid range.

10. **Bulk amorphous steels** are Fe-based multicomponent metallic glasses with high glass-forming ability. Such alloys have been developed recently by PONNAMBALAM ET AL. [14] and by LIU AND COWORKERS [15]. Although conventional steels with crystalline structure have been extensively utilised by industries, bulk amorphous steels have potential to supersede crystalline steels for some critical structural and functional applications, because of their unusual combinations of material properties. These include higher strength and hardness, better magnetic properties, and better corrosion resistance.

11. **Organic Glasses** consist of carbon-carbon chains which are so entangled that cooling of the melt prevents crystallisation. The chains in organic glasses can be crosslinked, with consequent changes in their properties. Increasing the degree of crosslinking, for example, increases the viscosity of

the melt and the glass-transition temperatures. Technological interest in organic glasses stems, for example, from the work on polymer elctrolytes such as polyethylene oxide (PEO) and polypropylene oxide (PPO) salt complexes and their potential for battery applications (see Chap. 27).

12. **Natural Glasses** glasses like all minerals are relics of the history of our planet. There is quite an abundance of natural glasses on earth [16]. Obsidian is perhaps the best-known example. Obsidian is a shiny natural glass, which is usually black or very dark green, but it can also be found in almost clear form. It is formed when lava cooles so quickly that no crystals can form. Ancient people throughout the world have used obsidian for arrowheads, knives, spearheads, and cutting tools of all kinds. Today, obsidian is used as a scalpel by doctors in very sensitive eye operations. Most obsidians are less than 65 million years old; it is alleged that devitrification occurs over longer periods. A typical composition of obsidian is 74 $SiO_2 \cdot 13.5 Al_2O_3 \cdot 1.6$ FeO/Fe_2O_3 1.4 CaO 4.3 Na_2O 4.5 K_2O 0.7 MnO. This makes obsidian a member of the aluminosilicate family.

Examples of other natural glasses are **fulgarites**, which are created when lightning strikes soil; impact glasses or **impactites** have been formed during meteorite impact event by 'shock transformation' or by melting of minerals and rocks due to the absorbed energy.

Scientifically, the most intriguing and hotly debated natural glasses are **tektites** [17]. They resemble obsidian in appearance and chemical composition; however, they have a very low water content, a low alkali content, and they always contain pure silica glass. There are many tektite-strewn fields, areas over which related tektites are found; the more studied ones are the Australian tektites on land and the associated microtektites in deep-see deposits (Indian Ocean, Philippine Sea, Pacific Ocean), the moldavites of central Europe, and the Lybian desert glasses. It has been suggested that many of the characteristics of these glasses, in particular their homogeneity, indicate that these glasses have been molten and subsequently vitrified somewhere in outer space. It was concluded that the Australian tektites are of lunar volcanic origin and not the result of a terrestric meteorite impact.

The moldavites are of bottle-green, translucent colour. They are about 15 million years old and acknowledged to be fused ejecta associated with an impact of a meteorite of about one kilometer in diameter that formed the Ries crater in Southern Germany. The result of this impact was a shower of moldavites that fell in the valley of the Moldavian river near Prague in the Czech Republic.

References

1. G. Tammann, *Kristallisieren und Schmelzen*, Barth, Leipzig, 1903; *Aggregatzustände*, 2nd edn., Voss, Leipzig, 1923; *Der Glaszustand*, Voss, Leipzig, 1933;

2. F.E. Luborsky (Ed.), *Amorphous Metallic Alloys*, Butterworth Monographys in Materials, 1983
3. H.A. Davies, *Metallic Glass Formation*, Ch. 2 in [2]
4. R.W. Cahn, *Alloys Rapidly Quenched from the Melt*, Chap. 25 in: *Physical Metallurgy*, R.W. Cahn, P. Haasen (Eds.), Elsevier Science Publisher (1983)
5. W. Vogel, *Glass Chemistry*, Springer-Verlag, 1985
6. J.E. Shelby, *Introduction to Glass Science and Technology*, The Royal Society of Chemistry, Cambridge, RSC paperbacks, 1997
7. R.H. Doremus, *Glass Science*, John Wiley and Sons, New York, 1994
8. A.K. Varshneya, *Fundamentals of Inorganic Glasses*, Academic Press, Inc., 1994
9. A.W. Imre, S. Voss, H. Mehrer, Phys. Chem. Chem. Phys. **4**, 3219 (2002)
10. S.R. Ovshinsky, Phys. Rev. Lett. **21**, 1450 (1968)
11. W. Clement, R.H. Willens, P. Duwez, Nature **187**, 869 (1960)
12. H.H. Liebermann, *Sample Preparation: Methods and Process Characterization*, Ch. 2 in [2]
13. A. Peker, W.L. Johnson, Appl. Phys. Lett. **63**, 2342 (1993)
14. V. Ponnambalam et al., Appl. Phys. Lett. **83**, 1131 (2003); and J. Mater. Res. **19**, 1320 (2004)
15. Z.P. Lu, C.T. Liu, J.R. Thompson, W.D. Porter, Phys. Rev. Lett. **92**, 245503-1 (2004)
16. L.D. Pye, J.G. O'Keefe, V.D. Frechette (Eds.), *Natural Glasses*, J. Non-Cryst. Solids **67**, (1963)
17. J.G. O'Keefe, *Tektites*, The University of Chicago Press, 1963

29 Diffusion in Metallic Glasses

29.1 General Remarks

Metallic glasses differ from crystalline metallic alloys (Fig. 29.1) by the absence of translational atomic order. They are also called amorphous alloys or glassy metals and can be produced by a variety of techniques which usually involve rapid vitrification from the melt. The cooling occurs so rapidly that the atoms are frozen in their liquid configuration. There are also clear indications that local order exists in most amorphous metallic alloys, but no long-range order (Fig. 29.2).

Metallic glasses are of considerable technological importance. They reveal unique magnetic, mechanical, electrical, and corrosion properties which result from their amorphous structure [1–4]. Metallic glasses containing magnetic elements such as Fe, Ni, or Co are very soft ferromagnetic materials and have magnetic losses which are lower than those measured in any other crystalline alloy. They are widely used in electromagnetic devices such as relays, transformers, and other inductive devices. Metallic glasses are exceptionally hard and have high tensile strengths. Metallic glasses have electrical resistivities in the range of 100 to 300 $\mu\Omega$ cm, which is three or four times higher than that of iron. Metallic glasses owe many of their favourable properties to the fact that they do not contain lattice defects such as dislocations and grain boundaries.

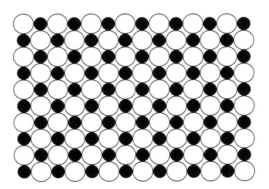

Fig. 29.1. Structure of an ordered binary crystalline solid (schematic)

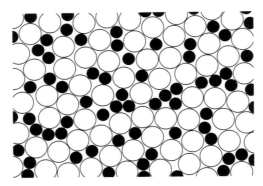

Fig. 29.2. Structure of a binary metallic glass (schematic)

They can be produced as homogeneous, metastable materials in composition ranges, where the equilibrium phase diagram requires heterogeneous phase mixtures of crystalline phases.

Historical Remarks: The first liquid metal alloy vitrified by cooling from the molten state to the glass transition was Au-Si as reported by DUWEZ AND COWORKERS in 1960 [5]. These authors made the discovery as a result of developing rapid quenching techniques for chilling metallic melts at very high cooling rates of 10^5 to 10^6 Ks^{-1}. The work of TURNBULL and of CHEN [6–8] was another crucial contribution to the field and illustrated the similarities between metallic and silicate glasses. This work clearly demonstrated the existence of a glass transition in rapidly quenched Au-Si glasses as well as other glass-forming alloys such as Pd-Si and Pd-Cu-Si, synthesised initially by the Duwez group. Already around 1950, TURNBULL AND FISHER had predicted that as the ratio between the glass-transition temperature, T_g, and the liquidus temperature, T_l, of an alloy increased from $T_g/T_l \approx 1/2$ to $2/3$, homogeneous nucleation of crystals in the undercooled melt should become very sluggish on laboratory time scales [6]. This *Turnbull criterion* for the suppression of crystallisation in undercooled melts is still today one of the best 'rules of thumb' for predicting the glass-forming ability of a liquid.

The field of metallic glasses gained momentum in the early 1970s when continuous casting processes for commercial manufacture of metal glass ribbons such as melt spinning were developed [10]. During the same period CHEN [9] used simple suction casting methods to form millimeter diameter rods of ternary Pd-Cu-Si alloys at cooling rates in the range of 10^3 Ks^{-1}. If one arbitrarily defines the 'millimeter scale' as 'bulk', then the Pd-based ternary glasses were the first examples of *bulk metallic glasses*. Experiments on Pd-Ni-P alloy melts, using boron oxide fluxing to dissolve heterogeneous nucleants into a glassy surface coating, showed that, when heterogeneous nucleation was suppressed, this ternary alloy with a reduced glass-transition temperature of $T_g/T_l \approx 2/3$ would form bulk glass ingots of centimeter size at cooling rates in the range of 10 Ks^{-1} [11, 12]. At the time, this work was perceived by many to be a laboratory curiosity.

During the late 1980s INOUE AND COWORKERS investigated the fabrication of amorphous aluminium alloys. In the course of this work, Inoue's team studied ternary alloys of rare earth materials with aluminium and ferrous metals. They found exceptional glass forming ability in rare-earth-rich alloys, e.g., in La-Al-Ni [13]. From there, they studied similar quaternary materials (e.g., La-Al-Cu-Ni) and developed alloys that formed glasses at cooling rates of under 100 Ks^{-1} with critical thicknesses ranging up to 1 centimeter. A similar family with the rare-earth metal partially replaced by the alkaline-earth metal Mg (Mg-Y-Cu, Mg-Y-Ni, ...) [14] along with a parallel family of Zr-based alloys (e.g., Zr-Cu-Ni-Al) [15] were also developed. These multicomponent glass-forming alloys demonstrated that bulk-glass formation was far more ubiquitous than previously thought and not confined to exotic Pd-based alloys. Building on the work of Inoue, JOHNSON AND COWORKERS [16, 17] developed a family of ternary and higher order alloys of Zr, Ti, Cu, Ni, and Be. These alloys were cast in the form of fully glassy rods of diameters ranging up to 5 to 10 centimeters. No fluxing is required to form such bulk metallic glasses by conventional metallurgical casting methods. The glass-forming ability and processability is comparable to that of many silicate glasses. Metallic glasses can now be processed by common methods available in a foundry [18].

Families of Metallic Glasses: The number and diversity of metallic glasses are continually increasing. We make no attempt to present a comprehensive list because of the complexity in ternary, quaternary, and higher order alloys. We simply mention several families of alloy systems in which glass formation from the melt occurs readily (see also Chap. 28).

Metallic glasses that require rapid cooling with rates of about 10^6 Ks^{-1} are denoted as *conventional metallic glasses*. For conventional metallic glasses the 'nose' of the nucleation curve of the TTT diagram lies in the range of 0.1 to 1 milliseconds (see Chap. 28). They are usually produced by melt-spinning for laboratory and commercial manufacture in the form of thin ribbons or sheets of about 40 µm thickness.

The first class of this type were alloys of late transition metals (LTM) (including group VIIB, group VIII, and noble metals) and metalloids (M) such as Si, B, and P. Metallic glasses of the type LTM-M are perhaps technogically still the most important ones. Many glasses based on Fe, Co, and Ni and on B and P with excellent soft magnetic properties belong to this group. It was at one time believed that the glass formation range is centered around a deep eutectic at about 20 at. % metalloid. Examples are $Au_{80}Si_{20}$, $Pd_{80}Si_{20}$, $Pd_{80}P_{20}$, or $Fe_{80}B_{20}$. When further solute species are added (late transition metals or metalloids), the glass-forming ability may increase further. Examples are $Fe_{40}Ni_{40}B_{20}$ and $Pd_{40}Ni_{40}P_{20}$.

A second group of conventional metallic glasses consists of alloys of early transition metals (ETM) and late transition metals (LTM). The former have high melting temperatures and addition of a LTM generally leads to a rapid decrease of the liquidus temperature down to an eutectic. The liquidus temperature then remains relatively low across one or more intermetallic phases of relatively low stability. Examples of this type are Zr-Co, Zr-Cu, Zr-Ni, Zr-Fe, and Nb-Ni alloys.

Most of the binary alloy systems of rare earth metals with late transition and group IB metals have also deep eutectics. They have been shown to be readily glass forming, if the composition is centered around the eutectic composition. Examples are La-Au, La-Ni, Gd-Fe, and Gd-Co alloys.

Bulk metallic glasses exhibit TTT diagrams with a crystallisation 'nose' in the range between 1–100 seconds or more. These alloys have an exceptional glass-forming ability and undercooled melts, which are relatively stable [18]. This permits diffusion studies even in the undercooled melt of bulk metallic glass-forming alloys. By contrast, conventional metallic glasses undergo crystallisation before the glass-transition temperature is reached and thus can be studied only below the glass-transition temperature. High glass-forming ability was recently found for bulk metallic glasses based on copper [21]. Applications of bulk metallic glasses benefit from their excellent elastic properties and the good formability in the supercooled liquid state. *Bulk amorphous steels* is a recent development [19, 20] with potential to replace conventional steels for some critical structural or functional applications.

29.2 Structural Relaxation and Diffusion

Glasses are thermodynamically metastable in a twofold sense: (i) They can undergo *crystallisation*, during which the material transforms to (a) crystalline phase(s). (ii) The properties of a glass may depend on its thermal history (see Chap. 28). Upon reheating a glass to the glass-transformation range, the glass properties may change due to a process which is called *structural relaxation*.

Structural relaxation of an amorphous material leads to a more stable amorphous state. Structural relaxation is accompanied by a number of changes in physical properties. Clearly, the extent of property changes for a given material depends on its thermal history and on the method of glass production. Changes due to structural relaxation are understandable by considering the volume (or enthalpy)-*versus*-temperature diagram of Fig. 29.3. The volume can be altered by a heat treatment, which allows equilibration of the structure to that pertaining to the heat treatment temperature. A fast cooled glass has a higher fictive temperature, a larger volume, and a lower density. The volume difference is sometimes denoted as the *excess volume*. If we reheat such a sample to a temperature within the transformation range, but below the original fictive temperature, the sample will readjust to the

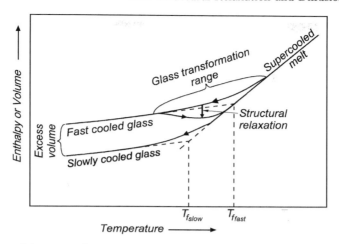

Fig. 29.3. Schematic illustration of structural relaxation in the V-T (or H-T) diagram of a glass-forming material

structure appropriate for the new temperature. Its volume will decrease. Although the changes in density occurring during structural relaxation are not particularly large (typically less than 1 %), they can be important for viscosity and ductility, as well as for magnetic, elastic, electric, and diffusion properties. A review of the effects of structural relaxation on various properties of metallic glasses has been given by CHEN [22]. In this section, we concentrate on structural relaxation of diffusion properties.

If structural relaxation occurs during the diffusion annealing of a sample, the diffusivity depends on time. Under such conditions the thin-film solution of Fick's second law (Chap. 3) remains valid, if the diffusivity D is replaced by its time average given by

$$\langle D(t) \rangle = \frac{1}{t} \int_0^t D(t') \mathrm{d}t' . \tag{29.1}$$

Equation (29.1) can be verified by showing that the thin-film solution with the time-averaged diffusivity $\langle D(t) \rangle$ is a solution of Fick's second law with the time-dependent (instantaneous) diffusivity

$$D(t) = \langle D \rangle + t \frac{\mathrm{d}\langle D \rangle}{\mathrm{d}t} . \tag{29.2}$$

The time-averaged diffusivity is the quantity that is accessible in a tracer experiment. Figure 29.4 displays time-averaged diffusivities for ^{59}Fe diffusion in amorphous $Fe_{40}Ni_{40}B_{20}$ [23]. In this example, the time-averaged diffusivity decreases by about half an order of magnitude with increasing annealing time.

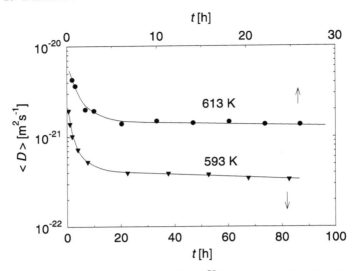

Fig. 29.4. Time-averaged diffusivities $\langle D \rangle$ of ^{59}Fe in as-cast $Fe_{40}Ni_{40}B_{20}$ as functions of the annealing time according to HORVATH AND MEHRER [23]

If a sufficient number of $\langle D \rangle$ values for various annealing times is measured, the instantaneous, time-dependent diffusivity can be deduced via Eq. (29.2).

Figure 29.5 displays instantaneous diffusivities for various as-cast metallic glasses determined in this way [24]. The main feature of Fig. 29.5 is the continuous decrease of $D(t)$ to a plateau value. In the following this plateau value is denoted as D_R and attributed to the relaxed amorphous state. The features described above are common to many diffusion studies on metallic glasses. The diffusivity decreases during diffusion annealing as a result of structural relaxation. This effect may be described by the relationship

$$D(t,T) = D_R(T) + \Delta D(t,T). \qquad (29.3)$$

The diffusivity enhancement, $\Delta D(t,T)$, drops to zero upon sufficient annealing and the diffusivity in the relaxed state, $D_R(T)$, depends on temperature only. Usually, within the experimental accuracy the temperature dependence of D_R can be described by an Arrhenius relation (see below).

The diffusivity enhancement in conventional metallic glasses is correlated with the excess free volume present in the as-quenched material (Fig. 29.3). This excess volume anneals out during structural relaxation and leads to an increase in density. Atoms can move more easily through a more open (less dense) structure than through a more dense structure. As a consequence, the diffusivity decreases during an annealing treatment at a temperature below the fictive temperature of the as-quenched glass. Sometimes the excess volume is also said to be due to 'quasi-vacancies' envisaged as localised defects being stable over several jumps [25]. In the language of quasi-vacancies the latter are

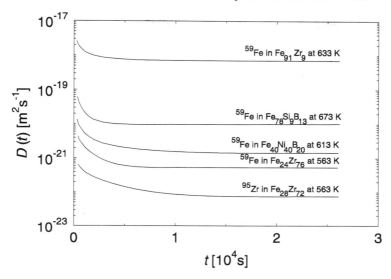

Fig. 29.5. Instantaneous diffusivities $D(t)$ of several conventional metallic glasses as functions of annealing time according to HORVATH ET AL. [24]

mobile during structural relaxation and remove the diffusivity enhancement. In contrast to self-diffusion in crystalline metals, which occurs via vacancies present in thermal equilibrium, quasi-vacancies in an as-quenched amorphous alloy are present in supersaturation and anneal out when they become mobile. As a result, the diffusivities slow down until they have reached their relaxed-state values.

The diffusivity enhancement depends on the material, its thermal history, and on the technique of glass production. According to Fig. 29.3 different fictive temperatures lead to different amounts of structural relaxation. For a given material with low fictive temperatures the diffusivity enhancement may be insignificant. As a consequence, some conflicting results in the literature about the magnitude of structural relaxation effects in diffusion are likely due to different techniques of alloy production such as melt-spinning, splat cooling, or co-evaporation.

29.3 Diffusion Properties of Metallic Glasses

Temperature Dependence: Diffusion measurements on **conventional metallic glasses** are usually carried out below the glass-transition temperature due to the limitations imposed by incipient crystallisation of the glass at higher temperatures. The diffusion coefficients in the structurally relaxed glassy state follow an Arrhenius-type temperature dependence

$$D_R = D^0 \exp\left(-\frac{\Delta H}{k_B T}\right), \tag{29.4}$$

thus yielding pre-exponential factors D^0 and activation enthalpies ΔH. Examples of Arrhenius plots for both metal-metal and metal-metalloid type amorphous alloys are shown in Fig. 29.6. The temperature range in which diffusion measurements have been performed is often limited to 200 K or less. At high temperatures the onset of crystallisation and at low temperatures the very small diffusivity prevents meaningful measurements.

The error margins imposed on the diffusion parameters are relatively large, being of the order of 0.2 eV for the activation enthalpy and about one order of magnitude for the pre-exponential factor. It was shown that the observed Arrhenian temperature dependence within these error bars is compatible with a narrow height distribution of jump barriers in the disordered structure of an amorphous alloy [26–28]. Another reason for the 'surprising' linearity of the Arrhenius plots are compensation effects between site and saddle-point disorder [29]. The most likely reason, however, is the collectivity of the atom-transport mechanism leading to an averaging of disorder effects in the atomic migration process (see below).

For **bulk metallic glasses** it is possible to carry out diffusion measurements in a temperature range that covers both the **undercooled melt** re-

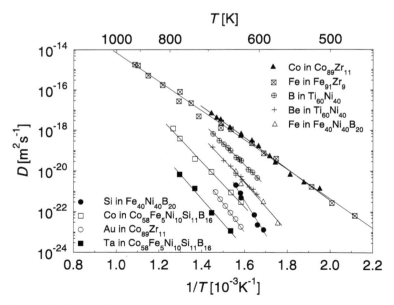

Fig. 29.6. Arrhenius diagram of self- and impurity diffusion in relaxed metal-metalloid and metal-metal-type conventional metallic glasses according to FAUPEL ET AL. [37]

gion and the glassy state. We illustrate diffusion in bulk metallic glasses for an alloy that has attracted great interest – namely the five component alloy $Zr_{46.75}Ti_{8.25}Cu_{7.5}Ni_{10}Be_{27.5}$ (commercially denoted as 'Vitreloy4'). Centimeter-size rods of this alloy can be produced by casting techniques [18] and the TTT diagram in the range of the glass transition and crystallisation is known from the work of BUSCH AND JOHNSON [30]. The temperature dependence of diffusion for a variety of elements in Vitreloy4 is displayed in Fig. 29.7. The following diffusers have been studied; Be [31], Ni [32, 33], Co [31, 34], Fe [31], Al [35], Hf [36] and the data have been assembled in two reviews [37, 54]. An important feature of Fig. 29.7 is that the diffusivities of several elements can be split into two different linear Arrhenius regions below and above a 'kink temperature'. The kink temperature correspond to the transition between the glassy and supercooled liquid states. The activation enthalpies and pre-exponential factors in the supercooled liquid state are higher than those below the kink temperature. In addition, the kink temperature separating the glassy and the supercooled region is higher for elements, which diffuse faster in the amorphous state.

It has been demonstrated that the diffusion times applied at low temperatures were too short to reach the metastable state of the undercooled liquid at these temperatures [33, 39]. A test of this interpretation of the nonlinear Arrhenius behaviour is shown in Fig. 29.8, in which the diffusivities

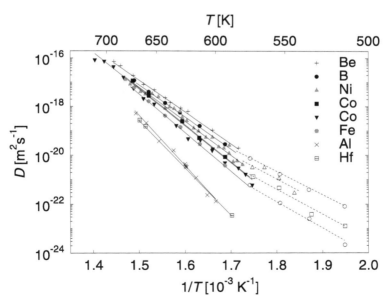

Fig. 29.7. Arrhenius diagram of tracer diffusion of Be, B, Fe, Co, Ni, Hf in the bulk metallic glass $Zr_{46.75}Ti_{8.25}Cu_{7.5}Ni_{10}Be_{27.5}$ (Vitreloy4) according to FAUPEL ET AL. [37]

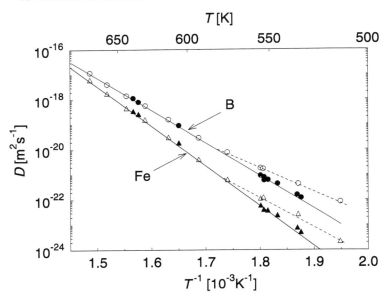

Fig. 29.8. Arrhenius diagram of tracer diffusion of B and Fe in $Zr_{46.75}Ti_{8.25}Cu_{7.5}Ni_{10}Be_{27.5}$ (Vitreloy4) according to FAUPEL ET AL. [37]. *Open symbols*: as-cast material from [31]; *filled symbols*: pre-annealed material from [39]

of Fe and B in 'as-cast' and 'pre-annealed' Vitreloy4 are displayed. For sufficiently long annealing times the material finally relaxes into the supercooled liquid state (see also Fig. 29.3). Open symbols in Fig. 29.8 represent diffusivities in the as-cast material, full symbols represent diffusivities measured after pre-annealing between 1.17×10^6 s and 2.37×10^7 s at 553 K, i.e. below the calorimetric glass-transition temperature. The diffusivities obtained after extended pre-annealing below 550 K are smaller than those of the as-cast material, whereas in the high-temperature region the diffusivities of the as-cast and the pre-annealed material coincide. Furthermore, the diffusivities in the relaxed material can be described by one Arrhenius equation, which also fits the high-temperature data of the as-cast material. This provides evidence that the kink in the temperature dependence of the diffusivity is not related to a change in the diffusion mechanism but depends on the thermal history of the material. It is caused by incomplete relaxation to the state of the undercooled liquid.

Correlation between D^0 and ΔH: Reported values of the activation enthalpy in conventional metallic glasses and supercooled glass melts, in general, range from 1 to 3 eV for different diffusers (excluding hydrogen). The pre-exponential factors D^0 show a wide variation from about 10^{-15} to 10^{13} m^2 s^{-1} [37]. This variation is much larger than the one reported for crys-

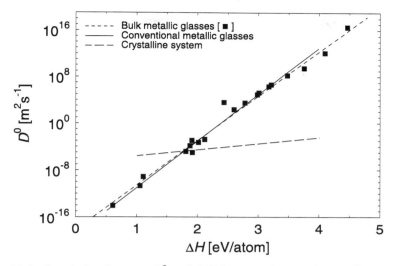

Fig. 29.9. Correlation between D^0 and ΔH for amorphous and crystalline metals according to [37]. *Solid line*: conventional metallic glasses; *dotted line*: bulk metallic glasses; *dashed line*: crystalline metals

talline metals and alloys (about 10^{-6} to 10^2 m^2 s^{-1}). The experimental values of D^0 and ΔH have been found to obey the following correlation:

$$D^0 = A \, \exp\left(\frac{\Delta H}{B}\right). \tag{29.5}$$

A and B are constants. This relationship has a universal character in the sense that it is valid not only for metallic glasses but also for self- and impurity diffusion in crystalline metals and alloys involving both interstitial and substitutional diffusion (see [37] and [38] for references). The values of D^0 and ΔH in both conventional metallic glasses and in the undercooled liquid state of bulk metallic glasses do follow the same relationship as shown in Fig. 29.9. However, the fitting parameters for metallic glasses ($A \approx 10^{-19}$ to 10^{-20} m^2 s^{-1}, $B \approx 0.055$ eV) and crystalline metals ($A \approx 10^{-7}$ m^2 s^{-1}, $B \approx 0.41$ eV) are quite different (Fig. 29.9).

The fact that the parameters A and B differ considerably for crystalline and amorphous metals indicates that the diffusion mechanism of metallic glasses is different from the interstitial or vacancy mechanisms operating in crystals.

Pressure Dependence: Studies of the pressure dependence of diffusion and the activation volumes deduced therefrom have been key experiments for elucidating diffusion mechanims of crystalline solids. For vacancy-mediated diffusion the activation volume equals the sum of the formation and migration volumes of the vacancy (see Chap. 8). The major contribution to the activation volume of self-diffusion for metallic elements comes from the formation volume, which typically lies between 0.5 and 1 atomic volumes. For interstitial diffusion no defect formation is involved and the activation volume equals the migration volume of the interstitial, which is small. A small activation volume implies a weak pressure dependence of the diffusion coefficient.

Measurements of the pressure dependence of diffusion in metallic glasses can be grouped into two categories [37]:

1. Systems with almost no pressure dependence: activation volumes close to zero were reported for metallic glasses, which mainly contain late transition elements and for tracers of similar size as the majority component. A typical example is displayed in Fig. 29.10. Small activation volumes allow vacancy-mediated diffusion to be ruled out and have been taken as evidence for a diffusion mechanism, which does not involve the formation of a defect.
2. Systems with significant pressure dependence: activation volumes comparable to those of vacancy-mediated diffusion in crystalline solids were mainly reported for diffusion in Zr-rich Co-Zr and Ni-Zr metallic glasses. They have tentatively been attributed to the formation of diffusion-mediating defects which are delocalised. On the other hand, molecular-dynamics simulations for Ni-Zr glasses suggest that diffusion takes place by thermally activated collective motion of chains of atoms (see Fig. 29.12 and Chap. 6). It has been proposed that the migration volume of chain-like motion is associated with a significant activation volume [41].

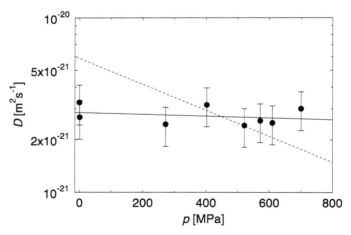

Fig. 29.10. Pressure dependence of Co diffusion in $Co_{81}Zr_{19}$ at 563 K according to [37]. The *dashed line* would corresponds to an activation volume of one atomic volume

29.3 Diffusion Properties of Metallic Glasses

Isotope Effects: Isotope effect measurements proved to be useful in deducing atomic mechanisms of diffusion in crystals (Chap. 9). Such studies have been performed on metallic glasses as well.

Almost vanishing isotope effects have been reported for Co diffusion in various relaxed, conventional metallic glasses by FAUPEL AND COWORKERS [40, 42–45]. The small isotope effects can be attributed to strong dilution of the mass dependence of diffusion due to the participation of a large number of atoms in a collective diffusion process. Isotope effect experiments are also reported for the deeply undercooled liquid state of bulk metallic glasses EHMLER ET AL. [46, 47] (Fig. 29.11). The magnitude of the isotope effect parameter is similar to the isotope effects found for (relaxed) conventional metallic glasses. This lends support to the view that the diffusion mechanism does not change at the calorimetric glass transition and demonstrates the collective nature of diffusion processes in metallic glasses [37].

So far, we have mentioned isotope effects in structurally relaxed metallic glasses. On the other hand, as-cast metallic glasses contain excess volume quenched-in from the liquid state. Magnitudes of the isotope effect parameter comparable to values observed for crystalline metals were reported for the as-quenched metal-metalloid glass $Co_{76.7}Fe_2Nb_{14.3}B_7$ [48]. Such observations suggest that during diffusion annealing of unrelaxed glasses quenched-in quasi-vacancies serve as diffusion vehicles until they have annealed out.

Atomic Mechanisms: Experiments and computer simulations show that diffusion mechanisms in metallic glasses contrast with diffusion in crystals. It requires other concepts, based on thermally activated highly collective processes.

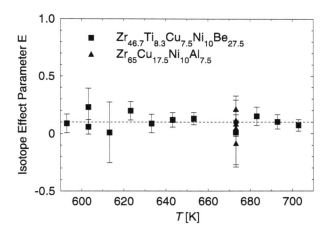

Fig. 29.11. Isotope effect parameter as function of temperature for Co diffusion in bulk metallic glasses according to [37]; data taken from EHMLER ET AL. [46, 47]

Molecular dynamics simulation has contributed significantly to the understanding of diffusion processes in undercooled melts and in metallic glasses. The following picture emerges from simulations in agreement with the experimental facts discussed above and in [37]:

With decreasing temperature transport of matter changes from liquid-like viscous flow via atomic collisions to thermally activated transport characteristics of solids. According to the simulations and to mode-coupling theory [52] this change-over occurs at a critical temperature T_C. Already well above T_C the dynamics starts to become heterogeneous in the form of collective motion of chains and rings of atoms. Upon cooling below T_C the calorimetric glass temperature is reached at T_g. Starting below T_C, but well above T_g, diffusion follows the classical Arrhenius relationship.

It is important to note that the linear Arrhenius behaviour observed within experimental accuracy in the supercooled liquid state is due to the limited temperature range of the experiment. Diffusivity measurements performed over wide temperature ranges in the liquid state can be described by a Vogel-Fulcher-Tammann type temperature dependence and not by an Arrhenius relation. Molecular-dynamics simulations [49–51] and mode-coupling theory [52] predict a downward curvature at higher temperatures. If an activation enthalpy is attributed to diffusion in the undercooled liquid state, it should be considered as an effective one. It is strongly increased by structural changes with temperature occurring in the undercooled liquid above the glass transition. A 'true' activation enthalpy can only be attributed to the slope of an Arrhenius line, if the structure does not change with temperature.

Mainly chain-like displacements of atoms have been observed in molecular dynamics simulations. Collective atomic motion in a chain-like manner leads to total displacements of the order of one nearest-neighbour distance. Such

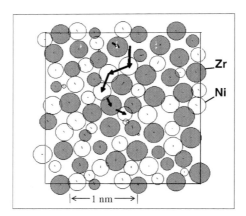

Fig. 29.12. Chain-like collective motion of atoms in a Co-Zr metallic glass according to molecular dynamics simulations by TEICHLER [55]

displacement chains typically involve 10 to 20 atoms, where each atom moves only a small fraction of the nearest-neighbour distance. This mechanism has been already discussed and illustrated in Chap.6. We repeat the previous figure for convenience in Fig. 29.12. With increasing temperature the total jump length, the jump length of single atoms, and the number of atoms involved in such displacements chains increases. With further increasing temperature these collective events become more and more frequent and finally merge into viscous flow.

29.4 Diffusion and Viscosity in Glass-forming Alloys

Viscosity measures the resistance of a melt to shear deformation. The *Stokes-Einstein relation* relates the viscosity to a viscosity diffusion coefficient (see also Chap.30) via

$$D_\eta = k_B T/(6\pi r \eta) \qquad (29.6)$$

k_B is the Boltzmann constant, r some atomic radius, and T the absolute temperature. For liquids the Stokes-Einstein relation is well appreciated. In an undercooled melt a decoupling of viscosity and diffusivity occurs around the critical temperature T_c of the mode-coupling theory. This decoupling is attributed to the arrest of liquid-like atomic motion, which leads to a discrepancy between viscosity diffusivity as compared to diffusivities deduced from tracer experiments. Near the caloric glass-transition temperature the difference can be several orders of magnitude. Molecular dynamics simulations on a binary Lenard-Jones mixture by MÜLLER-PLATE AND ASSOCIATES [56] have revealed a breakdown of the Stokes-Einstein relation at a temperature between the thermodynamic melting temperature T_m and the critical temperature T_c.

An experimental comparison between tracer diffusion and viscosity has been performed for a $Pd_{43}Cu_{27}Ni_{10}P_{20}$ alloy. This alloy was chosen because it has a glass-forming melt with high stability against crystallisation. This made it possible to measure tracer diffusivities from the glassy state through the supercooled melt up to the equilibrium melt. Diffusion of ^{32}P and ^{57}Co in $Pd_{43}Cu_{27}Ni_{10}P_{20}$ has been studied by ZÖLLMER ET AL. [57] and BARTSCH ET AL. [58] using tracer methods. Viscosity data are available for the supercooled and the equilibrium melt. Figure 29.13 shows a comparison between tracer diffusion of P and Co and the viscosity diffusion coefficient calculated from Eq. (29.6) using the atomic radius of Co (0.125 nm). The dashed lines indicate the melting temperature T_m, the critical temperature T_c, and the caloric glass-transition temperature T_g at a heating rate of 20 K min^{-1}.

In the equilibrium melt tracer diffusivities and viscosity diffusivities coincide as expected for liquid-like atomic transport. In the undercooled melt both available tracer diffusivities are about one order of magnitude higher

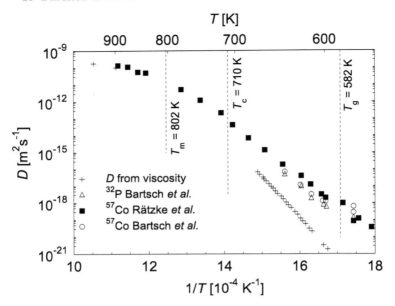

Fig. 29.13. Tracer diffusion coefficients of P and Co in comparison with viscosity diffusion coefficients of the alloy $Pd_{43}Cu_{27}Ni_{10}P_{20}$ according to BARTSCH ET AL. [58]

at 640 K than the diffusivity calculated from the viscosity; near the glass-transition this discrepancy is around two orders of magnitude. If one would use the atomic radii of other alloy elements instead of those for Co, the discrepancy would not be affected significantly. Although in bulk metallic glasses and their supercooled melts a considerable size dependence has been observed (see, e.g., Fig. 29.7), it is still under debate, whether the viscosity is determined solely by one element of the alloy. Obviously, the diffusivities of the metal atoms Co and the metalloid atoms P are decoupled from the viscosity of the supercooled melt.

References

1. F.E. Luborski, *Amorphous Metallic Alloys*, Butterworth & Co, Ltd., 1983
2. H. Mehrer, *Metallische Gläser*, Technische Mitteilungen **78**, 554 (1985)
3. R. Boll, H.R. Hilzinger, Elektrotechnische Zeitschrift **102**, 1096 (1981)
4. G. Herzer, Physikalische Blätter **5**, 43 (2001)
5. W. Clement, R.H. Willens, P. Duwez, Nature **187**, 869 (1960)
6. D. Turnbull, J.C. Fisher, J. Chem. Phys. **17**, 71 (1949); ibid. **18**, 198 (1950)
7. H.S. Chen, J. Chem. Phys. **48**, 2560 (1968)
8. H.S. Chen, Acta Metall. **22**, 1021 (1969)
9. H.S. Chen, Acta Metall. **22**, 1505 (1974)

10. H.H. Liebermann, *Sample Preparation: Methods and Process Characterisation*, Chap. 2 in [1]
11. A.L. Drehman, A.L. Greer, D. Turnbull, Appl. Phys. Lett. **41**, 716 (1982)
12. H.W. Kui, A.L. Greer, D. Turnbull, Appl. Phys. Lett. **45**, 615 (1984)
13. A. Inoue, T. Zhang, T. Masumoto, Mater. Trans., JIM, **33**, 965 (1990)
14. A. Inoue, Mater. Trans., JIM, **36**, 866 (1995)
15. A. Inoue, T. Zhang, T. Masumoto, Mater. Trans., JIM, **31**, 177 (1990)
16. A. Peker, W.L. Johnson, Appl. Phys. Lett. **63**, 2342 (1993)
17. X.H. Lin, W.L. Johnson, J. Appl. Phys. **78**, 6514 (1995)
18. W.L. Johnson, Mat. Res. Soc. Symp. Proc. **554**, 311 (1999)
19. V. Ponnambalam et al., Appl. Phys. Lett. **83**, 1131 (2003); and J. Mater. Res. **19**, 1320 (2004)
20. Z.P. Lu, C.T. Liu, J.R. Thompson, W.D. Porter, Phys. Rev. Lett. **92**, 245503-1 (2004)
21. D. Xu, G. Duan, W.L. Johnson, Phys. Rev. Lett. **92**, 245504-1 (2004)
22. H.S. Chen, *Structural Relaxation in Metallic Glasses*, Chap. 11 in [1]
23. J. Horvath, H. Mehrer, Cryst. Latt. Def. and Amorph. Mat. **13**, 1 (1986)
24. J. Horvath, K. Pfahler, W. Ulfert, W. Frank, H. Kronmüller, Mat. Sci. Forum **15–18**, 523 (1987)
25. W. Frank, J. Horvath, H. Kronmüller, Mater. Sci. Eng. **97**, 415 (1988)
26. H. Kronmüller, W. Frank, Radiat. Eff. Def. Solids **108**, 81 (1989)
27. A. van den Beukel, Acta Metall. Mater. **42**, 1273 (1994)
28. W. Frank, Defect and Diffusion Forum **143–147**, 695 (1997)
29. Y. Limoge, J.L. Bocquet, Phys. Rev. Lett. **65**, 60 (1990)
30. R. Busch, W.L. Johnson, Mater. Sci. Forum **269–272**, 577 (1998)
31. P. Fielitz, M.-P. Macht, V. Naundorf, G. Frohberg, J. Non-Cryst. Solids **250–252**, 674 (1999)
32. K. Knorr, M.-P. Macht, K. Freitag, H. Mehrer, J. Non-Cryst. Solids **250**, 669 (1999)
33. K. Knorr, M.-P. Macht, H. Mehrer, Mat. Res. Soc. Symp. Proc. **554**, 269 (1999)
34. H. Ehmler, K. Rätzke, F. Faupel, J. Non-Cryst. Solids **250–252**, 684 (1999)
35. E. Budke, P. Fielitz, M.-P. Macht, V. Naundorf, G. Frohberg, Defect and Diffusion Forum **143–147**, 825 (1997)
36. Th. Zumkley, V. Naundorf, M.-P. Macht, Z. Metallkd. **91**, 901 (2000)
37. F. Faupel, W. Frank, H.-P. Macht, H. Mehrer, V. Naundorf, K. Rätzke, H. Schober, S. Sharma, H. Teichler, *Diffusion in Metallic Glasses and Supercooled Melts*, Review of Modern Physics **75**, 1 (2003)
38. V. Naundorf. M.-P. Macht, A.S. Bakai, N. Lazarev, J. Non-Cryst. Sol. **224**, 122 (1998); ibid **250–252**, 679 (1999)
39. Th. Zumkley, M.-P. Macht, G. Frohberg, Scripta Mater. **45**, 471 (2001)
40. F. Faupel, W. Hüppe, K. Rätzke, Phys. Rev. Lett. **65**, 1219 (1990)
41. H. Schober, Phys. Rev. Lett. **88**, ... (2002)
42. W. Hüppe, F. Faupel, Phys. Rev. B **46**, 120 (1992)
43. A. Heesemann, K. Rätzke, F. Faupel, J. Hoffmann, K. Heinemann, Europhys. Lett. **29**, 221 (1995)
44. K. Rätzke, F. Faupel, J. Non-Cryst. Solids **181**, 261 (1995)
45. A. Heesemann, V. Zöllmer, K. Rätzke, F. Faupel, Phys. Rev. Lett. **84**, 1467 (2000)
46. H. Ehmler, A. Heesemann, K. Rätzke, F. Faupel, U. Geyer, Phys. Rev. Lett. **80**, 4919 (1998)

47. H. Ehmler, K. Rätzke, F. Faupel, J. Non-Cryst. Solids **250–252**, 684 (1999)
48. K. Rätzke, F. Faupel. Phys. Rev. B **45**, 7459 (1992)
49. H. Teichler, Defect and Diffusion Forum **143–147**, 717 (1997)
50. H. Teichler, Phys. Rev. B **59**, 8473 (1999)
51. M. Kluge, H.R. Schober, Defect and Diffusion Forum **194–199**, 849 (2001)
52. W. Götze, A. Sjölander, Rep. Progr. Physics **55**, 241 (1992)
53. H.R. Schober, Physica A **201**, 14 (1993)
54. F. Faupel, K. Rätzke, *Diffusion in Metallic Glasses and Supercooled Melts*, in: *Diffusion in Condensed Matter – Methods, Materials, Models*, P. Heitjans, J. Kärger (Eds.), Springer-Verlag, 2005
55. H. Teichler, J. Non-cryst. Solids **293**, 339 (2001)
56. P. Bordat, F. Affouard, M. Descamps, F. Müller-Plate, J. Phys.: Condensed Matter **15**, 5397 (2003)
57. V. Zöllmer, K. Rätzke, F. Faupel, J. Mater. Res. **18**, 2688 (2003)
58. A. Bartsch, K. Rätzke, F. Faupel, Appl. Phys. Lett. **89**, 121917 (2006)

30 Diffusion and Ionic Conduction in Oxide Glasses

30.1 General Remarks

Oxide glasses are the best known class of non-crystalline materials and comprise a large number of glass families. The most important ones are already mentioned in Chap. 28. When one considers the world-wide commercial use of oxide glasses, the need for an understanding of their structural elements and properties is obvious. The crystallography, chemistry, and physics of oxide glasses encompass a vast body of information. In the present chapter, we limit ourselves to some basic foundations. For more information, we refer to textbooks on glass by VOGEL [1], SHELBY [2], DOREMUS [3], and VARSHNEYA [4].

Most oxide glasses including silicate, germanate, borate, and many phosphate glasses are ionic conductors. Some phosphate and chalcogenide glasses are electronic conductors. Oxide glasses which contain transition metal elements are mixed conductors. Given the wide diversity of oxide glasses, we confine ourselves to illustrate some aspects of diffusion and ionic conduction by typical examples which concern vitreous silica, soda-lime silicate glasses, single alkali borate glasses, and features of the so-called mixed-alkali effect. Diffusion data for a large number of oxide glasses can be found, for example, in an early review by FRISCHAT [5] and in a more recent collection by JAIN AND HSIEH [6]. A coverage of the literature on ionic conductivity in oxide glasses can be found in INGRAM'S review [7].

Structure of Network Glasses: GOLDSCHMIDT, who is considered as the founder of modern crystal chemistry, suggested in the 1920s empirical rules for glass formation [8]. Like for crystalline structures he proposed that relations of the ionic sizes play a decisive role. He postulated ratios of cation to anion radii from 0.2 to 0.4 as a condition of glass formation. Indeed, the oxides SiO_2, B_2O_3, P_2O_5, GeO_2, and some other compounds fulfill this condition (Table 30.1).

An overwhelming number of oxide glasses are silicate glasses. The basic building block of crystalline silicates is the $SiO_{4/2}$ tetrahedron, a structural unit with a silicon atom in the center of four oxygen atoms[1]. In the case of

---
[1] We denote this unit as $SiO_{4/2}$ tetrahedron since each of the four O atoms is shared by two Si atoms.

Table 30.1. Ionic radii for typical glass-forming oxides or compounds according to VOGEL [1]

Compound	Cation radius	Anion radius
SiO_2	$r_{Si} = 0{,}039$ nm	$r_O = 0.14$ nm
B_2O_3	$r_B = 0.02$ nm	$r_O = 0.14$ nm
P_2O_5	$r_P = 0.034$ nm	$r_O = 0.14$ nm
GeO_2	$r_{Ge} = 0.044$ nm	$r_O = 0.14$ nm
BeF_2	$r_{Be} = 0.034$ nm	$r_F = 0.136$ nm

vitreous silica the same $SiO_{4/2}$ tetrahedra, which are regularly connected in crystalline silicates, are connected irregularly and form a disordered three-dimensional network. The structure of vitreous silica is readily described by silicon-oxygen tetrahedra linked at all four corners. Each oxygen is shared by two silicon atoms, which occupy the centers of the linked tetrahedra to form a continuous random network. Disorder is obtained in this structure by allowing variability in the Si-O-Si bond angle connecting adjacent tetrahedra.

The network hypothesis proposed by ZACHARIASEN [9] and enforced by the X-ray diffraction work of WARREN [10] in the 1930s represented an important step forward to our present understanding of the structure of glasses. According to this classical network idea, the following rules hold for the formation of three-dimensional network glasses:

1. An oxide or compound tends to form a glass, if it forms polyhedral groups as smallest building units. Examples are SiO_2, B_2O_3, GeO_2, P_2O_5, As_2S_3, and BeF_2.
2. Polyhedra should not share more than one corner.
3. Anions such as O^{2-}, S^{2-}, and F^- should not bind more than two central atoms of a polyhedron. In simple glasses, anions form bridges between two polyhedra.
4. The number of corners of a polyhedron must be smaller than six.
5. At least three corners of a polyhedron must connect with neighbouring polyhedra.

Depending on the glass-forming ability, an oxide may be called a glass (or network) former, a glass (or network) modifier, or an intermediate (conditional) oxide (see Table 30.1). Network-former ions, such as Si, B, P, and Ge,

Table 30.2. Examples of network former, network modifier, and intermediate ions

Network-former ions	Network-modifier ions	Intermediate ions
Si, Ge, B, P, Sb	Li, Na, K, Rb, Cs	Al, Bi, Mo, S
As, In, Tl	Ca, Ba, Pb, Sn	Se, Te, V, W

usually have coordination numbers of 3 or 4. Network-modifier ions, such as Na, K, Rb, and Ca, have coordination numbers generally not larger than 6. Intermediates may either enforce the network (coordination number 4) or further loosen the network (coordination number 6–8), but cannot form a glass alone.

A large number of properties of network glasses can be understood on the basis of the Zachariasen-Warren concepts. An increase of modifier cations breaks bridges or modifies the fundamental network. The increasing mobility of the building units accounts for decreasing viscosity and liquidus temperature, as well as for an increasing ionic conductivity and diffusivity of modifier cations.

Viscosity of Glass-forming Melts: Viscosity is a melt property. It measures the resistance of a liquid to shear deformation. The rate of shear deformation, $d\epsilon_{xy}/dt$, is related to the shear stress, τ_{xy}, via Newton's law of viscosity

$$\tau_{xy} = \eta \frac{d\epsilon_{xy}}{dt}, \qquad (30.1)$$

where η is the coefficient of viscosity, or simply the viscosity. When stress is written in units of Pa, the appropriate unit of the viscosity is Pa s. The old unit for η, based on the cgs system, was dyne s cm^{-2}. This unit, which is termed the Poise (symbol P), is used in all literature prior to 1970 and is still often used in glass technology. Since 1 Pa s = 10 P the conversion of units is straightforward.

Viscosity is the inverse of fluidity. A melt with a large fluidity will flow readily, whereas a melt with large viscosity has a high resistance to flow. The viscosity of a glass-forming melt plays a major rôle in determining the ease of glass formation. Glasses are most easily formed if the viscosity either is very high at the melting temperature of the crystalline phase or increases very rapidly with decreasing temperature. In either case, crystallisation is impeded by the kinetic barrier to atomic rearrangement which results from a high viscosity.

Viscosity is one of the most important properties in glass technology. It plays an enormous rôle in all stirring processes, in the buoyancy of bubbles during fining processes, during glass forming, and for nucleation and growth of crystalline phases. As pointed out in Chap. 28, a glass-forming melt acts as a liquid at high temperatures and turns into a glassy solid upon cooling. The viscosity-temperature $(\eta - T)$ relationship of a typical glass is shown in Fig. 30.1. In this figure a number of specific viscosities have been designated as reference points. These particular viscosities have been chosen because of their importance in various aspects of commercial or laboratory processing of glass-forming melts. Melting usually occurs at viscosities of 1 to 10 Pa s for commercial glasses, but can occur at lower viscosities for non-silicate, and in particular for non-oxide glasses.

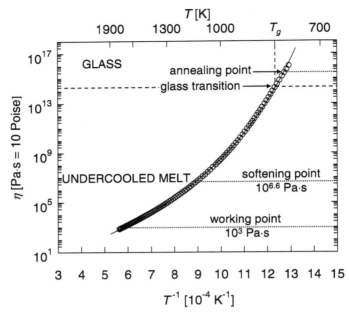

Fig. 30.1. Viscosity of a soda-lime-silicate glass (standard glass I of the Deutsche Glastechnische Gesellschaft, DGG). Particular viscosity points are indicated

Formation of a glass object from a melt requires shaping a viscous lump by some process involving deformation of the material. The melt must be fluid enough to allow flow under reasonable stresses but viscous enough to retain its shape after forming. In commercial forming methods, the melt is typically delivered to a processing device at a viscosity of 10^3 Pa s, which is known as the *working point*. Once formed, an object must be supported until the viscosity reaches a value sufficiently high to prevent deformation under its own weight, which ceases at a viscosity of $10^{6.6}$ Pa s. This viscosity is termed as the *softening point*. The temperature range between the working and softening points is denoted as the *working range*. Once an object is formed, the internal stresses which result from cooling can be reduced by annealing. The *annealing point* is usually considered to be at 10^{12} to $10^{12.4}$ Pa s. The glass transformation temperature, T_g, can be determined from measurements of the heat capacity or the thermal expansion coefficient during reheating of a glass. The temperature is somewhat dependent on the property measured and on the heating rate used in the measurement (Chap. 28). As a result, different studies will report slightly different values of T_g for supposedly identical glasses. Usually, the viscosity corresponding to T_g for common glasses has a value of about $10^{11.3}$ Pa s. A detailed discussion of viscosity-temperature as well as viscosity-composition relations can be found in a review on viscous flow and relaxation [11].

Vogel-Fulcher-Tammann Equation: A relatively good fit to viscosity data over the entire viscosity range is obtained by the Vogel-Fulcher-Tammann (VFT) equation

$$\eta = \eta_0 \exp\left(\frac{B}{T - T_0}\right), \qquad (30.2)$$

where η_0, B, and T_0 are constants. The VFT temperature T_0 for a given glass is always considerably lower than the value of T_g for that glass. While the VFT equation represents viscosity data over a wide temperature range quite well, it should be used with caution for temperatures at the lower end of the glass transformation region. It usually overestimates the viscosity in this regime. The VFT relation is in good agreement with experimental data above the transformation regime, but a theoretical justification of this equation is missing.

The viscosity can also be fitted, over limited temperature ranges, by an Arrhenius expression of the form

$$\eta = \eta_0(T) \exp\left[\frac{\Delta H_\eta(T)}{k_B T}\right], \qquad (30.3)$$

where $\eta_0(T)$ and $\Delta H_\eta(T)$ are pre-factor and activation enthalpy for viscous flow, respectively. The activation enthalpy for viscous flow is much lower for the fluid melt than for the high viscosity melt in the glass-transformation region. The temperature dependence between these limiting regions is decidedly non-Arrhenian; $\Delta H_\eta(T)$ and $\eta_0(T)$ decrease continually with increasing temperature.

Fragility of Melts: The degree of curvature of the Arrrhenius diagram of the viscosity of various melts can vary over wide ranges due to the variations in the value of T_0 relative to T_g. ANGELL has proposed to use this curvature as a basis for the classification of glass-forming melts [13]. Glasses which exhibit a near Arrhenian behaviour over their entire viscosity range (i.e. $T_0 \ll T_g$) are termed as *strong melts*, whilst those which exhibit a large degree of curvature are denoted as *fragile melts*. The concept of fragile-strong melt behaviour is summarised in a *fragility diagram* (Fig. 30.2). In general, strong melts have well-developed structural units like $SiO_{4/2}$ tetrahedra in silicate melts, at least partially covalent bonds, and only gradually dissociate with increasing temperature. Strong melts usually display only small changes in heat capacity upon passing through the glass transition region. Fragile melts are characterised by less well-defined short-range order and high configurational degeneracy. Their structures disintegrate rapidly with increases in temperature above T_g. Fragile melts are usually characterised by larger changes in the heat capacity at T_g. For example, conventional metallic

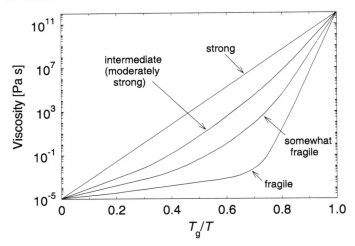

Fig. 30.2. Schematic fragility diagram for various melts

glasses usually have fragile melts, whereas bulk metallic glasses have strong melts (Chap. 29).

Stokes-Einstein Relation: The Irish scientist GEORGE STOKES (1819–1903) showed that a particle of radius r moving with a velocity v in a viscous medium experiences the frictional force $F = 6\pi r \eta v$. Assuming that the same relation applies to particles at the atomic scale yields via the Nernst-Einstein relation (see Chap. 11)

$$D_\eta = \frac{k_B T}{6\pi r \eta}. \tag{30.4}$$

This equation relates the viscosity of a fluid to some viscosity diffusion coefficient D_η. We use the index η to distinguish D_η from diffusion coefficients obtained from diffusivity measurements by the way of Fick's laws. Equation (30.4) was suggested for the first time by EINSTEIN [12] and is called the *Stokes-Einstein relation*. The viscosity diffusion coefficient and Fickian diffusion coefficients of the components of a glass melt in its supercooled state can be very different.

30.2 Experimental Methods

In principle, most of the methods described in Part II of this book for determining diffusivities can be used for studying glasses as well. The most reliable method is the radiotracer technique. As an example, Fig. 30.3 shows concentration depth profiles in borate glass after diffusing either ^{22}Na or ^{86}Rb from a thin layer deposited at the surface. The fitted thin-film solutions of

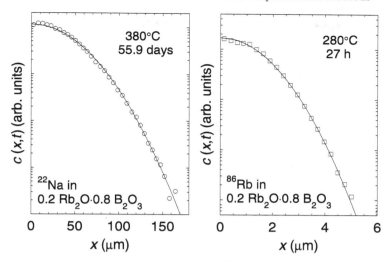

Fig. 30.3. Diffusion penetration profiles of ^{22}Na obtained by grinder sectioning (*left*) and of ^{86}Rb obtained by sputter sectioning (*right*) according to IMRE ET AL. [14]

Fick's second law confirm that diffusion coefficients can be obtained with good precision from such data.

Many oxide glasses are ionic conductors. Thus, a conductivity measurement is a powerful tool to deduce a conductivity diffusion coefficient of the species responsible for ion conduction (see below). Impedance spectroscopy is the most common method for conductivity measurements on ion-conducting solids (see Chap. 16). Examples of conductivity spectra measured by impedance spectroscopy are displayed in Fig. 30.4. This figure shows for a soda-lime silicate glass the real part of the complex conductivity as a function of the frequency ν for various temperatures. The plateau values at low frequencies represent dc conductivities σ_{dc}. The latter increase Arrhenius-activated with increasing temperature.

Typically, the monovalent cations are the most mobile species in oxide glasses followed by divalent cations, which are orders of magnitude slower (see below). For this reason, the majority of diffusion data on oxide glasses pertain to diffusion of monovalent modifier cations. If only one type of mobile cation is present in the glass, its diffusion coefficient can be deduced from σ_{dc} via the Nernst-Einstein equation (see Chap. 11), which we repeat here for convenience:

$$D_\sigma = \frac{k_\mathrm{B} \sigma_{dc} T}{N_{ion} q^2} . \qquad (30.5)$$

N_{ion} is the number density of mobile ions and q the charge of each ion. D_σ can be obtained from σ_{dc} if N_{ion} is known, say from the knowledge of the

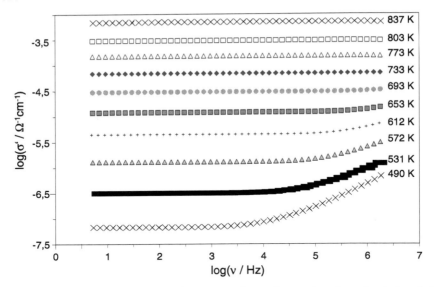

Fig. 30.4. Conductivity (real part) of a soda-lime silicate glass (standard glass I of DGG) *versus* frequency for various temperatures according to TANGUEP-NIJOKEP AND MEHRER [15]

molecular weight M and the mass density ρ using $N_{ion} = \rho N_A / M$, where N_A is the Avogadro number[2].

One needs to be careful in using the Nernst-Einstein relation, when several kind of ions contribute to the dc conductivity, unless the total conductivity can be shown to be dominated by one type of mobile ions. Also one has to distinguish D_σ from the tracer diffusion coefficient D^*. Both quantities are different and their ratio, $H_R = D^*/D_\sigma$, is termed the Haven ratio (see Chap. 11). The Haven ratio is usually smaller than unity due to correlation and collectivity effects in the atomic jump process.

Oxygen self-diffusion in glasses can be measured by means of the enriched stable isotope ^{18}O using $^{18}O/^{16}O$ isotopic exchange. In this procedure, samples of the glass are annealed in an ^{18}O atmosphere and the change in ^{18}O content is measured, for example, by secondary mass spectroscopy (SIMS). Determinations of the depth distribution of ^{18}O can also be done by nuclear reaction analysis (NRA). Then an energetic proton beam is used to induce the nuclear reaction $^{18}O(p,\alpha)^{15}N$ and the spectrum of the emitted α-particles is measured (Chap. 13). Nuclear magnetic relaxation (NMR) techniques and in particular measurements of the spin-lattice relaxation rate can be applied to self-diffusion studies of glasses as well (Chap. 15).

---

[2] The molecular formula of the glass should then be written in such a way that it contains one atom of the mobile ion.

Permeation of gases through glasses is of technological interest. If the diffusing species is a gas, it is possible to expose one face of a glass membrane (thickness Δx) to a known pressure of the gas, whilst the other side of the membrane is connected to a mass spectrometer. In one method, the permeating gas is continuously removed from the spectrometer side by pumping and maintaining a pressure difference Δp across the membrane and the steady state diffusion flow of gas, J, through the membrane is measured. From such experiments the permeability K can be determined via

$$K = \frac{J \Delta p}{\Delta x}. \tag{30.6}$$

If one assumes that the gas concentration in the glass, C, is given by *Henry's law*, $C = C_s p$, where C_s is the ambient pressure solubility and p the gas pressure, the permeability can be written as

$$K = DC_s, \tag{30.7}$$

where D is the diffusivity of the gas. Usually, diffusivity and solubility are both Arrhenius activated.

30.3 Gas Permeation

A number of gases permeate through glasses at rates which can have serious consequences for practical applications. Helium can readily permeate through many glasses used for vacuum tubes. Hydrogen permeation can result in coloration of glasses by the reduction of ions to a lower valency or to the metallic state and by the reaction with optically active defects. Oxygen permeating through the wall of an electric lamp can react with filament material, causing failure of the bulb.

Permeation rates of gases in vitreous silica are shown in Fig. 30.5. The data indicate that the permeability decreases as the atomic or molecular diameter of the diffusing species increases. The permeability decreases in the order He > H_2 > Ne > N_2 > O_2 > Ar > Kr. The trend for He, Ne, and hydrogen isotopes in all glasses is similar as in vitreous silica.

Permeation varies linearly with the partial pressure of the gas for pressures up to many atmospheres. The effect of glass composition on helium permeation has been studied for a variety of oxide glasses, including silicate, borate, germanate, and phosphate compositions. In general, He permeation decreases in silicate glasses with increasing modifier content. For example, the permeability of soda-lime silicate glasses, depending on the temperature, is two to four orders of magnitude lower than in pure vitreous silica (see, e.g., [4]). Helium permeates through vitreous silica most readily, since it is the most open-structured glass. The modifier ions occupy 'interstitial sites' of the network, thus blocking diffusion paths for He atoms.

In general, permeability data reflect the characteristics of 'interstitial spaces' in glasses. However, care is needed when the permeating species reacts with the glass network. For example, H_2 in vitreous silica reacts with the silica network to form hydroxyls and silan groups [16]. It has also been suggested that water diffuses in vitreous silica as molecules reacting with the network to form immobile hydroxyl ions.

30.4 Examples of Diffusion and Ionic Conduction

Below the glass-transition region, where the network structure is essentially rigid, self-diffusion of network formers is very slow. In comparison, self-diffusion of modifier cations is faster as they can move through the 'interstitial' channels of the network. Thus, it is the movement of modifier cations which determines many properties of the glass such as electrical conductivity, corrosion resistance, and dielectric break down. For this reason, the majority of diffusion data on oxide glasses refers to diffusion and ionic conduction of the modifier cations (see [6]). In what follows, diffusion and ion conduction of vitreous silica, soda-lime-glass, and borate glasses are used to illustrate typical features. In addition, the so-called mixed-alkali effect is described.

Vitreous Silica and Quartz: Vitreous silicon dioxide is an important technological material. It is the most refractory glass in commercial use and has high corrosion resistance, a low coefficient of thermal expansion, and good UV transparency. Apart from laboratory use, optical mirrors, high-efficiency

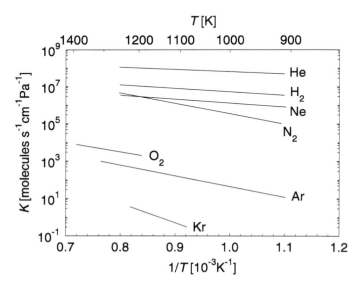

Fig. 30.5. Permeability of gases through vitreous SiO_2 according to SHELBY [2]

30.4 Examples of Diffusion and Ionic Conduction

lamps, optical fibers and dielectric films in microelectronic devices represent important applications (see also Chap. 28).

As already mentioned, vitreous SiO_2 as well as quartz crystals contain $SiO_{4/2}$ tetrahedra, which are linked together in three dimensions. In glassy silica, these tetrahedra form a random network, whereas in the crystal they are linked in an ordered fashion. Crystalline SiO_2 exists in many modifications as temperature and pressure varies. At ambient pressure trigonal low-quartz transforms around 575 °C to hexagonal high-quartz, which at 870 °C transforms to hexagonal trydimite. At 1470 °C cubic cristobalite is formed, which melts at about 1700 °C.

A comparison of diffusion in glassy and crystalline SiO_2 may be of special interest (Fig. 30.6). Diffusion of ^{22}Na in vitreous silica prepared from quartz[3] was studied between 170 and 1200 °C applying the residual activity

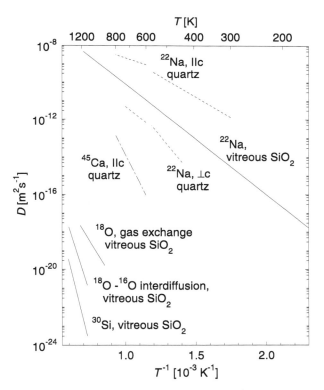

Fig. 30.6. Diffusion in vitreous silica and in quartz (for references see text)

---

[3] Despite its defined chemical composition one distinguishes in the literature different types of vitreous silica with respect to preparation method, raw material, and impurity content. These differences can lead to differences between the diffusion results of various groups. For example, glassy silica prepared from natural

method [17]. Diffusion of the stable isotope ^{30}Si has been measured in the temperature range 1110 to 1410 °C using SIMS [18]. Self-diffusion of oxygen was studied using a gas phase isotope exchange reaction [19]. SIMS has also been used to profile the interdiffusion of network oxygen in a vitreous $Si^{18}O_2 - Si^{16}O_2$ thin-film structure [20]. The diffusivity values are lower, but with a higher activation enthalpy (4.7 eV) than those reported in [19] and approach the diffusivity of network oxygen uncomplicated by gas phase exchange reactions. Figure 30.6 confirms that diffusion of the network former Si is very slow, whereas Na diffusion is relatively fast. The Si activation enthalpy of about 6 eV is close to the energy necessary to break Si-O bonds. The energy of a Si-O bond is about 2.9 eV [25]. For each $SiO_{4/2}$ tetrahedron the four half-bonds represent an energy of 5.8 eV. The main barrier for the movement of Si atoms seems to be indeed the Si-O bond energy. According to this reasoning, one could expect for oxygen diffusion an activation enthalpy of about half of that of Si diffusion. One experimental value of about 2.43 [19] seems to support this reasoning. However, other authors report values as low as 0.85 eV [26] and as high as 3.08 eV [27]. In view of this large scatter, it is likely that different diffusion mechanisms operate for oxygen and silicon.

Figure 30.6 also shows ^{22}Na diffusion in crystalline quartz parallel and perpendicular to its crystallographic axis [21–23] and ^{45}Ca diffusion in one direction [24]. The transition between high- and low-quartz at 575 °C influences Na diffusion. Since high-quartz has a hexagonal structure, which is of higher symmetry than the trigonal one of low-quartz, diffusion in high-quartz has a lower activation enthalpy. Figure 30.6 reveals a strong anisotropy of Na diffusion in quartz as well. Diffusion parallel to the axis is much faster than perpendicular to it. It is also remarkable that Na diffusion in vitreous silica lies between the Na diffusivities parallel and perpendicular to the crystallographic axis of low- and of high-quartz. Na and Ca have nearly the same ionic radii. Nevertheless, the tracer diffusivity of Ca is (at 600 °C) almost seven orders of magnitude lower than that of Na. This reflects the stronger linkage of Ca to the glass network.

Soda-Lime Silicate Glass: Silicate glasses form the largest class of oxide glasses (see Chap. 28). Most of them are used as window and container glasses. Soda-lime glasses are mainly ternary glasses often with some further minor additions. They usually contain about 10 to 20 mol % alkali oxides, primarily in the form of Na_2O, 5 to 15 mol % CaO and 70 to 75 mol % SiO_2. Use of dolomite as a source of CaO often implies that considerable MgO is also present in the glass. For special purposes some of the soda is replaced by K_2O or, less commonly, by Li_2O. Replacement of CaO and/or MgO by SrO and BaO occurs occasionally in the production of glasses.

---

quartz crystals either by electric melting or by plasma sputtering in a H_2 and O_2 plasma reveal nearly the same Na diffusivites, whereas glasses synthesised from $SiCl_4$ display a lower Na diffusivity presumably due to a distinctly lower content of hydroxyl groups [5].

30.4 Examples of Diffusion and Ionic Conduction   533

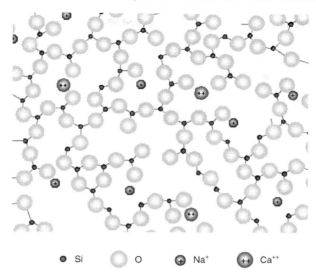

Fig. 30.7. Structure of a soda-lime silicate glass (schematic in two dimensions)

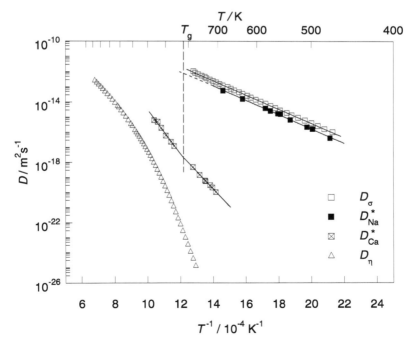

Fig. 30.8. Viscosity diffusion coefficient, D_η, tracer diffusivities, D^*_{Na}, D^*_{Ca}, and charge diffusion coefficient D_σ, of soda-lime silicate glass (standard glass I of DGG) according to TANGUEP-NIJOKEP AND MEHRER [15]

The structure of a soda-lime silicate glass (Fig. 30.7) is readily described by the structural rules of Zachariasen. The silicon-oxygen tetrahedron, with a coordination number of 4, serves as the basic building block. If modifier cations are introduced – by melting SiO_2, Na_2O, and CaO to form a soda-lime glass – some Si-O-Si bridges are broken. Oxygen atoms occupy free ends of separated tetrahedra thus forming non-bridging oxygen (NBO) units. The NBO units are the anionic counterparts of the alkali- or alkaline-earth ions. The cations (Na^+ or Ca^{2+}) are mainly incorporated at the severance sites of the network. Usually, every alkali ion has a neighbouring NBO, while every alkaline-earth ion has two neighbouring NBO units. This structure provides stronger network linkage at the alkaline-earth sites. Thus the divalent alkaline earth ions are less mobile than the monovalent alkali ions. The replacement – of alkali ions by alkaline-earth ions – reduces the ionic contributions to the electrical conductivity and improves the chemical durability of the glass.

Figure 30.8 and 30.9 illustrate mass transport properties in two similar soda-lime silicate glasses produced by the Deutsche Glastechnische Gesellschaft (DGG) as standard glass I and II for physical and chemical testing. The composition of standard glass I (in mole fractions) is: 71.8 %

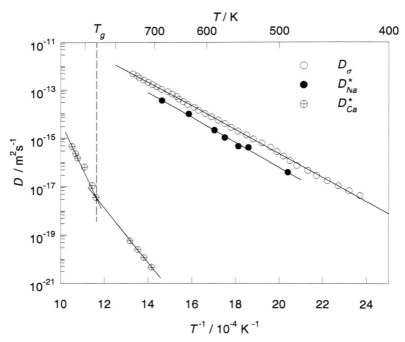

Fig. 30.9. Tracer diffusivities, D^*_{Na}, D^*_{Ca}, and charge diffusion coefficient, D_σ, of soda-lime silicate glass (standard glass II of DGG) according to TANGUEP-NIJOKEP AND MEHRER [15]

SiO_2, 14.52 % Na_2O, 7.22 % CaO, 6.24 % MgO, and some minor additions. The composition of standard glass II is: 71.37 % SiO_2, 13.19 % Na_2O, 10.43 % CaO, 5.01 % MgO, and some minor addition. Both glasses differ mainly in their content of alkaline-earth oxides. For standard glass I viscosity data for the undercooled melt are available from measurements performed at the Physikalisch Technische Bundesanstalt (Braunschweig, Germany). These data have been used to calculate the viscosity diffusion coefficient, D_η, from the Stokes-Einstein relation Eq. (30.4) using 0.042 nm for the ionic radius of Si (Fig. 30.8). D_η can be readily described by Vogel-Fulcher-Tammann behaviour. Also shown are tracer diffusivities of ^{22}Na and ^{45}Ca and the charge diffusion coefficient, D_σ, measured in the glassy state, which are all well represented by Arrhenius relations. At the glass-transition temperature, Ca diffusion is 6 orders of magnitude slower than Na diffusion and at lower temperatures the difference is even larger. This confirms the expectation that divalent Ca ions have a much stronger linkage to the network than Na ions. In addition, this large difference together with the fact that conductivity diffusion and Na tracer diffusion have the same activation enthalpy show that the electrical conductivity of soda-lime silicate glasses is due to the motion of Na ions.

Figure 30.10 shows the Haven ratios, $H_R = D^*_{Na}/D_\sigma$, of both standard glasses based on the assumption that only Na ions are mobile. The Haven ratios are: $H_R = 0.45$ for standard glass I and $H_R = 0.33$ for standard glass II. Both Haven ratios are temperature-independent within the experimental errors, indicating that the mechanism of Na diffusion does not change with temperature.

Alkali Borate Glasses: The structure of vitreous boric oxide (B_2O_3) differs considerably from that of vitreous silica. Although boron occurs in triangular as well as tetrahedral coordination in crystalline compounds, only the triangular state is formed in vitreous boric oxide. The $BO_{3/2}$ units are connected at all three corners via B-O-B bonds to form a network. It is also believed that vitreous boric oxide contains a certain amount of so-called boroxol groups consisting of three boron-oxygen triangles joined together. In contrast to vitreous silica, the basic building block of the vitreous boron oxide network is planar rather than three-dimensional. A three-dimensional structure is obtained by 'crumpling' the network. Since the primary bonds exist only within a plane, bonds in a third dimension are weak and the structure is easily disrupted. One consequence of this weakly bound structure is the low glass-transition temperature of vitreous boric oxide (about 260 °C), which is much lower than that of vitreous silica (about 1100 °C).

The arrangement of atoms (or ions) in an alkali borate glass is illustrated in Fig. 30.11. Whereas addition of alkali oxides to vitreous silica results in the formation of NBO units (see above), the effect of alkali-oxide addition to boric oxide cannot be explained on the basis of NBO formation. The addition of alkali oxide forces some of the boron to change from trigonal to tetrahedral configuration. Formation of two boron-oxygen tetrahedra consumes the additional oxygen provided by one alkali oxide molecule. If alkali ions are introduced into the trigonally coordinated network of vitreous B_2O_3, tetrahedrally coordinated $BO_{4/2}^-$ units are formed, which are the anionic counterparts of the alkali ions. Each Na_2O (or Rb_2O) molecule creates two $BO_{4/2}^-$ units. Only at concentrations larger than about 25 mol % alkali oxide, non-bridging oxygens appear. As evidenced by the glass-transition temperatures, the addition of alkali oxides enhances the stability of the glassy borates considerably at least below 25 mol % alkali content (Fig. 30.12).

Figure 30.13 shows an Arrhenius diagram of the dc conductivity (times temperature) of sodium borate glasses [28]. The conductivity is Arrhenius activated and increases many orders of magnitude when the alkali content increases from 4 to 30 mol %. In ion conducting glasses, the conductivity is determined by the number density of mobile ions and by their mobility. As a result, glasses which contain significant concentrations of monovalent ions are poor insulators, while glasses that are free of monovalent ions are excellent insulators. Figure 30.14 shows the effect of Li_2O, Na_2O, K_2O, and

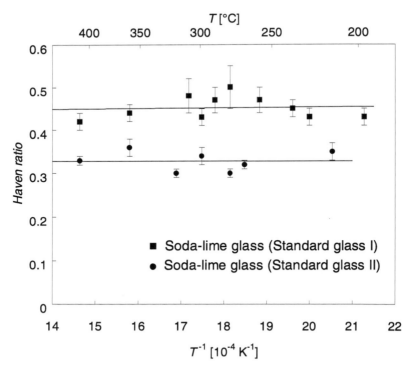

Fig. 30.10. Haven ratios of soda-lime silicate glasses according to TANGUEP-NIJOKEP AND MEHRER [15]

30.4 Examples of Diffusion and Ionic Conduction

Rb_2O additions on the conductivity of borate glass. Whereas the conductivity increases 5 to 6 orders of magnitude, the alkali content varies much less. This indicates that the mobility of ions increases significantly. The latter conclusion is supported by Na tracer diffusion studies, which show a very similar increase with Na_2O content [30]. In Fig. 30.14 also a decrease of conductivity for corresponding glasses containing the same alkali concentrations is observed in the order of increasing ionic radii: Li > Na > K > Rb. The smallest alkali ion entails the highest conductivity.

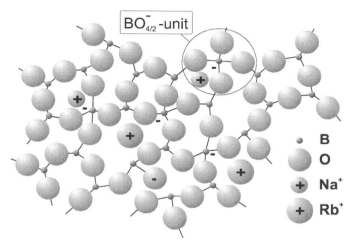

Fig. 30.11. Structure of sodium-rubidium borate glass (schematic in two dimensions)

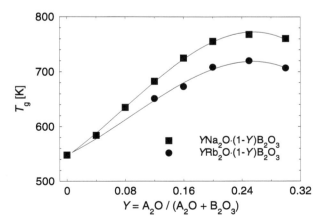

Fig. 30.12. Glass-transition temperatures of alkali borate glasses according to BERKEMEIER ET AL. [28]

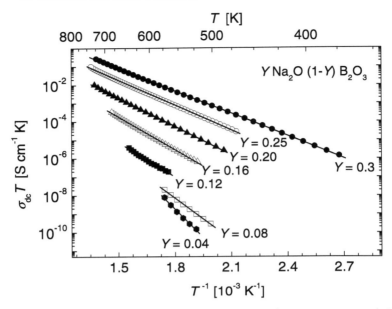

Fig. 30.13. Arrhenius diagram of the dc conductivity (times temperature) for Y Na$_2$O (1-Y)B$_2$O$_3$ glasses according to BERKEMEIER ET AL. [28]

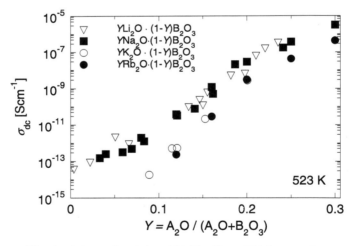

Fig. 30.14. Electrical dc conductivity of Li, Na, K, and Rb borate glasses according to BERKEMEIER ET AL. [28]

Mixed-Alkali Effect: Glasses containing two or more alkali oxides display the so-called mixed-alkali effect, which is one of the old but still very interesting features of ionic conduction and diffusion in glass. Figure 30.15 shows as a typical example the conductivity diffusion coefficient of a sodium-

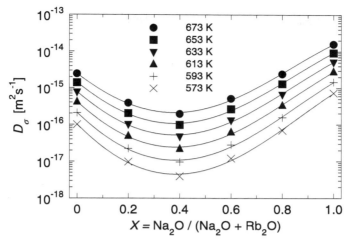

Fig. 30.15. Charge diffusion coefficient D_σ of mixed 0.2 [X Na$_2$O (1-X)Rb$_2$O] 0.8 B$_2$O$_3$ glasses according to IMRE ET AL. [29]

rubidium borate glass system according to IMRE ET AL.[29]. Further examples can be found, e.g., in a paper by GAO AND CRAMER [31]. If sodium ions are gradually replaced by rubidium ions the conductivity in Fig. 30.15 does not follow a linear mixing rule between the end-members. Instead, it passes through a deep minimum, which for this particular glass system is located near $X = 0.4$. Such conductivity minima are the best-known fingerprints of the mixed-alkali effect, which has been observed for many other mixed-alkali glasses. The depth of the mixed-alkali minimum decreases with increasing temperature. As a consequence the activation enthalpy ΔH of D_σ passes through a maximum for an intermediate composition (Fig. 30.16). Furthermore, the mixed-alkali effect decreases with decreasing total alkali content and vanishes for low total alkali contents [30].

In mixed-alkali silicate glasses containing 30 mol % alkali oxide the conductivity departs by as much as a factor of 10^3 to 10^6 from a linear mixing rule of the end compositions [32]. Besides dramatic departures from linearity in the conductivity, other transport properties of mixed-alkali glasses, such as tracer diffusion, viscosity, and internal friction display characteristic features of the mixed-alkali effect.

Of particular interest is self-diffusion of the alkali ions in a mixed-alkali glass. Systematic studies of tracer self-diffusion in mixed-alkali glasses are relatively rare [8], because tracer studies are very laborious and time-consuming. Figure 30.17 shows the tracer diffusivities of ^{22}Na and ^{86}Rb in sodium-rubidium borate glasses [14]. This figure reveals further typical aspects of mixed-alkali behaviour:

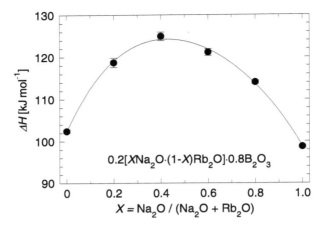

Fig. 30.16. Composition dependence of the activation enthalpy for conductivity diffusion of Fig. 30.15 [29]

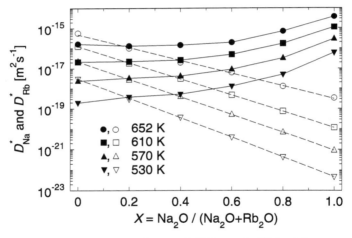

Fig. 30.17. Composition dependence of ^{22}Na and ^{86}Rb diffusion in mixed $0.2[X\,Na_2O(1-X)Rb_2O]0.8\,B_2O_3$ glasses according to IMRE ET AL. [14]. Na diffusion: *full symbols*; Rb diffusion: *open symbols*

1. The tracer diffusivity of the majority ion is higher than that of the minority ion regardless of the size relationship of the ions. On the other hand, this difference is much more pronounced if the minority ion is the larger ion. The Rb diffusivity on the Na-rich side is about 4 orders of magnitude lower than the Na diffusivity (Fig. 30.17).
2. The diffusivities of the two ions cross, when plotted as functions of the mixed-alkali composition. The crossover occurs usually at non-equiatomic

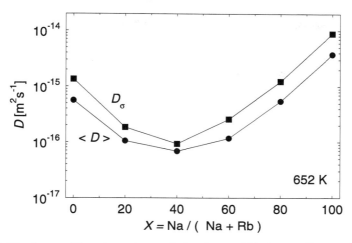

Fig. 30.18. Composition dependence of the charge diffusion coefficient, D_σ, and of the mean tracer diffusion coefficient, $\langle D \rangle$, of mixed Na-Rb borate glasses $0.2[\text{X Na}_2\text{O}\ (1\text{-X})\text{Rb}_2\text{O}]0.8\ \text{B}_2\text{O}_3$

compositions and the crossover composition is almost independent of temperature.

3. Similar observations were reported for Na-Cs silicate glasses by TERAI [33] and JAIN ET AL. [34], for Na-K silicate glasses by ESTROPIEV [35] and FLEMING AND DAY [38], for Na-Rb germanate glasses by ESTROPIEV [36], and for Na-Rb silicate glasses by MCVAY AND DAY [37].

4. One may define a hypothetical 'mean tracer diffusion coefficient'. For a Na-Rb borate glass it is given by

$$\langle D \rangle = X D^*_{Na} + (1-X) D^*_{Rb}, \tag{30.8}$$

where D^*_{Na} and D^*_{Rb} are the tracer diffusivities of Na an Rb ions, respectively. The as-calculated values of $\langle D \rangle$, due to the crossover in the diffusivities, produce a minimum near the minimum of D_σ (Fig. 30.18). One may also consider a 'common Haven ratio' defined via $\langle H \rangle = \langle D \rangle / D_\sigma$. From Fig. 30.18 it is obvious that the common Haven ratio passes through a maximum at an intermediate composition.

Understanding the mixed-alkali effect is one of the longstanding challenges in glass science [39, 40, 7]. It has been the subject of many studies, with several suggested explanations [41–47]. In the author's opinion the understanding of the trend in the tracer diffusivities is the key to this puzzle. Any proposed model of the mixed-alkali effect must somehow involve strong interactions between different alkalis such that a large number of majority alkali ions are immobilised or their mobility is at least considerably reduced, when minority ions are added. A widely accepted explanation of the mixed-alkali effect has not yet been put forward. Recent progress is summarised in reviews by DIETERICH AND MAASS [48] and by BUNDE ET AL. [49].

References

1. W. Vogel, *Glass Chemistry*, Springer-Verlag, 1985
2. J.E. Shelby, *Introduction to Glass Science and Technology*, The Royal Society of Chemistry, Cambridge, RSC paperbacks, 1997
3. R.H. Doremus, *Glass Science*, Wiley, New York, 1994
4. A.K. Varshneya, *Fundamentals of Inorganic Glasses*, Academic Press, Inc., 1994
5. G.H. Frischat, *Ionic Diffusion in Oxide Glasses*, Trans Tech Publication, Aedermannsdorf, Diffusion and Defect Monograph Series, 1975
6. H. Jain, C.H. Hsieh, *Diffusion in Oxide Glasses*, in: *Diffusion in Semiconductors and Non-Metallic Solids*, D.L. Beke (Vol. Ed.), Landolt-Börnstein, New Series, Group III: Condensed Mattter, Vol. 33, Subvolume B1, Springer-Verlag, 1999
7. M.D. Ingram, Phys. Chem. Glasses **28**, 215 (1987)
8. V.M. Goldschmidt, *Geochemische Verteilungsgesetze der Elemente*, Skr. Nor. Vidensk Akad. K1, 1; Mat. Naturvidensk. K1, 8. 7 (1926)
9. W.J. Zachariasen, J. Am. Ceram. Soc. **54**, 3841 (1932)
10. B.E. Warren, Z. Kristallogr. Mineral. Petrogr. **86**, 349 (1933)
11. D. Uhlmann, N. Kreidl, *Viscous Flow and Relaxation*, in: *Glass Science and Technology*, D. Uhlmann, N. Kreidl (Eds.), Academic Press, New York, 1985
12. A. Einstein, Z. f. Elektrochemie **17**, 235 (1908)
13. C.A. Angell, J. Non-Cryst. Solids **102**, 205 (1988)
14. A.W. Imre, S. Voss, H. Mehrer, Defect and Diffusion Forum **237–240**, 370 (2005)
15. E.M. Tanguep-Nijokep, H. Mehrer, Solid State Ionics **177**, 2839 (2006)
16. J.E. Shelby, in: *Treat. on Mat. Sci. and Tech.*, Vol. 17, M. Tomozawa, R.E. Doremus (Eds.), p.1, Academic Press, New York, 1979
17. G.H. Frischat, J. Amer. Ceram. Soc. **52**, 625 (1968)
18. G. Brebec, R. Seguin, C. Sella, J. Bevenot, J.C. Martin, Acta Metall. **26**, 327 (1980)
19. R. Haul, G. Dümbgen, Z. Elektrochemie **66**, 636 (1962)
20. J.C. Mikkelsen, Jr., Appl. Phys. Lett. **45**, 1187 (1984)
21. G.H. Frischat, Phys. Stat. Sol. **35**, K47 (1969)
22. G.H. Frischat, Ber. Dt. Keram. Ges. **47**, 238 (1970); and **47**, 313 (1970)
23. G.H. Frischat, J. Amer. Ceram. Soc. **53**, 357 (1970)
24. G.H. Frischat, Ber. Dt. Keram. Ges. **47**, 364 (1970);
25. H.F. Wolf, *Semiconductors*, Wiley Interscience, 1971
26. K. Mühlenbach, H.A. Schäffer, Can. Mineralogist **15**, 179 (1977)
27. E.W. Sucov, J. Amer. Ceram. Soc. **46**, 14 (1963)
28. F. Berkemeier, S. Voss, A.W. Imre, H. Mehrer, J. Non-Cryst. Sol. **351**, 3816 (2005)
29. A.W. Imre, S. Voss, H. Mehrer, Phys. Chem. Chem. Phys. **4**, 3219 (2002)
30. S. Voss, F. Berkemeier, A.W. Imre, H. Mehrer, Z. Phys. Chem. **218**, 1353 (2004)

31. Y. Gao, C. Cramer, Solid State Ionics **176**, 921 (2005)
32. R.J. Charles, J. Amer. Ceram. Soc. **48**, 432 (1965)
33. R. Terai, J. Non-Cryst. Sol. **5**, 121 (1971)
34. H. Jain, N.L. Peterson, H.L. Downing, J. Non-Cryst. Sol. **55**, 283 (1985)
35. K.K. Estrop'ev, in: *Structure of Glass*, E.B. Uvarov (Ed.), Vol. 2 (Consultant Bureau, New York, 1960)
36. K.K. Estrop'ev, V.K. Pavlovskii, in: *Structure of Glass*, E.A. Porai-Koshits (Ed.), Vol. 7 (Consultant Bureau, New York, 1966)
37. G.L. McVay, D.E. Day, J. Am. Ceram. Soc. **53**, 508 (1970)
38. J.W. Fleming, D.E. Day, J. Am. Ceram. Soc. **55**, 186 (1972)
39. J.O. Isard, J. Non-Cryst. Sol. **1**, 235 (1969)
40. D.E. Day, J. Non-Cryst. Sol. **21**, 343 (1969)
41. J.R. Hendriksen, P.J. Bray, Phys. Chem. Glasses **14**, 43 and 107 (1972)
42. G. Tomandl, H. Schaeffer, J. Non-Cryst. Sol. **73**, 179 (1985)
43. W.C. LaCourse, J. Non-Cryst. Sol. **95–96**, 905 (1987)
44. C.T. Moynihan, A.V. Lesikar, J. Amer. Ceram. Soc. **64**, 40 (1981)
45. M. Tomozawa, V. McGahay, J. Non-Cryst. Sol. **128**, 48 (1991)
46. A. Bunde, M.D. Ingram, P. Maass, J. Non-Cryst. Sol. **131–133**, 1109 (1991)
47. R. Kirchheim, J. Non-Crystalline Solids **272**, 84 (2000)
48. W. Dieterich, P. Maass, Chem. Physics **284** 439 (2002)
49. A. Bunde, W. Dieterich, P. Maass, M. Meyer, *Ionic Transport in Disordered Materials*, Ch. 20 in: *Diffusion in Condensed Matter – Methods, Materials, Models*, P. Heitjans, J. Kärger (Eds.), Springer-Verlag, 2005

Part VII

Diffusion along High-Diffusivity Paths and in Nanomaterials

31 High-diffusivity Paths in Metals

31.1 General Remarks

In contrast to gases and liquids, crystalline solids exhibit several structurally different paths by which atomic diffusion can take place. In the preceding parts of this book we have considered lattice diffusion (also denoted as bulk diffusion). The defects which aided diffusion through the crystal were atomic defects such as vacancies or interstitials. Grain boundaries, dislocations and free surfaces entered only to help attain the equilibrium concentration of point defects. For metals it has long been known [1] that the jump rates of atoms along dislocations, grain boundaries, and at free surfaces are much higher than in the lattice[1]. Because the diffusivity is high in these regions, the terms *high-diffusivity paths* or *diffusion short circuits* were coined to describe this very fast diffusion phenomena illustrated in Fig. 31.1. Lattice (or bulk) diffusion is characterised by its diffusion length $\sqrt{Dt}$. The deep penetrating diffusion fringes near the free surface, around the grain-boundary, and the dislocation line illustrate the effects of high-diffusivity paths.

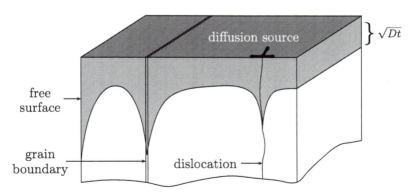

Fig. 31.1. Schematic illustration of high-diffusivity paths in a solid

---

[1] In ionic solids and ceramics, a space charge layer can form around extended defects and at surfaces, which can modify the picture developed here for metals.

High-diffusivity paths are of interest for several reasons. First, there is the question what is their influence on measurements of lattice diffusion coefficients. Second, with properly designed experiments it is possible to determine the diffusivity along diffusion short-circuits, to deduce their characteristic parameters, to learn more about their structure, and how atoms move in them. Furthermore, there are kinetic processes which are limited by high-diffusivity paths. For example, grain-boundary diffusion in polycrystalline materials plays a key rôle at temperatures below 0.6 T_m (T_m = melting temperature). Examples of processes controlled by grain-boundary diffusion are diffusional (Coble) creep, discontinuous precipitation, diffusion induced grain-boundary migration, recrystallisation, and sintering.

Diffusion short-circuits are important also in various areas of modern technology. Many structural changes in thin-film microelectronic, optoelectronic, and magnetic storage devices are controlled by short-circuit diffusion. These devices are often based on multilayer thin-film structures with film thicknesses comparable to the diffusion distances at the operating temperatures. The lifetime as well as the efficient performance of such devices depend largely on the physical integrity of the thin-film structures. Due to their high density of dislocations and grain boundaries and due to the possibility of large composition and/or stress gradients, thin-film structures are highly vulnerable. Because of the high density of grain boundaries in polycrystalline thin films, grain-boundary diffusion is often the dominating transport process at the temperatures of device operation. Detrimental effects and even device failure can be the result. Diffusion-failure of devices may also occur due to intermixing and compound formation mainly via grain-boundary and dislocation pipe diffusion between different layers. The efficiency of diffusion barriers used to prevent undesirable intermixing between layers of a thin-film device is also influenced by the short-circuit diffusion characteristics of the barrier layer. Therefore, an understanding and controlling of these processes is important to ensure the integrity and to improve the stability of thin-film devices [2].

Grain-boundary and dislocation-pipe diffusion is considered in detail in Chaps. 32 and 33. In the present chapter, we discuss some general features of high-diffusivity paths. We illustrate the spectrum of various diffusivities that can occur in crystalline solids and we mention empirical rules for grain-boundary diffusion. Finally, we address the question how the influence of high-diffusivity paths on the measurement of lattice diffusion coefficients can be avoided.

31.2 Diffusion Spectrum

Lattice diffusion represents the most severe constraint to atomic migration, leading to the lowest diffusivities and the highest activation enthalpies in a given material. Due to their distorted structure, dislocation cores have

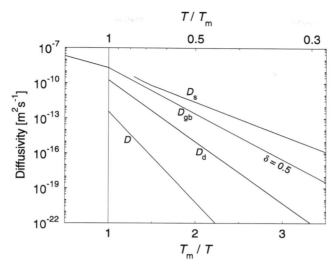

Fig. 31.2. Schematic illustration of the diffusion spectrum for metals in a reduced temperature scale; T_m denotes the melting temperature

smaller constraints for atomic motion than the lattice. High-angle grain boundaries with their less densely packed structure enable fast atomic diffusion. Surfaces offer the least constraints to the motion of diffusing defects and as a result, both adatoms and vacancies can promote the diffusion process. The diffusion spectrum for metals [1] is shown in Fig. 31.2. The diffusion data represent typical averages for a variety of metals and for the various high-diffusivity paths.

Experience has revealed the following hierarchy between lattice (or bulk) diffusivity D, dislocation pipe diffusivity D_d, grain-boundary diffusivity D_{gb}, and the diffusivity at surfaces D_s:

$$D \ll D_d \leq D_{gb} \leq D_s. \tag{31.1}$$

For the corresponding activation enthalpies the following inequalities hold:

$$\Delta H > \Delta H_d \geq \Delta H_{gb} > \Delta H_s. \tag{31.2}$$

Grain boundaries and dislocations are the most frequently encountered internal short circuits. They can be treated by similar phenomenological approaches as discussed in Chaps. 32 and 33.

31.3 Empirical Rules for Grain-Boundary Diffusion

Grain-boundaries provide the most prominent high-diffusivity paths. The grain-boundary diffusivity usually follows an Arrhenius type of temperature dependence

$$D_{gb} = D_{gb}^0 \exp\left(-\frac{\Delta H_{gb}}{k_B T}\right), \qquad (31.3)$$

where ΔH_{gb} is the activation enthalpy and D_{gb}^0 the pre-exponential factor of grain-boundary diffusion.

Typically, grain-boundary diffusion in metals is four to six orders of magnitude faster than lattice diffusion. The difference increases with decreasing temperature due to the smaller activation enthalpy of grain-boundary diffusion as compared to lattice diffusion. The pre-exponential factors D_{gb}^0 are not very much different from those of lattice diffusion. As illustrated in Fig. 31.2, the grain-boundary diffusivity approaches a common value for all metals of about 10^{-9} to 3×10^{-9} m^2 s^{-1} near the melting temperature. Even then the grain-boundary diffusivity is significantly larger than the diffusivity in the lattice. The grain-boundary diffusivity in metals near T_m is comparable to self-diffusion in liquid metals [7].

A large body of grain- and interphase boundary diffusion data for metals has been accumulated by GUST AND ASSOCIATES [5, 6]. As for lattice diffusion, self-diffusion is the most basic diffusion process for grain boundaries. For metals, the ratios of activation enthalpies of lattice and grain-boundary self-diffusion, $\Delta H_{gb}/\Delta H$, lie between 0.4 and 0.6. Table 31.1 summarises empirical correlations between grain-boundary self-diffusion properties for fcc, bcc, and hcp metals derived either by BROWN AND ASHBY [8] or by GUST ET AL. [9]. The physical justification of such correlations is rather obscure. At the atomic level there is no obvious link between the diffusion jump and the process of melting. Nevertheless, such empirical rules do work and are useful as a guide for the systematics of grain-boundary data. The obvious value of such rules is that they allow diffusion rates to be estimated for solids, for which little or no data are available. They should be used with a clear appreciation of the possible errors involved; but in some instances these are small. For lattice diffusion similar empirical correlations are described in Chap. 8.

Table 31.1. Empirical correlation between grain-boundary self-diffusion for fcc, bcc, and hcp metals and the melting temperature T_m. Left: according to BROWN AND ASHBY [8]. Right: according to GUST ET AL. [9]. For the grain-boundary width a value of $\delta = 0.5$ nm was assumed

Structure	D_{gb}^0 [m^2 s^{-1}]	ΔH_{gb} [J mol^{-1}]	D_{gb}^0 [m^2 s^{-1}]	ΔH_{gb} [J mol^{-1}]
fcc	1.89×10^{-5}	$83.0 \times T_m/\text{K}$	1.94×10^{-5}	$74.4 \times T_m/\text{K}$
bcc	0.67×10^{-3}	$97.6 \times T_m/\text{K}$	1.84×10^{-5}	$86.7 \times T_m/\text{K}$
hcp	0.55×10^{-4}	$89.8 \times T_m/\text{K}$	0.3×10^{-4}	$85.4 \times T_m/\text{K}$

31.4 Lattice Diffusion and Microstructural Defects

In experimental studies focused on lattice diffusion the influence of high-diffusivity paths is unwanted. If possible, it should be completely eliminated or at least reduced to a tolerable extent. Two major factors determine the quality of diffusion measurements, the method used and the extent to which the sample material is characterised. The importance of specifying the material cannot be overstated. The measured diffusivity depends apart from the chemistry of the sample on its microstructure. Dislocations and grain boundaries can lead to apparent values of the diffusivity that are usually larger than the true lattice value.

The most accurate method for diffusion studies is the radiotracer sectioning technique [3], described in Chap. 13. The tracer – initially deposited at the front face of the sample, may rapidly reach the side surfaces by surface diffusion or by evaporation and re-deposition and then diffuse inward (Fig. 31.1). Surface diffusion fringes along the side surfaces can be eliminated in careful tracer studies of lattice diffusion studies. To keep the diffusion one-dimensional, one removes about $6\sqrt{Dt}$ from the side edges of the sample. In this way, an influence of lateral diffusion fringes from the free surfaces of the sample is avoided.

The use of single crystals avoids grain-boundary diffusion. For polycrystals, distortion by 'grain-boundary tails' is one of the best known reasons for deviations from Gaussian or error function type penetration profiles expected for pure lattice diffusion. Of course, the influence of grain-boundary diffusion is smaller in coarse-grained than in fine-grained polycrystals. If single crystals are not available, polycrystals with large grain size should be used in lattice diffusion studies. Criteria for separating bulk and grain-boundary diffusion given by HARRISON [4] are discussed in Chap. 32. The condition of well-specified material also implies that its microstructure remains unchanged during the diffusion anneal. Grain growth in polycrystalline samples must be avoided by pre-annealing treatments at high enough temperatures.

Dislocations are almost inevitably present even in most single crystals. Extended pre-annealing at temperatures not much below the melting temperature T_m of the material can reduce the dislocation density. However, it will usually not remove all dislocations. Well-annealed single crystals of metals still contain about 10^{10} dislocations per m^2. This value corresponds to an average distance between dislocations of about 10 µm. A modification of Harrison's criteria [4] permits a separation of lattice and dislocation pipe diffusion. Usually dislocations densities of about 10^{10} m^{-2} can be tolerated if diffusion experiments are performed at temperatures above about $0.6 T_m$.

For some materials, single crystals can be grown practically dislocation-free due to elaborate crystal-growth techniques. The most prominent example are silicon single crystals. Silicon wafers cut from dislocation-free single crystals provide the base material for most microelectronic devices. Dislocations and grain boundaries usually act as sinks and sources for vacancies and self-interstitials. In this way dislocations help attain the equilibrium concentration of point defects. Point-defect equilibrium in metals is established in very short time intervals (typically several ms) compared to annealing times typical of diffusion experiments. In dislocation-free single crystals the free surfaces are the only sources and sinks for intrinsic point defects. As a consequence, super- or undersaturations of point defects can occur and have an influence on the diffusion kinetics. Consequences therefrom are discussed in the Chaps. 24 and 25 devoted to diffusion in semiconductors.

References

1. N A. Gjostein, *Short Circuit Diffusion*, In: *Diffusion*, H.I. Aaronson (Ed.), Am. Soc. for Metals, Metals Park, Ohio (1973)
2. S.P. Murarka, *Diffusion Barriers in Semiconductor Devices/Circuits*, in: *Diffusion Processes in Advanced Technological Materials*, D. Gupta (Ed.), William Andrew, Inc. 2005
3. S.J. Rothman, *The Measurement of Tracer Diffusion Coefficients in Solids*, Ch. 1 in: *Diffusion in Crystalline Solids*, G.E. Murch, A.S. Nowick (Eds.), Academic Press, Inc. 1984
4. L.G. Harrison, Trans. Farad. Soc. **57**,1191 (1961)
5. I. Kaur, W. Gust, *Grain and Interphase Boundary Diffusion*, Chap. 12 in: *Diffusion in Solid Metals and Alloys*, H. Mehrer (Vol. Ed.), Landolt-Börnstein, Numerical Data and Functional Relationships in Science and Technology, New Series Vol. III/26, Springer-Verlag, 1990. p.630
6. I. Kaur, W. Gust, L. Kozma, in two volumes: *Handbook of Grain and Interphase Boundary Diffusion Data*, Ziegler Press, Stuttgart, 1989
7. I. Kaur, Y. Mishin, W. Gust, *Fundamentals of Grain and Interphase Boundary Diffusion*, John Wiley and Sons, Ltd., 1995
8. A.M. Brown, M.F. Ashby, Acta Metall. **28**, 1085 (1980)
9. W. Gust, S. Mayer, A. Bögel, B. Predel, J. Physique **46** (C4), 537 (1985).

32 Grain-Boundary Diffusion

32.1 General Remarks

Diffusion along grain boundaries (grain-boundary diffusion) is a transport phenomenon of fundamental and technological importance. The fact that grain boundaries in metals provide high-diffusivity paths was already known in the 1930s mostly from indirect evidence. By 1950, the fast grain-boundary diffusion was well documented by autoradiographic images [1], from which the ratio between diffusivity along grain-boundaries and in the lattice was estimated by LE CLAIRE to be several orders of magnitude [2].

At about the same time, HOFFMAN AND TURNBULL in the metallurgy group at the General Electric Research Laboratory applied the radiotracer method (see Chap. 13) to study self-diffusion in silver [3]. They investigated diffusion of radioactive ^{110m}Ag in silver mono- and polycrystals, sectioned the samples into thin layers parallel to the source surface and measured the specific activity per layer, which is proportional to the concentration of the isotope[1]. In that way they obtained accurate concentration-depth curves. For monocrystals they deduced lattice self-diffusion coefficients by fitting the thin-film solution of Fick's second law to the experimental profiles. In contrast, the profiles measured on polycrystals clearly revealed long penetration 'tails' which were correctly attributed to the effect of diffusion along grain boundaries . For the quantitative description of profiles with 'tails' a theoretical model describing the coupled grain-boundary and lattice diffusion in a polycrystal was developed by J.C. FISHER [7], another member of the same laboratory. His work resulted in what is nowadays known as the 'FISHER model', a model which became a cornerstone of grain-boundary diffusion.

During the following decades the techniques for grain-boundary diffusion experiments have been considerably improved and extended to a wider temperature range and to a broad spectrum of materials. On the theoretical side, the Fisher model has been subject to careful mathematical analysis and extended to new situations encountered in diffusion measurements. For

---

[1] Already in the pioneering work of ROBERTS-AUSTEN on diffusion of gold in lead [4] sectioning experiments had been undertaken in combination with chemical assaying techniques. In the 1920s, VON HEVESY [5, 6] introduced the radiotracer sectioning technique to study solid-state diffusion (see Chap. 1).

a broad overview of both fundamentals and recent achievements the reader is referred to the textbook on grain-boundary diffusion by KAUR, MISHIN, AND GUST [8]. A collection of experimental data can be found in [9] for metals and alloys and for non-metals in [10]. Progress in the area of grain-boundary diffusion is summarised in several reviews [11–15], of which the one by HERZIG AND MISHIN [16] is the most recent.

Knowledge and understanding of grain-boundary diffusion is vital in materials science. Diffusion along grain boundaries often controls the evolution of the microstructure and properties of materials at elevated temperatures. In processes such as Coble creep, sintering, diffusion induced grain-boundary motion, discontinuous reactions, recrystallisation and grain growth, diffusion along grain-boundaries plays a prominent rôle. Grain-boundary diffusion ia important in thin-film interconnections and multilayer devices [17].

In this chapter, we first remind the reader of some basics of grain-boundary structure. Then we give a condensed review of the phenomenological theory of grain-boundary diffusion based on the Fisher model, which is strictly applicable to bicrystals. We consider grain-boundary diffusion in polycrystals and the pertinent A-, B-, and C-type of diffusion regimes. We mention foreign atom diffusion in the presence of segregation into the grain boundary. Finally, we report on some ideas about diffusion mechanisms in grain boundaries.

32.2 Grain Boundaries

Grain boundaries were already observed in the 19th century in optical micrographs of polycrystalline metal samples, which had been polished and etched to make the grain structure visible. Since HARGREAVES AND HILLS [18] grain boundaries are known to be transition regions between two neighbouring perfect crystals (grains), which are in contact with each other but differ in crystallographic orientation.

Grains in polycrystals without texture are oriented randomly. As a consequence, a wide range of ostensibly different grain boundaries exists. The nature of a boundary depends on the misorientation of the two adjoining grains and on the orientation of the boundary plane relative to them. The lattice orientation of any two grains can be made to coincide by rotating one of them through a suitable angle about a single axis. In general, the axis of rotation is randomly oriented with respect to both grains and to the grain-boundary plane. Grain boundaries are characterised by not less than five macroscopic parameters[2]: three for the rotation axis and two for the orientation of the grain-boundary plane. There are two families of boundaries

---

[2] There is an additional microscopic parameter, which is not relevant for the present discussion. It accounts for a miccroscopic translation between the two grains along the grain-boundary plane.

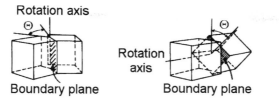

Fig. 32.1. Tilt boundary (*left*) and twist boundary (*right*)

that are relatively simple. These are *tilt boundaries* and *twist boundaries* illustrated in Fig. 32.1. For tilt boundaries the axis of rotation is parallel to the grain-boundary plane, whereas for twist boundaries the rotation axis is perpendicular to the boundary. Particular categories which are used to describe boundaries include:

(i) *Low-angle boundaries* with misorientations smaller than about 15 degrees, which consist of an array of discrete and clearly recognisable lattice dislocations (see below).
(ii) *High-angle boundaries* with misorientations larger than 15 degrees, where a lattice dislocation structure is no longer evident.
(iii) *Special boundaries* such as twins with particularly good lattice matching.
(iv) *General boundaries* which rather represent the average type of grain boundaries found in polycrystalline materials.

The topic of grain boundaries and their atomic structure has been reviewed by GLEITER [20], BALLUFFI [21], and in the anthology of WOLF AND YIP [22].

32.2.1 Low- and High-Angle Grain Boundaries

The simplest grain boundary is a low-angle symmetric tilt grain boundary shown in Fig. 32.2. This boundary can be considered as an array of parallel edge dislocations. The regions between the dislocations fit almost perfectly into both adjoining crystals whereas the dislocation cores are regions of poor fit in which the crystal structure is highly distorted. The energy of a low-angle boundary equals the total energy of the dislocations within the unit area of the boundary. This depends on the spacing L_d of the dislocations. For the array of Fig. 32.2 the spacing is given by

$$L_d = \frac{b}{\sin \Theta} \approx \frac{b}{\Theta}, \tag{32.1}$$

where b is the Burgers vector of the dislocations and Θ the angle of misorientation across the boundary. At small values of Θ, the dislocation spacing is large and the grain boundary energy γ is proportional to the number density of dislocations in the boundary, $1/L_d$, i.e.

$$\gamma \propto \Theta. \tag{32.2}$$

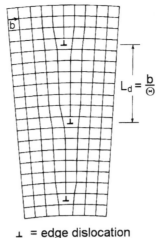

⊥ = edge dislocation

Fig. 32.2. Low-angle tilt boundary after BURGERS [19]

Low-angle twist boundaries can be considered as a planar network of screw dislocations [23].

As the misorientation Θ between the two grains increases, the strain fields of dislocations progressively cancel each other so that γ increases slower than Eq. (32.2) predicts. In general, when the misorientation exceeds 10 to 15 degrees the dislocation spacing is so small that the dislocation cores overlap. It is then impossible to identify individual dislocations. At this stage, the grain-boundary energy is almost independent of misorientation. High-angle boundaries contain large areas of poor fit and have a relatively open structure (Fig. 32.3). The bonds between the atoms are broken or highly distorted and consequently the grain-boundary energy is relatively high. Correlations between the macroscopic parameters of grain boundaries and their energy have been explored by atomistic computer simulations. For a review, we refer the reader to the already mentioned anthology of WOLF AND YIP [22]. The

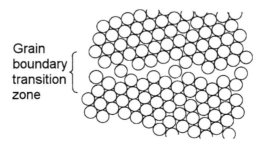

Fig. 32.3. Random high-angle grain boundary (schematic)

grain-boundary energy plays a central rôle in grain-boundary diffusion and in segregation of foreign atoms to the boundary. As a rule of thumb, high-angle grain-boundary energies are often found to be about one third of the energy of the free surface. In low-angle boundaries, however, most of the atoms fit very well into both lattices so that there is little free volume and the interatomic bonds are only slightly distorted. The regions of poor fit are restricted to dislocation cores.

32.2.2 Special High-Angle Boundaries

Not all high-angle boundaries have an open disordered structure. Special high-angle boundaries have significantly lower energies than random high-angle boundaries. Special boundaries occur at particular misorientations of the grains and orientations of the boundary plane which allow the adjoining lattices to fit together with relatively little distortion of the interatomic bonds.

The simplest special high-angle boundary is the boundary between twins. If the boundary is parallel to the twinning plane, the atoms in the boundary fit perfectly into both grains. The result is a coherent twin boundary illustrated in Fig. 32.4. In fcc metals the twinning plane is a close-packed {111} plane. Twin orientations in fcc metals correspond to a misorientation of 70.2 degrees around a ⟨110⟩-axis. A coherent twin boundary is a symmetric tilt boundary between the twin-related crystals. The atoms in such a boundary are essentially in undistorted positions and the energy of a coherent twin boundary is very low in comparison to the energy of a random high-angle boundary.

If the twin boundary does not lie exactly parallel to the twinning plane the atoms do not fit perfectly into each grain and the boundary energy is higher. Such boundaries are denoted as incoherent twin boundaries. The energy of a twin boundary is very sensitive to the orientation of the grain-boundary plane. If the boundary energy is plotted as a function of the boundary orientation (right part of Fig. 32.4) a sharp cusped minimum is obtained at the position of the coherent boundary.

Low grain-boundary energies are also found for other large-angle boundaries. A two-dimensional example is shown in Fig. 32.5. This is a symmetrical

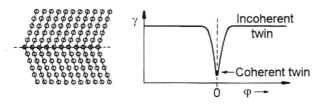

Fig. 32.4. A coherent twin boundary (*left*). Twin-boundary energy γ as a function of the orientation ϕ of the grain-boundary plane (*right*)

558   32 Grain-Boundary Diffusion

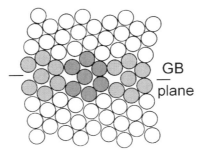

Fig. 32.5. A special large-angle boundary according to GLEITER [24]

tilt grain boundary between grains with a misorientation of 38.2 degrees. The boundary atoms fit rather well into both grains leaving little free volume. Moreover, a small group of atoms is repeated at regular intervals along the boundary.

High-resolution transmission electron microscopy (HREM) can be used to resolve the atomic structure of a grain boundary (see, e.g., [26]). Interfaces suitable for HREM are tilt boundaries, whose tilt axis coincides with a low-index zone axis. Figure 32.6 shows the HREM micrograph of a (113) [113] symmetric tilt boundary in a gold bicrystal. This grain boundary is periodic and several grain-boundary units along the boundary can be identified. This image also illustrates that the grain-boundary width δ (see below) is of the order of 0.5 nm.

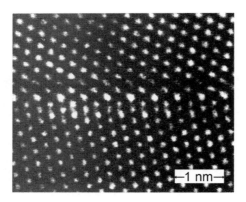

Fig. 32.6. A high-resolution transmission electron microscope image of a (113)[113] symmetric tilt boundary in gold according to WOLF AND MERKLE [25]

32.3 Diffusion along an Isolated Boundary (Fisher Model)

Most of the mathematical treatment of grain-boundary diffusion is based on the model first proposed by FISHER [7]. The grain boundary is represented by a semi-infinite, uniform, and isotropic slab of high diffusivity embedded in a low-diffusivity isotropic crystal (Fig. 32.7). The grain boundary is described by two physical parameters: the grain-boundary width δ and the grain-boundary diffusivity D_{gb}. The latter of course depends on the grain-boundary structure discussed above. It is usually much larger than the lattice diffusivity D in the adjoining grains, i.e. $D_{gb} \gg D$. The grain-boundary width is of the order of an interatomic distance. $\delta \approx 0.5\,\text{nm}$ is a widely accepted value (see above).

In a tracer diffusion experiment a layer of tracer atoms (either self- or foreign atoms) is deposited at the surface. Then, the specimen is annealed at constant temperature T for some time t. During the annealing treatment the labeled atoms diffuse into the specimen in two ways:

(i) by lattice diffusion directly into the grains and
(ii) much faster along the grain boundary.

Atoms which diffuse along the grain boundary eventually leave it and continue their diffusion path in the grains, thus giving rise to a lattice diffusion zone around the grain boundary. The total concentration of the diffuser in the specimen is the result of two contributions: a concentration c, established

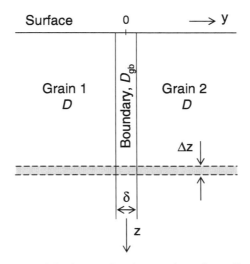

Fig. 32.7. FISHER's model of an isolated grain boundary. D: lattice diffusivity, D_{gb}: diffusivity in the grain boundary, δ: grain-boundary width

either directly by in-diffusion from the source or by leaking out from the grain boundary and the concentration inside the grain boundary, c_{gb}.

Mathematically, this diffusion problem can be described by applying Fick's second law to diffusion inside the grains and inside the grain-boundary slab. For composition-independent diffusivities we have:

$$\frac{\partial c}{\partial t} = D \left(\frac{\partial^2 c}{\partial y^2} + \frac{\partial^2 c}{\partial z^2} \right) \quad \text{for} \quad |y| \geq \delta/2,$$

$$\frac{\partial c_{gb}}{\partial t} = D_{gb} \left(\frac{\partial^2 c_{gb}}{\partial y^2} + \frac{\partial^2 c_{gb}}{\partial z^2} \right) \quad \text{for} \quad |y| < \delta/2. \quad (32.3)$$

In Eqs. (32.3) the coordinate system was chosen in such a way that the xz-plane is the symmetry plane of the grain boundary. Then, the concentration field depends on the variables y and z. Continuity of the concentrations and of the diffusion fluxes across the interfaces between grain-boundary and grains require the following boundary conditions:

$$c(\pm\delta/2, z, t) = c_{gb}(\pm\delta/2, z, t) \quad (32.4)$$

and

$$D \left[\frac{\partial c(y, z, t)}{\partial y} \right]_{|y|=\delta/2} = D_{gb} \left[\frac{\partial c_{gb}(y, z, t)}{\partial y} \right]_{|y|=\delta/2}. \quad (32.5)$$

These conditions apply for self-diffusion. In the case of foreign atom diffusion, grain-boundary segregation requires a modification, which is discussed later in this chapter.

Since the grain-boundary width is very small ($\delta \approx 0.5\,\text{nm}$) and $D_{gb} \gg D$, one can simplify the problem (see, e.g., [8]) and arrive at the following set of two coupled equations:

$$\frac{\partial c}{\partial t} = D \left(\frac{\partial^2 c}{\partial y^2} + \frac{\partial^2 c}{\partial z^2} \right) \quad \text{for} \quad |y| \geq \delta/2, \quad (32.6)$$

$$\frac{\partial c_{gb}}{\partial t} = D_{gb} \frac{\partial^2 c_{gb}}{\partial z^2} + \frac{2D}{\delta} \left(\frac{\partial c}{\partial y} \right)_{y=\delta/2} \quad \text{for} \quad |y| < \delta/2. \quad (32.7)$$

The first equation represents direct diffusion from the source into the lattice. In the second equation, the first term on the right-hand side represents the concentration change due to diffusion in the grain boundary. The second term describes the concentration change due to leakage of the diffusing species through the 'walls' of the grain-boundary slab into the grains. The mathematical problem reduces to the solution of Eqs. (32.6) and (32.7) after suitable initial and boundary conditions have been chosen.

It is convenient to introduce normalised variables, which correspond to the spatial coordinates y, z and to time t, respectively:

32.3 Diffusion along an Isolated Boundary (Fisher Model)

$$\xi \equiv \frac{y - \delta/2}{\sqrt{Dt}},$$

$$\eta \equiv \frac{z}{\sqrt{Dt}},$$

$$\beta \equiv \frac{(\Delta - 1)\delta}{2\sqrt{Dt}} \approx \frac{\delta D_{gb}}{2D\sqrt{Dt}}. \tag{32.8}$$

In these abbreviations

$$\Delta \equiv \frac{D_{gb}}{D} \tag{32.9}$$

is a dimensionless parameter, which equals the ratio of grain-boundary and lattice diffusivity. In physical terms, the variable ξ accounts for the extent of lateral lattice diffusion from the grain boundary into the grains. The quantity η accounts for the influence of direct lattice diffusion from the source into the grains; the smaller η the stronger is this influence. Whereas the physical meaning of ξ and η are obvious this is according to the author's experience rather less so for the parameter β, also called the *Le Claire parameter*. It is a measure of the extent to which grain-boundary diffusion is enhanced relative to lattice diffusion. Loosely speaking, one can consider β as the ratio of the transport capacity inside the grain-boundary slab, $c_{gb}D_{gb}\delta$, to the transport capacity along the grain-boundary fringe, $cD\sqrt{Dt}$, which has a width $\sqrt{Dt}$.

As we shall see below, diffusion profiles in bi- or polycrystals usually consist of a near-surface part dominated by lattice diffusion and a deep penetrating grain-boundary tail. Grain-boundary tails of the concentration field tend to level out as β decreases. Then, it becomes more difficult to reveal the influence of enhanced diffusion along grain boundaries in experiments. A question in this context is, what are the optimum conditions for the determination of grain-boundary diffusivities? The quantity β is relevant for this question. This can be seen from Fig. 32.8, in which isoconcentration contours are plotted for various values of β. The dotted line corresponds to the limiting case, $D_{gb} = D$, for which preferential grain-boundary diffusion is absent. The isoconcentration contours illustrate that the penetration of the diffuser along the grain boundary is much greater than anywhere else in the crystal. The larger the value of β, the more pronounced is the lateral diffusion fringe along the grain boundary. For an accurate determination of D_{gb} from sectioning experiments (see below) β must be at least 10. The annealing conditions must be chosen accordingly.

The solution for diffusion along an isolated grain-boundary slab embedded in a crystal can be written as follows:

$$c(\xi, \eta, \beta) = c_1(\eta) + c_2(\xi, \eta, \beta) \tag{32.10}$$

in the grains and

$$c(\eta) = c_{gb}(\eta) \tag{32.11}$$

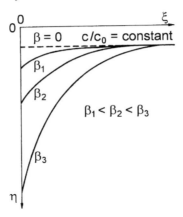

Fig. 32.8. Isoconcentration contours for various values of the LE CLAIRE parameter β

inside the boundary. In Eq. (32.10) the first term represents in-diffusion into the grains from the external source. The second term represents the leaking-out contribution from the grain boundary. The direct grain-boundary contribution, c_{gb}, can be neglected when $\sqrt{Dt} \gg \delta$; studies of the direct grain-boundary diffusion require $\sqrt{Dt} < \delta$. These distinctions are also related to the kinetic regimes B and C of diffusion in polycrystals, discussed later in this chapter.

Constant Source Solution: Let us at first consider the case of a constant source (also called infinite or inexhaustible source), with the diffuser concentration kept constant at the surface and zero everywhere inside the sample at the beginning. The initial and boundary conditions are:

$$c(y, 0, t) = c_0 \qquad \text{for} \quad t > 0,$$
$$c(y, z, 0) = 0 \qquad \text{for} \quad z > 0, \qquad (32.12)$$
$$c(y, \infty, 0) = 0.$$

c_o is the concentration of the diffuser at the surface in the source.

An approximate solution of the diffusion problem formulated in Eqs. (32.6), (32.7), and (32.12) was given already by FISHER [7]. An exact solution has been worked out three years later by WHIPPLE [27] using the Fourier-Laplace transformation method (see Chap. 3). We shall not go through the long and rather tedious mathematical exercise of deriving it. A transparent derivation of this solution can be found, e.g., in a textbook by ADDA AND PHILIBERT [28]. The first term of Eq. (32.10) is a complementary error function

$$c_1 = c_0 \operatorname{erfc}(\eta/2) \qquad (32.13)$$

32.3 Diffusion along an Isolated Boundary (Fisher Model)

and represents direct in-diffusion into the grains from the inexhaustible source. The second term in Eq. (32.10) represents the leakage contribution from the grain boundary into the grains. It is given by

$$c_2(\xi, \eta, \beta) = \frac{c_0 \eta}{2\sqrt{\pi}} \int_1^{\Delta} \frac{\exp(-\eta^2/4\sigma)}{\sigma^{3/2}} \mathrm{erfc}\left[\frac{1}{2}\left(\frac{\Delta-1}{\Delta-\sigma}\right)^{1/2}\left(\xi + \frac{\sigma-1}{\beta}\right)\right] d\sigma, \quad (32.14)$$

where σ is an integration variable. Note that the time variable is included in η and also in the Le Claire parameter β. At a fixed temperature $\beta \propto 1/\sqrt{t}$, i.e. β decreases with increasing time (see Eqs. 32.8).

Instantaneous Source (or Thin-Film) Solution: For an instantaneous source initial and boundary conditions are expressed by:

$$c(y, z, 0) = M\delta(z),$$
$$c(y, z, 0) = 0 \quad \text{for} \quad z > 0,$$
$$c(y, \infty, t) = 0,$$
$$\frac{\partial c(y, z, t)}{\partial z}\bigg|_{z=0} = 0. \quad (32.15)$$

$\delta(z)$ is the Dirac delta function and M the amount of diffuser deposited per unit area. This surface condition entails that the initial layer is completely consumed during the diffusion experiment.

An exact solution of the diffusion problem formulated in Eqs. (32.6), (32.7), and (32.15) has been worked out by SUZUOKA [29, 30], using the method of Fourier-Laplace transforms (see Chap. 3). The first term in Eq. (32.10) is

$$c_1(\eta) = \frac{M}{\sqrt{\pi Dt}} \exp\left(-\frac{\eta^2}{4}\right). \quad (32.16)$$

It describes lattice in-diffusion into the grains from a thin-film source. The second term in Eq. (32.10) represents the leakage contribution from the grain boundary:

$$c_2(\xi, \eta, \beta) = \frac{M}{\sqrt{\pi Dt}} \int_1^{\Delta} \left[\frac{\eta^2}{4\sigma} - \frac{1}{2}\right] \frac{\exp(-\eta^2/4\sigma)}{\sigma^{3/2}}$$
$$\mathrm{erfc}\left[\frac{1}{2}\left(\frac{\Delta-1}{\Delta-\sigma}\right)^{1/2}\left(\xi + \frac{\sigma-1}{\beta}\right)\right] d\sigma. \quad (32.17)$$

A comparison between the constant source solution, Eq. (32.14), and the instantaneous source solution, Eq. (32.17), shows that the latter can be obtained from Eq. (32.14) by a transformation through the operator $-\sqrt{Dt}\, \partial/\partial \eta$

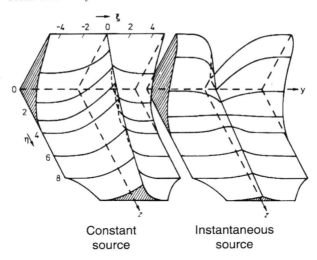

Fig. 32.9. Concentration contours for constant source (*left*) and a thin-film source solutions (*right*) for an arbitrary value of $\beta = 50$ according to SUZUOKA [30]

and replacing c_0 by M. Furthermore, in the case of the instantaneous source solution it can be shown that

$$\int_0^\infty c_2(\xi, \eta, \beta) \mathrm{d}\eta = 0. \tag{32.18}$$

A consequence of this equation is that the total amount of diffuser is given by the volume diffusion term $c_1(\eta)$, thus establishing that the total amount M of diffuser is conserved.

Figure 32.9 shows a comparison between the two types of diffusion sources for an arbitrarily chosen value of $\beta = 50$. For the thin-layer source the grain-boundary term c_2 is negative near the surface, indicating that in the near-surface region the crystal is supplying diffusing material to the grain boundary. The reason is that the source concentration decreases much more rapidly at the grain boundary than anywhere else. Thus, in the near-surface region the grain boundary behaves as a 'sink for the diffuser'. Beyond a certain depth the grain-boundary behaviour changes to that of a 'source for the diffuser', since then the direct volume diffusion from the source is negligible. In contrast, for an inexhaustible source the contribution c_2 is always positive. This implies that the grain boundary behaves as source of diffuser, irrespective of whether the near-surface region or the deeper regions are considered.

Average Concentrations in Thin Layers: Average concentrations are of prime interest for the analysis of grain-boundary diffusion experiments, which

32.3 Diffusion along an Isolated Boundary (Fisher Model)

are carried out by the radiotracer technique (see Chap. 13). Let us therefore discuss an expression for the average concentration in a thin layer, $\bar{c}$, at some depth z (Fig. 32.7). The total amount in a thin section between $z - \Delta z/2$ and $z + \Delta z/2$ and parallel to the free surface is given by the integral

$$\bar{c} = \frac{1}{L_x L \Delta z} \int_{-L_x/2}^{+L_x/2} \int_{-L/2}^{+L/2} \int_{z-\Delta z/2}^{z+\Delta z/2} [c(y,z,t) + c_{gb}(y,z,t)] \, dxdydz \qquad (32.19)$$

L_x and L are the dimensions of the bicrystal along the $x-$ and $y-$axes, respectively. The quantity $L_x L \Delta z$ is the section volume. For sake of simplicity, let us assume that the grain boundary lies in the center of the bicrystal and that the section is so thin that the concentration along the $z-$axis remains constant within a section. Then, the average concentration $\bar{c}$ is obtained by

$$\bar{c} = \frac{1}{L} \int_{-L/2}^{+L/2} [c(y,z,t) + c_{gb}(y,z,t)] \, dy. \qquad (32.20)$$

Furthermore, since c is an even function of y and c_{gb} practically constant within the boundary, we obtain

$$\bar{c}(z,t) = \frac{\delta}{L} \bar{c}_{gb}(z,t) + \frac{2}{L} \int_{\delta/2}^{L/2} c(y,z,t) dy. \qquad (32.21)$$

Let us for the moment neglect the amount of diffuser, c_{gb}, inside the boundary[3]. Then, we have

$$\bar{c}(z,t) = \frac{2}{L} \int_{\delta/2}^{L/2} c(y,z,t) dy. \qquad (32.22)$$

We know from Eq. (32.10) that the concentration field has two contributions, where c_1 represents bulk diffusion and c_2 is the grain-boundary leakage contribution given either by the thin-film solution (32.17) or by the constant-source solution (32.14). Since c_1 is constant in the xy-plane, the bulk diffusion contribution to the average concentration equals c_1 and we have

$$\bar{c}(z,t) = c_1(z,t) + \frac{2}{L} \int_{\delta/2}^{L/2} c_2(y,z,t) dy. \qquad (32.23)$$

---

[3] The same assumption is made in the next section for to type B kinetics in polycrystals. In type C kinetics the direct grain-boundary contributions is dominating.

In terms of the dimensionless variables ξ, η, β, and Δ, we get

$$\bar{c}(\eta, \beta) = c_1(\eta, t) + \frac{2\sqrt{Dt}}{L} \int_0^{(L-\delta)/2\sqrt{Dt}} c_2(\xi, \eta, \beta) d\xi. \tag{32.24}$$

For all practical purposes, the upper limit of the integral may even be replaced by infinity.

Substituting the thin-film solution Eq. (32.17) in Eq. (32.24) and using the standard formula

$$\int_x^\infty \text{erfc}(u) du = \frac{\exp(-x^2)}{\sqrt{\pi}} - x \, \text{erfc}(x), \tag{32.25}$$

we get

$$\bar{c}_2(\eta, \beta) = \frac{M}{L\sqrt{\pi}} \int_1^\Delta \left(\frac{\eta^2}{\sigma} - 2\right) \frac{\exp(-\eta^2/4\sigma)}{\sigma^{3/2}}$$
$$\left(\frac{\Delta - \sigma}{\Delta - 1}\right)^{1/2} \left[\frac{\exp(-Y^2)}{\sqrt{\pi}} - Y \text{erfc} Y\right] d\sigma, \tag{32.26}$$

where

$$Y = \frac{\sigma - 1}{2\beta} \left(\frac{\Delta - \sigma}{\Delta - 1}\right)^{1/2}. \tag{32.27}$$

Similarly, using the constant-source solution of Eq. (32.14) gives

$$\bar{c}_2(\eta, \beta) = \frac{c_0 \sqrt{Dt}}{L} \frac{2\eta}{\sqrt{\pi}} \int_1^\Delta \left(\frac{\eta^2}{\sigma} - 2\right) \frac{\exp(-\eta^2/4\sigma)}{\sigma^{3/2}}$$
$$\left(\frac{\Delta - \sigma}{\Delta - 1}\right)^{1/2} \left[\frac{\exp(-Y^2)}{\sqrt{\pi}} - Y \text{erfc} Y\right] d\sigma. \tag{32.28}$$

The factor $1/L$ in Eqs. (32.26) and (32.28) is important. For a bicrystal with dimensions L along x- and y-axes, $1/L$ represents the grain-boundary length per unit area of the bicrystal. An expression which is generally valid for a bicrystal or for polycrystals is:

$$\bar{c}(\eta, \beta) = c_1(\eta, \beta) + 2\lambda \sqrt{Dt} \int_0^\Lambda c_2(\xi, \eta, \beta) d\xi. \tag{32.29}$$

In Eq. (32.29) λ represents the grain-boundary length per unit area on the sample surface exposed to the diffuser.

For a bicrystal with a width L normal to the grain-boundary, Λ and λ are given by

$$\Lambda = \frac{L}{2\sqrt{Dt}} \quad \text{and} \quad \lambda = \frac{1}{L}. \tag{32.30}$$

For an array of uniformly spaced grain-boundaries with spacing d_s we have

$$\Lambda = \frac{d_s}{2\sqrt{Dt}} \quad \text{and} \quad \lambda = \frac{1}{d_s}, \tag{32.31}$$

and for a polycrystal having cubic grains with grain size d

$$\Lambda = \frac{d}{2\sqrt{Dt}} \quad \text{and} \quad \lambda = \frac{2}{d}. \tag{32.32}$$

For the more complicated case of a polycrystal with random distribution of grain size the reader may consult the textbook of KAUR, MISHIN, AND GUST [8].

Segregation of Foreign Atoms: Foreign atoms tend to segregate into grain boundaries. This process is called grain-boundary segregation and results in an excess concentration of the foreign element in the grain boundary. The mathematics of grain-boundary diffusion discussed so far has not taken into account segregation effects. This is justified for grain-boundary self-diffusion.

Diffusion of foreign elements (solutes) can be treated by the same mathematics, if grain-boundary segregation is taken into account in a suitable way. In the case of self-diffusion, it was assumed that the grain-boundary width δ is a purely geometric quantity and that the matching condition Eq. (32.4) simply expresses the continuity of concentration across the grain/grain-boundary interfaces. This assumption must be modified for solute atoms because they can segregate into the boundary. Accoording to GIBBS [31] segregation can be taken into account by introducing the *segregation factor* s. The matching condition then reads

$$sc(\pm\delta/2, z, t) = c_{gb}(\pm\delta/2, z, t) \tag{32.33}$$

Equation (32.33) rests on two assumptions:

1. The solute atoms in the grain boundary maintain local thermodynamic equilibrium with the solute atoms in the lattice adjacent to the interfaces. In other words, segregation is in local equilibrium at any depth z.
2. The grain-boundary segregation follows the *law of Henry*, which reads $c_{gb} = sc$. Henry's law implies that the segregation factor is a function of temperature only and not a function of c. Henry's law is applicable when both, c_{gb} and c, are small enough. This is usually the case, when impurity diffusion in a pure matrix is studied in radiotracer experiments.

Grain-boundary self-diffusion in alloys represents another interesting situation, which is not considered here. It may encompass also segregation effects of the alloy components. For details we refer the reader to the textbook of KAUR, MISHIN, AND GUST [8], to reviews by MISHIN AND HERZIG [15, 16], and by BERNARDINI AND GAS [33, 34].

32.4 Diffusion Kinetics in Polycrystals

As we have already seen, grain-boundary diffusion is a complex process which involves direct lattice diffusion from the source, diffusion along the grain boundary, leakage of the diffusant from the grain-boundary and subsequent lattice diffusion into fringes around the grain boundary. Depending on the relative importance of the various elementary processes, one can observe different diffusion kinetics (or diffusion regimes). Each regime prevails in a certain domain of annealing temperatures, annealing times, grain sizes, lattice and grain-boundary parameters. The knowledge of the diffusion regimes is important for designing diffusion experiments and for the interpretation of their results. This is because the shape of the diffusion profile depends on the dominating kinetic regime. Moreover, different diffusion parameters can be extracted in the various diffusion regimes, which therefore must be identified in dependable studies.

Figure 32.10 shows HARRISONS'S classification of the diffusion kinetics, which introduces three regimes called type A, B, and C [38]. This classification is the first and still the most widely used one for polycrystals. A more sophisticated classifications is proposed in [8, 14] and summarised in Chap. 34.

32.4.1 Type A Kinetics Regime

This kinetics is observed after diffusion anneals at high temperatures, or/and with long annealing times, or/and in materials with small grain size. Monte Carlo work by BELOVA AND MURCH [32] has shown that the lattice diffusion length, $\sqrt{Dt}$, need to be only a little larger than the spacing d between grain boundaries:

$$\sqrt{Dt} \geq d/0.8 \,. \tag{32.34}$$

Then, the diffusion fringes around neighbouring grain boundaries overlap and a diffusing atom may visit many grains and grain boundaries during a diffusion experiment. This results in an almost planar diffusion front with a penetration depth proportional to $\sqrt{t}$. From a macroscopic viewpoint the polycrystal obeys Fick's law for a homogeneous medium with some effective diffusion coefficient D_{eff}. The latter represents a weighted average of the lattice diffusivity D and grain-boundary diffusivity D_{gb}.

In the case of self-diffusion, HART [39] proposed an effective diffusivity for dislocated crystals. Modified for diffusion in polycrystals an approximate expression for the effective diffusivity is given by

32.4 Diffusion Kinetics in Polycrystals

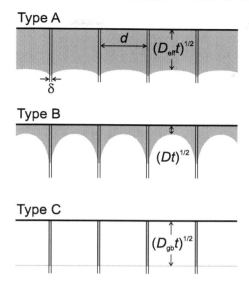

Fig. 32.10. Illustration of the type A, B, and C diffusion regimes in a polycrystal according to HARRISONS classification [38]

$$D_{eff} = gD_{gb} + (1-g)D. \qquad (32.35)$$

In Eq. (32.35) the quantity g is the fraction of atomic sites in the grain boundary of the polycrystal, which can be written as

$$g = \frac{q\delta}{d}, \qquad (32.36)$$

where q is a numerical factor depending on the grain shape. For example, $q = 1$ for parallel grain boundaries and $q = 3$ for cubic grains.

Diffusion profiles measured in the type A regime follow a Gaussian for instantaneous source conditions or an error function for constant source conditions. The only quantity which can be determined from the profile, is the effective diffusivity. Since usually $D_{gb} \gg D$, the effective diffusivity is larger than the lattice diffusivity. This explains why diffusion coefficients measured on polycrystals often can be larger than diffusion coefficients measured on monocrystalline samples. On the other hand, since the activation enthalpy of D is larger than that of D_{gb}, the enhancement due to grain boundaries becomes weaker with increasing temperature. Usually, at sufficiently high temperatures the effective diffusivity in coarse-grain polycrystals approaches the lattice diffusivity. In very fine-grained polycrystalline materials, the effective diffusivity can be dominated by the first term in Eq. (32.35) (see Chap. 34). Then, knowing the grain size one can estimate the product δD_{gb}.

For foreign atoms which segregate into the grain boundary, the Hart equation (32.35) has to be modified. Segregation of a diffuser can be taken into account by multiplying the fraction of atomic sites in the high-diffusivity path, g, by the segregation factor of the diffuser, s. For fine-grained polycrystals (nanocrystals), in which d is so small and/or segregation is so strong that $d \ll s\delta/2$, the effective diffusivity is given by [8]

$$D_{eff} = \tau D_{gb} + (1-\tau)D. \qquad (32.37)$$

τ and $1-\tau$ are the fractions of time that a diffusing foreign atom spends in the grain boundaries and in the grains, respectively. In the presence of segregation, these time fractions can be estimated as the corresponding volume fractions, g and $1-g$, multiplied by the pertinent concentrations, c and c_{gb}. Thus

$$\tau = \frac{gc_{gb}}{gc_{gb} + (1-g)c} \approx \frac{sg}{1+sg} = \frac{qs\delta/d}{1+qs\delta/d}. \qquad (32.38)$$

When $d \gg s\delta/2$, we have $\tau \approx sg$. Then, Eq. (32.37) reduces to

$$D_{eff} = sg D_{gb} + (1-sg)D. \qquad (32.39)$$

This equation was suggested already in 1960 by MORTLOCK [40]. For conventional polycrystals, say with $d > 50\,\mu\text{m}$, we have $\delta/d \approx 10^{-5}$. Then, even for large s-values $sg \ll 1$ is fulfilled and Eq. (32.39) simplifies further:

$$D_{eff} \approx sg D_{gb} + D. \qquad (32.40)$$

For a more detailed discussion of effective diffusion especially in nanomaterials we refer to Chap. 34 and a paper by BELOVA AND MURCH [41].

32.4.2 Type B Kinetics Regime

The conditions of type B kinetics are often encountered in diffusion experiments on polycrystals. This kinetics emerges after diffusion anneals at lower temperatures, or/and with relatively short annealing times, or/and in materials with sufficiently large grain size. Under such conditions the bulk diffusion length, $\sqrt{Dt}$, can become much smaller than the spacing d between grain boundaries. Simultaneously, the width of the grain-boundary fringes, which is given by $\sqrt{Dt}$, can be considerably larger than the grain-boundary width δ. In the case of solute diffusion with segregation an effective width, $s\delta$, must be considered. Thus, the conditions for type B kinetics are

$$s\delta \ll \sqrt{Dt} \ll d. \qquad (32.41)$$

In this regime, grain-boundary fringes develop by out-diffusion from the boundaries. In contrast to the type A regime, the lattice diffusion fringes of neighbouring grain boundaries do not overlap. Hence individual grain boundaries are isolated and the solutions for the boundary in bicrystals discussed in Sect. 32.3 can be used for polycrystals.

32.4 Diffusion Kinetics in Polycrystals

The left-hand part of the inequality Eq. (32.41) is equivalent to $\alpha \ll 1$ (see Eq. 32.45). It is also necessary for proper B regime conditions that the Le Claire parameter fulfills $\beta \gg 1$ (see Eq. 32.44). Then, the diffusion profile has the two-step character illustrated in Fig. 32.11. For self-diffusion the product, δD_{gb}, and for solute diffusion the triple product, $s\delta D_{gb}$, can be determined using the procedure described below.

Processing of Type B Grain-Boundary Diffusion Profiles: Most grain-boundary diffusion experiments are carried out by the radiotracer method in combination with a serial sectioning technique (Chap. 13). After the diffusion anneal, thin layers of the material are removed from the specimen and the radioactivity of each section is determined. This quantity is proportional to the average concentration per layer, $\bar{c}$, as a function of the penetration distance z. This function, called a concentration (or diffusion, or penetration) profile, contains the information about the grain-boundary diffusion parameters. It is subject to a mathematical treatment to extract these parameters.

In a grain-boundary diffusion experiment the overall penetration profile consists of two parts shown schematically in Fig. 32.11: first, a *near-surface part* due to the direct lattice diffusion from the surface. This part extends to about (3 to 4) $\times \sqrt{Dt}$, and the concentration in this region follows either a Gaussian or an error function depending on the surface condition. Second, a deep penetrating *grain-boundary tail* due to the simultaneous diffusion inside grain-boundaries, leakage from the boundary and lateral lattice diffusion into the adjacent grains. This tail is described by $\bar{c}_2$ of Sect. 32.3 and is considered in what follows.

As already mentioned, one is usually interested in the case $\Delta \equiv D_{gb}/D \gg 1$ with β remaining finite (see Eqs. 32.8 and 32.9). The variations of $\bar{c}_2$ computed from Eqs. (32.26) and (32.28) for a value of $\beta = 100$ (with $\Delta = 2 \times 10^6$ and $\sqrt{Dt} = 10\,\mu\text{m}$) are displayed in Fig. 32.12. For the instantaneous source the negative part of $\bar{c}_2$ near the source represents the behaviour of the grain boundary as a sink. The positive part at deeper penetration depths is the leakage contribution, when the boundary acts as source for diffusing atoms.

Of particular interest is the behaviour of $\bar{c}_2$ for $\eta > 3$ to 4, where direct lattice diffusion can be neglected. Figure 32.13 shows a plot of $\log \bar{c}_2$ versus $z^{6/5}$. In the tail region, the penetration profiles become straight lines when plotted in this way. An important aspect of Fig. 32.13 is that the slope $-\partial \log \bar{c}_2/\partial z^{6/5}$ is very nearly the same, irrespective of the type of diffusion source. As we shall see below, the slope of this line is used to determine an important grain-boundary diffusion parameter.

It should be noted, however, that the power 6/5 has no physical meaning and cannot be derived analytically. It simply offers a reasonably good mathematical approximation of the profile shape in the grain-boundary dominated profile range. Accurately measured experimental profiles usually do follow the $z^{6/5}$ rule over a wide concentration range. In practice, the linearity of

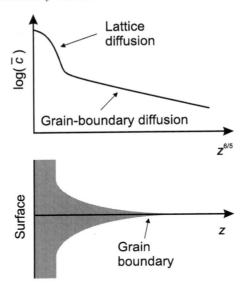

Fig. 32.11. Schematic illustration of a penetration profile in a bi-crystal for type B kinetics

the grain-boundary tail in the $\log \bar{c}$ versus $z^{6/5}$ plot is, on the one hand, a signature of predominant grain-boundary diffusion and, on the other hand, a criterion for the quality of an experiment.

Figure 32.14 shows experimental penetration profiles of Ag self-diffusion in polycrystalline silver, determined by serial sectioning and subsequent measurements of the activity per layer. The profiles are plotted as logarithm of the specific activity per layer as function of 6/5 power of the penetration distance. These profiles follow type B kinetics of grain-boundary diffusion for penetration depths larger than about 50 μm.

One can deduce the *triple-product*, $sD_{gb}\delta$, from slopes of diagrams like Fig. 32.14 as

$$sD_{gb}\delta = 1.322\sqrt{\frac{D}{t}}(-\partial \bar{c}/\partial z^{6/5})^{-5/3} \qquad (32.42)$$

for constant-source conditions and

$$sD_{gb}\delta = 1.308\sqrt{\frac{D}{t}}(-\partial \bar{c}/\partial z^{6/5})^{-5/3} \qquad (32.43)$$

for instantaneous-source conditions. If the lattice diffusion coefficient D is known from independent measurements, the triple product can be determined. In the case of self-diffusion the product $D_{gb}\delta$ is obtained.

Equations (32.42) and (32.43) require that the following conditions are met by the experiment:

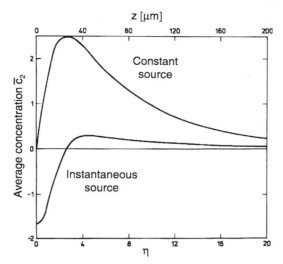

Fig. 32.12. Average thin-layer concentration at depth z of a diffuser entering via a grain boundary *versus* the normalised penetration depth $\eta = z/\sqrt{Dt}$ for $\beta = 100$ ($\Delta = 2 \times 10^6$ and $\sqrt{Dt} = 10\,\mu\mathrm{m}$) according to SUZUOKA [29]. The concentrations are expressed in units of $M/(L\sqrt{\pi})$ in the case of an instantaneous source and in units of $c_0\sqrt{Dt}/L$ for a constant source

(i) The parameter β defined by LE CLAIRE [36]

$$\beta = \frac{sD_{gb}\delta}{2D\sqrt{Dt}} = \alpha\Delta \qquad (32.44)$$

must be large enough (in practice $\beta > 10$). In the case of self-diffusion, we have $s = 1$ and β is given by Eq. (32.8).

(ii) The parameter

$$\alpha = \frac{s\delta}{2\sqrt{Dt}} \qquad (32.45)$$

must be small enough, in practice $\alpha < 0.1$.

Equations (32.42) and (32.43) are relations that are often used in the analysis of grain-boundary diffusion studies. If the Le Claire parameter β is smaller than 10^4 the numerical constants in Eq. (32.42) and (32.43) are slightly different. An analysis of the various solutions has been presented by LE CLAIRE [36]. A detailed practice-oriented discussion for various β ranges can be found in the textbook of KAUR, MISHIN, AND GUST [8].

We note that Eqs. (32.42) and (32.43) only provide the triple-product $sD_{gb}\delta$ or $D_{gb}\delta$ for self-diffusion. The individual values of s, D_{gb}, and δ remain unknown. Even in the case of self-diffusion ($s = 1$) one still needs to know the grain-boundary width δ to deduce the grain-boundary diffusivity D_{gb}.

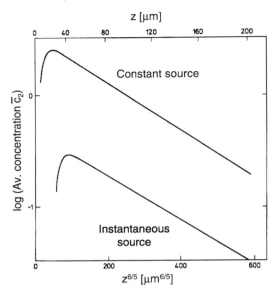

Fig. 32.13. Logarithm of the average thin layer concentration at depth z of the diffuser entering via a grain boundary *versus* $z^{6/5}$ (z = section depth, $\Delta = 2 \times 10^6$ and $\sqrt{Dt} = 10\,\mu m$) according to SUZUOKA [29]. The concentrations are expressed in units of $M/(L\sqrt{\pi})$ for an instantaneous source and in units of $c_0\sqrt{Dt}/L$ for a constant source

The usual assumption $\delta = 0.5\,\text{nm}$, already suggested by FISHER [7], proved to be a good guess. This value is consistent with determinations of δ by high-resolution electron microscopy and other techniques [8]. Furthermore, the combination of type B and type C regime measurements of grain-boundary self-diffusion (see below) showed that $0.5\,\text{nm}$ is a reasonable value for the grain-boundary width [35, 37].

32.4.3 Type C Kinetics Regime

Type C kinetics corresponds to conditions where lattice diffusion is practically 'frozen in'. Then, diffusion takes place along grain boundaries only, without any essential leakage into adjacent grains. This situation can be matched in diffusion anneals at sufficiently low temperatures and/or for very short diffusion times. In this regime we have

$$\sqrt{Dt} \ll s\delta. \qquad (32.46)$$

This criterion for type C kinetics is equivalent to $\alpha \gg 1$ (see Eq. 32.45). In practice, $\alpha > 10$ is sufficient. The concentration-depth profile is either a Gaussian function

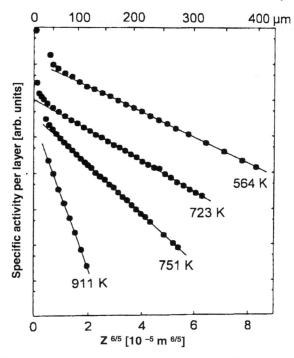

Fig. 32.14. Type B kinetics penetration profiles of self-diffusion in Ag polycrystals according to SOMMER AND HERZIG [35]

$$\bar{c}(z) \approx \bar{c}_{gb}(z) \propto \exp\left(-\frac{z^2}{4D_{gb}t}\right) \tag{32.47}$$

for an instantaneous source or an error function

$$\bar{c}(z) \approx \bar{c}_{gb}(z) \propto \mathrm{erfc}\left(\frac{z}{2\sqrt{D_{gb}t}}\right) \tag{32.48}$$

for a constant source at $z = 0$.

Grain-boundary diffusion studies in the type C regime are difficult. The major reason is that one has to detect the small amount of diffuser inside the boundaries. Type B experiments are less ambitious, since the lattice diffusion fringes around the grain boundaries enhance the average concentration in the grain-boundary tails. If type C kinetics profiles are measured in bi- or polycrystals, one can determine the grain-boundary diffusivity D_{gb}. For self-diffusion a combination of type B and type C kinetics experiments on the same material permits a determination of the grain-boundary product $D_{gb}\delta$ and of D_{gb}. Using this idea, grain-boundary self-diffusion studies in silver polycrystals yielded the already mentioned value $\delta \approx 0.5\,\mathrm{nm}$ for the grain-boundary width [35].

32.5 Grain-Boundary Diffusion and Segregation

Segregation occurs as a result of inhomogeneities in a solid. These inhomogeneities provide sites on which solute atoms have a lower Gibbs free energy. Such sites occur at free surfaces, grain boundaries, interfaces between different phases as well as at dislocations and stacking faults. All of these regions have solute concentrations that differ from each other and from that of the bulk material. For a detailed discussion of segregation effects the reader is referred to a review by HONDROS AND SEAH [45]. Theoretical studies based on atomistic computer simulations and experiments using atom probe field-ion microscopy are described by FOILES AND SEIDMAN [46]. In what follows, we consider segregation of solute atoms and its effects on grain-boundary diffusion.

Since the discovery of grain-boundary diffusion, the topic of segregation of foreign elements to grain boundaries has always been of interest. One of its effects was already noticed by one of the progenitors of physical metallurgy and solid-state diffusion: ROBERTS-AUSTEN stated in 1888 *'One thousandth part of antimony converts first rate best selected copper to the worst conceivable'*. We know nowadays that this was due to the influence of grain-boundary segregation on the mechanical behaviour of polycrystalline copper. Grain-boundary embrittlement is often caused by segregation of impurities. This effect is of considerable importance for engineering materials [44].

Solute Segregation at Grain Boundaries: Segregation to grain boundaries has been widely studied, not least for its rôle in controlling intergranular fracture in engineering materials. A useful empirical correlation between the solubility limit of a foreign atom in a host crystal, c_B, and the segregation factor was first noted by SEAH AND HONDROS [47]. As a 'rule of thumb' the segregation factor is inversely proportional to the the solubility limit of the segregant at the measuring temperature:

$$s = \frac{K}{c_B} \qquad (32.49)$$

Using this equation, the grain-boundary segregation factor can be estimated from handbooks of phase diagrams, when no measurements are available. In the dilute limit, Eq. (32.49) provides the correct theoretical description [45].

The solute segregation factors s introduced already in Eq. (32.33) can be determined directly when a sample of the segregated material can be fractured along grain-boundaries and the chemical composition is analysed using Auger spectroscopy (see Chap. 13) or some other technique. Unfortunately, such measurements of s are only possible for brittle materials such as ceramics or some intermetallic compounds.

Segregation Factors from Grain-boundary Diffusion: The segregation factor can be deduced also from grain-boundary diffusion studies. The idea is to combine type B and type C regime measurements. As pointed out above, grain-boundary diffusion experiments are usually performed in the type B regime. The profiles are plotted *versus* $z^{6/5}$ and the slope of the grain-boundary tail is determined (see, e.g., Eq. 32.43). For solute diffusion, this analysis yields the triple product $s\delta D_{gb}$, which is the only quantity that can be determined in type B conditions. While the grain-boundary width can be treated as a known constant ($\delta \approx 0.5$ nm), the grain-boundary diffusivity D_{gb} and the segregation factor s remain to be determined. Both quantities are temperature-dependent and may vary by orders of magnitude. Type C experiments directly provide the grain-boundary diffusivity D_{gb}. By combining values of D_{gb} from type C experiments at low temperatures with values of $s\delta D_{gb}$ extrapolated from type B regime measurements at higher temperatures, one gets

$$s\delta = \frac{(s\delta D_{gb})_{B-\text{regime}}}{(D_{gb})_{C-\text{regime}}}. \tag{32.50}$$

The triple product for solute diffusion usually follows an Arrhenius relation with an activation enthalpy Q_{gb} and a pre-factor $(s\delta D_{gb})_0$:

$$s\delta D_{gb} = (s\delta D_{gb})_0 \exp\left(-\frac{Q_{gb}}{k_B T}\right), \tag{32.51}$$

The segregation factor is an Arrhenius function as well. Segregation increases with decreasing temperature:

$$s = s_0 \exp\left(\frac{\Delta H_{seg}}{k_B T}\right). \tag{32.52}$$

The segregation enthalpy ΔH_{seg} is the gain in enthalpy that occurs when a solute atom is moved from a lattice site to a site in the grain boundary. s_0 is a pre-exponential related to the segregation entropy. Since the grain-boundary diffusivity of the solute also obeys an Arrhenius relation

$$D_{gb} = D_{gb}^0 \exp\left(-\frac{\Delta H_{gb}}{k_B T}\right), \tag{32.53}$$

we obtain for the activation enthalpy of the triple product

$$Q_{gb} = \Delta H_{gb} - \Delta H_{seg}. \tag{32.54}$$

If Q_{gb} and ΔH_{gb} are known from grain-boundary diffusion studies, one can determine the segregation enthalpy ΔH_{seg}.

Examples: Combined type B and C regime measurements of solute diffusion have been performed on few systems (see the reviews of MISHIN AND HERZIG [12, 15]). We give two examples:

Grain-boundary diffusion of radioactive Te in Ag has been studied over a wide temperature range from 378 to 970 K [48]. The penetrations profiles were carefully analysed, attributed to type B kinetics above 600 K, and triple

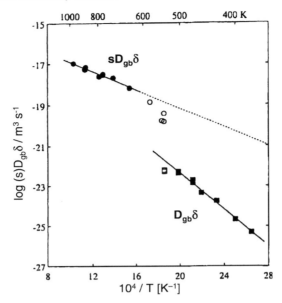

Fig. 32.15. Arrhenius diagram of the triple product $s\delta D_{gb}$ and of sD_{gb} from Te diffusion along grain boundaries in Ag according to HERZIG ET AL. [48]. Type B and C kinetics prevail above 600 K and below 500 K, respectively. The range 500–600 K corresponds to a transient regime

product values determined from $\log \bar{c}$ *versus* $z^{6/5}$ plots. Measurements below 500 K were attributed to type C kinetics and the profiles analysed by fitting Gaussian functions. Figure 32.15 shows an Arrhenius diagram of the $s\delta D_{gb}$ and δD_{gb} values determined in this way. The triple product follows the Arrhenius relation

$$s\delta D_{gb} = 2.34 \times 10^{-15} \exp\left(-\frac{43.5\,\text{kJ}\,\text{mol}^{-1}}{k_B T}\right) \text{m}^3\text{s}^{-1} \tag{32.55}$$

and the grain-boundary diffusion coefficient (assuming $\delta = 0.5$ nm) is described by

$$D_{gb} = 1.01 \times 10^{-4} \exp\left(-\frac{86.7\,\text{kJ}\,\text{mol}^{-1}}{k_B T}\right) \text{m}^2\text{s}^{-1}. \tag{32.56}$$

The ratio between $s\delta D_{gb}$ and sD_{gb} yields the segregation factor. The values for Te in Ag are displayed in Fig. 32.16 together with segregation factors for Au in Ag [49] obtained by the same procedure. The solubility of Te in the Ag lattice is very small. The high values for the segregation factor of Te in Ag (10^3 to 10^4) are consistent with the low lattice solubility of Te in Ag (see Eq. 32.49). The segregation enthalpy of Te is fairly high and equals

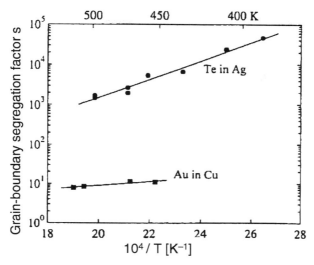

Fig. 32.16. Grain-boundary segregation factors for Te in Ag according to HERZIG ET AL. [48] and Au in Ag according to SURHOLT ET AL. [49] determined from combined type B and type C measurements

$\Delta H_{seg} = -43.3 \, \text{kJ} \, \text{mol}^{-1}$. Te atoms prefer to be incorporated on sites at silver grain boundaries. In contrast, the Au–Cu system is a system which forms a complete solid solution from the Au to the Cu side. Under such conditions, grain-boundary segregation is rather weak, which is reflected by a relatively small value of ΔH_{seg}.

32.6 Atomic Mechanisms of Grain-Boundary Diffusion

The atomic mechanisms of grain-boundary diffusion are still not well understood. It was an assumption for many years that atoms move in grain boundaries by exchanges with vacancies [8, 50, 51]. This assumption must be revised in view of recent modeling and simulation results for grain-boundary diffusion. Molecular dynamics simulations of grain-boundary diffusion in Ag [52, 53] and Cu [54–57] have pointed to a possible rôle of self-interstitials along with vacancies. The atomic structure of several symmetrical tilt grain boundaries in Cu and their interaction with vacancies and interstititals have been studied by MISHIN AND COWORKERS using molecular statics, molecular dynamics and other atomistic simulation methods [58–60].

The results can be summarised as follows: point defect formation energies in the grain boundary are on average lower than in the lattice but variations from site to site within the grain boundary are significant. The formation energies of vacancies and interstitials are close to each other, which makes both defects equally important for grain-boundary diffusion. Depending on the grain-boundary structure, grain-boundary diffusion can be dominated by vacancies or interstitials. The diffusion anisotropy depends on the grain-boundary structure, with diffusion in tilt boundaries along the tilt axis being either faster or slower than normal to the tilt axis.

Vacancies show effects such as delocalisation and instability at certain grain-boundary sites. Vacancies can move in grain-boundaries by single-atom exchanges, as they do in the lattice, but they can also move by collective jumps involving the simultaneous displacement of several atoms. Interstitial atoms can occupy relatively open positions between atoms, form split interstitials, or form highly delocalised displacement zones. Interstitials can move by the direct or indirect mechanism. Under the direct mechanism, an interstitial atom wanders along the grain boundary by jumping between neighbouring interstitial positions. Under the indirect mechanism, an interstitial atom displaces a neighbouring regular atom and takes its place. This process occurs by the simultaneous (collective) displacement of both atoms. Furthermore, an interstitial atom can initiate a chain of atomic displacements and push the terminal atom of the chain into an interstitial position. All atoms taking part in this chain process move in concert and not one after another. Interstitial dumbbells in grain boundaries always move by collective jumps of three or more atoms, as they do in the lattice. Ring mechanisms have also been found.

On the whole, it appears that atomistic mechanisms of diffusion in grain boundaries are different from lattice diffusion. A variety of mechanisms, most of which involve collective motion of two or more atoms, seem to play a rôle. Considering that the collective events often happen by displacements of chains or rings, we note an analogy with diffusion mechanisms in metallic glasses (see Chap.29).

References

1. R.S. Barnes, Nature **166**, 1032 (1950)
2. A.D. Le Claire. Philos. Mag. **42**, 468 (1951)
3. R.E. Hoffman, D. Turnbull, J. Appl. Phys. **22**, 634–639 (1951)
4. W.C. Roberts-Austen, Phil Trans. Roy. Soc. **A 187** (1896) 383–414
5. J. Groh, G. von Hevesy, Ann. Physik **63**, 85–92 (1920)
6. J. Groh, G. von Hevesy, Ann. Physik **65**, 216–222 (1921)
7. J.C. Fisher, J. Appl. Phys. **22**, 74 (1951)
8. I. Kaur, Y. Mishin, W. Gust, *Fundamentals of Grain and Interphase Boundary Diffusion*, John Wiley and Sons, Ltd., 1995
9. I. Kaur, W. Gust, *Grain- and Interphase Boundary Diffusion*, Chap. 12 in: *Diffusion in Solid Metals and Alloys*, H. Mehrer (Vol. Ed.), Landolt-Börnstein, Numerical Data and Functional Relationships in Science and Technology, New Series, Group III: Crystal and Solids State Physics, Vol. 26, Springer-Verlag, 1990, p. 630

10. G. Erdelyi, D.L. Beke, *Dislocation and Grain-boundary Diffusion in Non-metallic Systems*, Chap. 11 in: *Diffusion in Semiconductors and Non-metallic Systems*, D.L. Beke (Vol. Ed.), Landolt-Börnstein, Numerical Data and Functional Relationships in Science and Technology, New Series, Group III: Condensed Matter, Vol.33, Subvolume B1, Springer-Verlag, 1999, p.11–1
11. N.L. Peterson, Int. Metals Rev. **28**, 65 (1983)
12. Y. Mishin, Chr. Herzig, J. Bernardini, W. Gust, Int. Mater. Rev. **42**, 155 (1997)
13. D. Gupta, D.R. Campbell, P.S. Ho, in: *Thin Films: Interdiffusion and Reactions*, J.M. Poate, K.N. Tu, J.W. Mayer (Eds.), John Wiley and Sons, Ltd., 1978
14. Chr. Herzig, Y. Mishin, *Grain-Boundary Diffusion in Metals*, Chap. 4 in: *Diffusion in Condensed Matter*, J. Kärger, P. Heitjans, P. Haberlandt (Eds.), Fr. Vieweg und Sohn, Braunschweig/Wiesbaden 1998, p. 90
15. Y. Mishin, Chr. Herzig, Materials Science and Engineering **A 260**, 55 (1999)
16. Chr. Herzig, Y. Mishin, *Grain-Boundary Diffusion in Metals*, Chap. 8 in: *Diffusion in Condensed Matter – Methods, Materials, Models*, P. Heitjans, J. Kärger (Eds.), Springer-Verlag, 2005, p. 337
17. S.P. Murarka, *Diffusion Barriers in Semiconductor Devices/Circuits*, in: *Diffusion Processes in Advanced Technological Materials*, D. Gupta (Ed.), William Andrew, Inc. 2005
18. F. Hargreaves, R.J. Hills, J. Inst. Metals **41**, 257 (1929)
19. M. Burgers, Proc. Phys. Soc. (London) **452**, 23 (1940)
20. H. Gleiter, *Microstructure*, in: *Phyical Metallurgy*, R.W. Cahn, P. Haasen (Eds.), North-Holland Physics Publishing, 1983, p. 649
21. R.W. Balluffi, *Grain-Boundary Diffusion Mechanisms in Metals*, in: *Diffusion in Crystalline Solids*, G.E. Murch, A.S. Nowick (Eds.), Academic Press, Inc., 1984, p. 320
22. D. Wolf, S. Yip, *Materials Interfaces – Atomic-level Structure and Properties*, Chapman and Hall, 1992
23. W.T. Read Jr., *Dislocations in Crystals*, McGraw Hill, New York, 1953
24. H. Gleiter, Phys. Stat. Sol. (b) **45**, 9 (1971)
25. D. Wolf, K.L. Merkle, *Correlation between the Structure and Energy of Grain Boundaries in Metals*, p. 87 in [22]
26. J.H.C. Spence, *Experimental High-Resolution Electron Microscopy* (2nd edn.), Oxford University Press, 1988
27. R.T.P. Whipple, Philos. Mag. **45**, 1225 (1954)
28. Y. Adda, J. Philibert, *La Diffusion dans les Solides*, Vol. II, Presses Universitaires de France, Paris, 1966
29. T. Suzuoka, Trans. Jap. Inst. Metals **2**, 25 (1961)
30. T. Suzuoka, J. Phys. Soc. Japan **19**, 839 (1964)
31. G.B. Gibbs, phys. stat. sol. **16**, K27 (1966)
32. I.V. Belova, G.E. Murch, Philos. Mag. **81**, 2447 (2001)
33. J. Bernardini, Defect and Diffusion Forum **66–69**, 667 (1990)
34. J. Bernardini, P. Gas, Defect and Diffusion Forum **95–98**, 393–404 (1993)
35. J. Sommer, Chr. Herzig, J. Appl. Phys. **72**, 2758–2766 (1992)
36. A.D. Le Claire, Brit. J. Appl. Phys. **14**, 351 (1963)
37. P. Gas, D.L. Beke, J. Bernardini, Phil. Mag. Letters **65**, 133 (1992)
38. L.G. Harrison, Trans. Faraday Soc. **57**, 1191 (1961)
39. E.W. Hart, Acta Metall. **5**, 597 (1957)

40. A.J. Mortlock, Acta Metall **8**, 132 (1960)
41. I.V. Belova, G.E. Murch, in: *Nanodiffusion*, D.L. Beke (Ed.), J. of Metastable and Nanocrystalline Materials **19**, 23 (2004)
42. A.M. Brown, M.F. Ashby, Acta Metall. **28**, 1085 (1980)
43. W. Gust, S. Mayer, A. Bögel, B. Predel, J. Physique **46** (C4), 537 (1985)
44. C.L. Briant, *Interfacial Segregation, Bonding, and Reactions*, in: *Materials Interfaces – Atomic-Level Structure and Properties*, D. Wolf, S. Yip (Eds.), Chapman and Hall, London, 1992, p. 463
45. E.D. Hondros, M.P. Seah, *Interfacial and Surface Microchemistry*, in: *Physical Metallurgy*, R.W. Cahn, P. Haasen (Eds.), Elsevier Science Publishers B.V., 1983, p. 855
46. S.M. Foiles, D.N. Seidman, *Atomic Resolution Study of Solute-Atom Segregation at Grain Boundaries: experiments and Monte Carlo Calculation*, in: *Materials Interfaces – Atomic-Level Structure and Properties*, D. Wolf, S. Yip (Eds.), Chapman and Hall, London, 1992, p. 497
47. M.P. Seah, E.D. Hondros, Proc. Royal Soc. **A335**, 191 (1973)
48. Chr. Herzig, J. Geise, Y. Mishin, Acta Metall. Mater. **41**, 1683–1691 (1993)
49. T. Surholt, Y. Mishin, Chr. Herzig, Phys. Rev. B **50**, 3577–3590 (1994)
50. R.W. Balluffi, *Grain-Boundary Diffusion Mechanisms in Metals*, in: *Diffusion in Crystalline Solids*, G.E. Murch, A.S. Nowick (Eds.), Academic Press, Inc., 1984, p. 319
51. A.P. Sutton, R.W. Balluffi, *Interfaces in Crystalline Materials* Clarendon Press, Oxford, 1995
52. Q. Ma, C.L. Liu, J.B. Adams, R.W. Balluffi, Acta Metall. Mater. **41**, 143 (1993)
53. C.L. Liu, S.J. Plimpton, Phys. Rev. **B41**, 2712 (1995)
54. M. Nomura, S.-Y. Lee, J.B. Adams, J. Mater. Res. **6**, 1 (1991)
55. M. Nomura, J.B. Adams, J. Mater. Res. **7**, 3202 ((1992)
56. M. Nomura, J.B. Adams, J. Mater. Res. **10**, 2916 ((1995)
57. M.R. Sorenson, Y. Mishin, A.F. Voter, Phys. Rev. **B62**, 3658 (2000)
58. Y. Mishin, M.R. Sorensen, A.F. Voter, in: *Mass and Charge Transport in Inorganic Materials – Fundamentals to Devices*, P. Vincenzini, V. Buscaglia (Eds.), Techna, Faenza, 2000, p. 377
59. A. Suzuki, Y. Mishin, Interface Science **11**, 131 (2005)
60. Y. Mishin, *Atomistic Computer Simulation of Diffusion*, in: *Diffusion Processes in Advanced Technological Materials*, D. Gupta (Ed.), William Andrew, Inc., 2005, p. 113
61. W. Gust, S. Mayer, A. Bögel, B. Predel, J. Physique **46** (C4), 537 (1985)
62. I. Kaur, W. Gust, L. Kozma, in two volumes: *Handbook of Grain and Interphase Boundary Diffusion Data*, Ziegler Press, Stuttgart, 1989
63. Chr. Herzig, S. Divinski, Chap. 4 in: *Diffusion Processes in Advanced Technological Materials*, D. Gupta (Ed.), William Andrew, Inc., 2005

33 Dislocation Pipe Diffusion

Atomic migration in solids is more rapid along or close to dislocations than through the regular lattice (Fig. 31.1). Since practically all crystals contain dislocations, any measured diffusion rate will usually contain a dislocation contribution. This may be quite negligible for low dislocation densities, especially at high temperatures. However, it can become important at low temperatures because of the low activation enthalpy for dislocation diffusion relative to that of lattice diffusion (see Chap. 31). It may be even the dominant mode of transport in some diffusion-controlled processes observable at relatively low temperatures such as precipitation and metal oxidation. The study and understanding of dislocation diffusion is therefore a matter of importance.

It is common practise to denote the number of dislocations that penetrate the unit area, ρ_d, as the dislocation density. It corresponds to the total length of dislocation lines per unit volume of a crystal. For example, a typical dislocation density of a well-annealed metal is about 10^6 cm^{-2}. This corresponds to a dislocation length of 10 km per cm^3. In a heavily deformed metal the dislocation length can reach 10^5 to 10^6 km per cm^3.

The average distance Λ between dislocations depends on the dislocation arrangement. It is usually given by

$$\Lambda = \frac{K}{\sqrt{\rho_d}}, \tag{33.1}$$

where K is of the order of unity. For example, for a quadratic array of parallel dislocation $K = 1$, for a hexagonal array $K = \sqrt{3}/2$.

Following the classification of HARRISON [7] (see Chap. 32), three kinetic regimes of dislocation diffusion can be distinguished as in the case of grain-boundary diffusion. The occurrence of a particular regime depends on the average dislocation distance and the lattice diffusion length. Type A kinetics is observed for

$$\sqrt{Dt} > \Lambda. \tag{33.2}$$

In this case the diffusion fields of neighbouring dislocations heavily overlap. Type B kinetics prevails for

$$a \ll \sqrt{Dt} \ll \Lambda. \tag{33.3}$$

Then, the overlap of diffusion fields from neighbouring dislocations is negligible. Type C kinetics occurs for

$$a > \sqrt{Dt}, \qquad (33.4)$$

when diffusion is restricted to the dislocation core. The latter is characterised by the dislocation pipe radius a (see below).

The subject of dislocation diffusion has been reviewed by GIBBS AND HARRIS [1], GJOSTEIN [2], and BALLUFFI [3]. A collection of data for diffusion along dislocations in metals can be found in LE CLAIRE'S chapter in a data collection for metals [4] and in a chapter by ERDELYI AND BEKE in a data collection for non-metals [5]. A thorough mathematical analysis analogous to the analysis of grain-boundary diffusion in Chap. 32 has been given by LE CLAIRE AND RABINOVITCH [6]. The main features of their treatment and major results are summarised below.

33.1 Dislocation Pipe Model

The simplest model for discussing diffusion properties of dislocations has been introduced by SMOLUCHOWSKI [8] and is illustrated in Fig. 33.1. Dislocations are considered as cylindrical pipes of radius a. The diffusivity in the dislocation pipe, D_d, is larger than the lattice diffusivity, D, outside the pipe. A frequent assumption for the pipe radius is $a = 0.5$ nm.

More realistic models would take into account that the diffusion coefficient varies with the distance r from the dislocation core. However, there is no clear indication for a suitably simple form of $D_d(r)$. LUTHER [9] investigated the consequences of $D_d(r) \propto 1/r^2$. The advantage of Luther's approach is not very apparent. Therefore, we prefer to regard dislocation diffusion as being adequately represented by Smoluchowski's model. This approach is analogous to the Fisher model of grain-boundary diffusion.

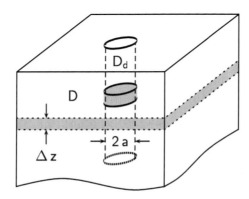

Fig. 33.1. Smoluchowski model of a dislocation pipe

33.1 Dislocation Pipe Model

The main features of dislocation diffusion can be illustrated for the case of isolated dislocations. If c and c_d represent the concentrations of the diffuser outside and inside the dislocation pipe, Fick's equations to be solved are:

$$\frac{\partial c}{\partial t} = D\left[\frac{1}{r}\frac{\partial}{\partial r}\left(r\frac{\partial c}{\partial r}\right) + \frac{\partial^2 c}{\partial z^2}\right] \quad \text{for} \quad r \geq a,$$

$$\frac{\partial c_d}{\partial t} = D_d\left[\frac{1}{r}\frac{\partial}{\partial r}\left(r\frac{\partial c_d}{\partial r}\right) + \frac{\partial^2 c}{\partial z^2}\right] \quad \text{for} \quad r < a. \quad (33.5)$$

In Eqs. (33.5) r and z are cylindrical coordinates. They denote the distance from the pipe axis and from the surface, respectively. The solutions are subject to the boundary conditions at $r = a$. There must be continuity of fluxes and concentrations:

$$\left(D\frac{\partial c}{\partial r}\right)_{r=a^+} = \left(D_d\frac{\partial c_d}{\partial r}\right)_{r=a^-},$$

$$c(r = a^+) = c_d(r = a^-). \quad (33.6)$$

For isolated dislocations the additional boundary conditions are

$$\frac{\partial c}{\partial r} \to 0 \quad \text{as} \quad r \to \infty. \quad (33.7)$$

As in the case of grain-boundary diffusion, two initial conditions at the surface $z = 0$ are considered: constant source (case I)

$$c(z = 0, r, t) = c_0 \quad (33.8)$$

and instantaneous source (case II)

$$c(t = 0) = 2M\delta(y). \quad (33.9)$$

In case I, the concentration at the surface is maintained at a value c_0 for all times $t \geq 0$. In case II, a very thin layer of $2M$ diffuser atoms per unit area is deposited on the surface.

Constant source conditions are appropriate, for example, if diffusion occurs from a vapour phase. Then c_0 is the concentration of the diffuser at the surface in equilibrium with the vapour. Constant source conditions are also appropriate for solubility-limited diffusion. Instantaneous source conditions simulate a conventional thin-film tracer diffusion experiment. However, it is only fully appropriate in practice if there is no rapid surface diffusion towards the dislocation to compensate the loss near $r = 0$ due to rapid diffusion down the dislocation pipe. In practice, neither the constant source condition nor the instantaneous source condition may exactly describe the situation prevailing at the surface. However, very likely they do represent the limits between which any experimental condition will lie.

While the diffusion fields for the constant source and the instantaneous source are different [6], we shall see below that the gradient of the log $\bar c$ versus z plot in the dislocation tail is the same for both cases. This is analogous to the Whipple and the Suzuoka solutions for grain-boundary diffusion (see Chap. 32). For diffusion of a solute that segregates to dislocations with an equilibrium segregation factor s, the second equation of (33.6) may be replaced by $sc = c_d$ (see also Chap. 32).

Exact solutions of the problem for isolated dislocations and for a hexagonal array of parallel dislocations all normal to and ending at the surface $z = 0$ of a semi-infinite solid have been worked out by LE CLAIRE AND RABINOVITCH [10–12] and summarised by the same authors [6]. For simplicity, we consider self-diffusion along dislocations, which implies $s = 1$. The solutions for the concentrations field around a dislocation, $c(r, z, t)$, are of the form

$$c(r,z,t) = c_1(z,t) + c_2(r,z,t) \quad \text{for} \quad r \geq a. \tag{33.10}$$

c_1 represents the standard expression for the concentration in the absence of the dislocation, under constant or instantaneous source conditions. c_2 is the additional concentration outside dislocations due the rapid diffusion down and out of them. The solutions for the diffusion fields are rather difficult to handle. We confine ourselves to the expressions for the average concentration, $\bar c(z,t)$, in a thin layer at some depth z after some time t.

33.2 Solutions for Mean Thin Layer Concentrations

Mean concentrations are of prime interest for the analysis of dislocation diffusion as for grain-boundary diffusion. Mean concentrations are measured in serial sectioning experiments. It is convenient to introduce abbreviations analogous to the normalised variables of grain-boundary diffusion:

$$\eta \equiv \frac{z}{\sqrt{Dt}},$$
$$\beta \equiv \frac{(\Delta - 1)a}{\sqrt{Dt}},$$
$$\alpha \equiv \frac{a}{\sqrt{Dt}},$$
$$\Delta \equiv \frac{D_d}{D}. \tag{33.11}$$

For the constant source (upper index I, surface concentration c_0), the solution for the mean concentration $\bar c^I$ is analogous to the Whipple solution of grain-boundary diffusion and can be written as, a sum of the complementary error function plus a dislocation tail:

$$\bar c^I(\eta) = c_0 \left[\text{erfc} \frac{\eta}{2} + \pi a^2 \rho_d Q^I \right]. \tag{33.12}$$

33.2 Solutions for Mean Thin Layer Concentrations 587

For the thin-layer source (upper index II), the solution for the mean concentration $\bar{c}^{II}$ is analogous to the Suzuoka solution of grain-boundary diffusion. It consits of a thin-film solution plus a dislocation tail:

$$\bar{c}^{II}(\eta) = \frac{M}{\sqrt{\pi D t}} \left[\exp\left(-\frac{\eta^2}{4}\right) + \pi a^2 \rho_d Q^{II} \right]. \tag{33.13}$$

In both cases, the contribution of dislocation diffusion is proportional to the total cross-sectional area of dislocations $\pi a^2 \rho_d$. In Eqs. (33.12) and (33.13) the quantities Q^I and Q^{II} are given by the following expressions [6]:

$$Q^I(\eta) = -\frac{16}{\pi^3}(\Delta - 1)^2 \int_0^\infty x^3 \exp(-x^2) \sin(\eta x) dx \int_0^\infty \frac{[\exp(-z^2) - 1]}{(\theta^2 + \phi^2)} \frac{dz}{z^3} \tag{33.14}$$

and

$$Q^{II}(\eta) = \frac{16}{\pi^{5/2}}(\Delta - 1)^2 \int_0^\infty x^4 \exp(-x^2) \cos(\eta x) dx \int_0^\infty \frac{[\exp(-z^2) - 1]}{(\theta^2 + \phi^2)} \frac{dz}{z^3}, \tag{33.15}$$

with

$$\theta = 2z Y_1(z\alpha) + (x^2 \beta - z^2 \alpha) Y_0(z\alpha),$$
$$\phi = 2z J_1(z\alpha) + (x^2 \beta - z^2 \alpha) J_0(z\alpha), \tag{33.16}$$

where J_1, J_0 and Y_1, Y_0 denote Bessel functions of the first and second kind of order one and zero, respectively. We note that constant source and instantaneous source solutions are related via

$$\bar{c}^{II}(\eta, t) = -\frac{M}{c_0 \sqrt{D t}} \frac{\partial}{\partial \eta} \bar{c}^I(\eta, t). \tag{33.17}$$

Fig. 33.2 compares the dependence of Q^I and Q^{II} on the normalised depth variable η. Both Q^I and Q^{II} first increase as η increases, pass through a maximum, and then decrease monotonically. Q^I is always positive and becomes zero at $\eta = 0$; Q^{II} has a zero and changes sign at the η-value for which Q^I is at its maximum, in accordance with Eq. (33.17). The value of Q^{II} is negative for smaller η-values because the finite amount of diffuser available under instantaneous source conditions is depleted around the dislocation pipe by the rapid diffusion down the pipe. At larger η-values, the mean concentration is enhanced in both cases as diffuser is leaking out of the dislocation pipe into its surrounding. The properties illustrated in Fig. 33.2 for dislocation pipe diffusion are qualitatively similar to the corresponding quantities for grain-boundary diffusion shown in Fig. 32.12.

When measurements are extended to values of the reduced penetration depth η such that the first terms of Eqs. (33.12) or (33.13) are negligible,

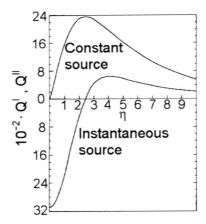

Fig. 33.2. Dislocation diffusion: mean thin-layer concentrations of the constant and instantaneous source solutions, Q^I and Q^{II}, for $\alpha = 10^{-2}$ and $(\Delta - 1) = 10^5$ according to LE CLAIRE AND RABINOVITCH [6]

dislocation tails can be observed in penetration profiles. The concentration in the tails is due to the material that has diffused down and out of the dislocations to depths well below those reached by lattice diffusion alone. The tail properties are determined by Q^I and Q^{II}. Figure 33.3 show logarithmic plots of Q^I and Q^{II} versus η for various values of the parameters α and $\alpha\beta$. These plots reveal the following features:

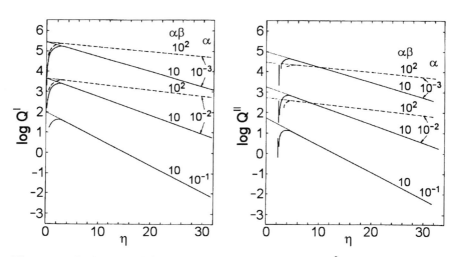

Fig. 33.3. Dislocation diffusion: constant source solution Q^I (*left*) and instantaneous source solution Q^{II} (*right*) versus η for $\alpha = 10^{-1}, 10^{-2}, 10^{-3}$, $\alpha\beta = 10$ (*full lines*) and $\alpha\beta = 10^2$ (*dashed lines*) according to LE CLAIRE AND RABINOVITCH [6]

1. Beyond η-values of 4 to 5 the plots are practically linear for $\alpha \leq 1$. This implies that plots of $\log \bar{c}$ are linear *versus* z for dislocation tails. We recall that plots of $\log \bar{c}$ for grain-boundary tails are linear *versus* $z^{6/5}$.
2. For given values of α and $\alpha\beta$, the slopes of the linear regions are practically the same for constant and instantaneous source conditions. The slopes can be represented as

$$\frac{\partial \ln Q^I}{\partial \eta} = \frac{\partial \ln Q^{II}}{\partial \eta} = -\frac{A(\alpha)}{\sqrt{\alpha\beta}}, \qquad (33.18)$$

where the quantity $A(\alpha)$ is given by

$$A^2(\alpha) = \frac{8}{\pi^2} \int_0^\infty \frac{\exp(-z^2)\,\mathrm{d}z}{z[J_0^2(z\alpha) + Y_0^2(z\alpha)]}. \qquad (33.19)$$

The properties of $A(\alpha)$ are illustrated in Fig. 33.4. $A(\alpha)$ is of the order of unity and a slowly varying function of α with very weak dependence on $\alpha\beta$.

3. In principle, the dislocation tail can provide an estimate of the dislocation density, if the experimental accuracy is sufficient to permit an extrapolation back to $z = 0$. The intersect with the ordinate, $\bar{c}_{int}$, is a measure of the dislocation density. For details we refer the reader to the review of LE CLAIRE AND RABINOVITCH [6].

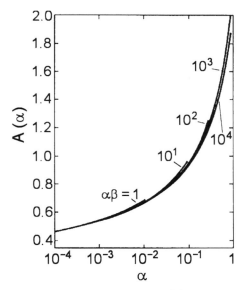

Fig. 33.4. Dislocation diffusion: The quantity $A(\alpha)$ is plotted as a function of α for various values of $\alpha\beta$ according to LE CLAIRE AND RABINOVITCH [6]

Type B kinetics: The solutions in Eqs. (33.12) and (33.13) correspond to isolated dislocations. They are appropriate if the average dislocation distance Λ is larger than the lattice diffusion length, i.e. for $\Lambda \gg \sqrt{Dt}$. Then, dislocations are sufficiently apart from each other and the individual diffusion fields do not overlap and type B kinetics prevails. According to Eq. (33.18) the measured slope of the dislocation tail is, for either surface condition, given by

$$\frac{\partial \ln \bar{c}}{\partial z} = -\frac{A(\alpha)}{\sqrt{(D_d/D - 1)a^2}} \qquad (33.20)$$

as long as $A(\alpha)$ is insensitive to variations in $\alpha\beta$. The quantity $(D_d/D - 1)a^2$ can be determined from the slope in the tail region of a plot of logarithm $\bar{c}$ *versus* penetration distance z with little uncertainty, because $A(\alpha)$ depends only weakly on α. In addition, the quantity $\Delta \equiv D_d/D$ can be determined from the slope of the tail, provided that the pipe radius a is known. Obviously the result for Δ depends on a. There are no undisputed measurements for a. A frequent assumption for metals is $a = 0.5$ nm [13, 14]. Larger values prevail in ionic crystals because of electrostatic effects.

Profiles of dislocation diffusion under type B kinetics conditions are similar to grain-boundary diffusion, with two major differences:

(i) As discussed above, diagrams of log $\bar{c}$ *versus* z are linear for dislocation tails. In contrast, grain-boundary tails are linear in plots of logarithm $\bar{c}$ *versus* $z^{6/5}$. In practice, a distinction between dislocation and grain-boundary diffusion on the basis of the shape of the penetration plots may be difficult. It requires measurements over at least two orders of magnitude in the tail region.

(ii) Because $A(\alpha)$ is only a slowly varying function of α, and thus an even more slowly varying function with time, the slopes of dislocation tails are almost independent of time. This is in marked contrast to grain-boundary tails, where the slopes in a plot of log $\bar{c}$ *versus* $z^{6/5}$ are proportional to $t^{-1/4}$ (see Chap. 32). This permits a relatively sensitive test to distinguish dislocation and grain-boundary diffusion tails. This test can, for example, also be used to distinguish diffusion along isolated dislocations from dislocations grouped in low-angle grain boundaries.

Type A kinetics: For long diffusion anneals and high dislocation densities type A kinetics condition prevail. This is the case for $\Lambda < \sqrt{Dt}$. Depending on the types of surface conditions either error function type or Gaussian penetration profiles evolve. Then, based on a dislocation volume fraction $g = \pi a^2 \rho_d$, an effective diffusivity, D_{eff}, can be measured (see Eq. 32.35), which according to HART [15] is given by

$$D_{eff} = D[1 + \pi a^2 \rho_d (D_d/D - 1)]. \qquad (33.21)$$

Type C kinetics: Type C kinetics is expected for $a > \sqrt{Dt}$, i.e. at low enough temperatures and/or for short diffusion anneals. The penetration profiles are linear in plots of log $\bar{c}$ versus penetration distance z^2. The slope is given by

$$\left(\frac{\partial \ln \bar{c}}{\partial z^2}\right)_{tail} = -\frac{1}{4 D_d t}. \qquad (33.22)$$

This equation can be used to determine D_d directly.

References

1. G.B. Gibbs, J.E. Harris, in: *Interfaces Conference*, Australian Institute of Metals, Butterworth, Melbourne, 1969
2. N.A. Gjostein, Chap. 9 in: *Diffusion*, Am. Soc. for Metals, Metals Park, Ohio, 1973
3. R.W. Balluffi, Phys. Stat. Sol. **42**, 11 (1970)
4. A.D. Le Claire, *Diffusion in Dislocations*, Chap. 11 in: *Diffusion in Solid Metals and Alloys*, H. Mehrer (Vol. Ed.), Landolt-Börnstein, Numerical Data and Functional Relationships in Science and Technology, New Series, Group III: Crystal and Solid State Physics, Vol.26, Springer-Verlag, 1990, p. 626
5. G. Erdelyi, D.L. Beke, *Dislocation and Grain-boundary Diffusion in Nonmetallic Systems*, Chap. 11 in: *Diffusion in Semiconductors and Non-metallic Systems*, D.L. Beke (Vol. Ed.), Landolt-Börnstein, Numerical Data and Functional Relationships in Science and Technology, New Series, Group III: Condensed Matter, Vol.33, Subvolume B1, Springer-Verlag, 1999, p. 11–1
6. A.D. Le Claire, A. Rabinovitch, *The Mathematical Analysis of Diffusion in Dislocations*, in: *Diffusion in Crystalline Solids*, G.E. Murch, A.S. Nowick (Eds.), Academic Press, Inc., 1984, p. 259
7. L.G. Harrison, Trans. Faraday Soc. **57**, 1191 (1961)
8. R. Smoluchowski, Phys. Rev. **87**, 482 (1952)
9. L.C. Luther, J. Chem. Phys. **43**, 2213 (1965)
10. A.D. Le Claire, A. Rabinovitch, J. Phys. C: Solid State Physics **14**, 3863 (1981)
11. A.D. Le Claire, A. Rabinovitch, J. Phys. C: Solid State Physics **15**, 3345 (1982), erratum 5727
12. A.D. Le Claire, A. Rabinovitch, J. Phys. C: Solid State Physics **16**, 2087 (1983)
13. H. Mehrer, M. Lübbehusen, Defect and Diffusion Forum **66–69**, 591 (1989)
14. Y. Shima, Y. Ishikawa, H. Nitta, Y. Yamazaki, K. Mimura, M. Isshiki, Y. Iijima, Materials Transactions, JIM, **43**, 173 (2002)
15. E.W. Hart, Acta Metall. **5**, 597 (1957)

34 Diffusion in Nanocrystalline Materials

34.1 General Remarks

'Nano' is the Greek word for 'dwarf'. In the International System of Units (SI) it is the decimal multiple 10^{-9} used as prefix of SI units. 'Nanoscience' refers to the range from one to several hundred nanometers and 'nanotechnology' are technologies in which atoms are manipulated in quantities of one to several thousand atoms. Nanoscience probably first gained attention in a 1959 lecture of the American Nobel laureate of 1965 in physics RICHARD FEYNMAN (1918–1988), who stated *'...that the day was not far off, when substances could be assembled at an atomic level'*. Although this day has not yet really come, nanotechnology involves at least manufacturing and characterisation of materials with crystal grains of nanometer size.

In nanocrystalline materials, new atomic structures and properties are generated by utilising the atomic arrangement in the cores of defects such as grain boundaries, interfaces, and dislocations. Depending on the type of defects utilised, nanocrystalline materials with different structures can be generated. These materials consist of a large volume fraction of defect cores and (strained) crystal lattice regions.

As an example, Fig. 34.1 shows the structure of a two-dimensional nanocrystalline material. The crystals are represented by periodic arrays of atoms in different crystallographic orientations (full circles). The atomic structures of the core regions of the boundaries between the crystallites are different because their structure depends on the crystal misorientations and on the boundary inclinations. The boundary core regions (open circles) are characterised by a reduced atomic density and by interatomic spacings deviating from those in the crystallites. Nanocrystalline materials are sometimes also denoted as nanophase materials or as nanometer-sized crystalline materials.

In this chapter, we consider mainly bulk nanocrystalline materials. Our understanding of nanocrystalline materials is documented in reviews, e.g., by SIEGEL AND HAHN [1], BIRRINGER AND GLEITER [2], GLEITER [3], GIALANELLA AND LUTTEROTTI [4], HEITJANS AND INDRIS [6], and CHADWICK [7].

Diffusion has attracted attention, largely because material transport belongs to the group of physical properties differing most from single-crystalline

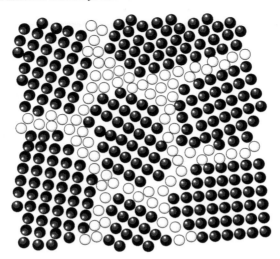

Fig. 34.1. Schematic view of a nanocrystalline material

or conventional polycrystalline materials. In addition, diffusion is a key property determining the suitability of nanocrystalline materials for applications. The present chapter is organised as follows: Sect. 34.2 reminds the reader of some important synthesis techniques of nanocrystalline materials. In Sect. 34.3 we consider theoretical aspects of diffusion in polycrystalline materials of various grain size. Sect. 34.4 illustrates diffusion in metallic nanomaterials by typical examples. Sect. 34.5 contains remarks about ionic conduction and diffusion in nanocrystalline ionic materials.

34.2 Synthesis of Nanocrystalline Materials

The synthesis of nanocrystalline materials is frequently carried out by assembling pre-generated small crystalline clusters created, for example, by inert-gas condensation, high-energy ball milling, and by chemical methods. Bulk materials are produced from nanocrystalline powders by consolidation and sintering. There are, however, also procedures like heavy plastic deformation, which do not require pre-generated clusters. We discuss some of the most important techniques and refer for details and for a more comprehensive list of synthesis techniques to reviews of GLEITER [3] and GIANELLI [4] and the anthology of NAIWA [5].

34.2.1 Powder Processing

Inert-gas Condensation: Inert-gas condensation is a direct extension of physical vapour deposition methods. In a UHV chamber backfilled with a low pressure of inert gas, typically He or Ar, the vapour phase of the material to be nanocrystallised is formed. Vapours are commonly produced in ovens, electron guns, sputter devices and laser evaporation sources.

For example, a substance (e.g., Cu) is vapourised in an oven. Then, the evaporated atoms transfer their kinetic energy to the inert gas and condense in the form of nanosized crystals. Direct evaporation into a gas produces an aerosol or a smoke of clusters. The mean cluster size can be controlled by varying the evaporation rate and the inert-gas pressure. Clusters with mean diameters as small as 3 to 4 nanometers can be generated by this process. The cluster size can be reduced and the size distribution sharpened by imposing a forced convective current to the inert gas to reduce cluster-cluster aggregation processes. The clusters are carried by the convective current of inert gas to the surface of a finger cooled with liquid nitrogen. Subsequently, the inert gas is removed and the particles can be scraped away and collected to be processed further. For instance, they can be directly compacted in the UHV chamber in a piston and anvil-like device using pressures of several GPa. The whole process leads to bulk nanocrystalline samples and can be carried out under well controlled atmosphere and clean conditions. This procedure was developed by GLEITER AND ASSOCIATES [2, 3]. It has been frequently reproduced and described in scientific and technical papers [4, 8] and has become a milestone in the history of nanostructural materials.

This processing technique also allows for great versatility. Composite materials can be produced by using two or more evaporation sources; oxide materials and other ceramic materials are obtained by mixing or replacing the inert gas with a reactive gas such as oxygen or nitrogen; the particle size can be controlled by the evaporation rate and the condensation gas pressure.

High-energy Milling: A few remarks on industrial application of milling techniques in conventional powder technology may be useful. High-volume, low-energy mills are used to process metallic powders and to modify their microstructure. For instance, industrial devices are employed in the production of oxide dispersion strengthened alloys. Milling is also used to achieve very fine-grained ceramic powders in metallic matrices, such as Mg- and Al-base light-weight alloys and Ni-base superalloys. Moreover, the formation of very fine mixtures of elemental and pre-alloyed metallic powders induced by milling promotes formation of stable solid solutions or intermetallic compounds. This occurs either directly during milling or upon heating milled powders to temperatures which trigger the reactions. The most popular laboratory mills, having common features to large scale devices, are planetary ball mills, vibratory mills, shaker mills, and attritor mills. Notes on the working principle of mills can be found, for instance, in [4, 9].

The reduction of grain size in powder samples to a few nanometers during heavy mechanical deformation in high-energy ball mills plays a significant rôle in the powder processing of nanostructured materials. In particular, this is true for bcc and hcp metallic powders as well as for intermetallic compounds.

During milling of metallic powders a large amount of cold-work energy can be stored in the materials. Values as high as 40 % of the enthalpy of melting are reported [10]. An important parameter is the minimum grain size, d_{min}, that can be reached by milling. Grain refinement is limited by the smallest separation between dislocations, d_c. When the grain size is in the range of d_c, dislocations become unstable and anneal out at the grain boundaries. This dynamic recovery prevents further grain refinement via plastic deformation. A linear relationship, $d_{min} \propto d_c$, has been proposed on the basis of data available for pure metals and binary alloys (see, e.g., Fe-Cu alloys [11]). d_c depends on the composition of the material. Finer grain size can be achieved by varying the composition, since the stacking fault energy and the modulus of the Burgers vector change accordingly, and with them the minimum grain size [13]. The ball milling method seems to be not applicable to fcc metals. If fcc metal powders are ball-milled, they sinter to larger particles up to millimeter size. Fcc metals are too soft for an effective energy storage during the milling process.

Mechanical milling may induce several transformations, such as the formation of nanostructures, still showing crystallinity. Partially or fully amorphous materials are also obtained. This is because a high amount of cold-work energy produced by ball milling can also promote amorphisation. This was reported for the first time for Ni-Nb powder mixtures [14]. Asymmetric interdiffusion of the components, resulting from their very different diffusivities [15], is one of the most important factors for this transformation, which can be regarded as an example of solid-state amorphisation [16, 17]. High-energy ball milling is also suitable to produce nanostructured alloys by mechanical alloying of the constituent powders.

Mechanical milling and mechanical alloying are comparatively inexpensive, require simple equipment, and are well suited to be scaled up to mass production of nanostructured powders. Actually, processing large powder feed-stocks can have the beneficial effect of reducing the concentrations of contaminants from milling media and atmosphere, which may reach intolerably high values in laboratory size equipment (see, e.g., [12]).

34.2.2 Heavy Plastic Deformation

Heavy plastic deformation methods also called severe plastic deformation (SPD) methods have attracted growing interest for the direct production of bulk nanostructured samples and provide interesting alternatives to powder processing routes. SPD methods have been reviewed by VALIEV ET AL. [18]. These methods overcome difficulties connected with residual porosity in compacted powder samples and impurities from the ball-milling process.

The basic idea of this approach, comprising various techniques, is to subject bulk specimen to large plastic deformation at relatively low temperatures. In this way, a uniform nanostructure can be induced and, thanks to the

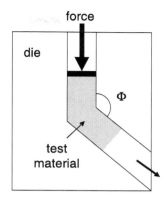

Fig. 34.2. Schematic drawing of an ECAP device used for severe plastic deformation according to VALIEV ET AL. [18]

specific design of SPD equipments, the intensive fracturing following strain-hardening is largely avoided. Another interesting aspect is the controlled deformation geometry. This allows one to obtain reproducible straining conditions, different from high-energy ball milling. In the latter case, the loading geometry is very complex and difficult to predict and model.

Several kinds of SPD methods are in use: severe plastic torsion straining, equal channel angular pressing (ECAP), and multiple forging [4, 18]. For example, during ECAP pressing a billet is pressed through a die using a facility with two channels which intersect at an angle Φ (Fig. 34.2). This permits intense plastic strain without changing the cross section area of the billet. Repeated deformation of the material in a defined way is also possible. In the case of hard materials, ECAP can be conducted at elevated temperatures. Typical sizes of the processed billets are 7 to 10 centimeter in length and not more than 2 centimeters in diameter.

The structure of SPD materials can be rationalised via transformations of a dislocated sample with critical dislocation density to a nanocrystalline material. The structures of SPD materials contain mainly high-angle grain boundaries. Soft metals such as Cu and Al can also be processed by SPD methods. Cu and Al pure metallic and composite compacts reach almost full density. Heavy plastic deformation methods can also be used for consolidating metallic and composite powders.

Whereas ECAP usually leads to microstrucures with grain sizes in the range of several 100 nanometers, high pressure torsion (HPT) can yield grain sizes that are in the range of 10 nanometers. Yet only small pellets can be sythesised via HPT.

An alternative technique to produce extremely fine-grained bulk nanocrystalline materials has been developed by WILDE AND COWORKERS [19]. It is based on repeated cold rolling with intermediate folding. This technique is capable of synthesising pure fcc metals in gram quantities with grain sizes below 10 nanometers. Such small grain sizes have not been achieved by any other SPD technique. This deformation procedure resembles the technique of ancient damascene sword making.

34.2.3 Chemical and Related Synthesis Methods

Chemical processing has great potential for the synthesis of nanocrystalline material. For example, sol-gel methods usually involve low preparation temperatures, preventing recrystallisation and coalescence of powder particles, and the possibility to control the composition through the selection of chemical reagents. However, full removal and decomposition of several chemicals, such as organic solvents, salts, organometallic precursors is not always easy to achieve. Therefore, the concentration of impurities in the final metallic phases can be rather high. For this reason, not many examples of fully metallic nanocrystalline materials processed by chemical methods can be found in the literature. Chemical routes are more frequently used to synthesise either fully ceramic systems or composites of nano-grained metallic phases dispersed into a ceramic matrix.

An interesting process for producing nanocrystalline Fe-Ni alloys has been reported by LEE ET AL. [23]. The idea of this process is to combine ball milling and chemical methods. Nanocrystalline precursors of Fe oxide and Ni oxide powders are synthesised by ball milling. The ball-milled oxides are then reduced in hydrogen atmosphere into powders of an Fe-Ni nanoalloy. The reduction process yields agglomerates of about 30 μm in diameter each composed of nanoalloy particles of about 30 nm in size. From these powders, nanocrystalline bulk material can be produced by pressureless sintering.

34.2.4 Devitrification of Amorphous Precursors

This route for synthesising nanomaterials is based on partial crystallisation of amorphous alloys, obtained by rapid solidification (see Chap. 29). Either fully nanocrystalline materials or composites of nanograins dispersed in an amorphous matrix can be obtained. High-strength Al-base alloys or ferromagnetic alloys can be prepared by crystallisation methods.

Aluminium–transition metal-rare earth alloys, for instance, Al-Ni-Y [20], can be quenched into the amorphous state and display very high mechanical strength. The strength can be further improved thanks to crystallisation of nanoparticles. The extremely high number density of the nanoparticles enchance the mechanical strength of Al-rich alloys or the saturation magnetisation of Fe-rich alloys. At the same time the stability of the amorphous-nanocrystalline composite structure is improved due to the overlap of diffusion fields that surround each nanoparticle [21]. It was also shown that severe plastic deformation of amorphous precursors ($Al_{88}Y_7fe_5$) can lead to nanocrystalline microstructures with refined grain size and enhanced stability against coarsening [22].

Soft-magnetic nanomaterials can be sythesised by crystallising amorphous precursors (see, e.g., [24]). For applications involving ac magnetic fields it is important to have a low coercive field. From a microstructural viewpoint the coercivity is high if magnetic domain walls are pinned in the material. Grain boundaries are usually very effective in hindering domain-wall movement. Therefore, a conventional approach to improve soft-magnetic properties is to increase the grain size. This is, however, only suitable for grain sizes in excess of about 100 nanometers. An inversion of the dependence of the coercive fields on the grain size occurs, when the thickness of the domain walls exceeds the average grain size. Two examples of such alloys are $Fe_{73.5}Si_{13.5}B_9Nb_3Cu_1$ (VITROPERM) and $Fe_{90}Zr_7B_3$ (NANOPERM) consisting of α-Fe nanocrystallites embedded in a residual amorphous matrix. The soft magnetic properties of such materials are comparable to, or even better than those of Fe-Si alloys, Fe-Ni permalloy, and completely amorphous Co and Fe based metallic glasses.

34.3 Diffusion in Poly- and Nanocrystals

Grain boundaries and interfaces are high-diffusivity paths (see Chap. 32). In nanomaterials a large amount of atoms is located in grain boundaries or interfaces (Fig. 34.1). For example, in material with an average grain size of 5 nm about 50 % of the atoms are located in boundaries. In material with 10 nm grains 20 % still lie in boundaries.

It is sometimes argued that grain-boundary diffusion plays no significant rôle at near-ambient temperature. This is not true for nanostructured materials. Let us consider grain-boundary self-diffusion in Cu as an example. Using experimental values for the grain-boundary diffusivity [26], we estimate that the diffusion length inside grain-boundaries, $\sqrt{D_{gb}t}$, at 700 K is about 8 μm for an annealing time t of 1 min. Near room temperature the corresponding diffusion length is still about one nanometer per minute. This means that in two hours an atom will travel about 10 nanometers inside the grain-boundary network, whereas the lattice diffusion length, $\sqrt{Dt}$, at room temperature is completely negligible [27]. The diffusion length $\sqrt{D_{gb}t}$ is comparable to the grain size in typical nanostructured Cu materials. Grain-boundary diffusion thus cannot be neglected near room temperature.

34.3.1 Grain Size and Diffusion Regimes

Diffusion in polycrystals is already considered in Chap. 32. Three main kinetic regimes were introduced and denoted as type A, B, and C. In the type C regime, diffusion takes place only inside the grain boundaries. In the B regime lattice diffusion fringes are formed by out-diffusion from the boundaries into the adjacent grains. In the type A regime the lattice diffusion fringes from adjacent boundaries overlap. For bi-crystals only type B and type C regimes are relevant.

Diffusion kinetics in poly- and nanocrystalline materials has been discussed in greater detail by KAUR, MISHIN, AND GUST [29]. These authors refined the classification of diffusion regimes A, B, and C by introducing further subregimes. In what follows, we summarise the essentials of their discussion. For convenience, we repeat two parameters relevant for grain-boundary diffusion, which had been already introduced in Chap. 32:

$$\alpha \equiv \frac{s\delta}{2\sqrt{Dt}} = \frac{s\delta}{2L_b} \quad \text{and} \quad \beta \equiv \frac{s\delta D_{gb}}{2D^{3/2}t^{1/2}} = \left(\frac{L_{gb}^B}{L_b}\right)^2. \quad (34.1)$$

s denotes the segregation factor of a solute and δ the grain-boundary width. In the following, it is useful to distinguish several characteristic lengths scales listed in Table 34.1.

Depending on the interrelations between the characteristic lengths scales, several kinetic regimes and subregimes of diffusion can be distinguished in polycrystals as illustrated in Fig. 34.3. In addition, polycrystals can be subdivided into three main classes according to the grain size: these classes are denoted as coarse-grained, fine-grained, and ultrafine-grained polycrystals. At a fixed temperature, polycrystals of each class show their own scenario of diffusion regimes in time. The scenarios for the various classes of polycrystals are listed in Table 34.2 and discussed below.

Coarse-grained Polycrystals: The following sequence of diffusion regimes occurs with increasing time: $C \to B_2 \to B_4 \to A$ (see Table 34.2, Fig. 34.3). Grain-boundary diffusion starts in the C regime, where the leakage of diffusing atoms from the boundaries to the grains is negligible and boundaries act as more or less parallel diffusion paths. If all boundaries are perpendicular to the surface, diffusion along boundaries gives rise to Gaussian or error-function diffusion profiles, depending on the initial conditions. In real polycrystals, the grain boundaries intersect at various angles with the surface. Nevertheless, the grain-boundary diffusion coefficient, D_{gb}, can be deduced from the mea-

Table 34.1. Characteristic length scales for diffusion in polycrystals

Lattice (or bulk) diffusion length	$L_b = \sqrt{Dt}$
Average grain size	d
Grain-boundary width	δ
Effective grain-boundary width	$s\delta$
Diffusion length inside boundaries (C regime)	$L_{gb}^C = \sqrt{D_{gb}t}$
Effective grain-boundary diffusion length (B regime)	$L_{gb}^B = \frac{\sqrt{sD_{gb}\delta}}{(4D)^{1/4}}t^{1/4}$

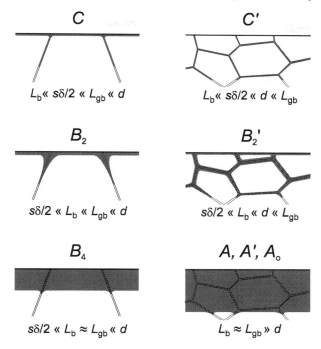

Fig. 34.3. Tracer distribution in various kinetic regimes and subregimes of diffusion in polycrystals according to KAUR, MISHIN AND GUST [29]

Table 34.2. Sequence of diffusion regimes and subregimes in polycrystals according to KAUR, MISHIN, AND GUST [29]. The sequences corresponds to a fixed temperature and increasing time

Class of polycrystal	Sequence of diffusion regimes in time
Coarse-grained	$C \to B_2 \to B_4 \to A$
Fine-grained	$C \to B_2 \to B_2' \to A'$
Ultrafine-grained	$C \to C' \to B_2' \to A'$
Special ultrafine-grained	$C \to C' \to A_0$

sured profile if the intersection angles are taken into account. Since $D_{gb} \gg D$, we have $L_{gb}^C \gg L_b$. Negligible leakage from the boundaries also means $\alpha \gg 1$ or $L_b \ll s\delta/2$.

The bulk diffusion length, L_b, grows with time and the condition for type C kinetics, $L_b \ll s\delta/2$, breaks down sooner or later and finally changes to the opposite, $L_b \gg s\delta/2$. Under the latter condition, the parameter α is small and the kinetics changes to the B_2 regime. Diffusion profiles in the B_2 regime assume a two-step shape, with a near-surface step caused by bulk in-diffusion from the surface with penetration depth $\approx L_b$ and a grain-boundary tail. The characteristic penetration depth of the grain boundary fringes is given by L_{gb}^B. From the grain-boundary tail of a diffusion profile the triple product $s\delta D_{gb}$ can be determined. Examples are shown in Fig. 32.14 of Chap. 32.

A comparison of the penetration distances L_{gb}^C and L_{gb}^B shows that with the change from the C to the B_2 regime the penetration law changes from $L_{gb}^C \propto \sqrt{t}$ to $L_{gb}^B \propto t^{1/4}$. From the condition $L_{gb}^C = L_{gb}^B$, the characteristic time and length, t' and L', of this transition are:

$$t' = \frac{s^2\delta^2}{4D} \quad \text{and} \quad L' = \frac{s\delta}{2}\sqrt{\frac{D_{gb}}{D}}. \qquad (34.2)$$

Note that in the B_2 regime we have preferential grain-boundary penetration, i.e. $L_{gb}^B \gg L_b$ and $\beta \gg 1$.

The lattice diffusion length, L_b, grows faster than L_{gb}^B (see Table 34.1). At some time L_b reaches the magnitude of L_{gb}^B. From the condition $L_b = L_{gb}^B$, the characteristic time and length, t'' and L'', of this intersection point are:

$$t'' = \frac{(s\delta D_{gb})^2}{4D^3} \quad \text{and} \quad L'' = \frac{s\delta D_{gb}}{2D}. \qquad (34.3)$$

These equations mark the end of the B_2 regime and the transition to the B_4 regime. In the B_4 regime diffusion proceeds with a nearly planar front (Fig. 34.3). Depending on the initial conditions, the diffusion profile is given by Gaussian or error function solutions.

Coarse-grained polycrystals in the present context means that the grain size d is much larger than L''. This implies that the grain-boundary diffusion fringes remain isolated for such a long time that diffusion along each individual boundary has enough time to evolve through all the regimes from C to B_4. Some time later, the lattice diffusion length reaches the average grain size ($d \approx L_b$) and then exceeds it. This marks the beginning of the A regime in which $L_b > d$. In the A regime the polycrystal behaves like an effective medium with some effective diffusivity. We consider the case of effective diffusion in Sect. 34.3.2.

Fine-grained Polycrystals: We now consider the evolution of diffusion regimes with increasing annealing time in polycrystals under conditions where the grain size lies in the range $L' \ll d \ll L''$. The following diffusion regimes develop with increasing time: $C \to B_2 \to B_2' \to A'$ (see Table 34.2, Fig. 34.3). Diffusion starts in grain boundaries (C regime) like for coarse-grained polycrystals. With increasing diffusion time, for $t \gg t'$ the B_2 regime is reached. After some time the effective grain-boundary diffusion length L_{gb}^B reaches and exceeds the average grain size. Then, a regime denoted as B_2' is reached, in which $L_{gb}^B \gg d \gg L_b \gg s\delta/2$. In the B_2' regime the tracer penetrates far beyond the average grain size, but nevertheless the diffusion fringes around the boundaries do not overlap.

Diffusion in the B_2' regime has been analysed by BOKSTEIN ET AL. [30] and by LEVINE AND MACCALLUM [31]. The main features of the diffusion kinetics in this regime are independent of grain shape and are (apart from geometrical factors) the same as in the B_2 regime for coarse-grained polycrystals. This is plausible from the following considerations: for $L_{gb}^B \gg d$ the diffusant concentration in the grain-boundaries encompassing the grains of the diffusion zone are almost the same. Since $d \gg L_b$, atoms that leave a grain boundary and travel into an adjacent grain behave very much the same as around an isolated planar boundary. Fisher's model appears to be also applicable if $L_b \ll d \ll L_{gb}^B$. However, the grain-boundary width should be replaced by an effective value according to $\delta \to g/A$, where g is the volume fraction of grain boundaries and A the grain-boundary area per unit volume. The ratio g/A differs from δ by a geometrical factor of the order of unity. For example in a cubic grain model $g \approx 3\delta/d$, $A = 6/d$ and then $g/A = \delta/2$. The triple product $s\delta D_{gb}$ is the only parameter that can be determined in this regime.

With further increase of the annealing time, the diffusion fringes around grain boundaries start to overlap and finally the bulk diffusion length becomes larger than the grain size. Then, we reach a regime with a planar diffusion front described by an effective diffusivity. This regime is denoted as A' to distinguish it from the A regime in coarse-grained polycrystals. In the present case, we have $d \ll L''$. In the type A regime of coarse-grained polycrystals, the effective diffusivity is dominated by the lattice diffusion coefficient, whereas in fine-grained material the effective diffusivity is dominated by diffusion along grain boundaries. However, the same expression for the effective diffusivity can be used (see, e.g., Eqs. 34.4 or 34.7).

Ultrafine-grained Polycrystals: Finally, let us consider diffusion in polycrystals with a grain size so small that $d \ll L'$. Ultrafine-grained polycrystals have the following scenario of diffusion regimes with increasing time: $C \to C' \to B_2' \to A'$ (see Table 34.2 and Fig. 34.3).

Again the diffusion process starts in the C regime, as long as $L_b \ll s\delta/2$ is fulfilled. With increasing time, the penetration depth inside the grain boundaries, L_{gb}^C, reaches and then exceeds the grain size. Then a new subregime C' occurs, not considered so far. It is characterised by $L_{gb}^C \gg d \gg s\delta/2 \gg L_b$. In this regime the diffusant penetrates deeply (compared to the grain size) inside the grain boundaries without considerable leakage into the grains. The moment t_0 of the transition $C \to C'$ can be estimated from the condition $L_{gb}^C = d$, which yields $t_0 = d^2/D_{gb}$. Using Eqs. (34.2) the transition time can also be written as $t_0 = (d/L')^2 t'$, which implies $t_0 \ll t'$ since in the present case $d \ll L'$. Therefore, the C' regime develops before the time t' at which the onset of the B_2' regime is expected. For times longer than t', which corresponds to the upper limit of the C' regime, the diffusion process reaches the B_2' regime, in which $L_{gb}^C \gg d \gg L_b \gg s\delta/2$. Some time later the lattice diffusion length exceeds the grain size and the diffusion kinetics enters the A' regime.

Strongly Segregating Solutes in Ultrafine-grained Polycrystals: In the previous discussion we had assumed that the grain size, although small, is still larger than $s\delta/2$. In ultrafine-grained polycrystals the effective grain-boundary width of strongly segregating solutes, $s\delta/2$, is larger than the grain size. Now we consider ultrafine-grained polycrystals in which the grain size is so small and/or segregation so strong that $d \ll s\delta/2$ is fulfilled. Then, the following scenario of diffusion regimes with increasing time emerges: $C \rightarrow C' \rightarrow A_0$ (see Table 34.2, Fig. 34.3). Figure 34.3 shows that as d approaches $s\delta/2$ the B'_2 regime shrinks and disappears for effective boundary widths, which are larger than the grain size. Thus after short C and C' regimes we arrive at an A type regime denoted as A_0. The effective diffusivity in the A_0 regime is given by a modified Hart-Mortlock equation discussed in Sect. 34.3.2.

34.3.2 Effective Diffusivities in Poly- and Nanocrystals

Theoretical aspects of diffusion in polycrystals with type A kinetics have been reconsidered by BELOVA AND MURCH [32] and MAIER AND ASSOCIATES [33, 34]. The authors of [32] discuss arrangements of grain-boundaries and grains illustrated in Fig. 34.4. Contributions to the effective diffusivity D_{eff} of poly- and nanocrystalline material arise from the lattice diffusivity D in the grains and the diffusivity D_{gb} inside the grain-boundaries. The volume fraction of grain boundaries is denoted as g.

Effective Self-diffusivity: Self-diffusion is the most basic diffusion process also in poly- and nanocrystalline materials. One can distinguish several limits for the effective diffusivity, depending on the arrangement of grains and grain boundaries:

The *upper limit* for the effective self-diffusivity is well-known for parallel arrangement of grain boundaries and grains in the diffusion direction

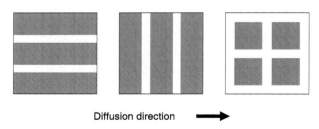

Fig. 34.4. Models representing grains (*dark*) and boundaries in a nanostructured material. *Left*: parallel arrangements of grains and grain boundaries in the diffusion direction. *Middle*: serial of arrangement of grains and grain boundaries. *Right*: grains represented as cubes

(Fig. 34.4, *left*). It is the Hart equation [35] already discussed in Chap. 32. We repeat it for convenience:

$$D_{eff}(\text{Hart}) = gD_{gb} + (1-g)D. \qquad (34.4)$$

In the solid-state diffusion literature the Hart equation has been the standard equation for interpreting the effective diffusivity measured in type A kinetics, where the bulk diffusion length, $L_b = \sqrt{Dt}$, is larger than the spacing between the grain boundaries.

The *lower limit* for the effective diffusivity corresponds to a series of alternating grains and grain boundaries in the diffusion direction (Fig. 34.4, *middle*). The pertaining diffusivity is:

$$D_{eff}(\text{series}) = \frac{D_{gb}D}{gD + (1-g)D_{gb}}. \qquad (34.5)$$

Equations (34.4) and (34.5) are exact for the geometries described.

For other arrangements and shapes of the grains, a number of different formalisms have been applied to describe the effective diffusivity. More than a century ago the famous scientist MAXWELL [36] developed a mean field approximation for the effective dc conductivity in a material composed of two phases with different conductivities. By analogy to this treatment, the effective diffusivity of a polycrystal is:

$$D_{eff}(\text{Maxwell}) = \frac{D_{gb}[(3-2g)D + 2gD_{gb}]}{gD + (3-g)D_{gb}}. \qquad (34.6)$$

Equation (34.6) is known as the *Maxwell equation* or as the *Maxwell-Garnett equation* [37]. In the limiting case, $D_{gb} \gg D$ and $g \ll 1$, the Hart equation yields $D_{eff}(\text{Hart}) \approx gD_{gb}$ and the Maxwell equation yields $D_{eff}(\text{Maxwell}) \approx 2gD_{gb}/3$.

Monte Carlo simulations by BELOVA AND MURCH have shown that the Maxwell-Garnett equation provides a fairly accurate description of the effective diffusivity [32]. These authors consider a simple phenomenological model represented by cubic arrangement of grains and grain boundaries (see Fig. 34.4, *right*). The results of the Monte Carlo simulation for $D_{gb}/D = 100$ and 1000 were compared with with Eqs. (34.4), (34.5), and (34.6). The comparison showed that for values of g that could reasonably be expected to correspond to nanocrystalline materials (g less than about 0.5) the Maxwell-Garnett equation (34.6) yields good results, whereas the Hart equation (34.4) and the series equation (34.5) are poorer approximations. These findings imply that the Hart equation should not be used, unless the experimental conditions for tracer penetration can be well described by parallel boundaries in the diffusion direction.

Effective Diffusivity of Solutes: Equations for the effective diffusivity of a solute in polycrystals, which correspond to the equations for the host atoms have also been derived. For parallel boundaries in diffusion direction the Hart equation was extended by MORTLOCK [38] to cover the case of effective solute diffusivity. For convenience we repeat Eq. (32.39):

$$D_{eff}(\text{Hart-Mortlock}) \approx sgD_{gb} + (1-sg)D. \tag{34.7}$$

Euation (34.7) is generally referred to as the Hart-Mortlock equation. All the diffusivities in this and subsequent equations refer to that of a solute with segregation factor s.

Unlike the Hart equation (34.4), the Hart-Mortlock equation (34.7) is not exact for solute diffusion even for parallel grain boundaries in the diffusion direction. In the presence of boundary segregation the fractions of time that diffusing atoms spend in grain boundaries and in grains determine the effective diffusivity. The time fractions can be calculated from the volume fractions, g and $1-g$, and the segregation factor (see Eq. 32.38). By combining Eqs. (32.37) and (32.38), we get:

$$D_{eff}(\text{corrected Hart-Mortlock}) = \frac{sgD_{gb} + (1-g)D}{1-g+sg}. \tag{34.8}$$

Equation (34.8) is the corrected Hart-Mortlock equation for solute diffusion.

For a series of grain boundaries and grains in the diffusion direction, the effective solute diffusivity is given by:

$$D_{eff}(\text{series-solute}) = \frac{sD_{gb}D}{gD_b + s(1-g)D_{gb}}. \tag{34.9}$$

This equation is the extension of (34.5) for solute atoms.

The Maxwell-Garnett equation (34.6) for self-diffusion has been modified by KALNINS ET AL. [33] to account for solute segregation. The modified Maxwell equation is

$$D_{eff}(\text{Maxwell-solute}) = \frac{sD_{gb}[(3-2g)D + 2sgD_{gb}]}{(1-g+sg)[gD+(3-g)sD_{gb}]}. \tag{34.10}$$

Like in the case of effective self-diffusion, Monte Carlo simulations for solute diffusion show that the modified Maxwell-Garnett equation (34.10) provides a much better description of the effective diffusivity than the corrected Hart-Mortlock equation (34.8) and the series equation (34.9) [32].

34.4 Diffusion in Nanocrystalline Metals

34.4.1 General Remarks

Since the pioneering work performed on nanocrystalline materials in the 1980s by GLEITER AND COWORKERS [39], diffusion in these materials has attracted permanent interest, largely because material transport in nanostructured materials belongs to the group of material properties differing most from their coarse-grained or single-crystalline counterparts. An overwiew of the early diffusion measurements on nanocrystalline metals and alloys was given in [40]. More recent developments can be found in a status report of WÜRSCHUM AND COWORKERS [28].

In the first diffusion studies in this field diffusivities in nanocrystalline Cu produced by inert-gas condensation and subsequent consolidation were found to be significantly faster than in grain boundaries of conventional polycrystals (see, e.g., [41, 42]). Soon after this initital era it was recognised that factors such as structural relaxation, grain growth, residual porosity, different types of interfaces, and perhaps triple junctions must be taken into account to obtain an unambigous assessment of diffusion in nanocrystalline metals. More recent studies taking structural relaxation and grain growth into consideration came to the conclusion that diffusivities in relaxed interfaces of nanocrystalline metals are similar to or only slightly higher than grain-boundary diffusivities obtained from conventional bicrystals or polycrystals.

Somewhat at variance with the finding that the grain boundary diffusivities of nanocrystalline materials are siminlar to those obtained from conventional polycrystals are the observations of super-plasticity [43] and increased strength and ductility [44] of nanostructured materials processed by severe plastic deformation. These properties have been attributed to the formation of non-equilibrium grain-boundaries with enhanced diffusivity [45]. However, so far the existence of such grain-boundary structures has not been established by experiments.

Most of the experimental techniques discussed in part II of this book have been applied to diffusion studies on nanocrystalline metals and alloys as well. These methods include radiotracer techniques, electron microprobe analysis, Auger electron and secondary ion mass spectrometry, Rutherford backscattering, and nuclear magnetic resonance. The nanostructured materials studied were prepared by various synthesis routes discussed above including inert-gas condensation and consolidation, severe plastic deformation, mechanical milling and compaction, and crystallisation of amorphous precursors. An overview of investigations available up to 2003 for metallic nanomaterials can be found in Table 1 of [28].

34.4.2 Structural Relaxation and Grain Growth

Since the conditions during the synthesis of nanocrystalline materials are far from thermodynamic equilibrium, the initial structure of grain-boundaries and interfaces of bulk samples may depend on their time-temperature history. For instance, for nanocrystalline metals prepared by inert-gas condensation and subsequent compaction or by severe plastic deformation structural relaxation effects have been reported, which lead to a decrease of the self-diffusivity in the boundaries in nano-Fe [46] and nano-Ni [48]. In both cases the grain-boundary diffusion coefficients in the relaxed state are similar or only slightly higher than the values expected for conventional grain boundaries.

The relaxed structure of nanocrystalline metals is prone to grain-boundary motion and grain growth. In this case the assessment of the diffusion behaviour is affected by the concomitant grain-boundary migration. The occurrence of grain growth during diffusion leads to a decrease of the interface fraction and, as a consequence of the growth-induced boundary migration to a slowing down of tracer diffusion, since tracer atoms are immobilised by incorporation in lattice sites of the crystallites. These complications may lead to deviations from diffusion profiles expected for type C kinetics.

34.4.3 Nanomaterials with Bimodal Grain Structure

In a number of nanocrystalline alloys, it has been possible to carry out diffusion measurements without complications caused by structural relaxation and grain growth. Despite the stable microstructure, the diffusion behaviour of nanocrystalline alloys may still be more complex than discussed in Sect. 34.3. One reason is the presence of several types of interfaces[1]. The existence of more than one type of boundaries may be a frequent feature of nanocrystalline materials, particularly when bulk samples are prepared from powders consisting of agglomerates of nanograins.

An interesting and well-studied example are nanocrystalline Fe-Ni alloys produced during hydrogen reduction of ball-milled oxide powders (see Sect 34.2). After sintering the microstructure of these nanoalloys remains stable up to fairly high temperatures of about 1100 K. Their structure is bimodal and consists of nanocrystalline grains of about 100 nm size clustered in agglomerates with an average size of 30 to 50 µm. In such a microstructure two types of interfaces exist: agglomerate boundaries and intra-agglomerate boundaries. Although this complexity was not included in the theoretical discussion of Sect. 34.3, we illustrate the state-of-the-art below by radiotracer diffusion in Fe-Ni nanoalloys. The analysis of the diffusion experiments in nano-material with a hierarchical microstructure is a sophisticated task. For a detailed discussion of the diffusion kinetics, taking into account fluxes from the agglomerate to the intra-agglomerate boundaries, we refer to a paper of DIVINSKI ET AL. [52].

Radiotracer experiments on nanocrystalline Fe-Ni alloys with bimodal microstructure are reported by DIVINSKI ET AL. [50, 51]. The data cover a wide temperature range and encompass diffusion in type A, B, and C kinetic regimes. Figure 34.5 shows examples of penetration profiles of ^{59}Fe

---

[1] Further reasons for a more complex behaviour, not considered here, can be the presence of intergranular amorphous phases in materials obtained by crystallisation of amorphous precursors and the occurrence of intergranular melting [28].

34.4 Diffusion in Nanocrystalline Metals

self-diffusion in nanocrystalline Fe-40 % Ni. The profiles are plotted as function of the penetration depth y either according to Gaussian penetration (y^2 axis, *left part*) or according to the Whipple-Suszuoka grain-boundary solution ($y^{6/5}$ axis, *right part*). The profile at the highest temperature corresponds to type A kinetics, the two profiles at lower temperatures reveal type B kinetics. For an unambiguous assessment of these profiles it is important to judge several parameters relevant for diffusion in polycrystals: using lattice diffusivities of conventional Fe-Ni alloys [49] (and $s = 1$ for self-diffusion) it can be shown that the parameter, $\alpha = s\delta/(2\sqrt{Dt})$, is always smaller or much smaller than unity [50]. This implies that considerable out-diffusion into the adjacent grains occurs for the profiles in Fig. 34.5 and excludes type C kinetics. Type A diffusion kinetics emerges when diffusion fringes from neighbouring boundaries overlap significantly, i.e. for $d/\sqrt{Dt} < 1$. Then, one expects diffusion profiles which are linear in a plot of logarithm of specific activity *versus* y^2. This is indeed the case for the 1013 K profile of Fig. 34.5. From such profiles an effective diffusivity can be deduced. On the other hand, if the grain-boundary fringes do not overlap, i.e. for $d/\sqrt{Dt} \gg 1$, type B kinetics is expected. Values between 40 and 80 are reported for the ratio between grain size and bulk diffusion length [50]. Under such conditions diffusion profiles should in general be composed of two parts (Chap. 32 and Sect. 34.3). The first part should correspond to direct in-diffusion from the surface. However, in the experiments shown in Fig. 34.5 the bulk penetration length is smaller

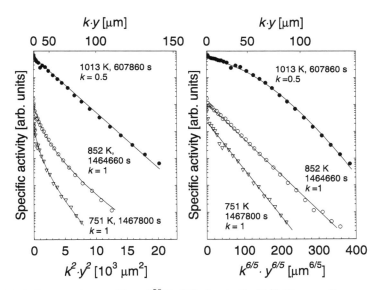

Fig. 34.5. Penetration profiles of ^{59}Fe diffusion in Fe-40 % Ni nanoalloys representing either type A or type B kinetics according to DIVINSKI ET AL. [50]: Fe diffusion plotted as function of y^2 (*left*). Fe diffusion plotted as function of $y^{6/5}$ (*right*)

than one µm. Since mechanical serial sectioning has been used, only the second part could be observed, which corresponds to boundary diffusion. The profiles at 852 and 751 K represent indeed Whipple-Suzuoka behaviour in the nanomaterial. The product δD_{gb} can be deduced from such profiles and D_{gb} is obtained if a value for $\delta \approx 0.5$ nm is assumed.

The effect of the bimodal microstructure has been revealed in experiments under type C conditions for the same material [51]. Figure 34.6 shows examples of penetration profiles of ^{59}Fe self-diffusion in a plot of the logarithm of the specific activity *versus* penetration distance squared. The existence of two types of interfaces – agglomerate and intra-agglomerate boundaries – manifests itself in two-stage diffusion profiles. Diffusivities in the grain-boundaries inside the agglomerates and diffusivities in the boundaries between the agglomerates have been deduced therefrom.

Figure 34.7 summarises grain-boundary diffusivities of Fe-Ni nanoalloys under type A and B [50], and type C kinetics conditions [51]. The results cover a relatively wide temperature interval. Data obtained in different dif-

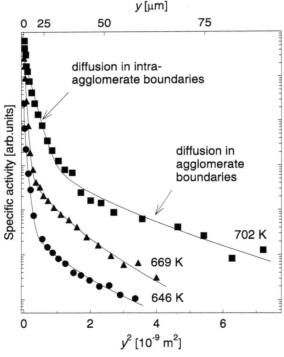

Fig. 34.6. Penetration profiles of ^{59}Fe diffusion in Fe-40 % Ni nanoalloys as function of y^2 representing type C kinetics according to DIVINSKI ET AL. [51]. Two types of grain boundaries contribute to the diffusion profiles

34.4 Diffusion in Nanocrystalline Metals 611

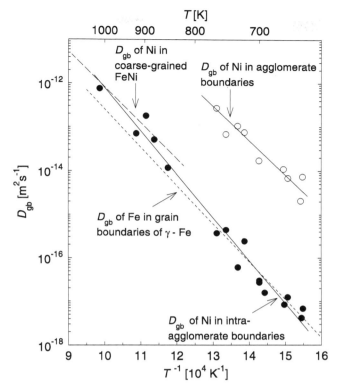

Fig. 34.7. Arrhenius diagram of Fe grain-boundary diffusion in Fe-40 % Ni nanoalloys according to DIVINSKI ET AL. [51]. *Open circles* and *solid line*: D_{gb} for agglomerate boundaries. *Filled circles* and *solid line*: D_{gb} for intra-agglomerate boundaries. For comparison, grain-boundary diffusion in conventional polycrystals is shown as *dashed lines*: Ni in Fe-Ni [53]; Fe in γ-iron [54]

fusion regimes are consistent, when a value of $\delta \approx 1$nm is assumed for the grain-boundary width. The grain-boundary diffusivity along well-relaxed intra-agglomerate boundaries has an activation enthalpy of about 190 kJ/mol and the diffusivities in the boundaries between the agglomerates is faster by about two orders of magnitude than that in the boundaries between the nanograins.

Grain-boundary diffusion of Ni has been measured in coarse-grained polycrystals of Fe-Ni alloys [53] and is also shown in Fig. 34.7. For Fe diffusion no data for grain-boundary diffusion in conventional polycrystals of Fe-Ni alloys are available. Therefore, the results on Fe-Ni nanoalloys are also compared with grain-boundary diffusion in coarse-grained γ-Fe [54]. This comparison seems to be justified, since bulk diffusion in γ-Fe and in conventional γ-Fe-Ni alloys are not much different [55]. The comparison indicates that the atomic mobilities in the intra-agglomerate boundaries of Fe-Ni nanoalloys are similar to those in large-angle boundaries of conventional polycrystals. This coincidence indicates that the grain-boundaries between the nanocrystallites had sufficient time to relax into a quasi-equilibrium state during the production process.

34.4.4 Grain Boundary Triple Junctions

In nanocrystalline materials a further aspect is the presence of many triple junctions. A triple junction is a linear defect that is formed when three grain boundaries join (Fig. 34.1). With decreasing grain size of nanocrystalline materials both the fractions of atoms located in grain boundaries as well as those located in triple junctions increase. It is well recognised that grain boundaries act as rapid diffusion paths in metals and can dominate mass transport at lower temperatures. The rôle of diffusion along triple junctions is not yet completely settled. It is, however, not unlikely that they can make an appreciable contribution to mass transport due to their more open structure compared to grain boundaries.

A mathematical model of triple junction diffusion analogous to the Fisher model for grain boundaries is available in the literature [56, 57]. Unfortunately, the rôle of triple junctions so far has been almost overlooked in the experimental diffusion literature. This is perhaps connected with the difficulty of separating the contribution of triple junction diffusion from the total diffusion flux. To the author's knowledge, only very few systematic studies are available. An example is diffusion of Zn in triple junctions of aluminium studied by PETELINE ET AL. [58]. The authors conclude that diffusivity along triple junctions at 280 °C is about three orders of magnitude faster than in grain boundaries.

An enhanced mobility at triple junctions might also be important for mechanisms of plastic deformation of nanostructured materials that are based on grain-boundary sliding [59]. Such mechanims, especially a rigid body rotation of nanograins under the application of an external shear stress, have been observed in molecular dynamics simulations [60]. For steric reasons, nanograin rotations need to involve considerable atomic transport, especially near triple junctions.

34.5 Diffusion and Ionic Conduction in Nanocrystalline Ceramics

Diffusion and ionic conduction in nanocrystalline ceramics has been reviewed by HEITJANS AND INDRIS [6] and by CHADWICK [7]. In this section, we focus on some selected diffusion and conductivity measurements in nanocrystalline ceramics. These examples comprise the classical oxygen ion conductor ZrO_2, the anion conductor CaF_2, and some composite materials.

34.5 Diffusion and Ionic Conduction in Nanocrystalline Ceramics

Ionic Conduction: The interest in nanocrystalline ion-conducting materials dates back to an observation of LIANG [61]. This author discovered for the composite LiF:Al$_2$O$_3$ that, when the insulator Al$_2$O$_3$ is added to the ion conductor LiF, the conductivity of the material increases by more than one order of magnitude (Fig. 34.8). In such systems, denoted as *dispersed ionic cinductors* (DIC), the enhanced conductivity has been attributed to conduction along interfacial regions between the ion-conducting grains and the grains of the insulator. Conventional DIC's are composites of microcrystalline materials, partially with sub-micrometer grains of the insulator. In principle, the conductivity enhancement may have different origins, such as the formation of space charge layers, an enhanced dislocation density, or the formation of new phases (see [6] for references). Similar results were reported for the composite CuBr:TiO$_2$ by KNAUTH AND ASSOCIATES [62–64]. These studies also showed that the conductivity enhancement is larger for 3 µm CuBr grains than for 5 µm grains.

An attractive explanation for a high conductivity along the interfaces has been suggested by MAIER in terms of the formation of a *space-charge layer* [65]. As discussed in Chap. 26, in ionic crystals the concentration of defects, e.g., cation and anion vacancies in the case of Schottky disorder, is equal in the bulk due to the constraint of charge neutrality even though the formation enthalpies of the defects are different. Near the grain-boundary or near an interface, this constraint is relaxed due to grain-boundary or interface charges and the concentrations of cation and anion vacancies can be different.

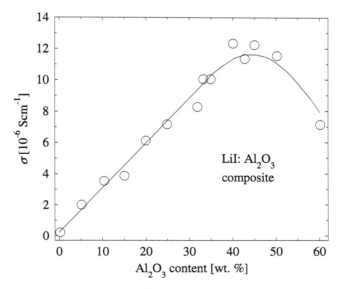

Fig. 34.8. Conductivity of LiI:Al$_2$O$_3$ composites according to LIANG [61]

This leads to the formation of a space charge layer. The unbalanced defect concentrations decay away in moving from the interface to the interior of the solid. The space charge layer can be treated by the classical Debye-Hückel theory [65]. This leads to a *Debye screening length*, L_D, given by

$$L_D = \sqrt{\frac{\epsilon_0 \epsilon_r k_B}{q^2 C_b T}}, \qquad (34.11)$$

where ϵ_0 and ϵ_r are the permittivities of vacuum and sample, respectively. C_b is the concentration of the majority carrier in the bulk and q its charge. For an ionic solid with $\epsilon_r = 10$ and a bulk carrier concentrationm of 10^{22} m^{-3} the Debye length is about 50 nm at 600 K. Thus, the effective space charge region is many times larger than the width of the boundary core, which for a grain boundary is typically 0.5 nm (see Chap. 32). The effect on the carrier concentration as the grain size decreases is illustrated in Fig. 34.9. The enhanced carrier concentration in material with grain sizes comparable or smaller than the Debye length translates into enhanced diffusivity and conductivity.

There are a number of investigations on dispersed ion conductors. We refrain from discussing all of them, since the results are far from beeing conclusive. Instead let us in the rest of this section focus just on the effect of particle size on diffusion and conduction.

A clearcut result has been reported by HEITJANS AND ASSOCIATES [66, 67] for conductivity studies in nanocrystalline CaF$_2$, which is a model substance for anionic conductors. The nanocrystalline material was prepared by inert-gas condensation with a particle size of 9 nm. As seen in Fig. 34.10, the overall conductivity in the nanocrystalline material was found to be four

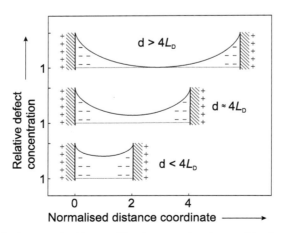

Fig. 34.9. Defect concentration profiles in nanostructures of ionic materials with dimension d. L_D is the Debye screening length

orders of magnitude higher than in polycrystals. As indicated by the solid line, the conductivity in the nanocrystalline material fits well to a space charge enhancement model [65, 67]. The enhanced conductivity is caused by the high number of grain boundaries. Analogous results have been obtained on nanocrystalline BaF_2 prepared by ball milling [68].

A fine example for the validity of the space charge model is provided by conductivity measurements on alternating thin films of CaF_2 and BaF_2 performed by MAIER AND COWORKERS [69]. The CaF_2–BaF_2 heterostructures were produced by molecular beam epitaxy, with layers in the nanometer regime. In agreement with the space charge model, the conductivity increases as the layer thickness decreases as shown in Fig. 34.11. For distances larger than 50 nm the conductivity is proportional to the number of interfaces. When the distance becomes smaller than the Debye screening length in the system (50 nm), the space charge layers of neighbouring interfaces overlap, which leads to an even stronger increase of the conductivity. At this point single interfaces loose their individual character and a nanoionic material with anomalous transport properties is generated.

Diffusion and Ionic Conduction in ZrO_2 and Related Materials:
A number of oxides shows fast oxygen ion conduction. Such materials have applications in solid electrolyte membranes in solid oxide fuel cells (SOFC) and as oxygen permeation membranes (see Chap. 27). Thus, there have been

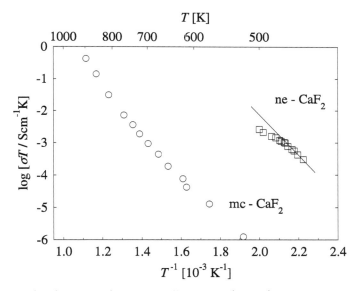

Fig. 34.10. Conductivity of nanocrystalline CaF_2 (*circles*) and of microcrystalline material (*diamonds*) according to HEITJANS AND ASSOCIATES [66, 67]. The solid has been calculated from the space charge layer model

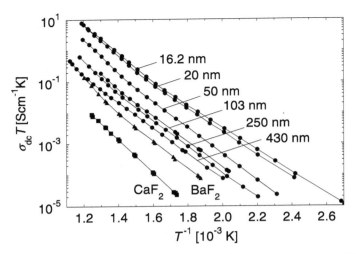

Fig. 34.11. Conductivity of CaF_2-BaF_2 layered heterostructures parallel to the layers of thickness L according to MAIER AND COWORKERS [69]

a number of studies of nanocrystalline zirconia. Common SOFC membranes usually consist of cubic stabilised ZrO_2. Pure ZrO_2 is monoclinic at normal temperatures and transforms at high temperatures to a tetragonal and then to a cubic structure. Addition of aliovalent dopants, such as yttrium (YSZ) and calcium (CSZ), stabilise at low concentrations the tetragonal phase and at higher concentrations the cubic phase. In addition to stabilise the cubic phase, the dopants are compensated by oxygen vacancies, which increase the conductivity.

Diffusion of oxygen in nanocrystalline monoclinic ZrO_2 has been studied by SCHAEFER AND ASSOCIATES [70]. Nanocrystalline powders were prepared by inert-gas condensation and *in situ* consolidation at ambient temperature and pressures of 1.8 GPa and subsequent pressureless sintering. Samples with a mass density of 97 % and an average grain size of 80 nm were obtained. The diffusion of ^{18}O has been investigated by SIMS profiling. The profiles could be attributed to three contributions: (i) diffusion in the grains, (ii) diffusion along the grain-boundaries, and (iii) diffusion due to residual pores in the sample. The grain-boundary diffusivity, D_{gb}, is reported to be 3 to 4 orders of magnitude larger than the diffusivity inside the grains, D. A comparison of the ^{18}O diffusion in the lattice and grain-boundary diffusivities in ZrO_2 with that of other oxide ceramics is shown in Fig. 34.12.

The available data for ZrO_2 are, however, not clearcut. Firstly, conductivity studies of bulk ZrO_2 showed that the grain-boundary diffusivity is *less* than the bulk diffusivity (see, e.g., [71, 72]). This has been attributed to the segregation of impurities into the grain boundaries forming blocking phases. However, blocking has also been proposed due to oxygen vacancy depletion

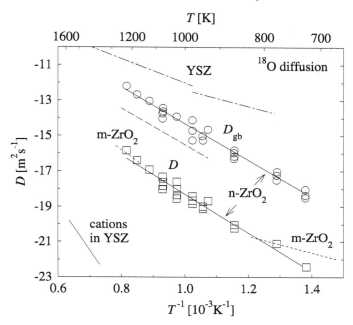

Fig. 34.12. Oxygen diffusion in ZrO$_2$ and YSZ. n-ZrO$_2$: nanocrystalline zirconia (*squares*: bulk diffusion, *diamonds*: interface diffusion); m-ZrO$_2$: microcrystalline zirconia; YSZ: yttrium stabilised zirconia (*dashed-dotted lines*). After SCHAEFER AND ASSOCIATES [70]

in the grain-boundary space charge layers [73]. Nanocrystalline YSZ with 30 to 50 nm grain size has been prepared by inert-gas condensation and the bulk and grain-boundary conductivities turned out to be similar to those of normal ceramics [74]. Similar results have been reported for nanocrystalline YSZ with a grain size of 90 nm [75].

Conclusion: Diffusion and ionic conduction in nanocrystalline ceramics is far from being well understood. This is mainly due to a lack of knowledge about the detailed microstructure, which is less well known than for nanocrystalline metals. The rôle of sample preparation has not been resolved for many systems. The complexity of these systems is determined, for example, by the number of phases involved, the deviation from purely cationic or anionic conduction, the average grain size and the width of the grain size distribution. More work is needed to avoid some of the complications found in early papers. Results of diffusion and conductivity are partly incompatible. For the case of ionic materials there appears to be a need for more studies of diffusion rather than conductivity measurements.

References

1. R.W. Siegel, H. Hahn, *Current Trends in Physics of Materials*, M. Yusouff (Ed.), World Scientific Publ. Co., Singapore, p. 403, 1987
2. R. Birringer, H. Gleiter, *Advances in Materials Science, Encyclopedia of Mat. Sci. and Eng.*, R.W. Cahn (Ed.), Pergamon Press, Oxford, p.339 (1988)
3. H. Gleiter, *Nanocrystalline Materials*, Progr. in Materials Science **33**, 223 (1989)
4. S. Gialanella, L. Lutterotti, *Nanocrystalline Metallic Materials*, p. 1 in [5]
5. H.S. Nalwa (Ed.) *Nanoclusters and Nanocrystals*, American Scientific Publishers, California, 2003
6. P. Heitjans, S. Indris, *Diffusion and Ionic Conduction in Nanocrystalline Ceramics*, J. Phys.: Condens. Matter **15**, R1257 (2003)
7. A.V. Chadwick, *Diffusion in Nanocrystalline Solids*, in: *Diffusion Fundamentals*, J. Kärger, F. Grindberg, P. Heitjans (Eds.), Leipziger Universitätsverlag 2005, p. 204
8. C. Suryanayarana, C.C. Koch, in *Non-equilibrium Processing of Materials*, C. Suryanayarana (Ed.), p. 313–345, Pergamon/Elsevier Science, Oxford, UK, 1999
9. C. Suryanayarana, Progr. in Materials Science **46**, 1 (2001)
10. H.J. Fecht, Nanostruct. Materials **6**, 33 (1995)
11. J. Eckert, J. Nanostruct. Materials **6**, 413 (1995)
12. Th. Sell, H. Mehrer, Z. Metallkd. **88**, 112 (1997)
13. J. Eckert, J.C. Holzer, C.E. Krill III, Scripta Metall. Mater. **27**, 1105 (1992)
14. C.C. Koch, O.B. Cavin, M.G. McCamey, J. Scarbrough, Appl. Phys. Lett. **43**, 1017 (1983)
15. A.W. Weeber, H. Bakker, Physica B **153**, 93 (1988)
16. R.B. Schwarz, W.L. Johnson, Phys. Rev. Lett. **51**, 415 (1983)
17. W.L. Johnson, Progr. Materials Science **30**, 81 (1986)
18. R.Z. Valiev, R.K. Islamgaliev, I.V. Alexandrov, *Bulk Nanostructured Materials from Severe Plastic Deformation*, Progr. Materials Science **27**, 4465 (2000)
19. G.P. Dinda, H. Rösner, G. Wilde, Materials Science and Engineering A **410–411**, 328 (2005)
20. A. Inoue, Progr. Materials Science **43**, 365 (1998)
21. G. Wilde, N. Boucharat, R.J. Herbert, H, Rösner, J.H. Perepezko, Advanced Engineering Material **5**, 125 (2003)
22. N. Boucharat, R. Herbert, H. Rösner, R. Valiev, G. Wilde, Scripta Materialia **53**, 823 (2005)
23. J.S. Lee, B.-H. Cha, Y.-S. Kang, Advanced Engineering Materials **7**, 467 (2005)
24. G. Herzer, J. Magn. Magn. Mater. **1122**, 258 (1992)
25. D.L. Beke (Ed.), *Nanodiffusion – Diffusion in Nanostructured Materials*, Trans Tech Publications, Special Issue of J. of Metastable and Nanocrystalline Materials **19** (2004)
26. T. Surholt, Chr. Herzig, Acta Mater. **45**, 3817 (1997)
27. K. Maier, Phys. Stat. Sol. (b) **44**, 567 (1977)
28. R. Würschum, S. Herth, U. Brossmann, Advanced Engineering Material **5**, 365 (2003)
29. I. Kaur, Y. Mishin, W. Gust, *Fundamentals of Grain and Interphase Boundary Diffusion*, John Wiley and Sons, Ltd., 1995

References 619

30. B.S. Bokstein, I.A. Magidson, I.L. Svetlov, Phys. Met. Metallogr. **6**, 81 (1958)
31. H.S. Levine, C.J. MacCallum, J. Appl. Phys. **31**, 595 (1960)
32. I.V. Belova, G.E. Murch, in: *Nanodiffusion*, D.L. Beke (Ed.), J. of Metastable and Nanocrystalline Materials **19**, 23 (2004)
33. J.R. Kalnin, E.A.Kotomin, J. Maier, J. Phys. Chem. Sol. **63**, 449 (2002)
34. J. Jamnik, J.R. Kalnin, E.A. Kotomin. J. Maier, Phys. Chem. Chem. Phys. **8**, 1310 (2005)
35. E.W. Hart, Acta. Metall. **7** 597 (1957)
36. J.C. Maxwell, *Treatise on Electricity and Magnetism, 3rd edition*, Clarendon Press, Oxford, 1904
37. J.C. Maxwell-Garnett, Philos. Trans. Roy. Soc. London **203**, 385 (1904)
38. A.J. Mortlock, Acta Metall. **8**, 132 (1960)
39. R. Birringer, H. Gleiter, H.-P. Klein, P. Marquardt, Phys. Rev. Lett. **A 102**, 365 (1984)
40. H. Gleiter, Phys. Stat. Sol. (b) **172**, 41 (1992)
41. J. Horvath, R. Birringer, H. Gleiter, Solid State Communications **62**, 319 (1987)
42. S. Schumacher, R. Birringer, R. Strau, H. Gleiter, Acta Metall. **37**, 2485 (1989)
43. S.X. McFadden, R.S. Mishra, R.Z. Valiev, A.P. Zhilyaev, A.K. Mukkerjee, Nature **398**, 684 (1999)
44. R.Z. Valiev, Nature **419**, 887 (2002)
45. R.Z. Valiev, R.K. Islamgaliev, I.V. Alexandrov, Progr. Mater. Science **45**, 103 (2000)
46. R. Würschum, U. Brossmann, H.-E. Schaefer, in: *Nanostructured Materials: Processing, Properties, and Applications*, C.C. Koch (Ed.), Noyes Publications, William Andrew Publishin, Norwich, NY, 2002, pp. 267
47. H. Tanimoto, P. Faber, R. Würschum, R.Z. Valiev, H.-E. Schaefer, Nanostruct. Mater. **12**, 681 (1999)
48. Y.R. Kolobov, G.P. Grabovetskaya, M.B. Ivanov, R.Z. Valiev, Scripta Mater. **44**, 873 (2001)
49. P. Million, J. Ruzickova, J. Velisek, J. Vrestal, Materials Science Engineering **50**, 43 (1995)
50. S.V. Divinski, F. Hisker, Y.-S. Kang, J.-S. Lee, Chr. Herzig, Z. Metallkd. **93**, 256 (2002)
51. S.V. Divinski, F. Hisker, Y.-S. Kang, J.-S. Lee, Chr. Herzig, Z. Metallkd. **93**, 265 (2002)
52. S. Divinski, J.-S. Lee, Chr. Herzig, J. of Metastable and Nanocrystalline Materials **19**, 55 (2004)
53. Y.S. Kang, J.-S. Lee, S.V. Divinski, Chr. Herzig, Z. Metallkd. **95**, 2 (2004)
54. P. Lacombe, P. Guiraldenq, C. Lemonie, 4th Colloque de Metallurgie, Saclay, Presses Universitaires de France,, 1992. p. 105
55. H. Bakker, *Self-diffusion in Homogeneous Binary Alloys and Intermediate Phases*, Chap. 4 in: *Diffusion in Solid Metals and Alloys*, H. Mehrer (Vol. Ed.), Landolt-Börnstein, Numerical Data and Functional Relationships in Science and Technology, New Series, Group III: Crystal and Solids State Physics, Vol. 26, Springer-Verlag, 1990
56. L.M. Klinger, L.A. Levin, A.L. Peteline, Defect and Diffusion Forum **143–147**, 1523 (1997)
57. A.L. Peteline, Mater. Science Forum **235–238**, 469 (1997)
58. A.L. Peteline, S. Peteline, O. Oreshina, Defect and Diffusion Forum **194–199**, 1265 (2001)

59. J.W. Cahn, Y. Mishin, A. Suzuki, Acta Mater. **54**, 4953 (2006)
60. D. Wolf, V. Yamakov, S.R. Philipot, A. Mukkerjee, H. Gleiter, Acta Mater. **53**, 1 (2005)
61. C.C. Liang, J. Electrochem. Soc. **120**, 1289 (1973)
62. J.M. Debierre, P. Knauth, G. Albinet, Appl. Phys. Lett. **71**, 1335–7 (1997)
63. P. Knauth, G. Albinet, J.M. Debierre, Ber. Bunsenges. Phys. Chem. **102**, 945 (1998)
64. P. Knauth, J. Electroceram. **5**, 111 (2000)
65. J. Maier, Solid State Ionics **23**, 59 (1987); J. Electochem. Soc. **134**, 1524 (1987); Progr. in Solid State Chemistry **23**, 171 (1995); Solid State Ionics **131**, 13 (2000)
66. W. Puin, P. Heitjans, Nanostruct. Mater. **6**, 885–8 (1995)
67. W. Puin, S. Rodewald, R. Ramlau, P. Heitjans, J. Maier, Solid State Ionics **131**, 159 (2000)
68. P. Heitjans, S. Indris, J. Mater. Science **39**, 5091 (2004)
69. N. Sata, K. Ebermann, K. Eberl, J. Maier, Nature **408**, 946 (2000)
70. U. Brossmann, R. Würschum, U. Södervall, H.-E. Schaefer, Nanostr. Mater. **12**, 871–4 (1999)
71. N.M. Beckmans, L. Heyne, Electrochem. Acta **21**, 303 (1976)
72. S.P.S. Badwal, A.E. Hughes, J. Eur. Ceram. Soc. **10**, 115 (1992)
73. X. Guo, W. Sigle, J. Fleig, J. Maier, Solid Stae Inics **154–155**, 555 (2002)
74. P. Mondal, H. Hahn, Ber. Bunsenges. Phys. Chem, **101**, 1765 (1997)
75. S. Jiang, J. Mater. Res. **12**, 2374 (1997)

List of Figures

2.1	Illustration of Fick's first law	28
2.2	Infinitesimal test volume. The in- and outgoing y-components of the diffusion flux are indicated by arrows. The other components (not shown) are analogous	30
2.3	Cartesian (*left*), cylindrical (*middle*), and spherical (*right*) coordinates	32
2.4	Diffusion direction in a single-crystal with principal diffusion axes x_1, x_2, x_3	34
3.1	Gaussian solution of the diffusion equation for various values of the diffusion length $2\sqrt{Dt}$	40
3.2	Gaussian solution of the diffusion equation and its derivatives	40
3.3	Solution of the diffusion equation for constant surface concentration C_s and for various values of the diffusion length $2\sqrt{Dt}$	43
3.4	Diffusion from a slab of width $2h$ for various values of $\sqrt{Dt}/h$	44
3.5	Absorption/desorption of a diffusing species of/from a thin sheet for various values of Dt/l^2	50
4.1	Schematic representation of unidirectional diffusion of atoms in a lattice	57
4.2	Example for a jump sequence of a particle on a lattice	60
4.3	Atomic jump process in a crystalline solid: the black atom moves from an initial configuration (*left*) to a final configuration (*right*) pushing through a saddle-point configuration (*middle*)	64
5.1	Vacancies in an elemental crystal	70
5.2	Arrhenius diagram of equilibrium concentrations of mono- and divacancies in metals (schematic)	73
5.3	Length and lattice parameter change *versus* temperature for Au according to SIMMONS AND BALLUFFI [23]	75
5.4	Equilibrium concentration of vacant lattice sites in Al determined by DD measurements according to [24]. DD data: + [25], • [27], × [26]. The concentration range covered by positron lifetime measurements is also indicated	76

5.5	Mean lifetime of positrons in Al according to SCHAEFER ET AL. [28]	78
5.6	Dumbbell configuration of a self-interstitial in an fcc lattice	80
5.7	Vacancies in a dilute substitutional alloy	81
5.8	Frenkel disorder in the cation sublattice of a CA ionic crystal	84
5.9	Schottky disorder in an CA ionic crystal	85
5.10	Electronic structure of semiconductors, with a defect with double acceptor character (*left*) and donor character (*right*)	89
6.1	Octahedral and tetrahedral interstitial sites in the bcc (*left*) and fcc (*right*) lattice	96
6.2	Direct interstitial mechanism of diffusion	96
6.3	Direct exchange and ring diffusion mechanism	97
6.4	Atom chain motion in an amorphous Ni-Zr alloy according to molecular dynamics simulations of TEICHLER [13]	98
6.5	Monovacancy mechanism of diffusion	99
6.6	Divacancy mechanism of diffusion in a close-packed structure	101
6.7	Interstitialcy mechanism of diffusion (colinear jumps)	101
6.8	Interstitial-substitutional exchange mechanisms of foreign atom diffusion. *Top*: dissociative mechanism. *Bottom*: kick-out mechanism	103
7.1	Encounter model: tracer displacement due to encounters with several different vacancies. The *numbers* pertain to tracer jumps promoted by a particular vacancy	111
7.2	Temporal correlation: bunching of tracer jumps within encounters	112
7.3	Example of a vacancy trajectory immediately after vacancy-tracer exchange in a two-dimensional lattice	115
7.4	*Left*: 'Five-frequency model' for diffusion in dilute fcc alloys. *Right*: 'Energy landscape' for vacancy jumps in the neighbourhood of a solute atom	118
7.5	Escape probabilities F_3 for the fcc, bcc, and diamond structure. For fcc and bcc lattices F_3 is displayed as function of ω_4/ω. For the diamond structure F_3 is a function of ω_4/ω_5. After MANNING [15]	120
7.6	'Four-frequency model' of solute diffusion in the bcc lattice	121
7.7	'Five frequency model' for jump rates in the diamond structure. ω_2: jump rate of vacancy-tracer exchange, ω_3 (ω_4): jump rate of vacancy jump from first (second) to second (first) nearest neighbour of solute atom, ω_5: jump rate of vacancy jump from second to third nearest neighbour of tracer	122
8.1	Arrhenius plot of diffusion for various elements in Pb; activation parameters from [1]	128

8.2	Pressure dependence of ^{198}Au diffusion in Au single crystals at constant temperature according to WERNER AND MEHRER [11]. Ω denotes the atomic volume of Au	134
8.3	Schematic illustration of the formation volume of a vacancy	136
8.4	Illustration of the formation volume of a self-interstitial	137
8.5	Illustration of the migration volume. *Upper part*: interstitial migration. *Lower part*: vacancy migration	137
8.6	Activation volumes for self-diffusion in Au *versus* temperature: *triangles* [8], *square* [9], *full circles* [11]	138
8.7	Self-diffusivities at the melting point, $D(T_m)$, for various classes of crystalline solids according to BROWN AND ASHBY [24]	142
8.8	Normalised activation enthalpies of self-diffusion, $\Delta H/(k_B T_m)$, for classes of crystalline solids according to BROWN AND ASHBY [24]...	144
8.9	Activation enthalpies of self-diffusion in metals, ΔH, *versus* melting temperatures, T_m, according to TIWARI ET AL. [25]	144
8.10	Activation enthalpies of self-diffusion in metals, ΔH, *versus* melting enthalpies, H_m, according to TIWARI ET AL. [25]	145
9.1	Schematic illustration of various isotope effect experiments. *Left*: isotope pair A^*/A^{**} in a solid element A. *Middle*: isotope pair B^*/B^{**} in a pure solid A. *Right*: isotope pairs A^*/A^{**} or B^*/B^{**} in a binary $A_x B_y$ compound	156
9.2	Simultaneous diffusion of the radioisotope pair ^{199}Au and ^{195}Au in monocrystalline Au according to HERZIG ET AL. [14] ..	157
9.3	Isotope effect parameters of self-diffusion in Au according to HERZIG ET AL. [14]	158
10.1	Schematic illustration of the Boltzmann-Matano method for a binary diffusion couple with starting compositions C_L and C_R	163
10.2	Molar volume of an A-B solid solution alloy (*solid line*) *versus* composition. The *dashed line* repesents the *Vegard rule*. The partial molar volumes, $\tilde{V}_A$ and $\tilde{V}_B$, and the molar volumes of the pure components, V_A and V_B, are also indicated	167
10.3	Composition profiles constructed according to the Sauer-Freise method. $V_{m,L}$ and $V_{m,R}$ are the molar volumes of the left-hand and right-hand end-members of the diffusion couple	167
10.4	Schematic illustration of a cross section of a diffusion couple composed of pure Cu and brass (Cu-Zn) prepared by SMIGELSKAS AND KIRKENDALL [8] before and after heat treatment. The Mo markers placed at the original contact surface moved towards each other. It was concluded that Zn atoms diffused faster outwards than Cu atoms move inwards $(D_{Zn}^I > D_{Cu}^I)$	168

List of Figures

10.5 Schematic velocity diagrams, pertaining to diffusion couples between the end-members $A_y B_{1-y}$ and $A_z B_{1-z}$ for $y > z$. On the A-rich side A diffuses faster and on the B-rich side B diffuses faster. Different situations are shown, which pertain to one stable Kirkendall plane (*upper part*), to an unstable plane (*middle part*), and to two stable Kirkendall planes, K_1 and K_3, and an unstable plane K_2 175

11.1 Schematic illustration of diffusion and drift 182

13.1 Typical ranges of the diffusivity D and the mean residence time $\bar{\tau}$ of direct and indirect methods for diffusion studies 211

13.2 Initial configurations for direct diffusion studies: **a**) Thin layer of A^* on solid A: tracer self-diffusion in pure elements. **b**) Thin layer of A^* or B^* on homogeneous A-B alloy: tracer self-diffusion of alloy components. **c**) Thin layer of C^* on solid A or on a homogeneous alloy: Impurity diffusion. **d**) Diffusion couple between metal-hydrogen alloy and a pure metal. **e**) Diffusion couple between pure end-members. **f**) Diffusion couple between two homogeneous alloys 214

13.3 Schematic illustration of the tracer method: The major steps – deposition of the tracer, diffusion anneal, serial sectioning, and evaluation of the penetration profile – are indicated 216

13.4 Penetration profile of the radioisotope ^{59}Fe in Fe$_3$Si obtained by grinder sectioning [15]. The *solid line* represents a fit of the thin-film solution of Fick's second law 219

13.5 Penetration profile of the radioisotope ^{59}Fe in Fe$_3$Al obtained by sputter sectioning [18]. The *solid line* represents a fit of the thin-film solution of Fick's second law 220

13.6 Ion-beam sputtering device for serial sectioning of diffusion samples 221

13.7 SIMS depth profiles of ^{30}Si measured before and after annealing at 925 °C for 10 days of a ^{28}Si isotope heterostructure. The initial structure consisted of a layer of ^{28}Si embedded in natural Si 224

13.8 Sputtering process at a surface of a solid 225

13.9 SIMS technique (schematic illustration) 226

13.10 Diffusion profiles for both stable isotopes ^{69}Ga and ^{71}Ga of natural Ga in AlPdMn (icosahedral quasicrystalline alloy) according to [38]. The *solid lines* represent fits of the thin-film solution 227

13.11 Schematic view of an electron microprobe analyser (EMPA) 228

13.12 Characteristic X-ray and Auger-electron production 229

13.13	Interdiffusion profile of a $Fe_{70}Al_{30}$–$Fe_{50}Al_{50}$ couple measured by EMPA according to SALAMON ET AL. [23]. *Dashed line*: composition distribution before the diffusion anneal	230
13.14	Schematic representation of Rutherford backscattering (RBS) and of nuclear reaction analysis (NRA)	231
13.15	Rutherford backscattering spectrometry: high-energy ion beam, electronics for particle detection and a schematic example of a RBS spectrum. The technique is illustrated for a thin layer of atoms with mass M deposited on a substrate of lower mass m ..	233
14.1	Schematic illustration of anelastic relaxation caused by reorientation of elastic dipoles (represented by *grey ellipses*)	238
14.2	Schematic illustration of Gorski-effect	238
14.3	Schematic illustration of anelastic behaviour. The strain response for an instantaneous stress-time function is shown in the *left half*. The stress response for an instantaneous strain-time function corresponds to the *right half*	240
14.4	Stress-strain relations for a periodically driven anelastic material at three different frequencies.......................	240
14.5	Internal friction, $Q^{-1} = \pi \tan \delta$, and frequency dependent modulus, M', as functions of $\omega \tau$	243
14.6	Octahedral interstitial sites in the bcc lattice	244
14.7	Diffusion coefficient for C diffusion in α-Fe obtained by direct and indirect methods: DIFF = in- and out-diffusion or diffusion-couple methods; IF = internal friction; EAE = elastic after effect, MAE = magnetic after effect	247
14.8	Mechanical loss spectrum of a Na_2O4SiO_4 at a frequency of 1 Hz according to ROLING AND INGRAM [18, 19]	249
15.1	Set-up for a NMR experiment (schematic)....................	255
15.2	Self-diffusion of ^6Li and ^7Li in liquid and solid Li studied by PFG-NMR according to FEINAUER AND MAJER [18]	258
15.3	Schematic iluustration of a T_1 measurement with an inversion-recovery (π-τ-$\pi/2$) pulse sequence	259
15.4	Temporal fluctuations of the local field – the origin of motional narrowing...	260
15.5	Schematic illustration of diffusional contributions (random jumps) to spin-lattice relaxation rates, $1/T_1$ and $1/T_{1\rho}$, and to the spin-spin relaxation rate $1/T_2$	261
15.6	Diffusion-induced spin-lattice relaxation rate, $(1/T_1)_{dip}$, of ^8Li in solid Li as a function of temperature according to HEITJANS ET AL. [8]. The B_0 values correspond to Larmor frequencies $\omega_0/2\pi$ of 4.32 MHz, 2.14 MHz, 334 kHz, and 53 kHz	263

List of Figures

15.7 Comparison of self-diffusivities for ^{6}Li in solid Li determined by PFG-NMR with spin-lattice relaxation results assuming a vacancy mechanism (*solid line*) and an interstitial mechanism (*dashed line*) according to MAJER [22] 264

15.8 Mössbauer spectroscopy. *Top*: moving source experiment; *bottom*: principles 265

15.9 Simplified, semi-classical explanation of the diffusional line-broadening of a Mössbauer spectrum. Q denotes the wave vector of the γ-rays 266

15.10 Mössbauer spectra for self-diffusion in polycrystalline Fe from a review of VOGL AND PETRY [27]. FWHM denotes the full-width of half maximum of the Mössbauer line. The spectrum at 1623 K pertains to γ-iron and the spectra at higher temperatures to δ-iron 267

15.11 Self-diffusion in γ- and δ-iron: comparison of Mössbauer (*symbols*) and tracer results (*solid lines*) according to VOGL AND PETRY [27] 268

15.12 Comparison between the dispersion relations of electromagnetic waves (EM waves) and neutrons 270

15.13 Neutron scattering geometry: in real space (*left*); in momentum space (*right*) 271

15.14 Energy spectrum of neutron scattering (schematic) 272

15.15 QENS spectrum of a Na monocrystal at 367.5 K according to GÖLTZ ET AL. [33]. *Dashed line*: resolution functionof the neutron spectrometer 273

15.16 *Top*: Self-correlation function G_s for a one-dimensional lattice. *Top*: The height of the *solid lines* represents the probability of occupancy per site. Asymptotically, the envelope approaches a Gaussian. *Bottom*: Incoherent contribution $S_{inc}(Q,\omega)$ to the dynamical structure factor and quasi-elastic linewidth $\Delta\Gamma$ *versus* scattering vector Q. According to [32] 276

15.17 Quasielastic linewidth as a function of the modulus $Q = |\boldsymbol{Q}|$ for polycrystalline Na_2PO_4 according to WILMER AND COMBET [47]. *Solid lines*: fits of the Chudley-Elliot model...... 278

15.18 Self-diffusion of Na in three $xNa_2SO_4(1-x)Na_3PO_4$ rotor phases according to WILMER AND COMBET [47]..................... 279

15.19 Self-diffusion of Na metal. Dependence of the QENS line broadening in three major crystallographic directions. Theoretical curves have been calculated for a monovacancy mechanism assuming nearest-neighbour junps (*solid lines*) and $a\langle 111 \rangle$ jumps (*dotted line*). From VOGL AND PETRY [27] according to [33, 34] 280

16.1 Schematic illustration of an impedance bridge with sample and electrodes 286

List of Figures 627

16.2 Circuits for the complex impedance and Cole diagram 288
16.3 Complex impedance (Cole diagram) for three temperatures representing volume conduction of a sodium-borate glass [3] 288
16.4 Conductivity spectra of a sodium borate glass in a diagram of logarithm $\sigma' \times T$ versus logarithm of the frequency ν. The onset frequencies of dispersion for various temperatures are connected by a *straight line* 289
16.5 Arrhenius diagram of the charge diffusivity D_σ and the tracer diffusivity D^* of ^{22}Na for a sodium borate glass 289
16.6 Spreading resistance profiling (schematic) 290
16.7 Spreading resistance profile of Se in Si (*left*) and concentration depth profile of Se deduced therefrom (*right*) 291

17.1 Diffusion of ^{63}Ni in monocrystalline Ni. $T > 1200\,K$: data from grinder sectioning [11]; $T < 1200\,K$: data from sputter sectioning [12] 298
17.2 Self-diffusion of fcc metals: noble metals Cu, Ag, Au; nickel group metals Ni, Pd, Pt; group IV metal Pb. The temperature scale is normalised to the respective melting temperature T_m ... 300
17.3 Self-diffusion of bcc metals: alkali metals Li, Na, K (*solid lines*); group-V metals V, Nb, Ta (*dashed lines*); group-VI metals Cr, Mo, W (*solid lines*). The temperature scale is normalised to the respective melting temperature T_m 301
17.4 Self-diffusion in single-crystals of Ag: *squares* [15], *circles* [16], *triangles* [17]. Mono- and divacancy contributions to the total diffusivity are shown as *dotted* and *dashed lines* with the following Arrhenius parameters: $D_{1V}^0 = 0.046 \times 10^{-4}\,m^2\,s^{-1}$, $\Delta H_{1V} = 1.76\,eV$ and $D_{2V}^0 = 2.24 \times 10^{-4}\,m^2\,s^{-1}$, $\Delta H_{2V} = 2.24\,eV$ according to an analysis of BACKUS ET AL. [17] 304
17.5 Effective activation volumes, ΔV_{eff}, of Ag self-diffusion *versus* temperature in units of the atomic volume Ω of Ag: *triangle*, *square* [21], *circles* [22] 305
17.6 Experimental isotope-effect parameters of Ag self-diffusion: *full circles* [15], *triangles* [25], *full square* [26], *triangles on top* [27], *open circles* [28] 306
17.7 Self-diffusion in single crystals of Zn, In, and Sn parallel and perpendicular to their unique axes 307
17.8 Hexagonal close-packed unit cell with lattice paranmeters a and c. Indicated are the vacancy jump rates: ω_a is within the basal plane and ω_b oblique to it 308
17.9 Self-diffusion in the α-, γ- and δ-phases of Fe: *full circles* [30]; *open circles* [31]; *triagles* [32]; *squares* [33] 309
17.10 Self-diffusion in α- und β-phases of titanium: *circles* [34]; *tiangles* [35]; *squares* [36] 310

List of Figures

18.1 Diffusion of interstitial solutes C, N, and O in Nb. For comparison Nb self-diffusion is also shown 314
18.2 Diffusion of H in Pd, Ni and Fe according to ALEFELD AND VÖLKL [13] 322
18.3 Diffusion of H, D, and T in Nb according to [13] 323

19.1 Diffusion of substitutional impurities in Ag and self-diffusion of Ag (*dashed line*). Diffusion parameters from [1, 2] 328
19.2 Diffusion of several impurities in Al and self-diffusion of Al (*dashed line*) according to [1, 9] 333
19.3 Diffusion of impurities in Pb and self-diffusion of Pb (*dashed line*) according to [1, 2] 334
19.4 Diffusion of impurities in Na and self-diffusion of Na according to BARR AND ASSOCIATES [18–20] 335
19.5 Tracer diffusion of ^{113}Sn and ^{195}Au in dilute Au-Sn solid-solution alloys according to HERZIG AND HEUMANN [29] .. 338

20.1 Ideally ordered structures of some cubic intermetallics: B2 (*left*), D0$_3$ (*middle*), L1$_2$ (*right*) 342
20.2 Ideally ordered structures of titanium aluminides: L1$_0$ (*left*), D0$_{19}$ (*right*) 342
20.3 Self-diffusion of ^{64}Cu and ^{65}Zn in CuZn according to KUPER ET AL. [17] 345
20.4 Self-diffusion in B2 structure intermetallics from [7]. The diffusing species is *underlined* 346
20.5 Schematic illustration of six-jump vacancy cycles in the B2 structure. The *arrows* show vacancy jumps; the *numbers* indicate the jump sequence 348
20.6 Illustration of the triple-defect diffusion mechanism in the B2 structure. The *arrows* show vacancy jumps; the *numbers* indicate the jump sequence 349
20.7 Illustration of the antistructural-bridge (ASB) mechanism. The *arrows* show vacancy jumps; the *numbers* indicate the jump sequence 349
20.8 Defect site fractions in B2 NiAl as a function of composition at 0.75 T_m from [39] 350
20.9 Ni tracer diffusion in B2 NiAl at various compositions X according to FRANK ET AL. [48] and DIVINSKY AND HERZIG [49] 351
20.10 Self-diffusion of Fe and Al in Fe$_3$Al 352
20.11 Self-diffusion of Fe and Al and interdiffusion in Fe$_2$Al 353
20.12 Solute diffusion of Zn, In, Ni, Co, Mn, and Cr in Fe$_3$Al according to [23, 51]. Fe self-diffusion in Fe$_3$Al is also shown for comparison 354

List of Figures 629

20.13 Schematic illustration of the sublattice vacancy mechanism in the majority sublattice of an $L1_2$ structured intermetallic. *Full circles*: majority atoms; *open circles*: minority atoms 355
20.14 Defect site fractions in $L1_2$ structured Ni_3Al as a function of composition at 0.75 T_m from [39] 356
20.15 Self-diffusion in $L1_2$ structured Ni_3Al according to [39] 357
20.16 Self-diffusion in the $L1_2$ intermetallics Ni_3Ge, Ni_3Ga, and Ni_3Al. The temperature scale is normalised to the corresounding melting temperatures. For comparison self-diffusion in Ni is also shown. For references see [7] 358
20.17 Fe self-diffusion and Ge solute diffusion in three compositions of the $D0_3$ phase Fe_3Si according to GUDE AND MEHRER [54]. The temperatures are normalised to the corresponding liquidus temperatures. A slight influence of the paramagnetic-ferromagnetic transition can be seen for Fe diffusion in $Fe_{79}Si_{21}$ and $Fe_{82}Si_{18}$... 359
20.18 Ti and In diffusion along the two principal directions of γ-TiAl according to NAKAJIMA AND ASSOCIATES [63, 64]. Diffusion of Ga is shown for polycrystalline γ-TiAl according to HERZIG ET AL. [60] 361
20.19 Self-diffusion of ^{54}Mn and ^{63}Ni in polycrystalline, equiatomic NiMn according to PETELINE ET AL. [66] 362
20.20 Unit cell of tetragonal $MoSi_2$ with the lattice parameters a and c... 363
20.21 Mo, Si, and Ge diffusion along the two principal directions of $MoSi_2$ according to SALAMON AND MEHRER [68] 364
20.22 C15 type cubic Laves phase Co_2Nb. *Full circles* represent majority atoms and *open circles* minority atoms 365
20.23 Self-diffusion of Co and Nb in the cubic C15 Laves phase Co_2Nb according to DENKINGER AND MEHRER [73] 366

21.1 Icosahedral single-quasicrystal (Zn-Mg-Ho) with the habit of a dodecahedron. A dodecahedron has 12 pentagon-shaped faces and 20 vertices 372
21.2 The pseudo-Mackay cluster suggested for icosahedral Al-Pd-Mn [19]. It consists of a central cube (*filled grey circles*), an icosahedron (*filled black circles*), and an icosidodecahedron (*open circles*)... 373
21.3 Tracer diffusion in single-crystals of icosahedral Al-Pd-Mn according to MEHRER ET AL. [14]. Self-diffusion in Al is indicated as a *long-dashed line*............................... 375
21.4 Self-diffusion and diffusion of solutes in Al according to [14] 377

21.5 Self-diffusion of ^{65}Zn in icosahedral $Zn_{64.2}Mg_{26.4}Ho_{9.4}$ and $Zn_{60.7}Mg_{30.6}Y_{8.7}$ quasicrystals and in a related hexagonal phase (h-ZnMgY) according to GALLER ET AL. [34]. *Dashed lines*: self-diffusion in Zn parallel and perpendicular to its hexagonal axis; *dotted line*: Zn diffusion in icosahedral Al-Pd-Mn 378

21.6 Self-diffusion of ^{65}Ni, ^{60}Co, and ^{57}Co in decagonal Al-Ni-Co quasicrystals from [14, 34, 36]. Monte Carlo simulations of Al diffusion are also shown [40] 380

22.1 Diamond structure of Si and Ge (*right*) and zinc blende structure (*left*) 385

22.2 Resistivity of various materials and silicon of various doping levels 388

23.1 Self-diffusion of Si and Ge and of some metals (Cu, Au, Na) in a homologous temperature scale 395

23.2 Self-diffusion coefficients of Ge measured by tracer methods [7–9, 11, 12] 399

23.3 Doping dependence of Ge self-diffusion according to WERNER ET AL. [12] 401

23.4 Activation volumes of Ge self-diffusion according to WERNER ET AL. [12]. For comparison activation volumes of Au self-diffusion are also shown [14] 402

23.5 Self-diffusion coefficients of Si measured by tracer methods [15–23] 403

23.6 Si self-diffusion coefficients (*symbols*) compared with the self-interstitial and vacancy transport products, $C_I^{eq}D_I$ and $C_V^{eq}D_V$, according to BRACHT ET AL. [23] 405

24.1 Schematic phase diagram of a semiconductor and a foreign element (or a compound of a foreign element) illustrating the phenomenon of retrograde solubility 410

24.2 Equilibrium solubilities of substitutional (*open symbols*) and interstitial (*filled symbols*) Cu in Ge according to [10] 412

24.3 Diffusion of foreign atoms in Ge compared with Ge self-diffusion according to [3]: *Solid lines* represent diffusion of elements that are incorporated on substitutional sites and diffuse via the vacancy mechanism. *Long-dashed lines* represent diffusion of hybrid elements, which are mainly dissolved on substitutional sites; their diffusion proceeds by the dissociative mechanism via a minor fraction in an interstitial configuration (Au, Ag, Ni, Cu). The *short-dashed lines* represent diffusion of elements that diffuse via a direct interstitial mechanism (H, Li [16, 17]). The *short-dashed line on top* shows the diffusivity deduced for interstitial Cu 413

24.4 Diffusion of foreign atoms in Si compared with Si self-diffusion according to [4]: *Solid lines* represent diffusion of elements that are incorporated on substitutional sites and diffuse via the vacancy or interstitialcy mechanism (B, Al, Ge, P, As, Sb, As, C). *Long-dashed lines* represent diffusion of hybrid elements, which are mainly dissolved on substitutional sites: their diffusion proceeds via a minor fraction in an interstitial configuration (Au, Pt, S, Zn). The *short-dashed lines* indicate the elements (H [14], Li [16–18], Cu, Fe, Ni, O) that diffuse via the direct interstitial mechanism 414

24.5 Bond-centered configuration of the oxygen interstitial in the Si lattice according to FRANK ET AL. [2] 416

24.6 Vacancy enthalpy as a function of the coordination site away from a substitutional dopant (*full circle*) 419

25.1 Computer simulation of in-diffusion into a thick sample via the kick-out mechanism according to STOLWIJK [4]. The *left part* (**a**) shows the concentration of substitutional foreign atoms. The *right part* (**b**) shows the associated self-interstitial supersaturations. The values of the ratio $\alpha_j \equiv C_I^{eq} D_I / C_i^{eq} D_i$ are: $\alpha_1 = 1/25, \alpha_2 = 1/5, \alpha_3 = 1, \alpha_4 = 5, \alpha_5 = 25$, with the indices j listed in the diagrams 431

25.2 In-diffusion of Au into a thick dislocation-free Si specimen according to STOLWIJK ET AL. [13]. The *solid line* represents a fit of the kick-out mechanism. The *dashed line* is an attempt to fit a complementary error function predicted for a concentration-independent diffusivity 434

25.3 In-diffusion of Au into dislocation-free Si wafers for various annealing times at 1273 K according to [13]. The Au concentration C_{Au} is given in units of the solubility C_{Au}^{eq}; the penetration depth x as a fraction of the wafer thickness d. For one penetration profile of the two 4.27 h anneals $d \approx 300$ µm, otherwise $d \approx 500$ µm.. 435

25.4 Comparison of Au diffusion into a dislocation-free Si wafer (1.03 h at 1273 K) with the prediction of the kick-out mechanism (*solid line*) [13]. Circles: $x < 0$; squares $x > 0$................. 436

25.5 In-diffusion of substitutional Zn measured on dislocation-free (*crosses*) and highly dislocated (*squares*) Si samples; Zn was diffused simultaneously at 1115 °C for 2880 s into both samples, using metallic Zn as vapour source. The lower profile (*circles*) represents in-diffusion into a dislocation-free wafer also at 1115 °C using a Zn solution in HCl as diffusion source. After BRACHT ET AL. [17]... 437

25.6 Diffusion profiles of substitutional Zn (Zn_s) at 1021 °C for various diffusion times according to BRACHT ET AL. [17]. *Top*: Dislocation-free Si wafers; *solid lines* show calculated profiles based on the kick-out model using one set of parameters C_s^{eq} and $C_I^{eq} D_I$. *Bottom*: Highly dislocated Si; *solid lines* show fitting with complementary error functions 438
25.7 Comparison of Cu penetration profiles in almost dislocation-free Ge (1126 K, 900 s) and in a crystal with high dislocation density (1124 K, 780 s) according to [17] 441
25.8 Products of solubility × effective diffusivity × correlation factor for Cu, Ag, and Au diffusion in Ge with various dislocation densities compared to the Ge tracer diffusivity according to [17]. 442
25.9 Penetration profile of ^{60}Co in Nb after 10.3 days of annealing at 1422 K in double-logarithmic representation according to [34]. 444
26.1 Ionic conductivity of halide crystals. The corresponding activation enthalpies are listed. For comparison the conductivity of the fast ion conductor $RbAg_4I_5$ is also shown.............. 450
26.2 Examples of point defects in ionic crystals: Schottky defects, Frenkel defects, divalent cation impurity and cation vacancy, complex of divalent cation and cation vacancy, vacancy pair 451
26.3 Schematic diagram of charge diffusivity, D_σ, and tracer diffusivities of anions and cations, D_A^* and D_C^*, in alkali halides. *Parallel lines* in the extrinsic region correspond to different doping contents C_i 462
26.4 Conductivity of a NaCl single crystal doped with a site fraction of 1.2×10^{-5} Sr^{2+} ions according to BENIÈRE ET AL. [22] 463
26.5 Self-diffusion of ^{22}Na and ^{36}Cl in intrinsic NaCl. Also indicated is the product of charge diffusion coefficient D_σ and correlation factor $f_V = 0.781$. From LASKAR [15] 464
26.6 Diffusion of the homovalent impurities Cs, Rb F, I, and Br in NaCl according to [32–34]. Self-diffusion of Na and Cl is also indicated for comparison 466
26.7 Concentration dependence of the diffusion coefficient of divalent cations, D_2, relative to its saturation value, $D_2(sat)$, according to [39]. The curves refer to three values of the association enthalpy, ΔG_{iV_C}, normalised with $k_B T$ 467
26.8 *Left*: migration of interstitial Ag^+ ions via the direct interstitial and by the interstitialcy mechanism. *Right*: pathways for Ag^+ movements by the colinear (*double arrows*) and the non-colinear (*solid arrows*) interstitialcy mechanism 469
26.9 Tracer self-diffusion coefficients for the constituents of AgCl [41] and AgBr [24, 42]. D_σ was calculated from the ionic conductivity via the Nernst-Einstein relation [37] 470

List of Figures   633

26.10 Schematic illustration of the effect of divalent cationic impurity doping on the temperature dependence of the conductivity of AgCl 471
26.11 Schematic illustration of the effect of divalent cationic impurity doping on the isothermal conductivity of AgCl. The *dashed lines* represent the effect of anionic impurity doping 472

27.1 Electrical dc conductivity of several fast ion conductors. Some ordinary solid electrolytes and concentrated H_2SO_4 are shown for comparison 476
27.2 Crystal structure of α-AgI. *Large circles*: I^- ions; *filled small circles*: octahedral sites; *filled squares*: tetrahedral sites; *filled triangles*: trigonal sites. Octahedral, tetrahedral, and trigonal sites can be used by Ag^+ ions 478
27.3 Probabiliy distribution of Ag in α-AgI at 300 °C according to CAVA, REIDINGER, AND WUENSCH [42] 478
27.4 Cation pathway in an fcc anion sublattice according to FUNKE [19]. *Filled squares*: tetrahedral sites; *small filled circles*: octahedral sites 479
27.5 Fluorite structure (prototype CaF_2): *Filled circles* represent anions and *open circles* cations. *Diamonds* represent sites for anion interstitials 481
27.6 Perovskite structure 482
27.7 Sites for Na^+ ions in the conduction plane of β-alumina. *m*: mid-oxygen position, *br*: Beevers-Ross site, *abr*: anti-Beevers-Ross site. *Open circles*: O^{2-}, *grey circles*: O^{2-} spacer ions 483
27.8 Conductivities of some single crystal β-aluminas according to WEST [45] 484
27.9 Schematic illustration of ion solvation and migration in amorphous polymer electrolytes according to [62] 487
27.10 Tracer diffusion coefficients of ^{22}Na and ^{125}I in an amorphous PEO–NaI polymer electrolyte compared to the charge diffusivity, D_σ, according to STOLWIJK AND OBEIDI [62, 63]. The *dashed line* is shown for comparison: it represents the sum $D(^{22}Na) + D(^{125}I)$ 487

28.1 X-ray diffractogram of a crystal (*left*) and of a glass (*right*) 493
28.2 Volume (or enthalpy) *versus* temperature diagram of a glass-forming liquid 495
28.3 Differential Scanning Calorimetry (DSC) thermogram of a $0.2(0.8Na_2O\ 0.2\ Rb_2O)\ 0.8B_2O_3$ glass measured at a heating rate of 10 K/min from [9]. The glassy and undercooled liquid state are indicated. The strong exothermic signal (near 650 °C) corresponds to the crystallisation of the undercooled melt 496

List of Figures

28.4 Schematic time-temperature-transformation diagram (TTT diagram) for the crystallisation of an undercooled melt 497

29.1 Structure of an ordered binary crystalline solid (schematic) 503
29.2 Structure of a binary metallic glass (schematic) 504
29.3 Schematic illustration of structural relaxation in the V-T (or H-T) diagram of a glass-forming material 507
29.4 Time-averaged diffusivities $\langle D \rangle$ of ^{59}Fe in as-cast $Fe_{40}Ni_{40}B_{20}$ as functions of the annealing time according to HORVATH AND MEHRER [23] 508
29.5 Instantaneous diffusivities $D(t)$ of several conventional metallic glasses as functions of annealing time according to HORVATH ET AL. [24] 509
29.6 Arrhenius diagram of self- and impurity diffusion in relaxed metal-metalloid and metal-metal-type conventional metallic glasses according to FAUPEL ET AL. [37] 510
29.7 Arrhenius diagram of tracer diffusion of Be, B, Fe, Co, Ni, Hf in the bulk metallic glass $Zr_{46.75}Ti_{8.25}Cu_{7.5}Ni_{10}Be_{27.5}$ (Vitreloy4) according to FAUPEL ET AL. [37] 511
29.8 Arrhenius diagram of tracer diffusion of B and Fe in $Zr_{46.75}Ti_{8.25}Cu_{7.5}Ni_{10}Be_{27.5}$ (Vitreloy4) according to FAUPEL ET AL. [37]. *Open symbols*: as-cast material from [31]; *filled symbols*: pre-annealed material from [39] 512
29.9 Correlation between D^0 and ΔH for amorphous and crystalline metals according to [37]. *Solid line*: conventional metallic glasses; *dotted line*: bulk metallic glasses; *dashed line*: crystalline metals 513
29.10 Pressure dependence of Co diffusion in $Co_{81}Zr_{19}$ at 563 K according to [37]. The *dashed line* would corresponds to an activation volume of one atomic volume 514
29.11 Isotope effect parameter as function of temperature for Co diffusion in bulk metallic glasses according to [37]; data taken from EHMLER ET AL. [46, 47] 515
29.12 Chain-like collective motion of atoms in a Co-Zr metallic glass according to molecular dynamics simulations by TEICHLER [55] . 516
29.13 Tracer diffusion coefficients of P and Co in comparison with viscosity diffusion coefficients of the alloy $Pd_{43}Cu_{27}Ni_{10}P_{20}$ according to BARTSCH ET AL. [58] 518

30.1 Viscosity of a soda-lime-silicate glass (standard glass I of the Deutsche Glastechnische Gesellschaft, DGG). Particular viscosity points are indicated 524
30.2 Schematic fragility diagram for various melts 526

List of Figures 635

30.3 Diffusion penetration profiles of ^{22}Na obtained by grinder sectioning (*left*) and of ^{86}Rb obtained by sputter sectioning (*right*) according to IMRE ET AL. [14] 527
30.4 Conductivity (real part) of a soda-lime silicate glass (standard glass I of DGG) *versus* frequency for various temperatures according to TANGUEP-NIJOKEP AND MEHRER [15] 528
30.5 Permeability of gases through vitreous SiO_2 according to SHELBY [2] 530
30.6 Diffusion in vitreous silica and in quartz (for references see text) 531
30.7 Structure of a soda-lime silicate glass (schematic in two dimensions) 533
30.8 Viscosity diffusion coefficient, D_η, tracer diffusivities, D^*_{Na}, D^*_{Ca}, and charge diffusion coefficient D_σ, of soda-lime silicate glass (standard glass I of DGG) according to TANGUEP-NIJOKEP AND MEHRER [15] 533
30.9 Tracer diffusivities, D^*_{Na}, D^*_{Ca}, and charge diffusion coefficient, D_σ, of soda-lime silicate glass (standard glass II of DGG) according to TANGUEP-NIJOKEP AND MEHRER [15] 534
30.10 Haven ratios of soda-lime silicate glasses according to TANGUEP-NIJOKEP AND MEHRER [15] 536
30.11 Structure of sodium-rubidium borate glass (schematic in two dimensions) 537
30.12 Glass-transition temperatures of alkali borate glasses according to BERKEMEIER ET AL. [28] 537
30.13 Arrhenius diagram of the dc conductivity (times temperature) for Y Na_2O (1-Y)B_2O_3 glasses according to BERKEMEIER ET AL. [28] 538
30.14 Electrical dc conductivity of Li, Na, K, and Rb borate glasses according to BERKEMEIER ET AL. [28] 538
30.15 Charge diffusion coefficient D_σ of mixed 0.2 [X Na_2O (1-X)Rb_2O] 0.8 B_2O_3 glasses according to IMRE ET AL. [29] 539
30.16 Composition dependence of the activation enthalpy for conductivity diffusion of Fig. 30.15 [29] 540
30.17 Composition dependence of ^{22}Na and ^{86}Rb diffusion in mixed 0.2[X Na_2O(1-X)Rb_2O]0.8 B_2O_3 glasses according to IMRE ET AL. [14]. Na diffusion: *full symbols*; Rb diffusion: *open symbols* 540
30.18 Composition dependence of the charge diffusion coefficient, D_σ, and of the mean tracer diffusion coefficient, $\langle D \rangle$, of mixed Na-Rb borate glasses 0.2[X Na_2O (1-X)Rb_2O]0.8 B_2O_3 541

31.1 Schematic illustration of high-diffusivity paths in a solid 547
31.2 Schematic illustration of the diffusion spectrum for metals in a reduced temperature scale; T_m denotes the melting temperature 549

List of Figures

32.1 Tilt boundary (*left*) and twist boundary (*right*) 555
32.2 Low-angle tilt boundary after BURGERS [19] 556
32.3 Random high-angle grain boundary (schematic) 556
32.4 A coherent twin boundary (*left*). Twin-boundary energy γ as a function of the orientation ϕ of the grain-boundary plane (*right*) 557
32.5 A special large-angle boundary according to GLEITER [24] 558
32.6 A high-resolution transmission electron microscope image of a (113)[113] symmetric tilt boundary in gold according to WOLF AND MERKLE [25] 558
32.7 FISHER's model of an isolated grain boundary. D: lattice diffusivity, D_{gb}: diffusivity in the grain boundary, δ: grain-boundary width 559
32.8 Isoconcentration contours for various values of the LE CLAIRE parameter β 562
32.9 Concentration contours for constant source (*left*) and a thin-film source solutions (*right*) for an arbitrary value of $\beta = 50$ according to SUZUOKA [30] 564
32.10 Illustration of the type A, B, and C diffusion regimes in a polycrystal according to HARRISONS classification [38] 569
32.11 Schematic illustration of a penetration profile in a bi-crystal for type B kinetics 572
32.12 Average thin-layer concentration at depth z of a diffuser entering via a grain boundary *versus* the normalised penetration depth $\eta = z/\sqrt{Dt}$ for $\beta = 100$ ($\Delta = 2 \times 10^6$ and $\sqrt{Dt} = 10\,\mu\text{m}$) according to SUZUOKA [29]. The concentrations are expressed in units of $M/(L\sqrt{\pi})$ in the case of an instantaneous source and in units of $c_0\sqrt{Dt}/L$ for a constant source 573
32.13 Logarithm of the average thin layer concentration at depth z of the diffuser entering via a grain boundary *versus* $z^{6/5}$ (z = section depth, $\Delta = 2 \times 10^6$ and $\sqrt{Dt} = 10\,\mu\text{m}$) according to SUZUOKA [29]. The concentrations are expressed in units of $M/(L\sqrt{\pi})$ for an instantaneous source and in units of $c_0\sqrt{Dt}/L$ for a constant source 574
32.14 Type B kinetics penetration profiles of self-diffusion in Ag polycrystals according to SOMMER AND HERZIG [35] 575
32.15 Arrhenius diagram of the triple product $s\delta D_{gb}$ and of sD_{gb} from Te diffusion along grain boundaries in Ag according to HERZIG ET AL. [48]. Type B and C kinetics prevail above 600 K and below 500 K, respectively. The range 500–600 K corresponds to a transient regime 578

32.16 Grain-boundary segregation factors for Te in Ag according to HERZIG ET AL. [48] and Au in Ag according to SURHOLT ET AL. [49] determined from combined type B and type C measurements 579

33.1 Smoluchowski model of a dislocation pipe 584
33.2 Dislocation diffusion: mean thin-layer concentrations of the constant and instantaneous source solutions, Q^I and Q^{II}, for $\alpha = 10^{-2}$ and $(\Delta - 1) = 10^5$ according to LE CLAIRE AND RABINOVITCH [6] 588
33.3 Dislocation diffusion: constant source solution Q^I (left) and instantaneous source solution Q^{II} (right) versus η for $\alpha = 10^{-1}, 10^{-2}, 10^{-3}$, $\alpha\beta = 10$ (full lines) and $\alpha\beta = 10^2$ (dashed lines) according to LE CLAIRE AND RABINOVITCH [6] .. 588
33.4 Dislocation diffusion: The quantity $A(\alpha)$ is plotted as a function of α for various values of $\alpha\beta$ according to LE CLAIRE AND RABINOVITCH [6] 589

34.1 Schematic view of a nanocrystalline material 594
34.2 Schematic drawing of an ECAP device used for severe plastic deformation according to VALIEV ET AL. [18] 597
34.3 Tracer distribution in various kinetic regimes and subregimes of diffusion in polycrystals according to KAUR, MISHIN AND GUST [29] 601
34.4 Models representing grains (dark) and boundaries in a nanostructured material. Left: parallel arrangements of grains and grain boundaries in the diffusion direction. Middle: serial of arrangement of grains and grain boundaries. Right: grains represented as cubes...................................... 604
34.5 Penetration profiles of ^{59}Fe diffusion in Fe-40 % Ni nanoalloys representing either type A or type B kinetics according to DIVINSKI ET AL. [50]: Fe diffusion plotted as function of y^2 (left). Fe diffusion plotted as function of $y^{6/5}$ (right) 609
34.6 Penetration profiles of ^{59}Fe diffusion in Fe-40 % Ni nanoalloys as function of y^2 representing type C kinetics according to DIVINSKI ET AL. [51]. Two types of grain boundaries contribute to the diffusion profiles 610
34.7 Arrhenius diagram of Fe grain-boundary diffusion in Fe-40 % Ni nanoalloys according to DIVINSKI ET AL. [51]. Open circles and solid line: D_{gb} for agglomerate boundaries. Filled circles and solid line: D_{gb} for intra-agglomerate boundaries. For comparison, grain-boundary diffusion in conventional polycrystals is shown as dashed lines: Ni in Fe-Ni [53]; Fe in γ-iron [54] 611
34.8 Conductivity of LiI:Al$_2$O$_3$ composites according to LIANG [61] .. 613

34.9 Defect concentration profiles in nanostructures of ionic materials with dimension d. L_D is the Debye screening length .. 614

34.10 Conductivity of nanocrystalline CaF_2 (*circles*) and of microcrystalline material (*diamonds*) according to HEITJANS AND ASSOCIATES [66, 67]. The solid has been calculated from the space charge layer model 615

34.11 Conductivity of CaF_2-BaF_2 layered heterostructures parallel to the layers of thickness L according to MAIER AND COWORKERS [69]... 616

34.12 Oxygen diffusion in ZrO_2 and YSZ. n-ZrO_2: nanocrystalline zirconia (*squares*: bulk diffusion, *diamonds*: interface diffusion); m-ZrO_2: microcrystalline zirconia; YSZ: yttrium stabilised zirconia (*dashed-dotted lines*). After SCHAEFER AND ASSOCIATES [70] 617

Index

activation energy 133
activation enthalpy 127, 128, 133, 143, 242, 246, 260, 263, 297, 299, 300, 310, 321, 328, 330, 332, 345, 395, 416, 418, 450, 459, 462, 510, 512, 525, 535, 540, 549, 550, 577
  activation enthalpy of self-diffusion 132, 144
  activation enthalpy of solute diffusion 132
activation parameters 128, 130, 133, 299, 300, 313, 321, 398
  activation parameters and elastic constants 146
  model of Zener 146
activation volume 132, 135, 145, 298, 305, 332, 376, 401, 514
  activation volume and melting point 145
  activation volume of ionic crystals 140
  activation volume of ionic conduction 133
  activation volume of self-diffusion 135
  activation volume of solute diffusion 139
  effective activation volume 133
  formation volume of a divacancy 136
  formation volume of a monovacancy 136
  formation volume of a self-interstitial 136
  formation volume of Schottky pairs 140
  migration volume 137

  migration volume of the cation vacancy 140
AgCl and AgBr 469
alkali halide 458, 461, 467
anelastic relaxation 237, 456, 457
anelasticity 237, 239
anisotropic media 33
anisotropy ratio 308
Arrhenius diagram 127, 246, 287, 289, 301, 304, 317, 332, 333, 345, 374–376, 465, 471, 510–512, 536, 538, 578, 611
Arrhenius relation 127, 297, 316, 508
ARRHENIUS, SVANTE AUGUST 4
atomic jump process 55, 64
  Debye frequency 64
  saddle point 65
  simulation of atomic jump processes 66
attempt frequency 65, 129, 132, 143, 152, 297, 302, 316, 396
Auger-electron spectroscopy (AES) 230
  Auger electron 230
  Auger emission 229
axial flow 32, 38

B2 intermetallics 346
  antistructural-bridge (ASB) mechanism 350
  B2 Fe-Al 353
  B2 NiAl 351
  B2 order 361
  coupling between diffusivities 347
  six-jump-cycle (6JC) mechanism 347
  triple-defect mechanism 348
  vacancy-pair mechanism 350

BARDEEN, JOHN 10, 62, 105, 386
BEKE, DESZÖ 13
Bessel functions 51, 587
binary alloy 80
　Lomer equation 82, 100
　solute 81
　solute-vacancy pair 81, 118
　solvent 81
　vacancies in concentrated alloys 82
　vacancies in dilute alloys 81
binary intermetallics 341
　B2 (or CsCl) structure 342
　$C11_b$ structure 343
　DO_3 (or Fe_3Si) structure 342
　DO_{19} structure 343
　$L1_0$ (or CuAu) structure 343
　$L1_2$ (or Cu_3Au) structure 343
BOKSTEIN, BORIS 13
Boltzmann transformation 162
Boltzmann-Matano method 163, 165, 229
　Boltzmann-Matano equation 164
　Matano plane 162, 164
borate glasses 537
BRACHT, HARTMUT 15
BROWN, ROBERT 1, 5
　Brownian motion 1, 55

cartesian coordinates 32
centrifugal forces 181
charge diffusion coefficient 185, 285, 287, 461, 463, 487, 533–535, 539, 541
chemical diffusion 161
chemical diffusion coefficient 162, 183, 212
chemical potential 161, 170, 180, 184, 194
　thermodynamic activity 170
classical ion conductors 83
colinear and non-colinear jumps 468
collective correlation factor 201, 204
collective mechanism 97, 155, 159, 516
　direct exchange 97
　interstitialcy mechanism 98
　non-defect mechanisms 98
　ring mechanism 97
collective motion 580
configurational entropy 71

continuum theory of diffusion 27
correlation factor 62, 105, 106, 111, 112, 114, 132, 151, 158, 185, 195, 200, 308, 316, 329, 356, 397, 419, 428, 461, 465
　activation enthalpy of the correlation factor 123
　correlation factor for diamond lattice 121
　correlation factor of self-diffusion 115
　escape probability 119, 121, 122
　geometric correlation factor 153
　impurity form 123, 151
　recursion formula 113
　solute correlation factor 119, 123, 139
　solute correlation factor bcc 120
　vacancy trajectory 114
Cottrell atmospheres 181
crystallisation 496, 504, 506, 510, 523, 598
Cu_3Au rule 365, 366
　majority element 366
　minority element 366
cylindrical coordinates 32

DO_3 inremetallics 367
DO_3 intermetallics 357
　Cu_3Sn 358
　DO_3 Fe_3Si 358
　sublattice vacancy mechanism 358
DANIELEWSKI, MAREK 13
Darken equations 170, 171, 188, 193, 203, 215
　Darken-Dehlinger equations 171
　Darken-Manning equations 172, 203, 215, 356
　Manning factor 172
　vacancy-wind factor 172, 173
DAYANANDA, MYSORE 14
dc conductivity 285, 456, 476, 527, 538, 613, 615, 616
dielectric relaxation 457
differential dilatometry (DD) 74
diffusion and ionic conduction in oxide glasses 521
　alkali borate glasses 535
　annealing point 524

fragile melts 525
fragility diagram 525
fragility of melts 525
gas permeation 529
glass-forming oxides 522
glass-transition temperature 537
intermediate ions 522
mixed-alkali effect 538
network-former ions 522
network-modifier ions 522
non-bridging oxygen (NBO) 534
permeability 529
soda-lime silicate glass 532
softening point 524
strong melts 525
structure of network glasses 521
structure of soda-lime silicate glass 533
structure of sodium-rubidium borate glass 537
viscosity 524
viscosity of glass-forming melts 523
vitreous silica and quartz 530
working point 524
working range 524
diffusion coefficient 28, 57, 59
  diffusivity tensor 33
  principal diffusion coefficients 33
  principal diffusivities 33
diffusion couple 41, 214
diffusion entropy 129, 297, 300, 398
diffusion equation 30, 31, 37
diffusion in a plane sheet 47
  desorption and absorption 49
  eigenvalues 48
  out-diffusion from a plane sheet 48
  separation of variables 47
diffusion in a sphere 51
diffusion in metallic glasses 503
  bulk amorphous steels 506
  bulk metallic glasses 506
  chain-like collective motion of atoms 516
  conventional metallic glasses 505
  correlation between D^0 and ΔH 512
  diffusion and viscosity 517
  diffusivity enhancement 508
  equilibrium melt 517
  excess volume 506
  families of metallic glasses 505
  instantaneous diffusivity 508
  isotope effect parameter for bulk metallic glasses 515
  melt-spinning 505
  mode-coupling theory 517
  molecular-dynamics simulations 516
  pressure dependence 513
  quasi-vacancies 508
  structural relaxation and diffusion 506
  temperature dependence for bulk metallic glasses and their supercooled melt 510
  temperature dependence for conventional metallic glasses 509
  time-averaged diffusivity 507
  Turnbull criterion 504
diffusion in nanocrystalline materials 593
  agglomerate boundaries 608
  characteristic length scales for diffusion 600
  chemical and related synthesis methods 598
  coarse-grained polycrystals 600
  Debye screening length 614
  devitrification of amorphous precursors 598
  diffusion and ionic conduction in ZrO_2 and related materials 615
  diffusion in nanocrystalline metals 606
  dispersed ionic cinductors (DIC) 613
  effective diffusivity of solutes 605
  effective self-diffusivity 604
  fine-grained polycrystals 602
  grain size and diffusion regimes 599
  Hart equation 605
  Hart-Mortlock equation 606
  heavy plastic deformation (SPD methods) 596
  high-energy milling 595

inert-gas condensation 594
intra-agglomerate boundaries 608
ionic conduction in nanocrystalline
  ceramics 613
Maxwell equation (Maxwell-Garnett
  equation) 605
nanocrystalline monoclinic ZrO_2
  616
nanomaterials with bimodal grain
  structure 608
oxygen diffusion in ZrO_2 and YSZ
  617
powder processing 594
refined classification of diffusion
  regimes 600
space-charge layer 613
strongly segregating solutes in
  ultrafine-grained polycrystals
  604
structural relaxation and grain
  growth 607
structure of nanocrystalline material
  594
synthesis of nanocrystalline materials
  594
triple junctions 612
ultrafine-grained polycrystals 603
diffusion in quasicrystalline alloys 371
  decagonal quasicrystals 379
  diffusion anisotropy in decagonal
    Al-Ni-Co 381
  icosahedral Al-Pd-Mn 374
  icosahedral Zn-Mg-RE 378
  molecular dynamic simulations 380
  phasons 376, 380
  self-diffusion and diffusion of solutes
    in Al 377
  vacancy-mediated 376
diffusion length 39, 217, 599–602, 604,
  609
diffusion mechanism 95, 154, 186, 380,
  486, 512, 515
diffusion of foreign-atoms in semiconductors 409
  defect-controlled diffusion of hybrid
    solutes 420
  diffusion of foreign atoms in Ge 413
  diffusion of foreign atoms in Si 414

diffusion of hydrogen 415
diffusion of oxygen 415
dopant diffusion 416
dopant diffusion in Ge 417
dopant diffusion in Si 417
equilibrium solubilities 411
interstitial diffusion 414
interstitial-controlled diffusion of
  hybrid solutes 421
P diffusion in Si 419
retrograde solubility 410
solubility limit of oxygen 415
solubility of a foreign elements 409
diffusion of hydrogen 317
  absorption and desorption methods
    319
  electrochemical techniques 319
  Gorsky effect 319
  hydrogen in iron 322
  hydrogen in nickel 321
  hydrogen in niobium 322
  magnetic relaxation 320
  mechanical relaxation methods 319
  non-steady-state permeation 318
  nuclear magnetic relaxation (NMR)
    320
  nuclear reaction analysis (NRA)
    318
  palladium-hydrogen 321
  quasielastic neutron scattering
    (QENS) 320
  radiotracer method 318
  Snoek effect 319
  steady-state permeation 318
diffusion of impurities 327
  electrostatic model 331
  Friedel oscillations 332
  impurity diffusion in Al 332
  normal impurity diffusion 327
  Thomas-Fermi approximation 331
diffusion of interstitial solutes 313
  C, N, and O 313
  diffusion-couple methods 315
  in-diffusion experiments 315
  magnetic relaxation 315
  nuclear reaction analysis (NRA)
    315
  out-diffusion methods 315

Index 643

paired carbon interstititals 317
radiotracer method 315
Snoek-effect – internal friction
  methods 315
stable isotopes 315
steady-state methods 314
diffusion vehicle 63, 462, 515
diffusivity 28
diffusivity tensor 33
dilute substitutional alloys 327
direct methods 209, 246, 256
dislocation density 551, 583
dislocation pipe diffusion 583
  constant source 585
  dislocation density 583
  dislocation pipe radius 584
  dislocation tail 590
  dislocation tails in penetration
    profiles 588
  effective diffusivity (Hart equation)
    590
  instantaneous source 585
  mean thin-layer concentrations 586,
    588
  Smoluchowski model 584
  type A kinetics 583, 590
  type B kinetics 583, 590
  type C kinetics 584, 591
dispersion relation 270
dissociative diffusion in open metals
  333
  effective diffusivity 335
dissociative mechanism 102, 334, 336,
  391, 413, 425, 428, 430, 439, 445
divacancy 63, 72, 100, 130, 136
  Gibbs free energy of interaction 73
divacancy mechanism 100, 130, 154,
  158
DIVINSKI, SERGUEI 15
drift velocity 181

effective activation volume 305
Einstein relation 55, 60, 199
EINSTEIN, ALBERT 6, 55
Einstein-Smoluchowski relation 55,
  58, 60, 105, 199, 210, 213
electrical methods 285
electromigration 179, 192, 456

electron microprobe analysis (EMPA)
  227
EMPA equipment 228
X-ray emission 228
encounter model 263, 269, 277
energy spectrum of neutron scattering
  272
enthalpy of migration 65
entropy of migration 65
equation of continuity 29, 30
error function 42, 44, 47, 562, 575, 586
  complementary error function 42
  error function and approximation
    44
  Grube-Jedele solution 42
  probability integral 42
external driving force 179

FARADAY, MICHAEL 475, 480
fast ion conductor 83, 141, 450, 475
  AgI 477
  Beevers-Ross sites 483
  conduction plane of β-alumina 483
  crystal structure of α-AgI 478
  fast silver ion conductors 477
  fluorite structure 481
  halide ion conductors 480
  lanthanum gallates and LSGM 482
  Li_2SO_4 485
  lithium ion conductors 484
  lithium nitride 485
  NASICON 484
  open lattice structure 477
  oxide ion conductors 481
  PbF_2 480
  PEO–salt systems 486
  perovskite oxide ion conductors 482
  perovskite structure 482
  polymer electrolytes 485
  probabiliy distribution of Ag in
    α-AgI 478
  $RbAg_4I_5$ 479
  sodium β-alumina 482
  stabilised zirconia 481
  tracer diffusion in PEO–NaI 487
FEYNMAN, RICHARD 593
Fick's first law 28
Fick's laws in isotropic media 27

Fick's second law   30, 46, 162, 212, 507, 527, 560, 585
FICK, ADOLF   1, 2, 27
FISHER, J.C.   10, 553
five-frequency model   117, 202, 329, 337, 465
foreign diffusion coefficient   214
formation enthalpy   71, 72, 77, 86, 390, 396, 472
formation entropy   71
Fourier's law   29, 31
Frenkel disorder   69, 84, 452, 453
  Frenkel pair   84
  Frenkel product   84
  Frenkel-pair formation properties   85
FRENKEL, JAKOV IL'ICH   1, 8, 69
frequency factor   127

Gaussian solution   39, 40, 182, 456, 574
GEGUZIN, YAKOV E.   11
Gibbs free energy of activation   129, 132
Gibbs free energy of binding   73
Gibbs free energy of formation   71
Gibbs free energy of interaction   82
Gibbs free energy of migration   65, 131
Gibbs-Duhem relation   171, 197
glass-forming ability   497, 505
glasses   493
  aluminosilicate glasses   499
  amorphous semiconductors   499
  borosilicate glasses   499
  bulk amorphous steels   500
  bulk metallic glasses   500
  caloric glass-transition temperature   496
  DSC thermogram   496
  fictive temperature   495
  fulgarites   501
  glass families   498
  glass transformation region   495
  glass-transition temperature   495
  impactites   501
  inorganic glasses   494
  lead silicate glasses   499
  metallic glasses   494, 500, 505
  moldavites   501
  natural glasses   501
  non-silica-based glasses   499
  nose in the TTT diagram   497
  nucleation   497
  obsidian   501
  organic glasses   500
  soda-lime silicate glasses   498
  tektites   501
  undercooled melt   495, 496
  vitreous silica (SiO_2)   498
  volume-temperature diagram   494
Gorski effect   181
GRAHAM, THOMAS   1, 2, 317
  Graham's law   2
grain boundaries   554
grain-boundary diffusion   10, 221, 549, 551, 553, 568
  average concentration in thin-layer   573
  average concentrations in thin layers   564
  concentration contours   564
  constant source solution (whipple solution)   562
  diffusion mechanisms   579
  effective diffusivity   568, 570
  Fisher model (isolated boundary)   559
  grain boundaries   554
  grain-boundary diffusivity   559, 575
  grain-boundary parameters   554
  grain-boundary segregation   560
  grain-boundary tail   571
  grain-boundary width   559, 574
  Harrisons classification of diffusion in polycrystals   569
  high-angle boundaries   555
  HREM   558
  isoconcentration contours   562
  Le Claire parameter   561, 573
  low-angle boundaries   555
  molecular dynamic simulations   579
  segregation   576
  segregation enthalpy   577
  segregation factor   567, 577
  segregation factor in Ag   579
  segregation factors from grain-boundary diffusion   576
  segregation of foreign atoms   567

special high-angle boundary 557
thin-film solution (Suzuka solution) 563
tilt boundaries 555
triple product 571, 572, 578
twin boundary 557
twist boundaries 555
type A kinetics 568
type B kinetics 570
type B kinetics diffusion profile 571
type B kinetics penetration profiles of self-diffusion 575
type C kinetics 574
gravitational forces 181
GUPTA, DEVENDRA 13

Haven ratio 185, 288, 461, 470, 528, 535, 536, 541
HEITJANS, PAUL 14
HERZIG, CHRISTIAN 15
HEUMANN, THEODOR 15
HEVESY, GEORG KARL VON 4, 333, 553
   radiotracers 4
high-diffusivity paths 129, 547
   diffusion short circuits (high-diffusivity paths) 547
   diffusion spectrum 548
   dislocation density 551
   dislocation pipe diffusion 551
   dislocation pipe diffusivity 549
   grain-boundary diffusivity 549
   lattice (or bulk) diffusivity 549
   lattice diffusion and microstructural defects 551
   polycrystals 551
   rules for grain-boundary diffusion 549
   surface diffusion 551
hybrid solutes 102, 147, 334, 336, 390, 409, 411, 413, 420, 422, 425, 439

impedance spectroscopy 285, 286, 527
   Cole diagram 287
   complex conductivity 286
   complex impedance 286
   Conductivity spectrum 289
   impedance bridge 286
implantation 217

impurity diffusion 131, 214, 465, 510
impurity diffusion coefficient 214
indirect methods 209, 210, 246
interdiffusion 161, 168, 170, 197, 215, 353
   interdiffusion in ionic crystals 186
   interdiffusion coefficient 31, 161, 165, 166, 170, 172, 198, 215, 229
intermetallics 86
   antisite defect 87
   bound triple defects 88
   elements of disorder 87
   triple defect disorder 87
   triple defects 88
   vacancy-type defects 88
internal friction 237, 239, 241
interstitial diffusion 316
interstitial mechanism 95, 107, 264, 316, 411, 413, 422
   direct interstitial mechanism 96, 130
interstitial-substitutional diffusion 425
   combined dissociative and kick-out diffusion 425
   computer simulation of in-diffusion 431
   diffusion limited by the flow of interstitial solutes 429
   diffusion limited by the flow of intrinsic defects 427
   diffusion of Co in Nb 443
   dissociative diffusion in metals 443
   dissociative diffusion of Cu, Ag, and Au in Ge 440
   dissociative reaction 425
   effective diffusion coefficient i 429
   effective diffusion coefficient I+II 428, 432
   effective diffusion coefficient V 439
   flow product (or transport product) 427, 429
   flow product of Ge vacancies 441
   flow product of interstitial solutes in Ge 440
   interstitial flow product of Cu in Ge 441

kick-out diffusion in dislocation-free specimen 432
kick-out diffusion of Au and Pt in Si 434
kick-out diffusion of Zn in Si 435
kick-out reaction 425
law of mass action 426
local equilibrium 427
numerical analysis 430
source or sink for point defects 426
supersaturation of self-interstitials 431
tracer self-diffusion coefficient 428
interstitialcy mechanism 100, 107, 186, 395, 419, 421, 468, 469
intrinsic diffusion 168, 215
intrinsic diffusion coefficient 161, 168, 172, 197, 215
intrinsic gettering 415
ion-beam analysis 231
　nuclear reaction analysis (NRA) 231
　Rutherford backscattering (RBS) 231
　stopping power 234
ion-implantation 217
ionic compounds 83
ionic conductivity 179, 460
ionic crystals 83, 449
　anion sublattice 449
　anti-Frenkel defects 452
　cation sublattice 449
　cation vacancy 451
　charge neutrality 449
　colinear and non-colinear interstitialcy jumps 469
　common features of alkali halides 467
　conductivity and self-diffusion of AgCl and AgBr 469
　conductivity and self-diffusion of NaCl 462
　degree of association 456, 466
　doping effects 454
　doping effects for AgCl and AgBr 471
　electroneutrality 83
　experimental techniques 457
　extrinsic defects 454
　extrinsic region 455, 459
　Frenkel defect 451
　Frenkel product 452
　heterovalent impurities 466
　homovalent impurities 465
　impurity diffusion in NaCl 465
　impurity-vacancy complex 455
　intrinsic defects 452
　intrinsic region 455, 458
　ionic conduction 449
　ionic conductivity of halide crystals 450
　isotope effect and electromigration in NaCl 464
　jump rate of anion vacancies 458
　jump rate of cation vacancies 458
　Kröger-Vink notaion 454
　migration enthalpy of cation vacancies 462
　Schottky defect 451
　Schottky product 452
　silver halides AgCl and AgBr 468
　structural constraint 83
　tracer self-diffusion 458
　vacancy pair 451, 453, 460
irreversible thermodynamics 191
　diffusion in binary alloys 195
　entropy production 192
　isothermal diffusion 193
　Onsager reciprocity theorem 192
　Onsager transport equations 191
　thermodynamic forces 192
　tracer self-diffusion in element crystals 193
isotope effect 151, 298, 306, 316, 321, 323, 456, 464, 515
　effective isotope-effect parameter 154
　Einstein model 152
　incoherent tunneling 324
　kinetic energy factor 153, 399
　many-body effects 153
　non-classical isotope effect 323
　single-jump mechanisms 151
isotope effect experiment 155
　isotope pair 155, 157

isotope effect parameter   153, 155, 158, 306, 464
isotopically controlled heterostructure   223, 227, 402
   chemical vapour deposition (CVD)   223
   molecular beam epitaxy (MBE)   223

JAIN, HIMANSHU   11
JOST, WILHELM   10
jump rate   56, 58, 61, 65, 131, 246, 280, 308, 337, 547
jump rate of a solute atom   100
jump vector   280

KÄRGER, JÖRG   14
kick-out mechanism   102, 336, 391, 421, 425, 428, 430, 431, 436
Kirkendall effect   168–170, 172, 197, 297
   inert markers   168, 173
   Kirkendall markers (inert markers)   174
   Kirkendall plane   169, 174
   Kirkendall velocity   169
   microstructural stability of the Kirkendall plane   169
   stable Kirkendall plane   175
   two stable Kirkendall planes   176
   unstable Kirkendall plane   176
KIRKENDALL, ERNEST   8, 168
KLOTSMAN, SEMJON   13
KOIWA, MASAHIRO   13

L1$_2$ inremetallics   367
L1$_2$ intermetallics   355, 358
   L1$_2$ compounds Ni$_3$Ge and Ni$_3$Ga   356
   L1$_2$ Ni$_3$Al   355
   sublattice vacancy mechanism   355, 356
Laplace transformation   45
Laplace transform   45
lattice (or bulk) diffusion   547, 551
lattice diffusion   11, 212, 559, 568
LAUE, MAX VON   1, 8
Laves phases   364
   C15 type cubic Laves phase Co$_2$Nb   365

Index   647

Co-antisite atoms   365
law of mass action   335, 412, 426, 429, 456
LIDIARD, ALAN B.   12
LIMOGE, YVES   13
linear flow   32, 38
local equilibrium   427
Lomer equation   329
Lomer expression   117
LOO, FRANS VAN   14
Ludwig-Soret effect   180

magnetic relaxation   250, 320
MANNING, JOHN   12
mean residence time   62, 109, 211, 245, 260, 263, 276
mechanical spectroscopy   237
   complex elastic modulus   240
   elastic dipoles   237, 244
   forced oscillations   243
   Gorski relaxation   237, 248
   Gorski relaxation time   248
   hysteresis loop   239
   loss angle   241
   mean modulus   242
   mechanical loss in ion-conducting glasses   249
   point-defect relaxation   238
   relaxation strength   242
   relaxation time   245
   relaxation time of Snoek relaxation   246
   relaxed elastic modulus   239
   Snoek relaxation   237, 244
   standard linear solid   239
   strain relaxation time   239
   stress relaxation time   239
   techniques of mechanical spectroscopy   242
   three-point-bending   243
   torsional pendulums   243
   unrelaxed elastic modulus   241
   Zener relaxation   247
MEHRER, HELMUT   15
melting properties and diffusion   141
   activation enthalpy and melting properties   143
   diffusivities at the melting point   141

van Liempt rule 143
metallic glasses (glassy metals, amorphous alloys) 500, 503
migration enthalpy 146, 396, 472
migration entropy 146
MISHIN, YURI 14
mixed-alkali effect 538
mobility 170, 179, 181, 183
Mössbauer spectroscopy (MBS) 112, 253, 264, 279, 359
   Debye-Waller factor 268
   diffusional line-broadening 266, 268
   Doppler shift 265
   Mössbauer isotopes 265
   natural linewidth 266
   self-correlation function 268
mono- and divacancies 73, 136, 303
monovacancy 70, 99, 130, 136
monovacancy mechanism 99, 130, 154, 158, 302
MUNDY, JOHN 11
MURCH, GRAEME 13

NaCl 462, 464, 466
NAKAJIMA, HIDEO 13
nanocrystalline materials 593, 594
NERNST, WALTHER 475
Nernst-Einstein relation 182–184, 460, 470, 487, 527
Nernst-Planck equation 186, 188
   Nazarov-Gurov equation 189
network glasses 521
neutron scattering 271
non-steady-state diffusion 39
nuclear magnetic relaxation (NMR) 112, 253, 528
   Bloch equation 256
   field-gradient NMR (FG-NMR) 256
   free induction decay (FID) 259
   Koringa relation 262
   Larmor frequency 254
   local field 260
   motional narrowing 260
   NMR experiment 255
   nuclear electric quadrupole moment 259
   nuclear magnetic moment 259
   PFG-NMR 257
   spectral density function 262
   spin-lattice relaxation time 256, 258, 262
   spin-spin relaxation time 256
   Zeeman effect 254
nuclear reaction analysis (NRA) 528

ÖCHSNER, ANDREAS 14
octahedral sites 95, 244, 313
Ohm's law 29
Onsager matrix 187, 191, 192
ONSAGER, LARS 9, 191
order-disorder transition 341, 344
   Cu-Zn system 344
   Fe_3Al system 345
   Fe-Co system 344
oxide glasses 521

permeation 38
PERRIN, JEAN BAPTISTE 1, 7
PETERSON, NORMAN 11
phenomenological coefficients 191, 199, 200
PHILIBERT, JEAN 12
point defect 69, 396, 472, 475
point source 52
positron annihilation spectroscopy (PAS) 77, 359, 364
   mean lifetime of positrons 78
   trapping model 78
pre-exponential factor 127, 143, 147, 297, 300, 304, 316, 328, 332, 459, 510, 550
pressure dependence of diffusion 132

quantum effects 321
quartz 530
quasicrystals 371
   dodecahedron 372
   Penrose tiling 372
   pseudo-Mackay clusters 373
   single-quasicrystal 371
quasielastic neutron scattering (QENS) 253, 269, 271
   Chudley-Elliot random jump diffusion 276
   diffusional broadening of QENS signal 275
   incoherent scattering function 275

incoherent scattering function for random jump motion 276
linewidth of quasielastic line 275
quasielastic linewidth 272
quasielastic linewidth of polycrystalline sample 278

radial diffusion in a cylinder 50
random alloy model 172, 203, 205
random walk 55, 56, 58, 105
   distribution function 58
   mean square displacement 59, 106
   random walk on a lattice 60
   true random walk 61
   uncorrelated random walk 61
rapid quenching (RQ) 77
residual activity method 222
ROBERTS-AUSTEN, WILLIAM CHANDLER 3, 333, 553, 576
ROTHMAN, STEVEN J. 12

Sauer-Freise method 166
   partial molar volume 166
   Vegard rule 166
scattering cross section for neutrons 270
Schottky disorder 69, 85, 452, 453, 613
   Schottky pair 85
   Schottky product 86
   Schottky-pair formation properties 86
SCHOTTKY, WALTER 1, 8, 69
Schrödinger equation 31
second rank tensor 374
secondary ion mass spectroscopy (SIMS) 224
   SIMS instrument 224
   sputtering process 225
   TOF-SIMS instrument 225
secondary mass spectroscopy (SIMS) 528
SEEGER, ALFRED 16
segregation 567, 576
SEITH, WOLFGANG 14
self-diffusion 213, 214, 227, 257, 297, 300, 301, 314, 334, 335, 343, 357, 375, 378, 395, 458, 464, 550, 560, 599
   enriched stable isotope 227

self-diffusion in metals 297
   bcc metals – empirical facts 301
   Curie temperature 309
   fcc metals – empirical facts 299
   hexagonal close-packed and tetragonal metals 306
   magnetic transition 308
   metals with phase transitions 308
   mono- and divacancy interpretation 303
   monovacancy interpretation 302
   self-diffusion in group-IV transition metals 310
   self-diffusion in iron 309
   standard interpretation 303
self-diffusion in semiconductors 395
   activation volumes of Ge self-diffusion 402
   charge states of self-interstitials and vacancies 397
   concentration of self-interstitials 396
   concentration of vacancies 396
   doping dependence of Ge self-diffusion 400
   self-diffusion of Ge 399
   self-diffusion of Si 403
   self-interstitial transport product of Si 405
   vacancy transport product of Si 405
self-interstitial 63, 69, 79, 102, 136, 396, 404, 422, 579
   dumbbell configuration 80
   pure metal 79
   radiation damage 80
   semiconductor 80
semiconductors 88, 385
   background doping 89, 91
   band gap 89, 386
   charged defects 90
   diamond structure 385
   direct band gap III-V compounds 389
   doping elements 388
   electric-field effect 391
   electronic structure 89
   electrons and holes 387

fast diffusing metallic impurities   390
Fermi level   89
Fermi-level effect   391
germanium   385
  high perfection   391
  hybrid solutes   390
  impurities with deep levels   390
  intrinsic carrier density   89, 387
  ion-pairing   391
  low solubilities   391
  major contaminants O and C   390
  Moore's law   389
  n-type and p-type   387
  oxidation-enhanced or -retarded diffusion   391
  packing density   88
  self-doping   91, 391
  semiconductor age   386
  semiconductor single crystals   387
  shallow dopants   390
  silicon   385
  silicon dioxide   388
  silicon-carbide   389
  thermal defect concentrations   89
  under- or oversaturation of point defects   392
  zinc blende structure   385
serial sectioning   217
  grinder sectioning   219
  ion-beam sputter sectioning (IBS)   219
  lathe sectioning   219
  mechanical sectioning   218
SHEWMON, PAUL   12
silicate glasses   532, 533, 539
silver halide   468
sinks and sources for point defects   188, 426, 443, 552
site fraction of monovacancies   72
slab source   43
SMOLUCHOWSKI, MARIAN   6
solute diffusion   116, 327, 336
  solute diffusion coefficient   337
solvent diffusion   116, 327, 336
  linear enhancement factor   337
  partial correlation factor   337
  solvent diffusion coefficient   337

Soret effect   180
spherical coordinates   32
spherical flow   38
spreading resistance profiling   285, 290
  foreign atom concentration   292
  resistivity   292
  spreading resistance   291
  spreading resistance profile   290
steady-state diffusion   37
  hollow cylinder   38
  planar membrane   38
  spherical shell   38
Stokes-Einstein relation   517, 526, 535
STOLWIJK, NICOLAAS   15
STRAUMAL, BORIS   14
structural relaxation   506, 508
sum rules   198, 204
Summerfield scaling   288
supersaturation of point defects   428, 430, 437

TAMMANN, GUSTAV   494
temperature dependence of diffusion   127
temporal correlation   112
  bunching effect   112
tetrahedral sites   95, 313
thermodynamic factor   171, 184, 187, 197
thermoelectric devices   192
thermomigration   180
thermotransport   180
thin-film solution   39, 182, 217, 219, 220, 563, 587
  sandwich geometry   39
  thin-film geometry   39
tracer diffusion coefficient   171, 184, 200, 202, 212, 298, 518, 528, 533, 534, 539
tracer diffusion experiment   215, 507, 608
  evaporation losses of the matrix   222
  evaporation losses of tracer   221
tracer diffusion of cations and anions   456
tracer difuusion coefficient
  grain-boundary tail   221

tracer self-diffusion coefficient   213, 215, 303, 397, 441, 460, 469
tracer self-diffusion coefficient in HCP   307
transmission electron microscopy (TEM)   77
transport coefficient   187, 191
transport product   397, 398, 404, 420, 427
TTT diagram   497, 505, 511
TURNBULL, DAVID   10, 553

undersaturation of point defects   428, 430, 439
uniaxial intermetallics   360
  anisotropy ratio   361, 364
  diffusion asymmetry   363
  equiatomic NiMn   361
  $L1_0$ TiAl   360
  phase transition B2-$L1_0$ 362
  tetragonal $MoSi_2$ 363

V-T diagram   495, 507
vacancy   63, 69, 99, 194, 196, 421, 579
  pure metals   70

vacancy mechanism   98, 114, 130, 185, 264, 395, 396, 398, 413, 417, 418, 421, 441
vacancy mechanism of self-diffusion   108, 130
  mean residence time of a vacancy   109
  mean residence time of the tracer atom   109
  return probability   110
  rule of thumb   108, 116
  vacancy-tracer encounter   109
vacancy properties   74
  properties of monovacancy formation   79
vacancy source or sink   335
vacancy-impurity complex   330
vacancy-mediated solute diffusion   116, 130
vacancy-solute interaction   336
viscosity   495, 517, 525
viscosity diffusion coefficient   517, 526, 533, 535
vitreous silica   529, 531
Vogel-Fulcher-Tammann behaviour   487, 516, 525

Springer Series in
SOLID-STATE SCIENCES

Series Editors:
M. Cardona   P. Fulde   K. von Klitzing   R. Merlin   H.-J. Queisser   H. Störmer

91 **Electronic Properties and Conjugated Polymers III**
Editors: H. Kuzmany, M. Mehring, and S. Roth

92 **Physics and Engineering Applications of Magnetism**
Editors: Y. Ishikawa and N. Miura

93 **Quasicrystals**
Editor: T. Fujiwara and T. Ogawa

94 **Electronic Conduction in Oxides**
2nd Edition   By N. Tsuda, K. Nasu, A. Fujimori, and K. Siratori

95 **Electronic Materials**
A New Era in MaterialsScience
Editors: J.R. Chelikowski and A. Franciosi

96 **Electron Liquids**
2nd Edition   By A. Isihara

97 **Localization and Confinement of Electrons in Semiconductors**
Editors: F. Kuchar, H. Heinrich, and G. Bauer

98 **Magnetism and the Electronic Structure of Crystals**
By V.A. Gubanov, A.I. Liechtenstein, and A.V. Postnikov

99 **Electronic Properties of High-T_c Superconductors and Related Compounds**
Editors: H. Kuzmany, M. Mehring, and J. Fink

100 **Electron Correlations in Molecules and Solids**
3rd Edition   By P. Fulde

101 **High Magnetic Fields in Semiconductor Physics III**
Quantum Hall Effect, Transport and Optics   By G. Landwehr

102 **Conjugated Conducting Polymers**
Editor: H. Kiess

103 **Molecular Dynamics Simulations**
Editor: F. Yonezawa

104 **Products of Random Matrices** in Statistical Physics   By A. Crisanti, G. Paladin, and A. Vulpiani

105 **Self-Trapped Excitons**
2nd Edition
By K.S. Song and R.T. Williams

106 **Physics of High-Temperature Superconductors**
Editors: S. Maekawa and M. Sato

107 **Electronic Properties of Polymers**
Orientation and Dimensionality of Conjugated Systems
Editors: H. Kuzmany, M. Mehring, and S. Roth

108 **Site Symmetry in Crystals**
Theory and Applications
2nd Edition
By R.A. Evarestov and V.P. Smirnov

109 **Transport Phenomena in Mesoscopic Systems**
Editors: H. Fukuyama and T. Ando

110 **Superlattices and Other Heterostructures**
Symmetry and Optical Phenomena
2nd Edition
By E.L. Ivchenko and G.E. Pikus

111 **Low-Dimensional Electronic Systems**
New Concepts
Editors: G. Bauer, F. Kuchar, and H. Heinrich

112 **Phonon Scattering in Condensed Matter VII**
Editors: M. Meissner and R.O. Pohl

113 **Electronic Properties of High-T_c Superconductors**
Editors: H. Kuzmany, M. Mehring, and J. Fink

Springer Series in
SOLID-STATE SCIENCES

Series Editors:
M. Cardona   P. Fulde   K. von Klitzing   R. Merlin   H.-J. Queisser   H. Störmer

114 **Interatomic Potential and Structural Stability**
Editors: K. Terakura and H. Akai

115 **Ultrafast Spectroscopy of Semiconductors and Semiconductor Nanostructures**
By J. Shah

116 **Electron Spectrum of Gapless Semiconductors**
By J.M. Tsidilkovski

117 **Electronic Properties of Fullerenes**
Editors: H. Kuzmany, J. Fink, M. Mehring, and S. Roth

118 **Correlation Effects in Low-Dimensional Electron Systems**
Editors: A. Okiji and N. Kawakami

119 **Spectroscopy of Mott Insulators and Correlated Metals**
Editors: A. Fujimori and Y. Tokura

120 **Optical Properties of III–V Semiconductors**
The Influence of Multi-Valley Band Structures   By H. Kalt

121 **Elementary Processes in Excitations and Reactions on Solid Surfaces**
Editors: A. Okiji, H. Kasai, and K. Makoshi

122 **Theory of Magnetism**
By K. Yosida

123 **Quantum Kinetics in Transport and Optics of Semiconductors**
By H. Haug and A.-P. Jauho

124 **Relaxations of Excited States and Photo-Induced Structural Phase Transitions**
Editor: K. Nasu

125 **Physics and Chemistry of Transition-Metal Oxides**
Editors: H. Fukuyama and N. Nagaosa

126 **Physical Properties of Quasicrystals**
Editor: Z.M. Stadnik

127 **Positron Annihilation in Semiconductors**
Defect Studies
By R. Krause-Rehberg and H.S. Leipner

128 **Magneto-Optics**
Editors: S. Sugano and N. Kojima

129 **Computational Materials Science**
From Ab Initio to Monte Carlo Methods.   By K. Ohno, K. Esfarjani, and Y. Kawazoe

130 **Contact, Adhesion and Rupture of Elastic Solids**
By D. Maugis

131 **Field Theories for Low-Dimensional Condensed Matter Systems**
Spin Systems and Strongly Correlated Electrons
By G. Morandi, P. Sodano, A. Tagliacozzo, and V. Tognetti

132 **Vortices in Unconventional Superconductors and Superfluids**
Editors: R.P. Huebener, N. Schopohl, and G.E. Volovik

133 **The Quantum Hall Effect**
By D. Yoshioka

134 **Magnetism in the Solid State**
By P. Mohn

135 **Electrodynamics of Magnetoactive Media**
By I. Vagner, B.I. Lembrikov, and P. Wyder

136 **Nanoscale Phase Separation and Colossal Magnetoresistance**
The Physics of Manganites and Related Compounds
By E. Dagotto

137 **Quantum Transport in Submicron Devices**
A Theoretical Introduction
By W. Magnus and W. Schoenmaker

Made in the USA
Lexington, KY
12 June 2012